# Anaerobic Treatment of Domestic Wastewater

## *Present Status and Potentialities*

# Anaerobic Treatment of Domestic Wastewater

## *Present Status and Potentialities*

Edited by
Maria Concetta Tomei
*Water Research Institute (IRSA) of the Italian National Research Council (CNR)*

and

Juan Manuel Garrido
*Cross-disciplinary Research Centre in Environmental Technologies (CRETUS) and Chemical Engineering Department, Universidade de Santiago de Compostela*

IWA PUBLISHING

**Published by** 

**IWA Publishing**
**Unit 104–105, Export Building**
**1 Clove Crescent**
**London E14 2BA, UK**
Telephone: +44 (0)20 7654 5500
Fax: +44 (0)20 7654 5555
Email: publications@iwap.co.uk
Web: www.iwapublishing.com

First published 2024


*British Library Cataloguing in Publication Data*
A CIP catalogue record for this book is available from the British Library

ISBN: 9781789063462 (paperback)
ISBN: 9781789063479 (eBook)
ISBN: 9781789063486 (ePub)

Doi: 10.2166/9781789063479

This eBook was made Open Access in May 2024.

# Contents

## Chapter 3
## *Anaerobic membrane bioreactors* . . . *51*

*D. C. Stuckey*

doi: 10.2166/9781789063479_0001

# Introduction

The book presents the state of knowledge and the future perspectives of anaerobic processes applied as mainstream treatment of domestic wastewater, which in the last few decades, due to the development of effective technologies such as the high-rate bioreactors, became competitive with the conventional aerobic process even in moderate climatic regions.

The book consists of 12 chapters covering the main aspects related to the anaerobic treatment of domestic wastewater including engineering, microbiology, process monitoring and control, sustainability derived by energy and resource recovery, life-cycle assessment, and techno-economic analysis. Effective technologies namely upflow anaerobic sludge blanket bioreactors, membrane bioreactors, and granular and immobilized biomass systems, potentially able to achieve removal performance of high organics even at low temperatures, are presented in detail, and their status of application is reported.

For the nutrient content of treated effluents, two alternatives are considered: their recovery, which is preferable to maximize the process sustainability, and low impact post-treatment solutions. Operational strategies to recovery energy and water are also discussed for different final use options.

Concerning the drawback caused by the dissolved methane in the treated effluent, possible alternatives to enhance either its recovery within the process to maximize energy production or its use as an alternative carbon source to improve denitrification potential, avoiding in any case the emission of this powerful greenhouse gas, are critically analyzed.

Constraints such as removal of microbial indicators (i.e., helminth eggs, pathogenic microorganisms, viruses) and fate of micropollutants are discussed to evaluate possible interventions for enhancing their removal.

Feasibility of anaerobic treatment for domestic wastewater is also evaluated through life-cycle assessment and techno-economic analysis to define the real competitiveness with conventional activated sludge processes.

At the end, the book will provide the reader all the information required to evaluate anaerobic processes as a feasible alternative in treating domestic wastewater, expected advantages in terms of sustainability, and the still open research questions, placing special emphasis on hot topics such as the fate of microplastics and antibiotic resistance in anaerobic systems.

## WHY THIS BOOK?

In spite of the fact that the potential advantages of anaerobic domestic wastewater treatment have been well highlighted in numerous earlier studies, the application of the process is still limited to regions with favorable warm climatic conditions. This book can effectively contribute to the dissemination of the information on the current state of knowledge, which is very promising, and speed up the research activities required to complete the last steps toward an extended full-scale application, including the necessary post-treatment options, especially in temperate regions of the world.

Through a combination of theoretical knowledge and practical considerations, this book aims to provide readers, doctoral, and master's students on a water treatment topic, and even professionals or researchers in the water sector, with the knowledge necessary to know, analyze, design, and implement anaerobic bioreactors for sewage treatment. As we embark on this journey toward anaerobic wastewater treatment, we invite readers to explore the science, technology, and real-world applications that make this approach a reliable alternative in the search for more economical and sustainable solutions for domestic wastewater treatment.

doi: 10.2166/9781789063479_0003

# Chapter 1

# Anaerobic treatment of low-strength wastewater: applicability and hygienization potential

M. C. Tomei[1]*, V. Stazi[1], J. Calabria de Araújo[2] and C. Leite Madeira[3]

[1]Water Research Institute (IRSA), National Research Council (CNR), Roma, Italy
[2]Federal University of Minas Gerais, Belo Horizonte, Brazil
[3]Department of Civil Engineering, University of Texas at El Paso, El Paso, United States
*Correspondence: concetta.tomei@irsa.cnr.it

## ABSTRACT

Anaerobic treatment of domestic wastewater (DWW) is currently applied in warm climate regions because the process kinetics at these temperatures are competitive with the conventional activated sludge process, and the solution is advantageous given the possibility of energy and nutrient recovery and the reduced sludge production. However, the lower metabolic capacity of anaerobic bacteria than aerobic ones makes the process less efficient, especially for low concentrated wastewater as the domestic ones. This challenge can be overcome by applying high-rate bioreactors, such as anaerobic filters, upflow anaerobic sludge blanket reactors, expanded granular sludge bed reactors, and anaerobic baffled reactors, allowing the decoupling of hydraulic retention time (HRT) and sludge retention time (SRT). They operate with highly concentrated biomass, thus low HRTs (and consequently lower reactor volumes) can be applied to achieve process rates suitable for practical application. Another important aspect to evaluate the feasibility of the anaerobic treatment of DWW is the hygienization potential, which is gaining increasing attention in light of safe water reuse. In this chapter, an overview of the present state of knowledge of high-rate systems as well as multi-step treatment options, in combination with the anaerobic process, are presented and discussed. In addition, an extended analysis of pathogen removal in anaerobic processes treating low-strength wastewater (with or without post-treatment) is reported with the aim of providing an exhaustive contribution to this aspect, which is of relevance for water reuse and process sustainability.

**Keywords:** anaerobic process, low-strength wastewater, high-rate bioreactors, combined processes, pathogen removal, microbial indicators.

## 1.1 WHY ANAEROBIC TREATMENT OF LOW-STRENGTH WASTEWATER?

### 1.1.1 Advantages and potentialities in comparison to conventional activated sludge process

Recent trends in domestic wastewater (DWW) treatment no longer consider DWW as a waste to be treated or disposed of but as a source of energy and valuable products. This approach is in line with the United Nations sustainable development goals (SDGs) 6 and 12 and in particular with the following target goals: 'achieving the environmentally sound management of chemicals and all wastes throughout their life cycle; reducing waste generation through prevention, reduction, recycling and reuse; encourage companies to adopt sustainable practices' (SDG 12) and 'improve water quality by reducing pollution, eliminating dumping and minimizing release of hazardous chemicals and materials, halving the proportion of untreated wastewater and substantially increasing recycling and safe reuse globally' (SDG 6). Potentially recoverable resources in anaerobic treatment consist of water itself whose quality depends on the applied post-treatment, fertilizing nutrients (nitrogen and phosphorus as main nutrients but also potassium and sulfur), and energy (Bae *et al.*, 2014; Foresti *et al.*, 2006; Kujawa-Roeleveld & Zeeman, 2006; McCarty *et al.*, 2011).

The majority of wastewater treatment plants in moderate and cold climate regions are operated with aerobic processes, using different configurations based on the conventional activated sludge (CAS) process whereas anaerobic processes have a minor role being applied only for sludge digestion in large DWW treatment plants. CAS plants achieve high removal efficiencies of the influent organic load, but show main drawbacks related to the high energy demand for aeration and to the high production of sludge, which requires handling, treatment and disposal (Leitão *et al.*, 2006; Martinez-Sosa *et al.*, 2011; Smith *et al.*, 2012). Moreover, only a minor fraction of the wastewater energy content can be recovered through anaerobic sludge digestion, whereas the energy associated with the dissolved organic fraction is completely lost. As a result, in CAS the energy balance is negative, that is, more energy is consumed in wastewater treatment than gained with anaerobic sludge digestion (McCarty *et al.*, 2011).

Anaerobic treatment, owing to attractive advantages of energy saving, biogas recovery, and lower sludge production, has the potential of being a more sustainable technology (Bae *et al.*, 2014; Wen *et al.*, 1999) for DWW treatment and increasing investigations have focused on this topic in the last few decades. McCarty *et al.* (2011) have published an important reference study on the potential advantages in terms of energy recovery from anaerobic DWW treatment compared to a CAS system coupled with anaerobic sludge digestion. Assuming typical DWW at a chemical oxygen demand (COD) concentration of 500 mg/L, they estimated a doubling of $CH_4$ production in comparison to CAS and observed that the energy production greatly exceeds the energy required for plant operation. This important result means that anaerobic DWW treatment can be a net energy producer.

DWW can also be an important carrier medium for nutrients in the nutrient cycle. Recycling nitrogen and phosphorus present in sewage rather than wasting them could help to minimize the anthropogenic production of fertilizers. An average evaluation of potential nutrient and fertilizer recovery from the anaerobic DWW treatment has been reported in Verstraete *et al.* (2009), who estimated a potential recovery per $m^3$ sewage of 0.05 kg nitrogen, 0.01 kg phosphorus, and 0.1 kg organic fertilizers, calculated on the basis of 20% organic matter remaining after anaerobic digestion.

At first, anaerobic digestion kinetics were considered feasible only for high-strength wastewater at temperatures above 20–25°C. Thus, the first anaerobic bioreactors were used in tropical regions and designed for high organics' concentrated wastewater as the agro-food ones (Foresti *et al.*, 2006). However, over the past 50 years, technologies that are more efficient have been developed opening a wider spectrum of anaerobic process applications, including the possibility of treating DWW at lower temperatures. Since then, the application of the anaerobic process as the mainstream treatment unit of DWW significantly increased, particularly in developing countries such as Brazil, Colombia, Mexico, Egypt, and India, where this technology is considered to be a fully feasible treatment option characterized by a favorable cost–benefit ratio (Aiyuk *et al.*, 2006; McCarty *et al.*, 2011).

In the last few decades, the already mentioned increased interest in a more sustainable wastewater treatment approach pushed research activities toward high-performance technologies allowing the extension of anaerobic treatment applicability to low-strength wastewater such as DWW under temperate climate conditions, and this is one of the main challenges for anaerobic technology evolution.

### 1.1.2 Drawbacks of anaerobic DWW treatment

For a better understanding of some of the drawbacks of the anaerobic DWW treatment, it is useful for the reader to have a short presentation of the basic reactions occurring in the process. Anaerobic organic matter degradation is carried out in four steps in series: (1) hydrolysis; (2) acidogenesis; (3) acetogenesis; and (4) methanogenesis. In the first step, complex organic compounds such as proteins, carbohydrates, and lipids are hydrolyzed to obtain simpler soluble products such as amino acids, sugars, long-chain fatty acids, and glycerin by the action of extracellular enzymes excreted by fermentative bacteria. This hydrolytic step is considered the rate-limiting step of the overall process for streams containing lipids and/or a significant amount of particulate matter (Khanal, 2008). In a general simplified model of the anaerobic biodegradation process, it is assumed that the first step is performed by fermentative bacteria, whose catabolic phase produces acids and alcohols (acidogenesis), which are then readily utilized as substrates by acetogenic bacteria to yield acetate (acetogenesis). In the final step, methanogens obtain energy from converting acetate, carbon dioxide and hydrogen to methane (Schink, 1997).

As a rule of thumb, in common practice, aerobic-based technologies are suitable for the treatment of low-strength wastewater (biochemical oxygen demand (BOD) $\leq$1000 mg/L), whereas anaerobic ones can be better applied for the treatment of high-strength wastewater (BOD $>$4000 mg/L). Furthermore, aerobic systems, compared to anaerobic ones, are able to achieve higher removal efficiencies for soluble biodegradable organics and the produced biomass is generally characterized by good settleability, resulting in lower effluent suspended solid (SS) concentration. Thus, the effluent quality from an aerobic process is generally higher than that in an anaerobic process in terms of both residual organics and solids' content. In the presence of stringent effluent discharge limits, a post-treatment after the anaerobic step is necessary to achieve the required removals of organic matter, SSs, pathogenic microorganisms, and possibly nutrients. Several alternatives have been tested for post-treatment including aerobic reactors, physical–chemical processes, or nature-based alternatives such as wetlands and oxidation ponds (Chan *et al.*, 2009; El-Khateeb *et al.*, 2009; Kim *et al.*, 2011; Van Haandel *et al.*, 2006).

The lower efficiency of anaerobic processes is due to a lower metabolic capacity of anaerobic bacteria resulting in longer retention times required in comparison to aerobic ones (Van Haandel *et al.*, 2006). This critical aspect can be overcome with high-rate anaerobic systems, which have the ability to uncouple hydraulic retention time (HRT) and solid retention time (SRT) effectively. Therefore, quite low HRTs can be applied due to the high biomass concentration in the system (Daud *et al.*, 2018; Gömec, 2010).

As mentioned earlier, temperature strongly affects the rates of anaerobic conversion processes. Therefore, some essential improvements must be made in the conventional design of high-rate reactors to enable their applications under sub-optimal temperatures (Lettinga *et al.*, 2001). Many researchers (Kettunen & Rintala, 1997; Lettinga *et al.*, 1999; Matsushige *et al.*, 1990; Rebac *et al.*, 1995; Svensson, 1984; Wu *et al.*, 1993) have investigated the effect of temperature on the maximum substrate utilization rates of microorganisms. Most studies on the metabolic activity of mesophilic anaerobic methanogenic bacteria showed a negative effect with decreasing temperatures. Thus, full-scale applications of anaerobic DWW were restricted to wastewater with temperatures above 18°C (Lettinga *et al.*, 2001). To avoid lower microbial activity, external heating should be provided to maintain the required temperature, which will result in a consistent increase of operating costs (Aiyuk *et al.*, 2006; Foresti, 2002). However, a wide range of temperature conditions was tested showing that temperature may not be a limiting factor in anaerobic treatment applications if the appropriate process design is chosen (Nachaiyasit & Stuckey, 1997; Patel & Madamwar, 2002; Petropoulos *et al.*, 2017; Van Lier *et al.*, 1997). For example, well-established upflow anaerobic sludge blanket (UASB) and expanded granular sludge bed (EGSB) reactors are able to meet the requirements necessary for high-rate anaerobic treatment (Aiyuk *et al.*, 2004; Seghezzo *et al.*, 1998) if operated with a sufficiently high biomass concentration achieved with granular or immobilized biomass. Under unfavorable environmental conditions, however, granules could disintegrate leading to a decrease of the bioreactor performance (Connaughton *et al.*, 2006). Anaerobic membrane bioreactors (AnMBRs), by coupling membrane filtration with anaerobic treatment, could provide an alternative strategy for DWW treatment at low temperatures (Smith *et al.*, 2012). Regardless of short HRTs, they can retain the biomass in the reactor more effectively and can produce high-quality effluent in terms of SSs, COD, and pathogens. Moreover, it is possible to reuse and recycle the treated effluent for non-potable purposes (Ho & Sung, 2010). However, the applicability of large-scale AnMBRs for the treatment of DWW still remains an issue, as will be elaborated below.

## 1.2 HIGH-RATE SYSTEMS

### 1.2.1 Principle of operation and configuration of high-rate bioreactors

Anaerobic treatment in high-rate bioreactors is an advanced technology effective for environmental protection and resource preservation, depending on the required effluent standards. It can be applied as standalone technology or in combination with other treatment steps such as aerobic and physical or chemical pre/post-treatments. The peculiarity of high-rate bioreactors is the uncoupling of HRTs and SRTs. They operate with highly concentrated biomass, thus, relatively low HRTs can be applied (Daud *et al.*, 2018; Gömec, 2010; Lettinga *et al.*, 1997a, 1997b; Seghezzo *et al.*, 1998) reducing the bioreactor volume.

According to Iza (1991), good performance of high-rate anaerobic reactors is attributed to the following features:

(a) high biomass concentration achieved by different methods, that is, settling, attachment to solids (fixed or mobile), or recirculation;
(b) enhanced contact between biomass and wastewater obtained with the optimal configuration of the hydraulic regime in the bioreactor;
(c) adaptation procedures and enhanced growth strategies improving the biomass activity;
(d) effective separation of the biomass from the liquid phase.

Given the high SRT values, slowly growing microorganisms are not removed from the system while the bioreactor can be operated at short HRTs (Gömec, 2010), high-organic loadings can be applied in smaller reactors, and the high SRT provides generally good process stability (Aiyuk *et al.*, 2006).

Typical high-rate anaerobic bioreactors include: anaerobic filter (AF), UASB, EGSB, anaerobic baffled reactor (ABR), fixed film fluidized-bed, and expanded-bed reactors. An important step in the evolution of these technologies was the development of UASB and EGSB reactors. They were first designed for the treatment of industrial wastewater, then the application was extended to low-strength wastewater and, currently, they constitute the most powerful anaerobic treatment device for sewage (Aiyuk *et al.*, 2006; Bajpai, 2017; Bodkhe, 2009; Gömec, 2010; Iza, 1991; Liu *et al.*, 2018; Seghezzo *et al.*, 1998; Van Haandel *et al.*, 2006). A schematic representation of the most applied high-rate bioreactors for DWW treatment is reported in Figure 1.1, and briefly described in the following sections.

#### *1.2.1.1 Anaerobic filters*

Young and McCarty (1969) first proposed an AF bioreactor, which has been applied to both high- and low-strength wastewaters. An AF consists of one or more vertical filter beds containing inert media, such as rocks or plastics, which act as a support for biomass attachment. Moreover, they are characterized by a structure favoring the entrapment of suspended flocs or microorganisms. Influent is fed upward through a filling material, resulting in an enhanced contact between the attached biomass and the wastewater (Bajpai, 2017; Switzenbaum, 1983).

AFs can achieve high substrate removal efficiency operating at low HRTs, can tolerate shock loads, and require small volumes. The treated effluent has low SS concentration, thus it is not necessary to add a solid separation phase and sludge recycle. Moreover, efficient and quick recovery of the biological activity was observed when restarted after a stop period. All these features contribute to affordable construction, operation, and maintenance costs (Manariotis & Grigoropoulos, 2003, 2008; Martín *et al.*, 2010).

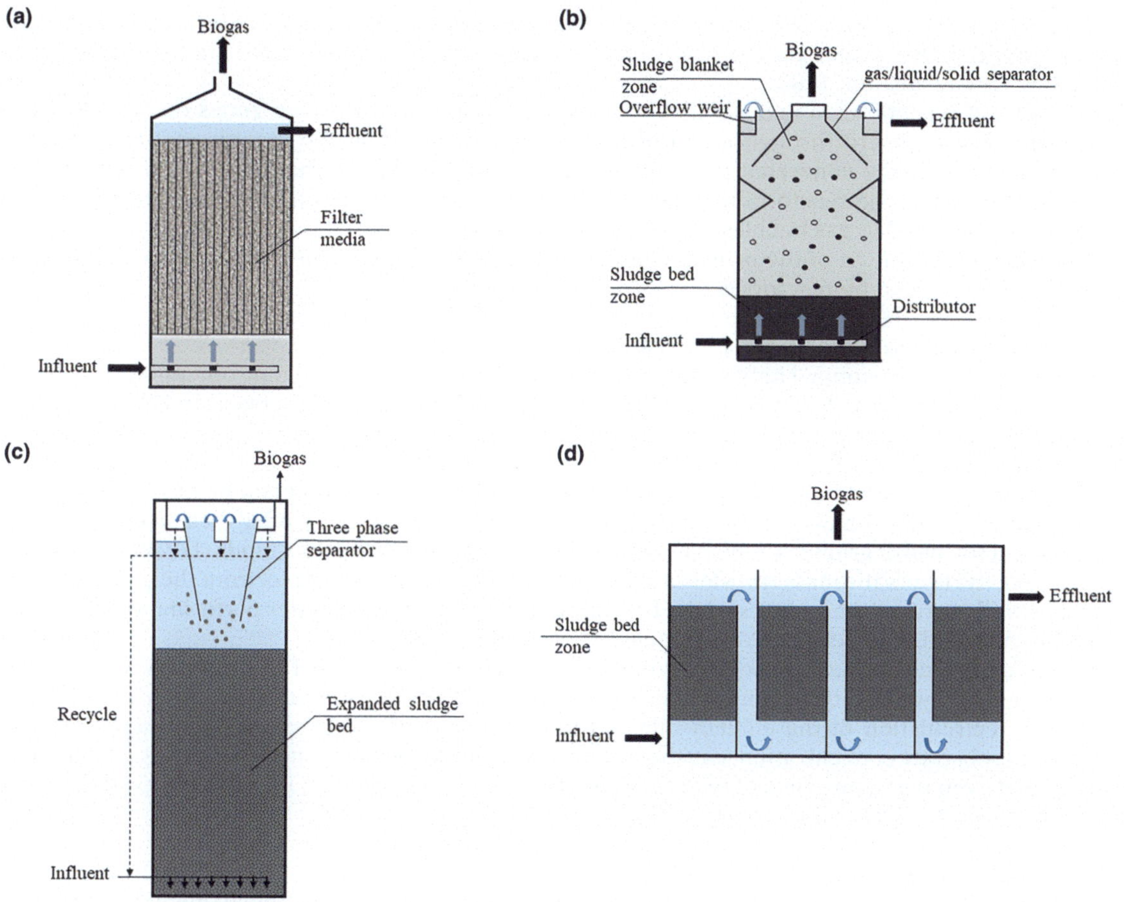


**Figure 1.1** Schematic representation of high-rate bioreactors and related streams: (a) AF, (b) UASB, (c) EGSB, and (d) ABR (*source*: from Stazi & Tomei, 2018).

Nowadays, AF is gaining more attention as an alternative technological solution for mainstream anaerobic treatment of low-strength wastewater under ambient conditions, especially in small decentralized facilities located in moderate climate regions (Manariotis & Grigoropoulos, 2006).

#### 1.2.1.2 Upflow anaerobic sludge blanket reactors

A UASB reactor was proposed by Lettinga and his coworkers in 1970s and it is recognized as one of the best anaerobic treatment devices developed so far. Currently, UASB reactors are extensively applied as consolidated technology for treating DWW in large full-scale plants covering a population of 1 million inhabitants (Chernicharo *et al.*, 2015). Simplicity, low investment and operation costs, and the long favorable experience gained in the treatment of a wide range of wastewater types are some features of these bioreactors (Elmitwalli, 2000; Yu *et al.*, 1997).

UASB reactors are fed in upflow mode operating with suspended biomass, and they also act as settling devices, as the SS retention is extremely important for the biological processes taking place in the system (Bajpai, 2017; Gömec, 2010). Under proper physical and chemical operating conditions, anaerobic sludge can flocculate and form granules with excellent settling properties, so that it is not susceptible to wash-out from the system. High active biomass concentrations can be established in the bioreactor with better exploitation of the reactor working volume (Chernicharo *et al.*, 2015; Seghezzo *et al.*, 1998; Switzenbaum, 1983).

A biological reaction area and a sedimentation section are the two main zones of a reactor. The wastewater flows upward through a bed of highly active sludge and the organic matter contained in the influent is finally converted to methane and carbon dioxide. A gas–solid–liquid separator is located at the top of the reactor and provides the separation of the biogas and the sludge particles entrapped or attached to gas bubbles in the liquid effluent (Bajpai, 2017; Gömec, 2010).

A detailed presentation of the principle of operation, technological features, evolution, and status of application of UASB bioreactors is discussed in Chapter 2.

#### 1.2.1.3 Expanded granular sludge bed reactors

Aiming to optimize the contact between sludge and wastewater, and to overcome operational problems such as preferential flows, dead zones, and hydraulic short circuits that might occur in UASB reactors, an advanced reactor configuration, namely EGSB bioreactor, has been proposed (Bajpai, 2017; Gömec, 2010; Rinzema *et al.*, 1993).

The solution consisted in applying a higher influent upflow velocity that can be obtained by using an appropriate height/diameter ratio and/or high effluent recirculation. Under these conditions, the expansion of the bed of granules, which is achieved by the upward flow of the fluids (DWW and biogas) at high velocity, favors the mass transfer of substrates and reaction products to/from the granules. This condition will increase the kinetics of the biological process leading to better performance compared to a conventional UASB reactor (Van Haandel *et al.*, 2006).

EGSBs can efficiently remove soluble pollutants, but a scarce removal of SSs is observed due to the high applied upflow velocity, which can cause the rupture of flocs and granules (Seghezzo *et al.*, 1998). Moreover, recirculation of the effluent dilutes the bulk phase of the bioreactor with a consequent decrease of substrate concentration, causing a decrease in the anaerobic kinetics. However, it has been verified that even in the presence of recirculation and at low temperatures, EGSBs could satisfactorily treat low-strength wastewater (Gömec, 2010).

#### 1.2.1.4 Anaerobic baffled reactors

An ABR was developed in the 1970s by McCarty and co-workers. It is a compartmentalized reactor composed of a series of UASB reactors in which alternating hanging and standing baffles are placed to direct the liquid flow upward and downward from one section to the next (Chinwetkitvanich & Ruchiraset, 2017). High concentrations of biomass are retained in the upflow region of each compartment resulting in high performances, whereas overall sludge production is characteristically low (Bajpai, 2017; Gömec, 2010). This configuration minimizes the risk of clogging and excessive sludge bed expansion that may occur in other systems, such as AFs and UASBs (Manariotis & Grigoropoulos, 2002).

The main advantage of an ABR is the possibility of separating acidogenesis and methanogenesis phases longitudinally down the reactor, allowing the two-phase operation without the use of complex control devices and at reduced costs. The two-phase mode can significantly increase acidogenic and methanogenic performance because the different bacterial groups can operate under optimized operational conditions. This is particularly important for methanogenic bacteria, which are not affected by the possible pH decrease derived from the accumulation of volatile acids produced in the acidogenic phase (Barber & Stuckey, 1999; Chinwetkitvanich & Ruchiraset, 2017; Rozzi *et al.*, 1994; Tomei *et al.*, 1994). Furthermore, at low temperatures, the compartmentalization might improve the hydrolysis of less biodegradable substrates in the first zone of the reactor where the pH is low. Other peculiarities of these bioreactors include high tolerance to hydraulic and organic shock loads as well as high biomass retention times and lower sludge production. Operational problems detected at pilot/full-scale include the requirement of shallow reactors for maintaining acceptable liquid and gas upflow velocities, and the difficulty in ensuring a uniform influent distribution (Tilche & Vieira, 1991).

Several studies have revealed the potential of ABRs for the application of the anaerobic process (Abbasi *et al.*, 2017; Gopala Krishna *et al.*, 2008; Manariotis & Grigoropoulos, 2002; Wang *et al.*, 2004). For example, Xi-quan and Zhao-hua (2008) successfully tested ABRs for the treatment of low-strength wastewater, with high efficiencies for COD removal in the front compartments, indicating that the number of compartments could be reduced in several of the investigated cases.

#### 1.2.1.5 Anaerobic membrane bioreactors

As mentioned earlier, modern high-rate anaerobic reactors offer effective strategies for biomass retention, that is, the biggest issue that limits the widespread application of anaerobic processes to low-concentrated streams. Meanwhile, some peculiarities of the process, that is, long start-up periods, difficulty in establishing optimal operating conditions suitable for all the involved bacterial groups, low efficiency at low temperatures, and other influent characteristics such as salinity, still represent critical conditions for extended applicability to the treatment of DWW. A promising alternative to resolve these drawbacks is the combination of membrane with anaerobic bioreactors, which is potentially able to provide a sustainable wastewater treatment characterized by complete biomass retention with the added benefits of lower sludge production, and high-quality effluent, also under critical conditions as for saline wastewater (Lin *et al.*, 2013; Watanabe *et al.*, 2017).

Grethlein proposed the AnMBR concept for the first time in 1978. He employed an external cross-flow membrane to treat DWW, and achieved 85–95% BOD reduction and 72% nitrate removal simultaneously. The first commercially available AnMBR was constructed in the early 1980s by Dorr-Oliver to treat high-strength whey processing wastewater. Since then, AnMBRs have been investigated for municipal and industrial wastewater treatment for a wide spectrum of contaminant loads (Liao *et al.*, 2006; Skouteris *et al.*, 2012). By the 2000s, studies on AnMBRs focused on several issues such as filtration characteristics, system performance, characterization of membrane foulants, and membrane fouling. The success of submerged aerobic membrane bioreactors (MBRs) in the early 2000s encouraged the exploration of submerged anaerobic membrane bioreactors (SAnMBRs) for wastewater treatment. In the later 2010s, SAnMBRs were studied in-depth, with the aim to improve biogas production, extend the application scope, and solve related technical problems (Lin *et al.*, 2013). Although this topic has been widely investigated, studies focused mainly on single treatment system, and there are still some challenging issues regarding AnMBR systems, such as membrane fouling, low flux, high capital, and operational costs (Chernicharo *et al.*, 2015; Lin *et al.*, 2013; Ozgun *et al.*, 2013).

Given the high potentialities of MBRs and the great research interest specifically for the treatment of DWW, Chapter 3 is dedicated to this technology.

### 1.2.2 Status of application

Table 1.1 shows several examples of applications of high-rate anaerobic bioreactors for DWW treatment at laboratory, pilot, and full scale. Their performances are evaluated in terms of COD removal and biogas production for different operating conditions: temperatures, HRTs, and organic loading rates (OLRs). These parameters are the most investigated because their variations strongly affect the process performance. Based on the data shown in Table 1.1, the removal efficiency of COD is 70% on average, with influent COD concentration in the range of 30–1500 mg/L. The highest value of COD reduction is 95%, and it was obtained for a lab-scale UASB ($T = 28°C$) and an ABR ($T = 35°C$) for influent COD of 1000 and 500 mg/L, respectively (Ghangrekar *et al.*, 1996; Langenhoff & Stuckey, 2000). In the first case, biogas production was 0.49–0.55 N $m^3/kg\ COD_{removed}$. Similar values were achieved with an ABR operating under similar conditions, that is, HRT 8 h, temperature in the range of 22–28°C, and influent COD in the range of 505–914 mg/L (Nasr *et al.*, 2009). However, the ABR achieved lower COD removal (≈70%) than the UASB, and achieved efficiencies of ≈80%, at HRT of 24 h. Satisfactory efficiencies for COD removal, ranging from 73 to 88%, were also obtained with AF and EGSB reactors at 35°C and HRT of 24 and 6 h, respectively, and at $T = 15°C$ and HRT in the range of 10–17 and 3.5–5.7 h, respectively (Kobayashi *et al.*, 1983; Li *et al.*, 2007).

**Table 1.1** Examples of application of high-rate anaerobic sewage treatment at different scale (since 2000).

| Bioreactor Configuration | Volume (L) | Operation Temperature (°C) | HRT (h) | OLR (kg COD/m³ day) | Influent COD (mg/L) | COD Removal (%) | Biogas Production (N m³/kg $COD_{removed}$) | $CH_4$ Production (N m³/kg $COD_{removed}$) | Reference |
|---|---|---|---|---|---|---|---|---|---|
| AF | 1.35 | 15–17 | 10–17 | 1–4 | 705 | 80 | | 0.15 | Martín *et al.* (2010) |
| UASB | 3.5 | 9 | 12 | | 310 | 37 | | | Bodík *et al.* (2000) |
| | 3.5 | 15 | 12 | | 310 | 48 | | | |
| | 4 | 13 | 8 | | 456 | 67 | | | Elmitwalli (2000) |
| | $20 \times 10^3$ | 20 | 6 | | 300 | 70 | | | Monroy *et al.* (2000) |
| | $100 \times 10^3$ | 20–25 | 12 | | 500 | 70–80 | | | |
| | $2200 \times 10^3$ | 20 | 20.3 | | 600 | 75–80 (BOD) | | | |
| | 15.7 | 25–13 | 4.7 | 1.6 | 312 | 64–70 | | 0.16–0.26 | Uemura and Harada (2000) |
| | $810 \times 10^3$ | 31 | 9.4 | | 549 | 75 | | | Florencio *et al.* (2001) |
| | 140 | 15 | 6 | 2.88 | 721 | 44 | | 0.09 | Mahmoud *et al.* (2004) |
| | 6.45 | 10 | 8.15 | 1.61 | | 80 | 0.136 | | Gömec (2005) |
| | $60 \times 10^3$ | 18–25 | 23–27 | | 1531 | 51 | | 0.25 | Halalsheh *et al.* (2005) |
| | 120 | 27 | 6 | | 195–816 | 53–57 | | | Leitão *et al.* (2006) |
| | 5.65 | 15 | 24 | 0.935–1.03 | 800–1000 | 83 | | 0.3 | Akila and Chandra (2007) |
| | 30 | 12–27 | 5.08 | | 30–700 | 70 | | | Gömec *et al.* (2009) |

*(Continued)*

**Table 1.1** Examples of application of high-rate anaerobic sewage treatment at different scale (since 2000). (*Continued*)

| Bioreactor Configuration | Volume (L) | Operation Temperature (°C) | HRT (h) | OLR (kg COD/m³ day) | Influent COD (mg/L) | COD Removal (%) | Biogas Production (N m³/kg $COD_{removed}$) | $CH_4$ Production (N m³/kg $COD_{removed}$) | Reference |
|---|---|---|---|---|---|---|---|---|---|
| EGSB | $205 \times 10^3$ | 16–19 | 1.5–5.8 | | 391 | 30 | | | Van der Last and Lettinga (1992) |
| | 4.7 | 15–25 | 3.5–5.7 | 1.6–4.5 | 383–849 | 73–88 | $0.28 \times 10^{-3}$–$0.58 \times 10^{-3}$ N m³/day | | Chu *et al.* (2005) |
| | 3 | 35 | 6 | 1.66 | 150 | 81 | Little or no biogas production | | Li *et al.* (2007) |
| | 18 | 13 | 0.75 | | 369 | 66 | | | Chunjuan *et al.* (2009) |
| ABR | 10 | 35 | 10 | | 500 | 95 | 0.003 N m³/day | | Langenhoff and Stuckey (2000) |
| | 10 | 20 | 10 | | 500 | 70 | 0.0022 N m³/day | | |
| | 10 | 10 | 10 | | 500 | 60 | 0.0019 N m³/day | | |
| | $3 \times 10^3$ | | 238 | | 564 | 58 | 0.39 mol/h | | Foxon *et al.* (2004) |
| | 100 | 18 | 9.5 | | 30–700 | 63 | | | Gömec *et al.* (2009) |
| | 15 | 22–28 | 8 | 2.1 | 505–914 | 67.5 | 0.455 | | Nasr *et al.* (2009) |
| | 15 | 22–28 | 12 | 1.3 | 505–914 | 75.6 | 0.35 | | |
| | 15 | 22–28 | 18 | 0.958 | 505–914 | 79.7 | 0.35 | | |
| | 15 | 22–28 | 24 | 0.669 | 505–914 | 82 | 0.35 | | |
| | 1000 | 18 | 12 | | 760 | 43 | | 0.24 | Hahn and Figueroa (2015) |

*Source:* Modified from Stazi and Tomei (2018).

Concerning full-scale applications of UASB reactors, Monroy *et al.* (2000) and Florencio *et al.* (2001) reported average COD removal of 75% operating at similar HRTs (12 and 9.4 h), influent COD concentrations (500 and 549 mg/L), and at temperatures of 20–25 and 31°C, respectively.

Biogas generation in sewage treatment is still an investigated topic and little data are available in the literature. Reported $CH_4$ yield values are between 0.09 and 0.3 N $m^3$/kg $COD_{removed}$, whereas for biogas composition typical data for the anaerobic digestion of DWW are 70–80% methane, 10–25% nitrogen, and 5–10% carbon dioxide and they are strongly influenced by the operating temperature (Noyola *et al.*, 2006).

A critical aspect of the anaerobic process applied to DWW treatment concerns the presence, in the treated effluent, of dissolved methane ($dCH_4$), which, if not properly recovered, can be released into the atmosphere. The emission of this greenhouse gas (GHG) with a global warming potential 28 times higher than $CO_2$, represents a performance decrease in terms of energy recovery and exerts a strong negative environmental impact. Losses of $dCH_4$ through anaerobic effluents can vary between 45 and 88% of the total $CH_4$ produced, depending on load, reactor type, and temperature. Consequently, strategies for efficient $dCH_4$ recovery or reuse within the same treatment process are required for making anaerobic treatment an attractive option in this field. Strategies for optimizing methane/energy recovery will be discussed in detail in Chapter 7.

## 1.3 MULTI-STEP TREATMENT SYSTEMS IN COMBINATION WITH THE ANAEROBIC PROCESS

UASB reactors have been the most applied treatment option for DWW in several tropical countries, whereas the employment of other high-rate bioreactors has increased in recent years. However, as already mentioned, due to the complexity of the composition of sewage and its high SS content, further research efforts are required for the extensive application of these reactors in moderate climate zones. The main possible drawbacks caused by low operating temperatures, as mentioned in the previous sections, include a decreased process efficiency, a higher content of volatile fatty acids and SSs in the effluent, a pH decrease, a lag-phase in gas production, and low methane content in the biogas (Gömec, 2010).

Several alternative options have been proposed to resolve the above-mentioned problems:

- pre-treatment to remove SSs;
- pre-treatment performed with a high-rate anaerobic bioreactor followed by a post-treatment;
- two-stage systems constituted by a first hydrolytic bioreactor followed by a methanogenic bioreactor;
- combination of two high-rate anaerobic bioreactors and anaerobic bioreactor/digester.

Settling or physical–chemical pre-treatment of DWW prevents the accumulation of solids in the bioreactor and can favor the formation of granular sludge (Vieira & Souza, 1986). It is known that hydrolysis at low temperature occurs at very low rates and represents the limiting step of the whole removal process. This is why at low temperatures the pre-treatment can improve the anaerobic process kinetics and the problems due to hydrolysis, in the case of pre-treated sewage, can be minimized (Foresti *et al.*, 2006).

Post-treatment of anaerobic effluent is advisable to guarantee the effluent quality in terms of organic matter and nutrients considering the intrinsic limitations associated with the anaerobic process kinetics, which are directly related to the substrate concentration, and the stringent discharge standards in many countries and regions, especially in Europe (Chernicharo *et al.*, 2015). In several cases, a post-treatment is required for the removal of residual COD, nutrients, and pathogens in the effluent. The choice of a post-treatment strongly depends on the characteristics of the anaerobic effluent and on local law standards for the reuse of the treated effluent or discharge to the environment

(Kujawa-Roeleveld & Zeeman, 2006). Several post-treatment methods have been proposed in the literature: some examples are waste stabilization ponds (Van Haandel & Lettinga, 1994), trickling filters (Chernicharo & Nascimento, 2001), dynamic membrane filters (Quek *et al.*, 2017), aerobic chambers (Da Silva *et al.*, 2017), activated sludge (Von Sperling *et al.*, 2001), and constructed wetlands (El-Khateeb *et al.*, 2009). Among them, aerobic post-treatment was demonstrated to be effective in enhancing sludge stabilization (Tomei & Carozza, 2015) and pathogen removal (Tomei *et al.*, 2016); whereas constructed wetlands are particularly suitable for decentralized wastewater treatment in rural areas (Álvarez *et al.*, 2008). These latter methods have been deeply investigated and widely utilized being characterized by low cost and easy operation. An extensive analysis of the applicability of constructed wetlands combined with anaerobic digesters is presented elsewhere (Álvarez *et al.*, 2008; De la Varga *et al.*, 2013; El-Khateeb *et al.*, 2009; Liu *et al.*, 2015).

In two-stage systems, the particulate organic matter is partially converted into soluble compounds in the first hydrolytic reactor. Thus, the effluent of the first reactor mostly contains dissolved organic matter, which is suitable to be digested in a second methanogenic reactor. Some investigated alternatives include the use of two UASBs in series, an AF, and a UASB, and a UASB-septic tank (ST) followed by a conventional UASB (Van Haandel *et al.*, 2006). A drawback of a two-stage system can be the high solid accumulation in the first reactor, which occurs with low hydrolysis rates, as in the case of low temperatures. According to Zeeman *et al.* (1997), a two-stage anaerobic process resulted in higher removal efficiency compared to one-phase UASB systems.

Sequential anaerobic systems consist in the combination of different high-rate anaerobic bioreactors in which a post-treatment takes place in the second reactor to polish off the effluent coming from the first reactor. A classic example is the ST with AF. Some other options are an UASB and AF, an UASB and EGSB, and an AF and anaerobic hybrid (AH) reactor (Foresti *et al.*, 2006; Van Haandel *et al.*, 2006). As an EGSB reactor is suitable for low-strength wastewater and low temperatures, it can be used as a polishing reactor of UASBs or any other anaerobic effluent, which needs intense mixing and effective contact between biomass and wastewater to achieve good performances. Influent should enter the EGSB reactor with substrates predominantly dissolved, either acidified or non-acidified, because high upflow velocity is unfavorable for retaining SSs (Van Haandel *et al.*, 2006). Even with flocculent sludge but constituted by dense biomass, pilot-scale polishing reactors for UASB sewage effluent showed good COD and SS removal efficiency (Kato *et al.*, 2003).

An AF followed by an AH reactor operated with granular sludge achieved very good performance for DWW treatment at low temperatures (Lettinga *et al.*, 2001). The AF retained the suspended COD, whereas the AH reactor consisting of a sludge bed in the lower part and filter material in the upper part, physically retained the biomass and enhanced the biological activity, contributing to further COD reduction (Kujawa-Roeleveld & Zeeman, 2006). Elmitwalli (2000) observed a removal efficiency of 71% for total COD when operating a combined AF–AH system treating sewage at 13°C with HRTs of 4 and 8 h, respectively.

Moreover, the removal of colloidal COD could be enhanced in a two-step system by the addition of a small amount of a cationic polymer before the second step, due to the increase in the size of the colloidal particles and consequent easy settling and better retention (Lettinga *et al.*, 2001).

Mahmoud (2002) investigated a combined UASB–anaerobic digester system for both sewage treatment and sludge stabilization at low temperatures. The SS of the influent is retained in the UASB operating at 15°C and the concentrated sludge is fed to the digester operating at 35°C. Digested sludge containing methanogens is recirculated to the UASB reactor to improve its methanogenic capacity and dispersed biodegradable solids attached to the sludge flocs are biodegraded in the digester. In this way, it is possible to prevent the accumulation of non-degraded solids in the sludge bed and to gain better removal efficiencies and sludge stabilization. Biogas produced in the digester should be reused for its heating. The performance of the UASB–digester system was compared with a one-stage UASB, showing substantially better removal performances (Mahmoud *et al.*, 2004).

## 1.4 MICROBIAL COMMUNITY IN ANAEROBIC REACTORS

Organic matter present in low-strength effluents is converted to biogas in anaerobic reactors through a multistage process involving at least four groups of microorganisms: hydrolytic bacteria (including polysaccharolytic, proteolytic, and lipolytic), fermentative bacteria, acetogenic bacteria (including proton-reducing and syntrophic), and methanogenic archaea (Kallistova *et al.*, 2014). These microbial groups carry out the four main steps leading to the conversion of organic macromolecules, such as proteins, lipids, and polysaccharides, to the primary end-products methane and carbon dioxide. These steps are named hydrolysis, acidogenesis, acetogenesis, and methanogenesis (Figure 1.2), and they are discussed below.

(1) *Hydrolysis:* Organic matter is present in low-strength effluents in the form of polymeric materials such as carbohydrates, lipids, and proteins. Hydrolysis is the first step in the decomposition of these materials and it determines the overall rate of the degradation process. Fermentative microbes excrete extracellular enzymes (e.g., cellulases, lipases, and proteases) which break down their respective substrates into soluble monomeric units that can be further metabolized in the subsequent steps.
(2) *Acidogenesis:* Monomeric units such as sugars, amino acids, and fatty acids produced during the hydrolysis are metabolized by different facultative and obligatory anaerobic fermentative bacteria to hydrogen, carbon dioxide, fatty acids, and alcohols, and the composition of these products depend on the environmental conditions and type of metabolism. The biological reactions taking place in this step are thermodynamically favorable and the fermentative bacteria have the highest growth rates among the groups involved in the anaerobic conversion of organic matter (Thauer *et al.*, 1977).

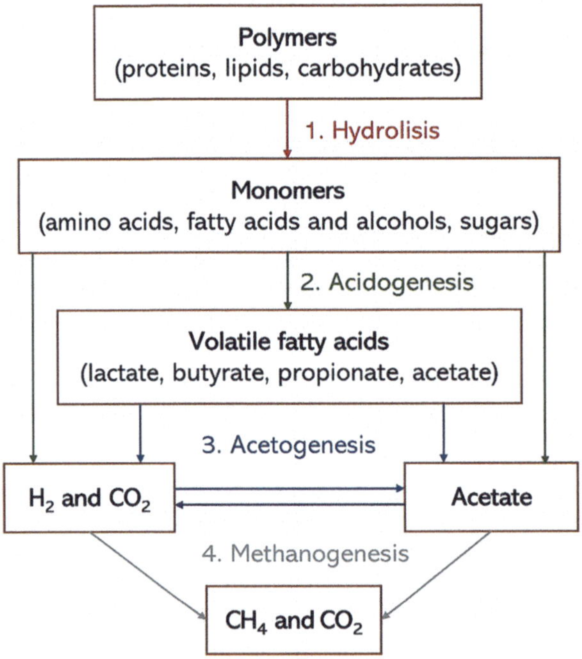


**Figure 1.2** Scheme of the main steps leading to the biological conversion of organic matter to biogas in anaerobic reactors.

(3) *Acetogenesis:* Fatty acids and alcohols produced in the acidogenesis are oxidized to hydrogen, carbon dioxide, and acetate by hydrogen-producing acetogenic bacteria. The conversion of fatty acids to acetate is usually performed by syntrophic communities of acetogens and methanogens because the latter can utilize the hydrogen produced by acetogens as an electron donor.

(4) *Methanogenesis:* Hydrogen and carbon dioxide on the one hand and acetate on the other, formed during either acidogenesis or acetogenesis steps are used as substrates by methanogenic hydrogenotrophic and acetoclastic archaea respectively, resulting in the production of methane. High sensitivity of the methanogenic consortium was observed for environmental factors such as pH, ammonia concentration, OLRs, and macro and micro-nutrients (Ma *et al.*, 2009), which could lead to system imbalance or failure.

## 1.5 PATHOGEN REMOVAL IN ANAEROBIC REACTORS

### 1.5.1 Occurrence of pathogens in wastewater

Human pathogens are agents able to cause disease, being a health hazard to wastewater treatment plant workers and the population living downstream from effluent discharge points and near sludge disposal sites. Their main source in DWW is fecal matter and urine contaminated by infected people (Cai & Zhang, 2013). Pathogens can enter wastewater treatment plants through animal wastes from meat-processing facilities and rats in sanitary sewers. In addition, pathogens may end up in wastewater due to inflow and infiltration from sites containing animal waste on the ground (Gerardi & Zimmerman, 2004).

The most commonly found pathogens in wastewater are bacteria, viruses, fungi, protozoa, and helminths. People can be exposed to these organisms through unsafe drinking water, aerosols from spray irrigation, and recreational use of reclaimed water, meaning that communities lacking access to proper sanitation are at greater risk of exposure to waterborne diseases. For instance, over 360,000 kids under the age of 5 are estimated to die of diarrhea annually (WHO, 2019), which is considered the second highest cause of infant mortality in developing countries (Ugboko *et al.*, 2020). Table 1.2 provides some examples of ubiquitous wastewater pathogens and the diseases associated with them.

### 1.5.2 Regulations on pathogens in wastewater

Several methods to identify and quantify different types of pathogens in wastewater have been established, and most of them involve direct detection using a microscope or culturing of pathogens using artificial media. Despite the low cost, these methods are not always feasible due to the time taken to perform each analysis. Recently, methods relying on the use of polymerase chain reaction (PCR) have been implemented to detect short sections of the genome of specific viral of bacterial pathogens. PCR-based methods are highly specific and sensitive; however, this technology is still costly and the environmental contaminants present in the samples can interfere with the analysis due to low extraction efficiency of the genetic material (Toze, 1999). In addition, high-throughput shotgun sequencing techniques were found to be a potential approach to investigate the distribution, diversity, and abundance of pathogens in wastewater and sludge samples (Cai & Zhang, 2013). Although these techniques can achieve high accuracy and avoid the potential bias characteristic of PCR-based methods, they do not provide a reliable volumetric quantification of pathogens and are relatively costly compared to direct detection or culture-based methods.

Thus, indicator organisms, typically fecal and total coliforms, have been adopted to determine the risk of the presence of pathogens in water and sludge samples (Rose, 2005). *Escherichia coli*, a member of the fecal coliform group, is considered a reliable indicator of fecal pollution due to its prevalence in human and animal feces and its rare incidence in non-polluted environments (Odonkor & Ampofo, 2013). Moreover, there are well-established simple, rapid, and inexpensive methods to detect *E. coli* in water, allowing for the implementation of routine monitoring of pathogens. Although

**Table 1.2** Common genera of pathogens reported in wastewater and diseases associated with them.

| Pathogen | Genus | Diseases |
|---|---|---|
| Bacteria | *Vibrio* | Cholera |
| | *Salmonella* | Salmonellosis |
| | *Shigella* | Shigellosis |
| | *Escherichia* | Urinary tract infection, gastroenteritis |
| | *Burkholderia* | Melioidosis |
| | *Campylobacter* | Gastroenteritis |
| Viruses | *Mastadenovirus* | Acute respiratory infection, gastroenteritis |
| | *Astrovirus* | Gastroenteritis |
| | *Calicivirus* | Gastroenteritis |
| | *Norovirus* | Gastroenteritis |
| | *Enterovirus* | Gastroenteritis |
| | *Rotavirus* | Gastroenteritis |
| Protozoa | *Cryptosporidium* | Cryptosporidiosis |
| | *Giardia* | Giardiasis |
| | *Entamoeba* | Amebiasis |
| Helminths | *Dracunculus* | Guinea worm disease |
| | *Fasciola* | Fascioliasis |
| Fungi | *Aspergillus* | Aspergillosis |
| | *Candida* | Candidiasis |
| | *Rhizopus* | Mucormycosis |

*Source:* Adapted from Magana-Arachchi and Wanigatunge (2020) and Olaolu *et al.* (2014).

*E. coli* has been used as an indicator organism for well over 60 years, but Wu *et al.* (2011) showed that *E. coli* is not strongly correlated with several specific pathogens, suggesting that monitoring multiple indicators might be a better approach for water quality assessment.

Although current regulations do not include the monitoring of microbiological contaminants for treated wastewater effluent, water reuse and resource recovery have become widespread, leading the authorities to propose minimum quality requirements to ensure protection to environment and human health. To address the impacts of water scarcity and droughts, the European Commission proposed a regulation for minimal requirements on water reuse (EU Legislation 2020/741). For the use of reclaimed water in irrigation, parameters such as total SS, turbidity, and *E. coli* must fall below the quality requirements for each quality class. In some situations, *Legionella* spp. and intestinal nematodes must also be monitored. Fecal coliform or *E. coli* are also water quality parameters for non-food crop irrigation in several states across the United States, Mexico, and Australia. Bacterial indicators such as total or fecal coliform, *E. coli*, and *Enterococci* have also been adopted as water quality parameters for urban reuse in several American states. In addition, coliphage, a viral indicator, *Giardia*, *Cryptosporidium*, and *Clostridium* are regulated in some states for water reuse for irrigation of food crops (EPA, 2012).

A study comparing 70 current worldwide regulations and guidelines on agricultural water reuse indicates that some of the common pathogen indicators are fecal coliform, *E. coli*, intestinal nematodes, total coliform, thermotolerant coliform, and *Enterococci* (Shoushtarian & Negahban-Azar, 2020). According to the authors, these microbial parameters and their thresholds are not adequate to ensure that the water reuse practices will not cause harm to human health. Although pathogen indicators are used as water quality parameters in water reuse regulations, it is important to notice that the bacterial community structure may be modified during conventional wastewater treatment processes and reuse site practices, which could lead to higher abundance of species that may be a public health concern. For instance, Kulkarni *et al.* (2018) observed a progressive increase in the abundance of *Legionella* spp. along the wastewater treatment train. Its abundance was reportedly higher at the sprinkler system pumphouse at the irrigation site than in the influent of the wastewater treatment plant, showing that guidelines for water reuse should include a broader assessment of the bacterial community to protect human health.

### 1.5.3 Pathogen removal mechanisms

Anaerobic reactors are primarily designed for COD removal and biogas production. However, they also have the ability to remove pathogens from wastewater, increasing the quality of the final effluent and helping to protect public health. The removal of pathogens occurs through different mechanisms in anaerobic reactors such as adsorption, predation, and competition between microorganisms.

(1) *Adsorption:* Pathogenic microorganisms can be removed from the liquid phase through adhesion to the surfaces of organic matter and SS in wastewater and subsequent removal via sedimentation and/or filtration. The adsorption of viruses to solid surfaces has been investigated by Armanious *et al.* (2016). According to the authors, the process is governed by electrostatic interactions. These interactions depend on the solution pH and ionic strength, which affects the surface charges of the viruses and the sorbents. Hydrophobic effects may also have a favorable contribution to adsorption, especially in systems where high apolarity of the virus and sorbent surfaces is observed. Extracellular polymeric substances excreted from anaerobic bacteria in the sludge can also interact with adjacent bacteria and viruses, leading to their aggregation to sludge through electrostatic and physical interactions (Mahmoud *et al.*, 2003).

(2) *Predation:* Organisms present in anaerobic reactors such as viruses, bacteria, and protozoa have different types of trophic interactions. Bacterivory, a process in which an organism feeds on bacteria as its main source of nutrients, and viral lysis are considered the two main causes of bacterial mortality (Miki & Jacquet, 2008). On the contrary, the authors also show that certain bacteria have active mechanisms to resist viral attacks, contributing to the elimination of viruses, and that viruses can be grazed by heterotrophic protozoa. All these interactions can lead to significant reduction in the number of specific organisms in anaerobic reactors, including pathogens. For instance, studies have shown that enteric viruses can be internalized by free-living protozoa, and human adenovirus can be internalized by ciliates in wastewater (Verbyla & Mihelcic, 2015).

(3) *Competition:* The anaerobic environment in the reactor favors the growth of certain microorganisms that are responsible for steps of the biodegradation of organic matter such as fermentation and methanogenesis. Thus, these microorganisms can outcompete pathogens for resources such as organic matter and nutrients, which can limit their growth and reduce their populations in the reactor. Although pathogen removal by competition has not been reported in anaerobic reactors treating low-strength wastewater, it has been observed during anaerobic sludge digestion, where competition for substrates led to a decline in the concentrations of enteric bacteria in a reactor under mesophilic temperatures (Smith *et al.*, 2005).

Overall, these mechanisms work together to remove pathogens from DWW in anaerobic reactors. The efficiency of pathogen removal can be affected by various factors, such as reactor design, operational conditions, and the characteristics of the wastewater being treated, which are discussed in the next section.

### 1.5.4 Pathogen removal efficiency

Anaerobic reactors rely on microbial activity to degrade organic matter, producing biogas in the process. Pathogens are also removed during the treatment of wastewater, but the effectiveness of this removal can vary depending on the type of anaerobic reactor used. Typical anaerobic systems do not produce effluents that comply with discharge standards for protection of receiving water bodies and for water reuse. Thus, a post-treatment step is usually employed to ensure better quality of the effluent, as shown in Table 1.3.

Studies have reported on the removal of pathogens such as coliforms (total and fecal), viruses, and helminth eggs in anaerobic reactors. In most studies shown in Table 1.3, the removal of pathogens in anaerobic reactors such as UASBs is usually below 1-log reduction. For instance, the virus removal ranged from 0.52 log for F-specific coliphage to 0.99 log reduction for RNA coliphage, whereas bacteria removal ranged from 0.67 log for fecal coliforms to 1.14 log reduction for *E. coli*. Temperature is known as a major factor affecting virus inactivation in anaerobic reactors, as the survival rates of viruses tend to decrease with increasing temperatures (Plaza-Garrido *et al.*, 2022). In the studies assessing viral removal presented in this chapter, the temperatures in UASB reactors were below 25°C, in the mesophilic range. To date, the mechanism for the removal of pathogenic bacteria in UASB reactors has not been widely investigated, but it could be related to their sorption to sludge. According to Espinosa *et al.* (2021), the number of *E. coli* in the sludge of an UASB reactor was one order of magnitude higher than in the clarified liquid effluent. Low removal rates were also observed for protozoa cysts and helminth eggs in UASB reactors. Yaya-Beas *et al.* (2015) showed that the removal rate of *Ascaris suum* eggs increased as the upflow velocity decreased in the reactor. At lower upflow velocities, the viscosity of the flocculent anaerobic sludge decreases, reducing its capacity to retain helminth eggs.

Because UASB reactors do not produce effluents with desirable characteristics for environmental discharge and water reuse, they are commonly combined with a post-treatment step. Processes such as sand filtration and downflow hanging sponge (DHS) reactors can reduce the pathogens concentrations via retention of the biomass in the media. For example, fecal coliforms and total coliphage removals of 1.86 log and 1.26 log reduction, respectively, were achieved in a DHS unit composed of two rectangular sheets in series. A sand filter was not as efficient, showing a fecal coliform removal of 0.44 log reduction, which could be due to the lower surface area of the media as compared with a DHS reactor. High-rate algal ponds have also been successfully used to improve the quality of the UASB effluent. Some of the mechanisms for pathogen removal in this process are direct and indirect sunlight-mediated inactivation, predation, starvation, high dissolved oxygen, and high pH values (Von Sperling, 2005).

AnMBRs showed a better performance than UASB reactors in terms of pathogen removal. Saddoud *et al.* (2006) demonstrated that bacteria, helminth eggs, and protozoa could be reduced several orders of magnitude by an anaerobic reactor with an ultrafiltration membrane. The mechanisms involved in the pathogen removal by the system are size exclusion, because pathogens, except for most viruses, are bigger than the pore size of ultra- or microfiltration membranes, and electrostatic repulsion, leading to 5–7 log reduction of fecal bacteria (Ottoson *et al.*, 2006). High removal of viruses smaller than the pore size can also be achieved, due to the formation of a biofilm on the membrane. However, it is important to notice that membrane fouling decreases the permeate flux, indicating that operation strategies must be developed to optimize viral removal and filtration performance (Zhang *et al.*, 2022).

## 1.6 CONCLUDING REMARKS

Feasibility of high-rate bioreactors for treating DWW has been demonstrated even at moderate climate regions. They, in fact, can provide a more sustainable treatment of DWW characterized by almost full recovery of energy and nutrient contents of the treated influent. Among the still critical issues to be investigated are the long start-up times, the high-nutrient contents of the effluent, which is

**Table 1.3** Pathogen removal in anaerobic systems (with or without post-treatment) treating low-strength wastewater.

| Treatment Units | Reactor Configuration | Influent | Retention Time | Temperature | Pathogens | | | | Reference |
|---|---|---|---|---|---|---|---|---|---|
| | | | | | Parameter | Influent | Effluent Anaerobic Reactor | Effluent Post-treatment | |
| UASB reactor followed by a DHS unit | UASB: 155 L column | Raw sewage | UASB: 6–30 h | 25°C | Total coliphage | $8.9 \times 10^3$ PFU/mL | $1.6 \times 10^3$ PFU/mL | $8.7 \times 10^1$ PFU/mL | Uemura *et al.* (2002) |
| | DHS: two rectangular sheets in series | | DHS: 2 h | | RNA coliphage | $3.0 \times 10^3$ PFU/mL | $3.1 \times 10^2$ PFU/mL | $2.9 \times 10^1$ PFU/mL | |
| | | | | | Fecal coliforms | $2.4 \times 10^5$ CFU/mL | $4.7 \times 10^4$ CFU/mL | $6.5 \times 10^2$ CFU/mL | |
| UASB followed by baffled pond system | Pilot-scale UASB reactor: 9000 L | Raw sewage | Pilot-scale UASB reactor: 5.5 h | 22.7°C | *E. coli* | $1.68 \times 10^9$ MPN/100 mL | $1.68 \times 10^8$ MPN/100 mL | $3.87 \times 10^6$ MPN/100 mL | von Sperling *et al.* (2002) |
| | Polishing pond: $32 \times 10^3$ L | | Polishing pond: 6.3 days | | Helminth eggs | 64.26/L | 16.16/L | 0.09/L | |
| Full-scale UASB reactor treating DWW followed by polishing pond | 27 MLD, UASB reactor and polishing pond | Sewage collected downstream from grit chamber | UASB: NA | NA | Fecal coliforms | $1.58 \times 10^7$ MPN/100 mL | $1.58 \times 10^6$ MPN/100 mL | $4.62 \times 10^5$ MPN/100 mL | Pant and Mittal (2007) |
| | | | | | Fecal streptococci | $3.36 \times 10^6$ MPN/100 mL | $3.30 \times 10^5$ MPN/100 mL | $0.92 \times 10^4$ MPN/100 mL | |
| | | | Polishing pond: 24 h | | Salmonella | 280 MPN/100 mL | 27 MPN/ 100 mL | 10 MPN/ 100 mL | |
| | | | | | Shigella | 37 MPN/100 mL | 5 MPN/ 100 mL | 3 MPN/ 100 mL | |
| | | | | | Vibrio | 440 MPN/100 mL | 59 MPN/ 100 mL | 16 MPN/ 100 mL | |
| UASB reactor | 25 L reactor with bed height varying from 0.3 to 0.7 m | Raw domestic sewage | 4 h | 17.1–28.6°C | *Ascaris suum* eggs | Removal between 34 and 100%, 30 and 91%, and 34 and 56%, when the sludge bed in the UASB reactor was 19–25%, 31–38%, and 38–44% of the total reactor height, respectively | | NA | Yaya-Beas *et al.* (2015) |
| UASB followed by sand filter | Pilot-scale UASB reactor: 250 L | Domestic septic tank effluent | UASB: 6 h | 19.0°C | Fecal coliforms | $5.1 \times 10^5$ CFU/mL | $1.1 \times 10^5$ CFU/mL | $4.0 \times 10^4$ CFU/mL | Lohani *et al.* (2020) |
| | Sand and gravel filter: 1000 L | | | | | | | | |

*(Continued)*

**Table 1.3** Pathogen removal in anaerobic systems (with or without post-treatment) treating low-strength wastewater (*Continued*).

| Treatment Units | Reactor Configuration | Influent | Retention Time | Temperature | Pathogens | | | | Reference |
|---|---|---|---|---|---|---|---|---|---|
| | | | | | Parameter | Influent | Effluent Anaerobic Reactor | Effluent Post-treatment | |
| UASB followed by high-rate algal ponds (HRAP) | UASB: 343 L | Raw sewage | UASB: 7 h | UASB: 23.6°C | *E. coli* | $9.54 \times 10^8$ MPN/100 mL | $6.91 \times 10^7$ MPN/100 mL | $4.90 \times 10^4$ MPN/100 mL | Espinosa *et al.* (2021) |
| | 2 HRAPs in parallel: 205 L | | HRAPs: 8 days | HRAPs: 23.3°C | Somatic coliforms | $4.5 \times 10^5$ PFU/100 mL | $1.82 \times 10^5$ PFU/100 mL | $1.20 \times 10^4$ PFU/100 mL | |
| | Sedimentation chamber: 30 L | | | | F-specific coliphage | $1.26 \times 10^5$ PFU/100 mL | $3.80 \times 10^4$ PFU/100 mL | $9.12 \times 10^2$ PFU/100 mL | |
| Full-scale UASB reactor | $2794 \times 10^3$ L UASB reactor | Sewage collected downstream from tank for sand and grease removal | 8–10 h | 21.7°C | *Giardia* spp. cysts | 808/L | 5268/L | NA | Valdez *et al.* (2022) |
| | | | | | *Cryptosporidium* spp. oocysts | 10.7/L | 32.0/L | | |
| AnMBR | 50 L anaerobic reactor with 1 m² ultrafiltration membrane | Raw sewage | NA | 37°C | Total coliforms | $84 \times 10^5$ CFU/100 mL | ND | | Saddoud *et al.* (2006) |
| | | | | | Fecal coliforms | $42 \times 10^5$ CFU/100 mL | ND | | |
| | | | | | Fecal streptococci | $4.5 \times 10^5$ CFU/100 mL$^{-1}$ | ND | | |
| | | | | | Salmonella | 940 MPN/L | ND | | |
| | | | | | Helminth eggs | 11/L | ND | | |
| | | | | | Protozoan | $1.621 \times 10^5$ cysts/L | ND | | |
| Submerged AnMBR | 350 L anaerobic reactor with 3.5 m² membrane | Raw sewage | NA | 20°C | Fecal coliforms | $4.3 \times 10^6$ MPN/100 mL | 49 MPN/100 mL | ND | Martinez-Sosa *et al.* (2011) |
| AnMBR | 35 L anaerobic reactor with 1.4 m² membrane | Raw sewage | 8–12 h | 35°C | Mycobacterium | Net growth rate of –4.8 to 0.5 in the effluent | | | Medina *et al.* (2023) |
| | | | | | *Acinetobacter* | Net growth rate of –4.7 to –3.0 in the effluent | | | |
| | | | | | *Arcobacter* | Net growth rate of –2.6 to 1.6 in the effluent | | | |
| | | | | | *Pseudomonas* | Net growth rate of –2.2 to –1.0 in the effluent | | | |
| Anaerobic macrophyte pond followed by constructed wetland | System surface area: 300 m² | Raw sewage from university campus | NA | 13–26°C | Total coliforms | $3.1 \times 10^6$ CFU/100 mL | $5.6 \times 10^4$ CFU/100 mL | $2.5 \times 10^3$ CFU/100 mL | Guedes-Alonso *et al.* (2020) |
| | System volume: 340 m³ | | | | Fecal coliforms | $1.1 \times 10^6$ CFU/100 mL | $1.5 \times 10^4$ CFU/100 mL | $9.6 \times 10^2$ CFU/100 mL | |

MPN, Most Probable Number; PFU, Plate Forming Unit; MLD, million liters per day; NA, not available; ND, not detected.

a positive aspect when these nutrients can be recovered in agriculture, but can be an issue if they have to be removed to respect the regulatory limits. Moreover, low methane production and high methane solubility can affect the real amount of methane recovered with consequent reduction of the energy production and the potential emission of this dangerous GHG in the atmosphere. This is why research efforts should be devoted to technological solutions able to maximize the recovery of the produced methane and achieve an effective positive energy balance in wastewater treatment.

Best results in facing treatment of DWW at low temperature were obtained with AnMBRs, which improve effluent quality over the anaerobic mainstream bioreactors, coupled with some of the highest energy recovery values. After the proved success of AnMBRs on DWW treatment at lab-scale, several pilot-scale studies have recently been conducted (Shin & Bae, 2018), but a complete picture of the real potentialities and exhaustive information on the energy balance are still not exhaustive. Additional research as well as quantitative environmental and economic evaluations for full-scale implementation of AnMBRs are required due to several reasons, such as membrane fouling, membranes sensitivity to toxicity, and low nitrogen and phosphorus removals. To the best of our knowledge, results on full-scale trials are not available. Most authors, indeed, worked with laboratory-scale bioreactors, but because the membrane performance cannot be directly scaled-up from laboratory to real plant, further studies are needed to facilitate the implementation of this technology at full-scale wastewater treatment plants.

Another promising solution is UASB bioreactors, a well-known technology applied at full scale for the treatment of DWW in warm climate regions. Suspended biomass UASB bioreactors cannot guarantee the same performance of AnMBRs in terms of effluent quality but their performance can be substantially improved by operating with immobilized or granular biomass and/or by adding a post-treatment depending on the required characteristics of the treated effluent. In comparison to AnMBRs, UASB is a simpler technology, extensively tested even at full scale, and characterized by lower costs of installation and maintenance, so to be considered as a valid alternative for DWW treatment depending on the specific required effluent standards.

Same considerations apply for pathogens' removal: also in this case, AnMBRs and UASBs coupled with a post-treatment (i.e. sand filtration, DHS reactors, and high-rate algal ponds) appear as feasible solutions able to achieve satisfactory performance for effluent hygienization. In addition, AnMBRs can also achieve high removal efficiencies for antibiotic resistant bacteria and antibiotic resistance genes, which will be discussed in Chapter 12.

## REFERENCES


Abbasi H. N., Lu X. and Xu F. (2017). Seasonal performance and characteristic of ABR for low strength wastewater. *Applied Ecology and Environmental Research*, **15**, 263–273, https://doi.org/10.15666/aeer/1501_263273

Aiyuk S., Amoako J., Raskin L., van Haandel A. and Verstraete W. (2004). Removal of carbon and nutrients from domestic wastewater using a low investment, integrated treatment concept. *Water Research*, **38**, 3031–3042, https://doi.org/10.1016/j.watres.2004.04.040

Aiyuk S., Forrez I., Lieven D. K., van Haandel A. and Verstraete W. (2006). Anaerobic and complementary treatment of domestic sewage in regions with hot climates – a review. *Bioresource Technology*, **97**, 2225–2241, https://doi.org/10.1016/j.biortech.2005.05.015

Akila G. and Chandra T. S. (2007). Performance of an UASB reactor treating synthetic wastewater at low-temperature using cold-adapted seed slurry. *Process Biochemistry*, **42**, 466–471, https://doi.org/10.1016/j.procbio.2006.09.010

Álvarez J. A., Ruíz I. and Soto M. (2008). Anaerobic digesters as a pretreatment for constructed wetlands. *Ecological Engineering*, **33**, 54–67, https://doi.org/10.1016/j.ecoleng.2008.02.001

Armanious A., Aeppli M., Jacak R., Refardt D., Sigstam T., Kohn T. and Sander M. (2016). Viruses at solid–water interfaces: a systematic assessment of interactions driving adsorption. *Environmental Science and Technology*, **50**, 732–743, https://doi.org/10.1021/acs.est.5b04644

Bae J., Shin C., Lee E., Kim J. and McCarty P. L. (2014). Anaerobic treatment of low-strength wastewater: a comparison between single and staged anaerobic fluidized bed membrane bioreactors. *Bioresource Technology*, **165**, 75–80, https://doi.org/10.1016/j.biortech.2014.02.065

Bajpai P. (2017). Anaerobic reactors used for waste water treatment. In: Anaerobic Technology in Pulp and Paper Industry, Springer, Singapore, pp. 37–53.

Barber W. P. and Stuckey D. C. (1999). The use of the anaerobic baffled reactor (ABR) for wastewater treatment: a review. *Water Research*, **33**, 1559–1578, https://doi.org/10.1016/S0043-1354(98)00371-6

Bodík I., Herdová B. and Drtil M. (2000). Anaerobic treatment of the municipal wastewater under psychrophilic conditions. *Bioprocess Engineering*, **22**, 385–390, https://doi.org/10.1007/s004490050748

Bodkhe S. Y. (2009). A modified anaerobic baffled reactor for municipal wastewater treatment. *Journal of Environmental Management*, **90**, 2488–2493, https://doi.org/10.1016/j.jenvman.2009.01.007

Cai L. and Zhang T. (2013). Detecting human bacterial pathogens in wastewater treatment plants by a high-throughput shotgun sequencing technique. *Environmental Science and Technology*, **47**, 5433–5441, https://doi.org/10.1021/es400275r

Chan Y. J., Chong M. F., Law C. L. and Hassell D. G. (2009). A review on anaerobic–aerobic treatment of industrial and municipal wastewater. *Chemical Engineering Journal*, **155**, 1–18, https://doi.org/10.1016/j.cej.2009.06.041

Chernicharo C. A. L. and Nascimento M. C. P. (2001). Feasibility of a pilot-scale UASB/trickling filter system for domestic sewage treatment. *Water Science and Technology*, **44**, 221–228, https://doi.org/10.2166/wst.2001.0227

Chernicharo C. A. L., van Lier J. B., Noyola A. and Bressani Ribeiro T. (2015). Anaerobic sewage treatment: state of the art, constraints and challenges. *Reviews in Environmental Science and Bio/Technology*, **14**, 649–679, https://doi.org/10.1007/s11157-015-9377-3

Chinwetkitvanich S. and Ruchiraset A. (2017). The anaerobic baffled reactor (ABR): performance and microbial population at various COD loading rates. *International Journal of Geomate*, **12**, 78–84.

Chu L. B., Yang F. L. and Zhang X. W. (2005). Anaerobic treatment of domestic wastewater in a membrane-coupled expended granular sludge bed (EGSB) reactor under moderate to low temperature. *Process Biochemistry*, **40**, 1063–1070, https://doi.org/10.1016/j.procbio.2004.03.010

Chunjuan D., Xiao L., Suyun C., Yanxia W., Qingliang Z. and Bingnan L. (2009). Anaerobic treatment of actual domestic wastewater with EGSB reactor at ambient temperature. Bioinformatics and Biomedical Engineering. ICBBE 2009. 3rd International Conference, IEEE, pp. 1–3.

Connaughton S., Collins G. and O'Flaherty V. (2006). Development of microbial community structure and activity in a high-rate anaerobic bioreactor at 18°C. *Water Research*, **40**, 1009–1017, https://doi.org/10.1016/j.watres.2005.12.026

Da Silva J. A., Sarti A., Maintinguer S. I., Kaiser I. M. and da Silva G. H. R. (2017). Performance of an anaerobic baffled reactor with an aerobic chamber treating low-strength wastewater. *Desalination and Water Treatment*, **100**, 1–10, https://doi.org/10.5004/dwt.2017.21713

Daud M. K., Rizvi H., Akram M. F., Ali S., Rizwan M., Nafees M. and Jin Z. S. (2018). Review of upflow anaerobic sludge blanket reactor technology: effect of different parameters and developments for domestic wastewater treatment. *Journal of Chemistry*, **2018**, 1–13.

De la Varga D., Díaz M. A., Ruiz I. and Soto M. (2013). Avoiding clogging in constructed wetlands by using anaerobic digesters as pre-treatment. *Ecological Engineering*, **52**, 262–269, https://doi.org/10.1016/j.ecoleng.2012.11.005

El-Khateeb M. A., Al-Herrawy A. Z., Kamel M. M. and El-Gohary F. A. (2009). Use of wetlands as post-treatment of anaerobically treated effluent. *Desalination*, **245**, 50–59, https://doi.org/10.1016/j.desal.2008.01.071

Elmitwalli T. A. (2000). Anaerobic treatment of domestic sewage at low temperature. Ph.D. thesis, Sub-department of Environmental Technology, Wageningen University, Wageningen, The Netherlands.

EPA. (2012). Guidelines for water reuse.

Espinosa M. F., Verbyla M. E., Vassalle L., Rosa-Machado A. T., Zhao F., Gaunin A. and Mota C. R. (2021). Reduction and partitioning of viral and bacterial indicators in a UASB reactor followed by high rate algal ponds treating domestic sewage. *Science of the Total Environment*, **760**, 144309, https://doi.org/10.1016/j.scitotenv.2020.144309

Florencio L., Kato M. T. and de Morais J. C. (2001). Domestic sewage treatment in full-scale UASBB plant at Mangueira, Recife, Pernambuco. *Water Science and Technology*, **44**, 71–77, https://doi.org/10.2166/wst.2001.0182

Foresti E. (2002). Anaerobic treatment of domestic sewage: established technologies and perspectives. *Water Science and Technology*, **45**, 181–186, https://doi.org/10.2166/wst.2002.0324

Foresti E., Zaiat M. and Vallero M. (2006). Anaerobic processes as the core technology for sustainable domestic wastewater treatment: consolidated applications, new trends, perspectives, and challenges. *Environmental Science and Technology*, **5**, 3–19.

Foxon K. M., Pillay S., Lalbahadur T., Rodda N., Holder F. and Buckley C. A. (2004). The anaerobic baffled reactor (ABR): an appropriate technology for on-site sanitation. *Water SA*, **30**, 44–50.

Gerardi M. and Zimmerman M. (2004). Wastewater pathogens.

Ghangrekar M. M., Asolekar S. R., Ranganathan K. R. and Joshi S. G. (1996). Experience with UASB reactor start-up under different operating conditions. *Water Science and Technology*, **34**, 421–428, https://doi.org/10.2166/wst.1996.0578

Gömec C. Y. (2005). A comparative evaluation of anaerobic treatment of low strength wastewaters at mesophilic and psychrophilic temperatures. PhD. Thesis, Istanbul Technical University, Istanbul, Turkey.

Gömec C. Y. (2010). High-rate anaerobic treatment of domestic wastewater at ambient operating temperatures: a review on benefits and drawbacks. *Journal of Environmental Science and Health, Part A: Toxic/Hazardous Substances and Environmental Engineering*, **45**, 1169–1184.

Gömec C. Y., Eroglu V. and Wilderer P. A. (2009). Identifying aceticlastic and hydrogenotrophic methanogens in psychrophilic and mesophilic granular sludges treating synthetic sewage by means of FISH and CSLM. *Journal of Rapid Methods and Automation in Microbiology*, **17**, 135–153, https://doi.org/10.1111/j.1745-4581.2009.00164.x

Gopala Krishna G. V. T., Kumar P. and Kumar P. (2008). Treatment of low-strength complex wastewater using an anaerobic baffled reactor (ABR). *Bioresource Technology*, **99**, 8193–8200, https://doi.org/10.1016/j.biortech.2008.03.016

Guedes-Alonso R., Montesdeoca-Esponda S., Herrera-Melián J. A., Rodríguez-Rodríguez R., Ojeda-González Z., Landívar-Andrade V., Sosa-Ferrera Z. and Santana-Rodríguez J. J. (2020). Pharmaceutical and personal care product residues in a macrophyte pond-constructed wetland treating wastewater from a university campus: presence, removal and ecological risk assessment. *Science of the Total Environment*, **703**, 135596, https://doi.org/10.1016/j.scitotenv.2019.135596

Hahn M. J. and Figueroa L. A. (2015). Pilot scale application of anaerobic baffled reactor for biologically enhanced primary treatment of raw municipal wastewater. *Water Research*, **87**, 494–502, https://doi.org/10.1016/j.watres.2015.09.027

Halalsheh M., Sawajneh Z., Zu'bi M., Zeeman G., Lier J., Fayyad M. and Lettinga G. (2005). Treatment of strong domestic sewage in a 96 $m^3$ UASB reactor operated at ambient temperatures. two stage versus one stage UASB reactors. *Bioresource Technology*, **96**, 577–585, https://doi.org/10.1016/j.biortech.2004.06.014

Ho J. and Sung S. (2010). Methanogenic activities in anaerobic membrane bioreactors (AnMBR) treating synthetic municipal wastewater. *Bioresource Technology*, **101**, 2191–2196, https://doi.org/10.1016/j.biortech.2009.11.042

Iza J. (1991). Fluidized bed reactors for anaerobic wastewater treatment. *Water Science and Technology*, **24**, 109–132, https://doi.org/10.2166/wst.1991.0221

Kallistova A. Yu., Goel G. and Nozhevnikova A. N. (2014). Microbial diversity of methanogenic communities in the systems for anaerobic treatment of organic waste. *Microbiology*, **83**(5), 462–483, https://doi.org/10.1134/S0026261714050142

Kato M. T., Florencio L. and Arantes R. F. M. (2003). Post treatment of UASB effluent in an expanded granular sludge bed reactor type using flocculent sludge. *Water Science and Technology*, **48**, 279–284, https://doi.org/10.2166/wst.2003.0413

Kettunen R. H. and Rintala J. A. (1997). The effect of low temperature (5–29°C) and adaptation on the methanogenic activity of biomass. *Applied Microbiology and Biotechnology*, **48**, 570–576, https://doi.org/10.1007/s002530051098

Khanal S. K. (2008). Anaerobic Biotechnology for Bioenergy Production: Principles and Applications. John Wiley & Sons Ltd, West Sussex, UK.

Kim J., Kim K., Ye H., Lee E., Shin C., McCarty P. L. and Bae J. (2011). Anaerobic fluidized bed membrane bioreactor for wastewater treatment. *Environmental Science and Technology*, **45**, 576–581, https://doi.org/10.1021/es1027103

Kobayashi H. A., Stenstrom M. K. and Mah R. A. (1983). Treatment of low strength domestic wastewater using the anaerobic filter. *Water Research*, **17**, 903–909, https://doi.org/10.1016/0043-1354(83)90164-1

Kujawa-Roeleveld K. and Zeeman G. (2006). Anaerobic treatment in decentralised and source-separation-based sanitation concepts. *Environmental Science and Biotechnology*, **5**, 115–139, https://doi.org/10.1007/s11157-005-5789-9

Kulkarni P., Olson N. D., Paulson J. N., Pop M., Maddox C., Claye E., Rosenberg Goldstein R. E., Sharma M., Gibbs S. G., Mongodin E. F. and Sapkota A. R. (2018). Conventional wastewater treatment and reuse site practices modify bacterial community structure but do not eliminate some opportunistic pathogens in reclaimed water. *Science of the Total Environment*, **639**, 1126–1137, https://doi.org/10.1016/j.scitotenv.2018.05.178

Langenhoff A. A. M. and Stuckey D. C. (2000). Treatment of dilute wastewater using an anaerobic baffled reactor: effect of low temperature. *Water Research*, **34**, 3867–3875, https://doi.org/10.1016/S0043-1354(00)00136-6

Leitão R. C., van Haandel A. C., Zeeman G. and Lettinga G. (2006). The effects of operational and environmental variations on anaerobibc wastewater treatment systems: a review. *Bioresource Technology*, **97**, 1105–1118, https://doi.org/10.1016/j.biortech.2004.12.007

Lettinga G., Field J., Van Lier J. and Zeeman G. (1997a). Advanced anaerobic wastewater treatment in the near future. *Water Science and Technology*, **35**, 5–12, https://doi.org/10.2166/wst.1997.0347

Lettinga G., Hulshoff P. L. W., Zeeman G., Field J., Lier J. B., van Buuren J. C. L., van Janssen A. J. H. and Lens P. (1997b). Anaerobic treatment in sustainable environmental production concepts. Proceedings of the 8th International Conference on Anaerobic Digestion, Vol. **1**, Sendai, Japan, pp. 32–39.

Lettinga G., Rebac S., Parshina S., Nozhevnikova A., van Lier J. B. and Stams A. J. M. (1999). High-rate anaerobic treatment of wastewater at low temperature. *Applied and Environmental Microbiology*, **65**, 1696–1702, https://doi.org/10.1128/AEM.65.4.1696-1702.1999

Lettinga G., Rebac S. and Zeeman G. (2001). Challenge of psychrophilic anaerobic wastewater treatment. *Trends in Biotechnology*, **19**, 363–370, https://doi.org/10.1016/S0167-7799(01)01701-2

Li X. M., Guo L., Yang Q., Zeng G. M. and Liao D. X. (2007). Removal of carbon and nutrients from low strength domestic wastewater by expanded granular sludge bed-zeolite bed filtration (EGSB-ZBF) integrated treatment concept. *Process Biochemistry*, **42**, 1173–1179, https://doi.org/10.1016/j.procbio.2007.04.001

Liao B. Q., Kraemer J. T. and Bagley D. M. (2006). Anaerobic membrane bioreactors: applications and research directions. *Critical Reviews in Environmental Science and Technology*, **36**, 489–530, https://doi.org/10.1080/10643380600678146

Lin H., Peng W., Zhang M., Chen J., Hong H. and Zhang Y. (2013). A review on anaerobic bioreactors: applications, membrane fouling and future perspectives. *Desalination*, **314**, 169–188, https://doi.org/10.1016/j.desal.2013.01.019

Liu R., Zhao Y., Doherty L., Hu Y. and Hao X. (2015). A review of incorporation of constructed wetland with other treatment processes. *Chemical Engineering Journal*, **279**, 220–230, https://doi.org/10.1016/j.cej.2015.05.023

Liu Y., Huang L., Dong G., Liu G., Wu X., Wang C., Liu X. and Wang L. (2018). Enhanced granulation and methane recovery at low load by downflow sludge circulation in anaerobic treatment of domestic wastewater. *Bioresource Technology*, **249**, 851–857, https://doi.org/10.1016/j.biortech.2017.10.091

Lohani S. P., Khanal S. N. and Bakke R. (2020). A simple anaerobic and filtration combined system for domestic wastewater treatment. *Water-Energy Nexus*, **3**, 41–45, https://doi.org/10.1016/j.wen.2020.03.004

Ma J., Mungoni L. J., Verstraete W. and Carballa M. (2009). Maximum removal rate of propionic acid as a sole carbon source in UASB reactors and the importance of the macro- and micro-nutrients stimulation. *Bioresource Technology*, **100**(14), 3477–3482, https://doi.org/10.1016/j.biortech.2009.02.060

Magana-Arachchi D. N. and Wanigatunge R. P. (2020). Ubiquitous waterborne pathogens. In: Waterborne Pathogens: Detection and Treatment, M. Narasimha, V. Prasad and A. Grobelak (eds), Butterworth-Heinemann, Kidlington, Oxford, UK, pp. 15–42.

Mahmoud N. (2002). Anaerobic pre-treatment of sewage under low temperature (15°C) conditions in an integrated UASB-Digester system. Ph.D. thesis, Wageningen University, Wageningen, The Netherlands.

Mahmoud N., Zeeman G., Gijzen H. and Lettinga G. (2003). Solids removal in upflow anaerobic reactors, a review. *Bioresource Technology*, **90**(1), 1–9, https://doi.org/10.1016/S0960-8524(03)00095-6

Mahmoud N., Zeeman G., Gijzen H. and Lettinga G. (2004). Anaerobic sewage treatment in a one-stage UASB reactor and a combined UASB–digester system. *Water Research*, **38**, 2348–2358, https://doi.org/10.1016/j.watres.2004.01.041

Manariotis I. D. and Grigoropoulos S. G. (2002). Low strength wastewater treatment using an anaerobic baffled reactor. *Water Environment Research*, **74**, 170–176, https://doi.org/10.2175/106143002X139884

Manariotis I. D. and Grigoropoulos S. G. (2003). Anaerobic treatment of low-strength wastewater in a biofilm reactor. *Journal of Environmental Science and Health, Part A*, **38**, 2057–2068, https://doi.org/10.1081/ESE-120023332

Manariotis I. D. and Grigoropoulos S. G. (2008). Restart of anaerobic filters treating low-strength wastewater. *Bioresource Technology*, **99**, 3579–3589, https://doi.org/10.1016/j.biortech.2007.07.048

Manariotis I. D. and Grigoropoulos S. G. (2006). Municipal-wastewater treatment using upflow anaerobic filters. *Water Environment Research*, **78**, 233–242, https://doi.org/10.2175/106143005X90029

Martín M. A., De la Rubia M. A., Martín A., Borja R., Montalvo S. and Sánchez E. (2010). Kinetic evaluation of the psychrophylic anaerobic digestion of synthetic domestic sewage using an upflow filter. *Bioresource Technology*, **101**, 131–137, https://doi.org/10.1016/j.biortech.2009.08.010

Martinez-Sosa D., Helmreich B., Netter T., Paris S., Bischof F. and Horn H. (2011). Anaerobic submerged membrane bioreactor (AnSMBR) for municipal wastewater treatment under mesophilic and psychrophilic temperature conditions. *Bioresource Technology*, **102**, 10377–10385, https://doi.org/10.1016/j.biortech.2011.09.012

Matsushige K., Inamori Y., Mizuochi M., Hosomi M. and Sudo R. (1990). The effects of temperature on anaerobic filter treatment for low-strength organic wastewater. *Environmental Technology*, **11**, 899–910, https://doi.org/10.1080/09593339009384942

McCarty P. L., Bae J. and Kim J. (2011). Domestic wastewater treatment as a net energy producer – can this be achieved? *Environmental Science and Technology*, **45**, 7100–7106, https://doi.org/10.1021/es2014264

Medina J. S., Zhang S., Wang C., Zhou J. and Hong P.-Y. (2023). Decreasing hydraulic retention time of anaerobic membrane bioreactor: effect on core genera and microbial contaminants removal. *Bioresource Technology Reports*, **22**, 101389, https://doi.org/10.1016/j.biteb.2023.101389

Miki T. and Jacquet S. (2008). Complex interactions in the microbial world: underexplored key links between viruses, bacteria and protozoan grazers in aquatic environments. *Aquatic Microbial Ecology*, **51**, 195–208, https://doi.org/10.3354/ame01190

Monroy O., Famá G., Meraz M., Montoya L. and Macarie H. (2000). Anaerobic digestion for wastewater treatment in Mexico: state of technology. *Water Research*, **34**, 1803–1816, https://doi.org/10.1016/S0043-1354(99)00301-2

Nachaiyasit S. and Stuckey D. C. (1997). Effect of low temperatures on the performance of an anaerobic baffled reactor (ABR). *Journal of Chemical Technology and Biotechnology*, **69**, 276–284, https://doi.org/10.1002/(SICI)1097-4660(199706)69:2<276::AID-JCTB711>3.0.CO;2-T

Nasr F. A., Doma H. S. and Nassar H. F. (2009). Treatment of domestic wastewater using an anaerobic baffled reactor followed by a duckweed pond for agricultural purposes. *The Environmentalist*, **29**, 270–279, https://doi.org/10.1007/s10669-008-9188-y

Noyola A., Morgan-Sagastume J. M. and Lopez-Hernandez J. E. (2006). Treatment of biogas produced in anaerobic reactors for domestic wastewater: odor control and energy/resource recovery. *Environmental Science and Biotechnology*, **5**, 93–114, https://doi.org/10.1007/s11157-005-2754-6

Odonkor S. T. and Ampofo J. K. (2013). *Escherichia coli* as an indicator of bacteriological quality of water: an overview. *Microbiology Research*, **4**, 2, https://doi.org/10.4081/mr.2013.e2

Olaolu D. T., Akpor O. B. and Akor C. O. (2014). Pollution indicators and pathogenic microorganisms in wastewater treatment: implication on receiving water bodies.

Ottoson J., Hansen A., Björlenius B., Norder H. and Stenström T. A. (2006). Removal of viruses, parasitic protozoa and microbial indicators in conventional and membrane processes in a wastewater pilot plant. *Water Research*, **40**(7), 1449–1457, https://doi.org/10.1016/j.watres.2006.01.039

Ozgun H., Dereli R. K., Ersahin M. E., Kinaci C., Spanjers H. and van Lier J. B. (2013). A review of anaerobic membrane bioreactors for municipal wastewater treatment: integration options, limitations and expectations. *Separation and Purification Technology*, **118**, 89–104, https://doi.org/10.1016/j.seppur.2013.06.036

Pant A. and Mittal A. K. (2007). Monitoring of pathogenicity of effluents from the UASB based sewage treatment plant. *Environmental Monitoring and Assessment*, **133**(1–3), 43–51, https://doi.org/10.1007/s10661-006-9558-1

Patel H. and Madamwar D. (2002). Effects of temperatures and organic loading rates on biomethanation of acidic petrochemical wastewater using an anaerobic upflow fixed-film reactor. *Bioresource Technology*, **82**, 65–71, https://doi.org/10.1016/S0960-8524(01)00142-0

Petropoulos E., Dolfing J., Davenport R. J., Bowen E. J. and Curtis T. P. (2017). Developing cold-adapted biomass for the anaerobic treatment of domestic wastewater at low temperatures (4, 8 and 15°C) with inocula from cold environments. *Water Research*, **112**, 100–109, https://doi.org/10.1016/j.watres.2016.12.009

Plaza-Garrido A., Limaico M. and Villamar-Ayala C. A. (2022). Influence of wastewater treatment technologies on virus removal under a bibliometric-statistical analysis. *Journal of Water Process Engineering*, **47**, 102642, https://doi.org/10.1016/j.jwpe.2022.102642

Quek P. J., Yeap T. S. and Ng H. Y. (2017). Applicability of upflow anaerobic sludge blanket and dynamic membrane-coupled process for the treatment of municipal wastewater. *Applied Microbiology and Biotechnology*, **101**, 6531–6540, https://doi.org/10.1007/s00253-017-8358-6

Rebac S., Ruskova J., Gerbens S., van Lier J. B., Stams A. J. M. and Lettinga G. (1995). High-rate anaerobic treatment of wastewater under psychrophilic conditions. *Journal of Fermentation and Bioengineering*, **80**, 499–506, https://doi.org/10.1016/0922-338X(96)80926-3

Rinzema A., van Veen H. and Lettinga G. (1993). Anaerobic digestion of triglyceride emulsions in expanded granular sludge bed reactors with modified sludge separators. *Environmental Technology*, **14**, 423–432, https://doi.org/10.1080/09593339309385310

Rose J. B. (2005). Reduction of Pathogens, Indicator Bacteria, and Alternative Indicators by Wastewater Treatment and Reclamation Processes. IWA Publishing, London, UK, Vol. **4**, https://doi.org/10.2166/9781780404370

Rozzi A., Di Pinto A. C., Limoni N. and Tomei M. C. (1994). Start-up and operation of anaerobic reactors with bicarbonate control. *Bioresource Technology*, **48**, 215–219, https://doi.org/10.1016/0960-8524(94)90149-X

Saddoud A., Ellouze M., Dhouib A. and Sayadi S. (2006). A comparative study on the anaerobic membrane bioreactor performance during the treatment of domestic wastewaters of various origins. *Environmental Technology*, **27**(9), 991–999, https://doi.org/10.1080/09593332708618712

Schink B. (1997). Energetics of syntrophic cooperation in methanogenic degradation. *Microbiology Molecular Biology Reviews*, **61**, 262–280.

Seghezzo L., Zeeman G., van Lier J. B., Hamelers H. V. M. and Lettinga G. (1998). A review: the anaerobic treatment of sewage in UASB and EGSB reactors. *Bioresource Technology*, **65**, 175–190, https://doi.org/10.1016/S0960-8524(98)00046-7

Shin C. and Bae J. (2018). Current status of the pilot-scale anaerobic membrane bioreactor treatments of domestic wastewater: a critical review. *Bioresource Technology*, **247**, 1038–1046, https://doi.org/10.1016/j.biortech.2017.09.002

Shoushtarian F. and Negahban-Azar M. (2020). Worldwide regulations and guidelines for agricultural water reuse: a critical review. *Water*, **12**, 971–983, https://doi.org/10.3390/w12040971

Skouteris G., Hermosilla D., López P., Negro C. and Blanco Á (2012). Anaerobic membrane bioreactors for wastewater treatment: a review. *Chemical Engineering Journal*, **198–199**, 138–148, https://doi.org/10.1016/j.cej.2012.05.070

Smith S. R., Lang N. L., Cheung K. H. M. and Spanoudaki K. (2005). Factors controlling pathogen destruction during anaerobic digestion of biowastes. *Waste Management*, **25**(4), 417–425, https://doi.org/10.1016/j.wasman.2005.02.010

Smith A. L., Stadler L. B., Love N. G., Skerlos S. J. and Raskin L. (2012). Perspectives on anaerobic membrane bioreactor treatment of domestic wastewater: a critical review. *Bioresource Technology*, **122**, 149–159, https://doi.org/10.1016/j.biortech.2012.04.055

Stazi V. and Tomei M. C. (2018). Enhancing anaerobic treatment of domestic wastewater: state of the art, innovative technologies and future perspectives. *Science of the Total Environment*, **635**, 78–91, https://doi.org/10.1016/j.scitotenv.2018.04.071

Svensson B. H. (1984). Different temperature optima for methane formation when enrichments from acid peat are supplemented with acetate or hydrogen. *Applied and Environmental Microbiology*, **48**, 389–394, https://doi.org/10.1128/aem.48.2.389-394.1984

Switzenbaum M. S. (1983). Anaerobic fixed film wastewater treatment. *Enzyme and Microbial Technology*, **5**, 242–250, https://doi.org/10.1016/0141-0229(83)90072-8

Thauer R. K., Jungermann K. and Decker K. (1977). Energy conservation in chemotrophic anaerobic bacteria. *Bacteriological Reviews*, **41**(1), 100–180, https://doi.org/10.1128/br.41.1.100-180.1977

Tilche A. and Vieira S. M. M. (1991). Discussion on reactor design of anaerobic filters and sludge bed reactors. *Water Science and Technology*, **24**, 193–206, https://doi.org/10.2166/wst.1991.0225

Tomei M. C., Di Pinto A. C., Limoni N. and Rozzi A. (1994). Calibration of an automatic instrument for bicarbonate monitoring. *Environmental Technology*, **15**, 363–372, https://doi.org/10.1080/09593339409385439

Tomei M. C. and Carozza N. A. (2015). Sequential anaerobic/aerobic digestion for enhanced sludge stabilization: comparison of the process performance for mixed and waste activated sludge. *Environmental Science and Pollution Research*, **22**, 7271–7279, https://doi.org/10.1007/s11356-014-3130-2

Tomei M. C., Mosca Angelucci D. and Levantesi C. (2016). Two-stage anaerobic and post-aerobic mesophilic digestion of sewage sludge: analysis of process performance and hygienization potential. *Science of the Total Environment*, **545–546**, 453–464, https://doi.org/10.1016/j.scitotenv.2015.12.053

Toze S. (1999). PCR and the detection of microbial pathogens in water and wastewater. *Water Research*, **33**(17), 3545–3556, https://doi.org/10.1016/S0043-1354(99)00071-8

Uemura S. and Harada H. (2000). Treatment of sewage by a UASB reactor under moderate to low temperature conditions. *Bioresource Technology*, **72**, 275–282, https://doi.org/10.1016/S0960-8524(99)00118-2

Uemura S., Takahashi K., Takaishi A., Machdar I., Ohashi A. and Harada H. (2002) Removal of indigenous coliphages and fecal coliforms by a novel sewage treatment system consisting of UASB and DHS units.

Ugboko H. U., Nwinyi O. C., Oranusi S. U. and Oyewale J. O. (2020). Childhood diarrhoeal diseases in developing countries. *Heliyon*, **6**(4), e03690, https://doi.org/10.1016/j.heliyon.2020.e03690

Valdez F. Q., de Leite L. S., Zanetoni Filho J. A., Tango M. D. and Daniel L. A. (2022). Detection and removal of *Giardia* spp. cysts and *Cryptosporidium* spp. oocysts by anaerobic reactors in Brazil. *Environmental Technology*, **43**(13), 2059–2068.

Van der Last A. R. M. and Lettinga G. (1992). Anaerobic treatment of domestic sewage under moderate climatic (Dutch) conditions using upflow reactors at increased superficial velocities. *Water Science and Technology*, **25**, 167–178, https://doi.org/10.2166/wst.1992.0149

Van Haandel A. C. and Lettinga G. (1994). Anaerobic Sewage Treatment: A Practical Guide for Regions with a Hot Climate. John Wiley & Sons Ltd, Chichester, UK.

Van Haandel A., Kato M. T., Cavalcanti P. F. F. and Florencio L. (2006). Anaerobic reactor design concepts for the treatment of domestic wastewater. *Environmental Science and Biotechnology*, **5**, 21–38, https://doi.org/10.1007/s11157-005-4888-y

Van Lier J. B., Rebac S. and Lettinga G. (1997). High-rate anaerobic wastewater treatment under psychrophilic and thermophilic conditions. *Water Science and Technology*, **35**, 199–206, https://doi.org/10.2166/wst.1997.0383

Verbyla M. E. and Mihelcic J. R. (2015). A review of virus removal in wastewater treatment pond systems. *Water Research*, **71**, 107–124, https://doi.org/10.1016/j.watres.2014.12.031

Verstraete W., Van de Caveye P. and Diamantis V. (2009). Maximum use of resources present in domestic 'used water'. *Bioresource Technology*, **100**, 5537–5545, https://doi.org/10.1016/j.biortech.2009.05.047

Vieira S. M. M. and Souza M. E. (1986). Development of technology for the use of the UASB reactor in domestic sewage treatment. *Water Science and Technology*, **18**, 221–228, https://doi.org/10.2166/wst.1986.0093

Von Sperling M. (2005). Modelling of coliform removal in 186 facultative and maturation ponds around the world. *Water Research*, **39**(20), 5261–5273, https://doi.org/10.1016/j.watres.2005.10.016

Von Sperling M., Freire V. H. and Chernicharo C. A. (2001). Performance evaluation of an UASB-activated sludge system treating municipal wastewater. *Water Science and Technology*, **43**, 323–328, https://doi.org/10.2166/wst.2001.0698

von Sperling M., Chernicharo C. A., Soares A. M. and Zerbini A. M. (2002). Coliform and helminth eggs removal in a combined UASB reactor–baffled pond system in Brazil: performance evaluation and mathematical modelling. *Water Science and Technology*, **45**(10), 237–242, https://doi.org/10.2166/wst.2002.0341

Wang J., Huang Y. and Zhao X. (2004). Performance and characteristics of an anaerobic baffled reactor. *Bioresource Technology*, **93**, 205–208, https://doi.org/10.1016/j.biortech.2003.06.004

Watanabe R., Nie Y., Wakahara S., Komori D. and Li Y. Y. (2017). Investigation on the response of anaerobic membrane bioreactor to temperature decrease from 25°C to 10°C in sewage treatment. *Bioresource Technology*, **243**, 747–754, https://doi.org/10.1016/j.biortech.2017.07.001

Wen C., Huang X. and Qian Y. (1999). Domestic wastewater treatment using an anaerobic bioreactor coupled with membrane filtration. *Process Biochemistry*, **35**, 335–340, https://doi.org/10.1016/S0032-9592(99)00076-X

WHO. (2019). Progress on drinking water, sanitation and hygiene 2000–2017. UNICEF.

Wu W., Thiele J. H., Jain M. K. and Keikus J. G. (1993). Metabolic properties and kinetics of methanogenic granules. *Applied Microbiology and Biotechnology*, **39**, 804–811, https://doi.org/10.1007/BF00164470

Wu J., Long S. C., Das D. and Dorner S. M. (2011). Are microbial indicators and pathogens correlated? A statistical analysis of 40 years of research. *Journal of Water and Health*, **9**(2), 265–278, https://doi.org/10.2166/wh.2011.117

Xi-quan H. and Zhao-hua L. (2008). Operational characteristics of an anaerobic baffled reactor treating low strength wastewater. Proceedings of the 2nd International Conference on Bioinformatics and Biomedical Engineering, Shanghai, China, pp. 3135–3139.

Yaya-Beas R. E., Ayala-Limaylla C., Kujawa-Roeleveld K., van Lier J. B. and Zeeman G. (2015). Helminth egg removal capacity of UASB reactors under subtropical conditions. *Water*, **7**(5), 2402–2421, https://doi.org/10.3390/w7052402

Young J. C. and McCarty P. L. (1969). The anaerobic filter for waste treatment. *Journal of the Water Pollution Control Federation*, **41**, R160–R173.

Yu H., Tay J. and Wilson F. A. (1997). Sustainable municipal wastewater treatment process for tropical and subtropical regions in developing countries. *Water Science and Technology*, **35**, 191–198, https://doi.org/10.2166/wst.1997.0343

Zeeman G., Sanders W. T. M., Wang K. Y. and Lettinga G. (1997). Anaerobic treatment of complex wastewater and waste activated sludge – application of an upflow anaerobic solid removal (UASR) reactor for the removal and pre-hydrolysis of suspended COD. *Water Science and Technology*, **35**, 121–128, https://doi.org/10.2166/wst.1997.0373

Zhang J., Wu B., Zhang J., Zhai X., Liu Z., Yang Q., Liu H., Hou Z., Sano D. and Chen R. (2022). Virus removal during sewage treatment by anaerobic membrane bioreactor (AnMBR): the role of membrane fouling. *Water Research*, **211**, 118055, https://doi.org/10.1016/j.watres.2022.118055

doi: 10.2166/9781789063479_0029

# Chapter 2

# Upflow anaerobic sludge blanket reactors

T. Bressani-Ribeiro[1,2]* and C. A. L. Chernicharo[1,2]

[1]Department of Sanitary and Environmental Engineering, Federal University of Minas Gerais, Belo Horizonte, Brazil
[2]Chernicharo & Bressani Consultancy and Training in Sanitation, Belo Horizonte, Brazil
*Correspondence: thiagobressani@ufmg.br

**ABSTRACT**

This chapter addresses the use of upflow anaerobic sludge blanket (UASB) reactors for mainstream sewage treatment in warm climate regions. The first section presents the status of the application, highlighting the development of the third-generation UASB reactors concerning the efforts of more than three decades of continuous implementation. Next, a brief background is provided based on functioning principles and typical configurations. The following section on design procedures summarizes the most relevant criteria, in which modular gas–liquid–solid (GLS) separators play a crucial role. Based on suggested standardized dimensions and integrated process control (gas pressure and hydrostatic scum withdrawal), modular GLS separators improve the reliability of the design of (third-generation) UASB reactors treating sewage. Performance prediction for chemical oxygen demand and biochemical oxygen demand removal is addressed, as well as for effluent total suspended solid concentration. A dedicated section on the operation of full-scale UASB reactors discusses the management of sludge, scum, biogas and diffuse emissions. Aspects related to design, construction and operation are explored, pointing out the main hurdles and consolidated solutions. The chapter ends with final reflections on the advances made so far and the remaining bottlenecks, stressing the key role of mainstream sewage treatment for circular economy models.

**Keywords:** mainstream sewage treatment, scum, sludge, biogas, fugitive emissions.

## 2.1 STATUS OF APPLICATION IN WARM CLIMATE REGIONS AT FULL SCALE

After promising results from upflow anaerobic sludge blanket (UASB) reactors treating high-strength wastewater from potato and beet processing in the mid-1970s, the first demonstration-scale UASB reactor (64 $m^3$) treating sewage was implemented in the city of Cali, Colombia (Lettinga, 2014). Tropical temperatures have aided in overcoming shortcomings previously observed in lab- and pilot-scale tests in The Netherlands, and the Colombian trial paved the basis for criteria and parameters for the design and operation of UASB reactors applied to sewage treatment. Following such a successful experience, full-scale reactors (~1,000 $m^3$) were implemented in Cali, launching the first generation of UASB reactors directly treating sewage (mainstream). Moreover, research projects were deployed in India, Ghana, the Middle East and Latin America, Brazil, Colombia and Mexico.

Between the 1980s and the 1990s, the design of the first-generation UASB reactors for sewage treatment was focused on the liquid phase, whose main driver was the performance for chemical oxygen demand (COD) removal. The critical aspects were related to the need for consolidation of the recommended hydraulic parameters (i.e. hydraulic retention time (HRT), upflow velocity) as well as proper influent distribution and effluent collection. In other words, the first-generation UASB reactors for sewage treatment were dedicated to deriving the key design principles that are still in use.

To the best of our knowledge, Brazil currently has the largest number of UASB reactors for mainstream sewage treatment. A survey led by the Brazilian National Water Agency (ANA, 2017) identified 1,047 anaerobic-based sewage treatment plants (STPs) (38% of the total STPs surveyed), which deal with wastewater from about 21.5 million inhabitants (30% of all sewage presently being treated in Brazil). A comprehensive survey in the Latin American region approximately 10 years ago (Noyola *et al.*, 2012) showed that UASB reactors treating municipal wastewater accounted for 17% out of a total of 2,734 facilities in six countries (Brazil, Colombia, Chile, Dominican Republic, Guatemala and Mexico). Those results have been deemed as a significant milestone towards the implementation of mainstream anaerobic sewage treatment, especially considering an UASB system as a newcomer in the field of municipal sewage treatment compared to the century-old activated sludge process (Chernicharo *et al.*, 2015). In fact, the technology is fully mature in Latin America where UASB-based treatment plants of up to 1 million population equivalent (Onça STP, Belo Horizonte, Brazil) have been in operation for more than 15 years. The Indian experience with UASB reactors for mainstream sewage treatment is also mature, considering the report of implementation of several full-scale systems with a total capacity of 598,000 $m^3$/day, under the Yamuna Action Plan launched in the early 1990s (Uemura & Harada, 2010). The performance of UASB reactors treating sewage is further explored in Section 2.3.2.

The widespread implementation of mainstream anaerobic sewage treatment has led to the development of the second-generation UASB reactors, incorporating the lessons learned from the first-generation installations. In this case, attention was also paid to the solid phase as it showed to directly affect the system performance for organic matter removal. Protocols for sludge sampling and withdrawal were developed. Although efforts were also made for scum management, the second-generation UASB reactors lack a comprehensive approach for scum formation control and withdrawal. This prompted acute operational hurdles such as scum solidification and the collapse of gas–liquid–solid (GLS) separators. Therefore, although several full-scale UASB reactors for mainstream sewage treatment are currently in operation in warm climate regions, their design typically comprises significant drawbacks ascribed to the second generation.

The flaws from the second-generation UASB reactors hinder the whole potential of mainstream anaerobic processes for carbon conversion and energy recovery. This has prompted the development of a third generation, currently fully mature. Besides the improvements related to the management of the liquid and solid phases ascribed to the first and second generations, respectively, a holistic view was addressed considering the gaseous phase. In the third-generation UASB reactors for sewage treatment, effective devices for biogas collection and energy recovery were incorporated into the designs. This represents for the first time a complete integration of liquid, solid and gaseous phases. Furthermore, the third-generation UASB reactors demonstrated how the management of scum and biogas is intrinsically correlated, paving the way for odour and corrosion control. The management of diffusive gaseous emissions (i.e. dissolved hydrogen sulphide and methane) was also taken into account, helping to assess the carbon footprint of UASB-based STPs.

## 2.2 BACKGROUND

### 2.2.1 Principle of functioning of UASB reactors

The process principles of UASB reactors for mainstream sewage treatment are consolidated; the basics can be found in von Sperling and Chernicharo (2005). A UASB reactor comprises four functional units in a single tank (van Lier *et al.*, 2010), namely (1) a primary settler for removing suspended solids from

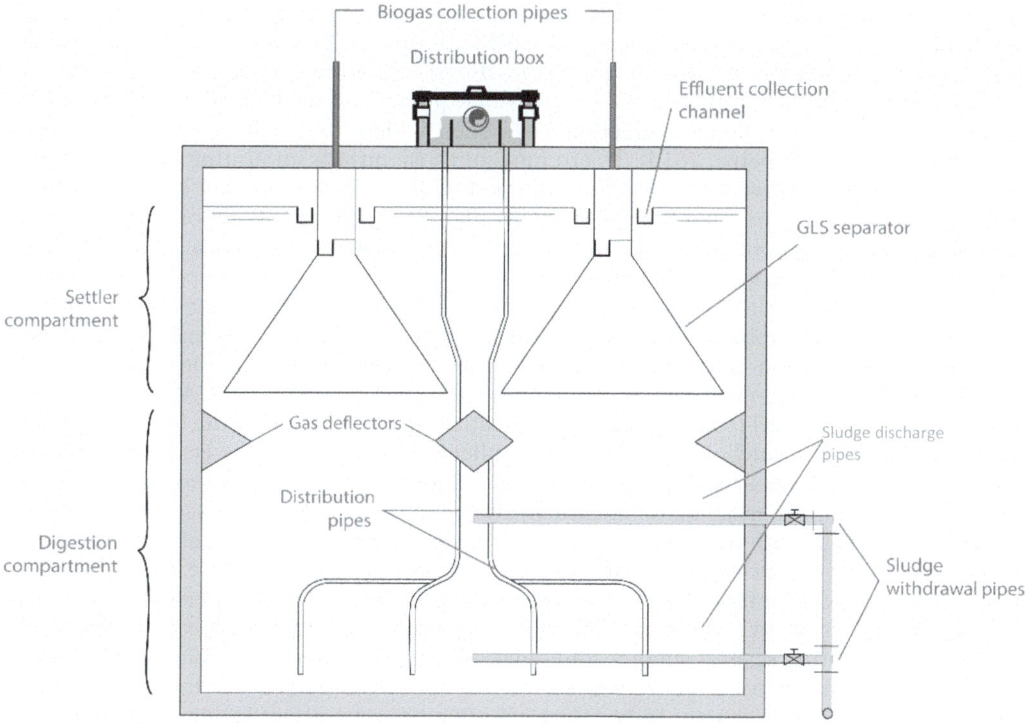


**Figure 2.1** Schematic representation of the main components of a UASB reactor treating sewage. *Source*: Adapted from Chernicharo and Bressani-Ribeiro (2019).

the influent, (2) a biological reactor (secondary treatment) for carbon removal through its conversion into methane, (3) a secondary settler for clarifying the effluent in the settling zone at the top of the reactor and (4) a sludge digester for thickening and digestion (or stabilization) of the retained organic solids and the sludge contributions from aerobic post-treatment systems. The main components of a UASB reactor for sewage treatment are shown in Figure 2.1.

In summary, the process is attained by an upward flux of wastewater passing firstly through a dense-sludge bed (3–5% TS) with a typical specific methanogenic activity of approximately 0.2 $gCOD_{CH_4}/g VS$ (volatile solids)/day (von Sperling & Chernicharo, 2005), followed by a more dispersed and light sludge (sludge blanket) close to the upper part of the so-called digestion compartment. A GLS separation device is situated at the cornerstone of the reactor, assuring (1) the return of the particles washed out from the sludge blanket to the digestion compartment; (2) the collection of the produced biogas and (3) the collection of the treated effluent.

### 2.2.2 Typical configurations of UASB reactors

Through the adaptation process of UASB reactors to treat low-concentration wastewater (such as sewage), an important outcome has been considered, which is the control of excessive upflow velocities in the digestion and settler compartments, especially during peak flows. Consequently, the height of the reactor is reduced (compared to reactors treating industrial wastewater) and its cross section is increased, as the design is mainly ruled by the hydraulic loading criteria. Additionally, to prevent higher upflow velocities in the settler compartment and to enable sludge to return to the digestion compartment, larger cross sections close to the settler compartment have been tested. In this case, the reactor adopts a variable section that is smaller when close to the digestion compartment and larger when close to the settler compartment. A UASB variation called RALF (fluidized bed anaerobic reactor (acronym in Portuguese)) was developed in Brazil following this assumption. Nevertheless, this significantly increases the construction complexity (e.g. inclined walls), and drawbacks associated with sludge withdrawal and biogas collection have been widely reported by sanitation companies (Chernicharo & Bressani-Ribeiro, 2019). To circumvent those pitfalls, mounting GLS separators and reduced upflow velocities prevent the need for enlarged surfaces at the top (van Lier *et al.*, 2010).

Full-scale UASB reactors treating sewage are typically circular (Figure 2.2a) or rectangular (Figure 2.2b). The first is especially considered for small-scale plants that generally have a single UASB reactor. Circular reactors are more economical from a structural standpoint (as they occupy a smaller perimeter than rectangular reactors with the same surface area), and alternative materials can be used for construction instead of reinforced concrete, such as glass fibre and polypropylene. Rectangular units are more suitable for medium- and large-scale STPs, especially when modulation is necessary, where a wall can serve two contiguous modules. Modulation is a crucial aspect for UASB reactors treating sewage, as the influent to an STP undergoes no equalization, thereby exposing the UASB reactor to flow variations that may be extremely high. For sewage collected in combined sewers (a typical condition in developed countries), large flow fluctuations (e.g. a 5–6 fold increase in flow during rain events) can occur. Even when using sewerage systems exclusively for sewage collection, excessive rainwater can enter the system through both infiltration and the irregular cross connections of rainwater (a typical condition in developing countries). Therefore, modulation aims to cope with these marked variations of the average flow, when one or more UASB reactors would only be used during peak flows to ensure the recommended HRTs. From an economic standpoint, implementing larger reactors may decrease costs compared to implementing several smaller units maintaining the same total volume. This is because shared walls can be removed for larger reactors. However, from an operational standpoint, modulation is also advisable as it allows a reactor to be taken out of operation without losing treatment capacity.

Open-air UASB reactors (without a top slab covering the settler compartment) can be used provided effluent collection design accounts for odour control (see Chapter 7). Other alternative configurations of UASB reactors for sewage treatment have been tested, namely Y-shaped UASB reactors, two-stage UASB reactors, combined UASB–digester systems and UASB reactors with double-stage biogas collection (Chernicharo & Bressani-Ribeiro, 2019). However, none of the configurations, except for hybrid anaerobic reactors, has been implemented effectively at full scale. Therefore, sparse, or even no information on adequate construction materials, maintenance, instrumentation and control

**Figure 2.2** Cylindrical UASB reactors in glass fibre (a) and rectangular UASB reactors in reinforced concrete (b). *Source*: Adapted from Chernicharo and Bressani-Ribeiro (2019).

needs is available. Furthermore, the third-generation UASB reactors are fully established considering cylindrical- or rectangular-shaped UASB reactors, as previously presented.

## 2.3 CURRENT DESIGN CRITERIA OF UASB REACTORS TREATING SEWAGE

### 2.3.1 Sizing parameters

As well documented in the classical literature (van Haandel & Lettinga, 1994; van Lier *et al.*, 2010; von Sperling & Chernicharo, 2005), the design of UASB reactors treating low-strength wastewater such as sewage (COD $< 1{,}000$ mg/L) in warm climate regions ($T > 20°C$) is controlled by the hydraulic loading instead of the organic loading criteria. For those conditions, the minimum required solid retention time (SRT) of 30 days (i.e. three times the doubling time of methanogens; van Lier *et al.*, 2010) has to be met. An exception is made for arid climate countries with limited water supply where the sewage concentration can attain up to 2,500 mg COD/L, whereas winter temperatures may drop to 15°C (Chernicharo *et al.*, 2015). In this case, the SRT should be the main design criterion for determining the UASB reactor volume.

The main design criteria of UASB reactors treating sewage are summarized in Table 2.1, supported by the full-scale experiences in Brazil. As a hydraulic-based design, the upflow velocity in the digestion

**Table 2.1** Main design criteria recommended for UASB reactors based on Brazilian experiences.

| Parameter | Unit | Value | | Comment |
|---|---|---|---|---|
| HRT | h | Temperature (°C) | $HRT_{avg}$ | Maximum peak flows should not last for more than 2–4 h to harm the buffer capacity of the settler compartment for solid retention. Adopting values close to the upper limits of each range ensures a larger capacity for sludge storage in the digestion compartment. |
| | | >25 | >6 | |
| | | 22–25 | 7–9 | |
| | | 18–21 | 8–10 | |
| | | 15–17 | 10–14 | |
| Upflow velocity at average flow | m/h | 0.5–0.7 | | Less than 1.1 m/h for the maximum peak flow. |
| Useful depth | m | 4–6 | | Minimum useful depths of digestion and settler compartments are 2.5 and 1.5 m, respectively. For the recommended upflow velocities and HRT, the UASB reactor depths should be between 4 and 6 m. |
| Feed inlet density | m² per feed point | 2.0–3.0 | | Minimum inlet pipe internal diameter should be 75 mm. |
| Angle of gas collector | Degree | ≥50 | | See below for standardized GLS separators. |
| Depth of the settler compartment | m | ≥1.5 | | |
| Velocity in the apertures to the settler compartment | m/h | <2.0–2.3 | | Less than 4.0–4.2 m/h for the maximum peak flow. |
| Surface loading rate in the settler compartment | m/h | 0.6–0.8 | | Less than 1.2 m/h for the maximum peak flow. This parameter is easier to determine than the HRT in the settler compartment, which should be between 1.5 and 2.0 h. |
| Overlap of the gas deflectors over the apertures for the settler compartment | m | 0.15–0.20 | | |

*Source:* Adapted from Chernicharo and Bressani-Ribeiro (2019).

and settler compartments is a crucial parameter. Therefore, for the average design flow definition, besides the per capita water consumption, rainfall infiltration coefficients that reflect local realities should be accounted for. Excessive velocities result in solid washout with the effluent, whereas low velocities can jeopardize the system performance owing to a worse contact between the influent substrate and microorganisms. The average influent sewage flow rate should be considered for calculations; nevertheless, maximum flow checks should be performed concerning the HRT, upflow velocity in the digestion compartment and surface loading rate in the settler compartment, as reported in Table 2.1. In the case of using pumping stations to feed UASB reactors, speed control of the pumps, using variable frequency drives or a minimum of three pumps (one standby and two in service, with an individual pumping capacity equal to half the maximum flow) is recommended, to avoid exceeding the maximum design flow rate.

The design of third-generation UASB reactors considers the use of modular GLS separators, such as the Étsus-1000. Such a device (Figure 2.3) is the result of a partnership between the Federal University of Minas Gerais (UFMG, Brazil) and the following Brazilian companies: Copasa (Companhia de Saneamento de Minas Gerais), Sabesp (Companhia de Saneamento Básico do Estado de São Paulo), Sanepar (Companhia de Saneamento do Paraná) and Fibrasa. A modular GLS separator is made of glass fibre, and serves as the solution for several operational problems of UASB reactors, namely (1) adequate management of the formation and removal of scum, (2) efficient collection of biogas without losses by leakages and (3) adequate collection of the treated effluent, avoiding the release of waste gases and the occurrence of hydraulic short circuits. The main advantages associated with the GLS separator Étsus-1000 are as follows:

- A modular structure, water and gas-tight, light and robust to facilitate transport and installation.
- Incorporation of scum removal connected to the biogas system, allowing for the control of odour nuisance and methane losses.
- Incorporation of an effluent collection system coupled with a level-checking device.
- Standard dimensions that allow for the design of UASB reactors for any incoming flow.

The conceived modulation for the GLS separator Étsus 1000 is reported in Table 2.2. Three different models were proposed as a function of the GLS length (dimensions $A$ and $B$, shown in

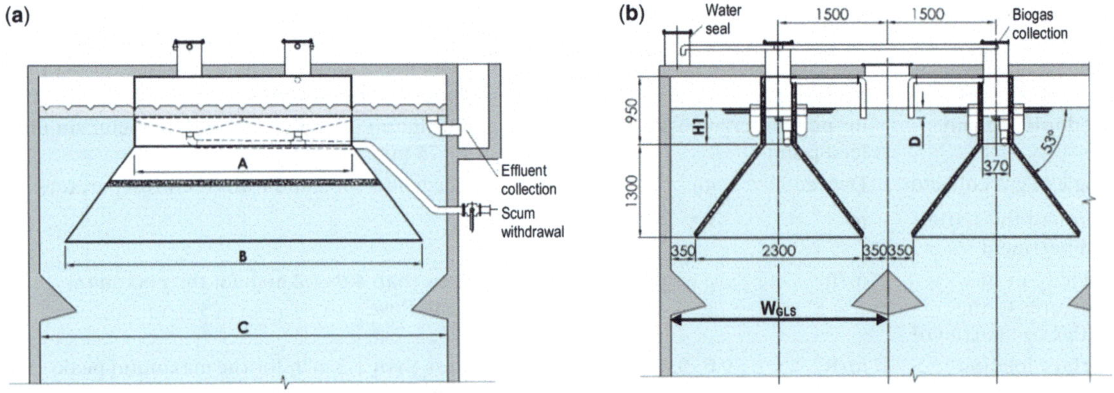


**Figure 2.3** Longitudinal (a) and cross-sectional (b) views of a standardized GLS separator (Étsus 1000) for UASB reactors treating sewage. *Source*: Adapted from Chernicharo and Bressani-Ribeiro (2019). $A$ and $B$ dimensions refer to the length of GLS separator. $C$ refers to the length of the UASB reactor. $W_{GLS}$ refers to the standard dimensions between the vertical axis of two GLS separators. $H_1$ refers to the submerged height of the gas chamber (adaptable). $D$ stands for the submerged height of the pressure-safe device (adaptable). Refer to the design example in Section 2.3.3 for further clarification.

**Table 2.2** Modulation of a standardized GLS separator (Étsus 1000) for UASB reactors treating sewage.

| GLS Separator Model | Number of Scum Withdrawal Points ($N_{scum\text{-}points}$) | Length of the GLS Separator (Upper Part), *A* (m) | Length of the GLS Separator (Lower Part), *B* (m) |
|---|---|---|---|
| 1 | 1 | 3 | 4.9 |
| 2 | 2 | 3–6 | 4.9–7.9 |
| 3 | 3 | 6–9 | 7.9–10.9 |

*Source:* Adapted from Chernicharo and Bressani-Ribeiro (2019).
*Notes:* (1) The length of the scum withdrawal channel should be the same as '*A*' but with convergent divisions to the $N_{scum}$ points (see Figure 2.3). (2) The values between the range limits of '*A*' and '*B*' can be incremented every 0.5 m up to the upper limit of the range

Table 2.2 and Figure 2.3). In summary, after determining the required area of a UASB reactor, the length of a chosen modular GLS could be adopted (dimension *A*) as well as the number of GLS to be implemented. The width of the reactor is a consequence of this design procedure, which is further exemplified (see Section 2.3.3). In practice, the three proposed modular GLS models can be applied to UASB reactors designed for any population equivalents (small-, medium- and large-scale STPs).

### 2.3.2 Performance of UASB reactors treating sewage

Organic carbon removal efficiencies of UASB reactors for mainstream sewage treatment in warm climate countries are in the range of 60–75% for biochemical oxygen demand (BOD) and 55–70% for COD, respectively, without a post-treatment step. Table 2.3 shows the efficiencies of full-scale UASB reactors for COD and total suspended solid (TSS) removal (Chernicharo & Bressani-Ribeiro, 2019). Underperforming systems can be mainly ascribed to the first- and second-generation reactors, which typically lack a thorough design control for scum withdrawal, besides engineering flaws incurred during the construction stage, and operational shortcomings reported by Chernicharo and Bressani-Ribeiro (2019).

It is worth stressing that the performance of a UASB-based system is highly dependent on the implementation of adequate operational routines particularly related to the preliminary treatment unit and the management of sludge and scum. A typical design and construction shortcoming refers to the lack of sludge dewatering capacity. In many cases where mechanical dewatering is taken out of service for maintenance (without sparing equipment), excessive sludge build-up occurs in the anaerobic reactors, leading to solid washout with the effluent and disturbing the operation of the entire treatment plant.

The COD removal efficiency of UASB reactors treating sewage can be estimated by empirical equations as proposed by Chernicharo and Bressani-Ribeiro (2019). Equations (2.1) and (2.2) predict COD and BOD removal efficiencies, suitable for well-operated full-scale systems, especially in the operational HRT range of 8–12 h. The original dataset derives from 25 full-scale UASB reactors treating sewage under tropical conditions (sewage temperature in the range of 20–27°C), with influent COD ranging from 300 to 1,400 mg/L and influent BOD varying between 150 and 850 mg/L:

$$E_{COD} = 48 \times (HRT^{0.14}) \tag{2.1}$$

where

$E_{COD}$ = UASB reactor efficiency for COD removal (%)
HRT = hydraulic retention time (h)
48 = empirical constant
0.14 = empirical constant

**Table 2.3** Performance of full-scale UASB-based STPs.

| Location | Effluent Concentration | | | Removal Efficiency | | | Population Equivalent (Inhabitants) | Reference |
|---|---|---|---|---|---|---|---|---|
| | COD (mg/L) | BOD (mg/L) | TSS (mg/L) | COD (%) | BOD (%) | TSS (%) | | |
| India | 202 | 60 | 150 | 63 | 67 | 70 | 93,500 | Pandey and Dubey (2014) |
| India | 139–567 | 57–159 | 72–452 | 29–75 | 45–79 | 40–70 | – | Khan *et al.* (2011) |
| Brazil | 283 | – | 132 | 58 | – | 49 | 3,047 | Silva *et al.* (2013) |
| Brazil | 114 | 38 | 132 | 79 | 84 | 59 | 70,000 | Rosa *et al.* (2012) |
| Brazil | 251 | 98 | 85 | 65 | 74 | 71 | 24,000 | Oliveira and Von Sperling (2011) |
| India | 515 | 115 | 113 | 41 | 50 | 47 | – | Mungray and Patel (2011) |
| India | 405 | 153 | 167 | 44 | 40 | 36 | – | Mungray and Patel (2011) |
| India | 145–250 | 55–75 | 160–240 | 45 | 60 | 34 | – | Walia *et al.* (2011) |
| Colombia | – | 60 | – | – | 77 | – | 320,000 | WERF (2010) |
| Brazil | 170 | 66 | 75 | 58 | 68 | 56 | 544,000 | Franco (2010) |
| Brazil | 247 | 97 | 112 | 62 | 67 | 54 | – | Van Lier *et al.* (2010) |
| India | 285 | 121 | 357 | 46 | 41 | 49 | – | Van Lier *et al.* (2010) |
| Brazil | 190 | 70 | 60 | 60 | 65 | 61 | 1,000,000 | Chernicharo *et al.* (2009) |
| Colombia | 144 | – | 81 | 58 | – | 65 | – | Peña *et al.* (2006) |
| Brazil | 181 | 75 | 127 | 64 | 74 | 51 | 24,719 | Baréa and Alem Sobrinho (2006) |
| Brazil | 106 | 69 | – | 72 | 72 | – | 150,000 | Carraro (2006) |
| Brazil | 161 | 66 | – | 77 | 78 | – | – | Tachini *et al.* (2006) |
| India | 403 | 130 | 380 | 47 | 50 | 7 | 55,000–570,000 | Sato *et al.* (2006) |
| Middle East | 221 | 83 | 63 | 71 | 70 | 85 | – | Nada *et al.* (2006) |
| India | – | – | – | 61 | 61 | 66 | – | Khalil *et al.* (2006) |
| Jordan | 632 | – | 180 | 58 | – | 62 | – | Halalseh *et al.* (2005) |
| Brazil | 237 | 64 | 127 | 60 | 69 | 52 | 3,808 | Busato (2004) |
| Brazil | 202 | – | 80 | 67 | – | 61 | 18,000 | Florencio *et al.* (2001) |
| Colombia | 177 | 69 | 72 | 66 | 78 | 69 | 9,000 | Peña *et al.* (2000) |
| Mexico | – | – | – | 70–80 | – | – | – | Monroy *et al.* (2000) |
| Median values | 212 | 70 | 127 | 62 | 68 | 58 | | |

*Source:* Adapted from Chernicharo and Bressani-Ribeiro *et al.* (2019).

$$E_{BOD} = 49.2 \times (HRT^{0.18}) \tag{2.2}$$

where

$E_{BOD}$ = UASB reactor efficiency for BOD removal (%)
HRT = hydraulic retention time (h)
49.2 = empirical constant
0.18 = empirical constant

As for predicting TSS effluent concentration of UASB reactors treating sewage (Equation (2.3)), key aspects should be considered before direct application, such as the existence of sludge and scum withdrawal routines, proper GLS separators and adequate upflow velocities in the digestion and settler compartments. Even under a proper HRT condition, the absence of strict sludge management can deteriorate the effluent quality:

$$\text{TSS} = 133 \times (\text{HRT}^{-0.36}) \tag{2.3}$$

where

TSS = TSS concentration in the effluent of UASB reactors (mg/L)
HRT = hydraulic retention time (h)
133 = empirical constant
0.36 = empirical constant

### 2.3.3 Design example of a third-generation UASB reactor for sewage treatment

Design example of a third-generation UASB reactor for 40,000 PE, based on the concept of modular GLS separators (see Section 2.3.1).

**Input data:**

- Population equivalent: PE = 40,000 inhabitants
- Average influent sewage flow rate: $Q_{s\text{-avg}} = 77.4$ L/s (278.6 $m^3/h$)
- Maximum hourly sewage influent flow rate: $Q_{s\text{-max}} = 133.0$ L/s (478.8 $m^3/h$)
- *Per capita* COD contribution: $QPCP_{COD} = 0.10$ kg/inhab/day
- Average influent COD concentration: $C_{COD} = 598$ mg/L (0.598 kg/$m^3$)
- Average influent BOD concentration: $C_{BOD} = 333$ mg/L (0.333 kg/$m^3$)
- COD removal efficiency (in terms of filtered COD): $E_{COD\text{-filtered}} = 80\%$
- Methane yield*: $Y_{CH_4} = 0.13$ N $m^3$ $CH_4$/g $COD_{removed}$
- Sludge yield: $Y_{sludge} = 0.15$ kg TS/kg $COD_{applied}$
- Sludge specific mass: $\gamma_{sludge} = 1{,}020$ kg TS/$m^3$
- Total solid concentration of the sludge removed by the UASB reactors: $C_{sludge} = 4\%$
- Temperature of the liquid: 20°C
  (*) Methane yield referred to as the amount recovered as biogas (does not take into account the loss of dissolved methane).

**Solution**:

(1) Assume a hydraulic retention time ($HRT_{adopted}$)
  - $HRT_{adopted} = 9.0$ h (see Table 2.1)

(2) Determine the total reactor volume ($V_t$)
  - $V_t = Q_{s\text{-avg}} \times \text{HRT} = 278.6\ m^3/h \times 9.0\ h = 2{,}507\ m^3$

(3) Assume a number of UASB reactors to be implemented ($N_r$)
  - $N_r = 4$ reactors

(4) Determine the volume of each UASB reactor ($V_r$)
  - $V_r = V_t/N_r = 2{,}507\ m^3/4$ reactors $= 627\ m^3$

(5) Assume a reactor useful height ($H_u$)
  - $H_u = 4.6$ m (adopting lower values leads to larger areas and, consequently, lower upflow velocities)

(6) Determine the area of each UASB reactor ($A_r$)
  - $A_r = V_r/H_u = 627\ m^3/4.6\ m = 136.3\ m^2$

(7) Determine the dimensions of each UASB reactor based on the available GLS separator dimensions (see Table 2.2 and Figure 2.3)
  (7.1) Assume a UASB reactor length ($L_{UASB}$) (preferably a multiple of 3.0 m, because of the GLS separator standard dimensions, $W_{GLS}$ in Figure 2.3b)
  - $L_{UASB} = 15.0$ m
  (7.2) Determine the number of GLS separators to be implemented ($N_{GLS}$)
  - $N_{GLS} = L_{UASB}/W_{GLS}$ (for standard dimensions of $W_{GLS}$, see Figure 2.3) = 15.0 m/3.0 m = 5
  (7.3) Determine the required UASB reactor width ($W_{UASB\text{-}req}$) separators to be implemented ($N_{GLS}$)
  - $W_{UASB\text{-}req} = A_r/L_{UASB} = 136.3\ m^2/15.0\ m = 9.1\ m$
  (7.4) Select a GLS separator length based on Table 2.2
  - Selected GLS separator: Model 3→ length of the lower GLS separator part ($L_{GLS\text{-}lower}$) = 8.4 m; length of the upper GLS separator part ($L_{GLS\text{-}upper}$) = 6.5 m; width of the lower GLS separator part ($W_{GLS\text{-}lower}$) = 2.30 m; width of the upper GLS separator part ($W_{GLS\text{-}upper}$) = 0.37 m
  (7.5) Assume a width for the aperture between the GLS separator and the UASB reactor wall ($W_{aperture\text{-}wall}$) (see Figure 2.3, Section 2.3.1)
  - $W_{aperture\text{-}wall} = 0.35$ m
  (7.6) Determine the resulting UASB reactor width ($W_{UASB}$)
  - $W_{UASB} = L_{GLS\text{-}lower} + 2 \times W_{aperture} = 8.4\ m + 2 \times 0.35\ m = 9.1\ m$
  (7.7) Verify the area and volume of each UASB reactor
  - $A_r = W_{UASB} \times L_{UASB} = 9.1\ m \times 15.0\ m = 136.5\ m^2$
  - $V_r = A_r \times H_u = 136.5\ m^2 \times 4.6\ m = 627.9\ m^3$

(8) Verify the HRT
  - $HRT = V_r \times N_r/Q_{s\text{-}avg} = 627.9\ m^3 \times 4/278.6\ m^3/h \approx 9.0\ h$

*Note:* The variables $H_u$, $L_{UASB}$ and $N_{GLS}$ can be managed to obtain different UASB reactor configurations.A schematic representation of two of the UASB reactor dimensions (plant and cross-section views) and the GLS separator configuration is shown in Figures 2.4 and 2.5.

(9) Verify the upflow velocities
  - For $Q_{s\text{-}avg}$: $v = Q_{s\text{-}avg}/(A_r \times N_r) = 278.6\ m^3/h/(136.5\ m^2 \times 4) = 0.51\ m/h$
  - For $Q_{s\text{-}max}$: $v = Q_{s\text{-}max}/(A_r \times N_r) = 478.8\ m^3/h/(136.5\ m^2 \times 4) = 0.88\ m/h$

(10) Determine the influent distribution system
  - Adopting an influence area ($A_i$) of 2.25 m² per distribution pipe (according to Table 2.1), the number of distribution pipes is:
    $N_d = A_r/A_i = 136.5\ m^2/2.25\ m^2 = 61$ distributors. Due to symmetry, adopt 60 distributors per reactor, as follows:
    - along the length of each reactor (15.0 m): 10 pipes
    - along the width of each reactor (9.1 m): 6 pipes

(11) Verify the velocities through the apertures to the settler compartment
  - Number of GLS separators ($N_{GLS}$): 20 (5 in each UASB reactor)
  - Lower area occupied by each GLS separator: $A_{GLS\text{-}lower} = (L_{GLS\text{-}lower} \times W_{GLS\text{-}lower}) = 8.4\ m \times 2.3\ m = 19.32\ m^2$
  - Area of each UASB reactor: $A_r = 136.5\ m^2$
  - Total area of the apertures: $A_{apertures} = A_r - A_{GLS\text{-}lower} = (N_r \times A_r) - (N_{GLS} \times A_{GLS\text{-}lower}) = (4 \times 136.5\ m^2) - (20 \times 19.32\ m^2) = 159.6\ m^2$
    - Verification of the velocities through the apertures:
  - For $Q_{s\text{-}avg}$: $v = Q_{s\text{-}avg}/A_{apertures} = 278.6\ m^3/h/159.6\ m^2 = 1.75\ m/h$
  - For $Q_{s\text{-}max}$: $v = Q_{s\text{-}max}/A_{apertures} = 478.8\ m^3/h/159.6\ m^2 = 3.00\ m/h$

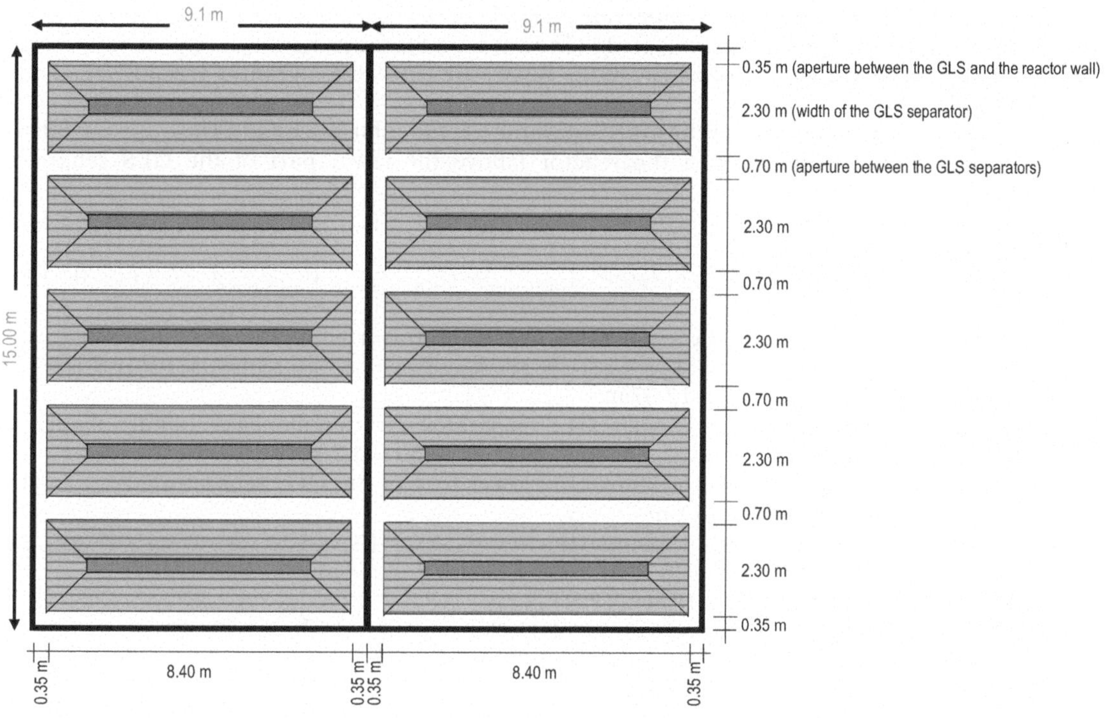


**Figure 2.4** Schematic representation of two of the UASB reactors dimensions (plant view) and the GLS separators (on the top of the reactors).

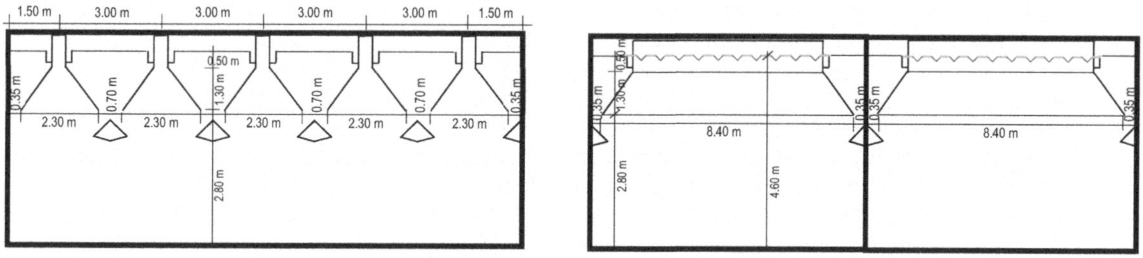


**Figure 2.5** Schematic representation of the UASB reactor dimensions (cross-section views).

(12) Verify the surface loading rates on the settler compartment

- Number of GLS separators ($N_{GLS}$): 20 (5 in each reactor)
- Upper area occupied by each GLS separator: $A_{GLS\text{-}upper} = (L_{GLS\text{-}upper} \times W_{GLS\text{-}upper}) = 6.5\ m \times 0.37\ m = 2.41\ m^2$
- Area of each UASB reactor: $A_r = 136.5\ m^2$
- Total area of the settler compartments: $A_{settlers} = A_r - A_{GLS\text{-}upper} = (N_r \times A_r) - (N_{GLS} \times A_{GLS\text{-}upper}) = (4 \times 136.5\ m^2) - (20 \times 2.41\ m^2) = 497.8\ m^2$
    - Surface loading rates on the settler compartment
- For $Q_{s\text{-}avg}$: $v = Q_{s\text{-}avg}/A_{settler} = 278.6\ m^3/h/497.8\ m^2 = 0.56\ m/h$
- For $Q_{s\text{-}max}$: $v = Q_{s\text{-}max}/A_{settler} = 478.8\ m^3/h/497.8\ m^2 = 0.96\ m/h$

(13) Verify the HRT on the settler compartment ($HRT_{settler}$)

- Number of GLS separators ($N_{GLS}$): 20 (5 in each UASB reactor)
- Height of the lower part of the GLS separator (inverted pyramid trunk): $H_{GLS\text{-}lower} = 1.30$ m
- Height of the upper part of the GLS separator (rectangular gas hood): $H_{GLS\text{-}upper} = 0.50$ m
- Height of the upper part of the reactor (above the lower part of the GLS separator): $H_{r\text{-}upper} = 1.80$
- Lower area of each GLS separator: $A_{GLS\text{-}lower} = (L_{GLS\text{-}lower} \times W_{GLS\text{-}lower}) = 8.4\ m \times 2.3\ m = 19.32\ m^2$
- Upper area of each GLS separator: $A_{GLS\text{-}upper} = (L_{GLS\text{-}upper} \times W_{GLS\text{-}upper}) = 6.5\ m \times 0.37\ m = 2.41\ m^2$
- Lower volume occupied by each GLS separator (inverted pyramid trunk): $V_{GLS\text{-}lower} = (H_{GLS\text{-}lower}/3) \times [A_{GLS\text{-}lower} + A_{GLS\text{-}upper} + SQR(A_{GLS\text{-}lower} \times A_{GLS\text{-}upper})] = (1.30/3) \times [19.32 + 2.41 + SQR(19.32 \times 2.41)] = 12.37\ m^3$
- Upper volume occupied by each GLS separator (rectangular gas hood): $V_{GLS\text{-}upper} = (L_{GLS\text{-}upper} \times W_{GLS\text{-}upper}) \times H_{GLS\text{-}upper} = (6.5 \times 0.37) \times 0.50 = 1.20\ m^3$
- Upper volume of each reactor (above the lower part of the GLS separator): $V_{r\text{-}upper} = A_r \times H_{r\text{-}upper} = 136.5 \times 1.80 = 245.7\ m^3$
- Total volume of the settler compartments: $V_{settlers} = (N_r \times V_{r\text{-}upper}) - N_{GLS} \times (V_{GLS\text{-}lower} + V_{GLS\text{-}upper}) = (4 \times 245.7\ m^3) - 20 \times (12.37 \times 1.20\ m^3) = 711.4\ m^3$
  - HRT of the settler compartment ($HRT_{settler}$):
- For $Q_{s\text{-}avg}$: $HRT_{settler} = V_{settler}/Q_{s\text{-}avg} = 711.4\ m^3/278.6\ m^3/h = 2.6\ h$
- For $Q_{s\text{-}max}$: $HRT_{settler} = V_{settler}/Q_{s\text{-}max} = 711.4\ m^3/497.8\ m^3/h = 1.4\ h$

(14) Estimate the COD removal efficiency

- According to Section 2.3.2: $E_{COD} = 48 \times (HRT^{0.14})$
- $E_{COD} = 48 \times (9.1^{0.14}) = 65\%$

(15) Estimate the BOD removal efficiency

- According to Section 2.3.2: $E_{BOD} = 49.2 \times (HRT^{0.18})$
- $E_{BOD} = 49.2 \times (9.1^{0.18}) = 73\%$

(16) Estimate the effluent COD and BOD concentrations

- $C_{effCOD} = C_{COD} - (E_{COD} \times C_{COD})/100 = 598 - (65 \times 598)/100 = 209$ mg COD/L
- $C_{effBOD} = C_{BOD} - (E_{BOD} \times C_{BOD})/100 = 333 - (73 \times 333)/100 = 90$ mg BOD/L

(17) Estimate the biogas flow rate ($Q_g$)

- Organic load removed: $OL_{removed} = Q_{s\text{-}avg} \times C_{COD} \times E_{COD\text{-}filtered} = 278.6\ m^3/h \times 24\ h/day \times 0.598\ kg/m^3 \times 0.80 = 3{,}198.8$ kg COD/day
- Daily methane recovered in biogas: $Q_{CH_4} = OL_{removed} \times Y_{CH_4} = 3{,}198.8$ kg COD/day $\times$ 0.13 N $m^3$ $CH_4$/g $COD_{removed}$ = 415.8 $m^3$ $CH_4$/day
- Adopting a methane content of 75%: $Q_g = Q_{CH_4}/0.75 = 415.8\ m^3\ CH_4/day/0.75 = 554.5\ m^3$ biogas/day

(18) Estimate the daily sludge production ($P_{sludge}$)

- Daily sludge mass production: $P_{sludge} = Q_{s\text{-}avg} \times C_{COD} \times Y_{sludge} = 278.6\ m^3/h \times 24\ h/day \times 0.598\ kg/m^3 \times 0.15$ kg TS/kg $COD_{applied}$ = 599.8 kg TS/day
- Daily sludge volumetric production: $V_{sludge} = P_{sludge}/(C_{sludge} \times \gamma_{sludge}) = 599.8$ kg TS/day/$(0.04 \times 1{,}020$ kg TS/$m^3) = 14.7\ m^3$/day

*Note*: A design spreadsheet can be found in Chernicharo and Bressani-Ribeiro (2019).

## 2.4 OPERATION OF FULL-SCALE UASB REACTORS FOR MAINSTREAM SEWAGE TREATMENT

### 2.4.1 Brief overview of the main operational challenges

The main operational challenges of UASB reactors for mainstream sewage treatment are associated with (1) the control of the preliminary treatment units, (2) the management of anaerobic sludge inside the digestion compartment, (3) the management of scum that accumulates inside the GLS separators, (4) the management of biogas and (5) the control of diffuse emissions. Preliminary treatment units are the core of the operation of UASB reactors, as raw domestic wastewater usually comprises components (debris of all kinds and sand) that should not be introduced into the biological process. Therefore, it is crucial to establish an operational routine of regular maintenance of the screens and grit chambers to ensure the effective removal of the coarse solids and grit prior to the UASB reactor. The following sections highlight the key aspects related to the integrated management of sludge, scum, biogas and diffuse emissions.

### 2.4.2 Management of sludge

Solid by-products (sludge and scum) produced in UASB reactors must be effectively managed to ensure process efficiency. Nevertheless, from a practical standpoint, sludge build-up and withdrawal control are often neglected, probably due to the low excess sludge production relative to aerobic processes. From the start-up of the system, the accumulation of sludge (biomass + inert solids) occurs after a few months of continuous operation, and the rate of accumulation depends on both biomass growth (microbial cell synthesis) and the presence of inert or non-biodegradable suspend solids in raw sewage. The latter can contribute significantly to the increase in the sludge yield coefficient and must, therefore, be included in the calculations of sludge production in UASB reactors. A special concern arises when sludge from post-treatment units is sent to digestion and thickening in the anaerobic reactor, or the facility receives exogenous contributions (e.g. leachate from landfills). In both cases, the sludge yield coefficient during the design phase should reflect augmented sludge production.

A key characteristic of well-operated UASB reactors refers to their high-solid retention capacity, ascribed to the combined effect of having a GLS separator, low-imposed upflow velocities and the high settleability of the sludge developing in the system (Leitão *et al.*, 2010). Therefore, high-solid retention times can be achieved with a high degree of sludge stabilization. Nevertheless, because of its compact volume (typically a function of a selected HRT – see Table 2.1), the reactor has a maximum capacity for sludge storage, determined mainly by the volume of the digestion compartment. The excess sludge must be withdrawn systematically to prevent exceeding such a maximum sludge storage capacity. A lack of a comprehensive sludge discharge routine can cause (Chernicharo & Bressani-Ribeiro, 2019):

- Excessive passage of sludge from the digestion compartment to the settler compartment, with a consequent increase in total solid concentration in the settler compartment.
- Solid washout with the effluent, leading to an increase in the effluent concentration of particulate COD and BOD.
- Overloading and eventual impairment of the post-treatment unit, particularly attached growth systems (e.g. biofilters, trickling filters and the like).
- Hurdle to manage the aerobic sludge generated in the post-treatment unit when it is sent for thickening and digestion in the UASB reactor.
- Excessive build-up of inert solids in the bottom of the reactor.
- Decrease in the usable volume of the reactor.

**Table 2.4** Main factors that can affect the management of sludge in UASB reactors treating sewage.

| Factors that can Affect the Operational Routine of Sludge Management | Non-compliance Source |
|---|---|
| Sludge sampling points, used to monitor total solid concentration along the height of the digestion compartment, are insufficient in number and/or installed in the wrong position. | Design and construction |
| Sludge withdrawal pipes in insufficient number or poorly distributed along the height of the digestion compartment. | Design and construction |
| Manuals or operating guidelines for UASB reactors and dewatering systems that do not provide adequate detail on the procedures to be followed by the operators. | Design |
| Lack of systematic removal of excess sludge from UASB reactors and drying beds because of logistical and administrative problems (e.g. service contract for sludge transport incompatible with the required frequency, preventing sludge withdrawal from the reactor). | Operational management |
| Failures in the designs of drying beds or mechanized dewatering systems, with capacity below the need to process the sludge produced in the STP or incompatible with the characteristics of the sludge to be dewatered (e.g. climatic conditions, time scale of operators, transport of sludge, availability of chemical products and spare parts, need for preventive maintenance, possibility of shutdowns because of corrective maintenance, and the like). | Design and operational management |
| Equipment of the mechanized dewatering system (e.g. sludge pumps, polymer dosers, centrifuges and the like) not in operation owing to difficulties with preventive and corrective maintenance. | Operational management |

*Source:* Adapted from Chernicharo and Bressani-Ribeiro *et al.* (2019).

Accordingly, the absence of and/or ineffective management of sludge in UASB reactors compromises the efficiency of the entire treatment system. Usually, the lack of an operational routine for systematic sludge withdrawals is associated with the factors described in Table 2.4. Those factors can be related to the design, construction and operation of UASB reactors (Chernicharo & Bressani-Ribeiro, 2019).

#### *2.4.2.1 Proper sludge withdrawal systems*

Two levels of sludge withdrawal pipes must be implemented in UASB reactors treating sewage to provide operational flexibility. A lower set of pipes, which allows the withdrawal of the more concentrated sludge that has better settling characteristics, and an upper set of pipes, which allows the withdrawal of the less concentrated sludge, with worse settling characteristics. The management of the sludge discharge system ensures the removal of the worst quality sludge (upper, less concentrated) and the preservation of the best-quality sludge (lower, more concentrated). Therefore, a dense-sludge bed is fostered at the bottom of the reactor, promoting a physical retention of suspended solids for further hydrolysis. On the contrary, sludge withdrawn from the bottom allows the removal of inert solids (e.g. sand) that enter the reactor. Moreover, the balance between lower and upper sludge discharge contributes to attain higher solid retention time, consequently, the withdrawal of more stabilized (digested) sludge. Besides, lower solids are lost to the settler compartment, directly reflecting the improvement of the effluent quality.

To prevent the undesired solid washout with the final effluent, the sludge mass within the reactor must be maintained between a minimum value that should match the amount of biomass sufficient to digest the influent organic load, and a maximum value that is dependent on the sludge retention capacity in the UASB reactor. The minimum mass is calculated from the organic load applied to the reactor and the specific methanogenic activity of the anaerobic sludge (Chernicharo & Bressani-Ribeiro, 2019).

The maximum mass of the sludge the reactor can store depends on close observation of the results of TSS and settleable solid trends in the effluent and the TS concentrations along the height of the digestion compartment (sludge profile). In other words, the maximum sludge storage capacity is defined as a threshold that triggers excessive loss of solids to the settler compartment and deterioration of the effluent quality. If no data series is available for the definition of the maximum mass, a possibility is ensuring the TS concentration at the highest point of the digestion compartment (located immediately before the gas deflector) always below 0.5%. The removal of the upper sludge favours the maintenance of low TS concentrations at the interface of the digestion and settler compartments.

The effective volume of sludge to be discarded is a function of the adopted withdrawal strategy and should be determined from the estimated sludge production and the mass of sludge to be withdrawn. One recommended strategy is to discard at least 50% of the excess sludge mass from the upper pipeline (1.0–1.5 m above the reactor bottom – see Figure 2.1). To compensate for the higher volumes of sludge extraction resulting from this practice (less concentrated sludge), a sludge homogenization tank should be implemented before the dewatering unit. The thickening of the sludge allows the supernatant to be returned to the UASB reactor and the remaining fraction to the dewatering unit.

Pre-programmed and continuous removal of the sludge is recommended to enable the reactor to function as steadily as possible. The definition of the percentage of sludge to be discarded from the upper and lower pipelines, as well as the frequencies of the pre-programmed extractions should be established according to the capacity and operational routine of the dewatering system (drying beds or mechanized).

### 2.4.3 Management of scum

The production and accumulation of scum inside the GLS separators of UASB reactors depend essentially on the composition of the raw sewage. Although oil and grease are an essential part of scum and should be removed to the maximum extent prior to the reactor, the presence of debris in the sewage is at the cornerstone of scum formation, as light inert materials tend to float inside the reactor (e.g. hair, cotton swabs, dental floss, cigarette butts, plastics, etc.) and become entrapped in the scum layer. Therefore, special care should be taken to design the preliminary treatment units, where grease traps and fine screens (preferably <6 mm) are advisable. Besides containing oil, grease and debris not removed in the preliminary treatment unit, 'floating sludge' can also entrap in the scum layer formed inside the GLS separator, which is caused by excess sludge in the digestion compartment and hydraulic overload in the reactor.

The lack of efficient scum removal devices, and the absence of a systematic protocol for scum removal, promotes the build-up of a solid and thick scum layer inside the GLS separator (Figure 2.6). This is one of the main operational hurdles reported in full-scale UASB reactors treating sewage. The scum layer prevents the release of the biogas, which escapes to the settler compartment of the reactor. A cascade of malfunctioning is triggered, as gas bubbles tend to drag solids (solid washout), deteriorating effluent quality and compromising the operation of post-treatment units. Besides, it contributes to fugitive emissions of hydrogen sulphide and methane. Therefore, odour nuisance and greenhouse gas emissions can become an issue. Furthermore, as the effective biogas recovery (capture) decreases, so does the renewable energy generation potential of the plant.

From a construction standpoint, the thickening of the scum layer imposes high-structural stresses on the walls of the GLS separator. This, in turn, can eventually cause the rupture of the GLS separator, particularly when manufactured from less-resistant materials, such as plastic canvas, metal sheets and the like. Moreover, as the scum layer gets thicker, the usable volume of the gas chamber decreases, changing the operating pressure. Accordingly, the scum must be removed periodically, as further addressed.

#### *2.4.3.1 Proper scum withdrawal systems*

The design improvements of the third-generation UASB reactors for sewage treatment incorporate scum collection channels inside the GLS separator to promote the hydrostatic removal of the scum. This removal procedure is based on the change in the water level inside the GLS separator, which allows

**Figure 2.6** Thick and solidified scum layer inside different GLS separators of UASB reactors treating sewage (without proper removal devices and operational routines). *Source*: Chernicharo and Bressani-Ribeiro (2019).

the scum to flow to the internal channel of the separator, and to be directed to the withdrawal pipe, outside the reactor. The change in the water level inside the GLS separator is achieved by increasing or reducing the pressure in the gas line between the separator and the water seal located on top of each UASB reactor (see Figure 2.3). A fluid scum layer is key to a high-performing hydrostatic removal device. Hence, scum should be removed frequently (two–three times a week) to avoid solidification.

Different configurations of hydrostatic scum removal devices have been tested, with satisfactory results in demonstration- and full-scale UASB reactors (Chernicharo & Bressani-Ribeiro, 2019). Scum removal efficiencies between 75 and 90% have been observed by Rosa *et al.* (2012), Santos (2014) and Díaz Flórez (2016), considering proper frequencies of discharge to avoid excessively thickening or solidification. Table 2.5 shows the main design, construction and operational guidelines for the adequate functioning of a GLS separator equipped with hydrostatic scum removal devices.

### 2.4.4 Management of biogas

The biogas produced in UASB reactors treating sewage comprises mainly methane ($CH_4$), nitrogen and carbon dioxide, whereas hydrogen sulphide ($H_2S$), oxygen, hydrogen, carbon monoxide, ammonia, siloxanes, water and particulate matter are present in smaller proportions (Noyola *et al.*, 2006). Methane has a calorific value of 9.9 kWh/N $m^3$ and its concentration defines the potential of recovering energy from the biogas. In addition, methane has a global warming potential (GWP) 28 times higher than that of carbon dioxide over a 100-year horizon (IPCC, 2014). Currently, most of the UASB-based treatment plants in warm climate regions do not store and/or use the biogas produced. Generally, the captured biogas is conveyed to an open flare, where it is destroyed thermally to reduce the rates of methane and hydrogen sulphide emitted to the atmosphere. Hydrogen sulphide is harmful to human health at high levels and is perceptible to humans from 0.47 to 100 ppb, having an unpleasant smell (Metcalf & Eddy, 2014; Park *et al.*, 2014). In addition, hydrogen sulphide can contribute to the corrosion of materials. Obviously, such biogas emissions need to be mitigated and controlled (Brandt *et al.*, 2019).

**Table 2.5** Guidelines for design, construction and operation of hydrostatic scum removal devices.

| Guidelines for Adequate Functioning of Hydrostatic Scum Removal Devices | Responsibility |
|---|---|
| GLS separator must be connected to a water seal to allow the control of internal pressures and, consequently, the adjustment of the water levels in relation to the scum collecting channel. | Design and construction |
| Scum collection channel should preferably be positioned adjacent to one of the walls of the GLS separator and not at the centre. The positioning of the channel next to one of the sides allows a larger water level and superior efficiency in the removal of scum. | Design and construction |
| Provision of inspection points and valves on the scum outlet pipes of each GLS separator, aiming to identify clogging points and avoiding uneven hydraulic head loss. | Design |
| Provision of extraction nozzles along the length of the scum collection channel to allow more adequate hydraulic flow conditions. | Design |
| Whenever possible, provisioning of automated valves to facilitate the scum removal operation. | Design |
| Ensuring gas compartment tightness to allow the establishment of small pressures in the biogas line (10–15 cm w c). | Design, construction and operation |
| Correct positioning of the scum collection channel at the time of construction, as installation at levels other than those specified in the design can prevent proper GLS pressurizing and operation of the device. | Construction |
| Optimum frequency of scum removal can be calibrated during start-up, based on visual inspection, which can be performed from the opening of the hermetic inspection hatches (or through transparent displays when available), before and after the withdrawal operation trials. The operational practice will enable establishing the optimal removal routine (frequency and time of withdrawal, degree of valve opening, volumes removed and the like). | Operation |
| Amount of scum accumulated inside the GLS separator should be checked for each specific condition, considering the characteristics of the raw sewage, the efficiency of the preliminary treatment units and the constructive characteristics of the reactor. | Operation |

*Source:* Adapted from Chernicharo and Bressani-Ribeiro *et al.* (2019).

Simply burning biogas in flares leads to the loss of its chemical energy. The design of third-generation UASB reactors comprises biogas recovery and possibly utilization, particularly for thermal and electricity generation (Possetti *et al.*, 2018). Heat recovery from biogas combustion has been studied as an alternative for the drying and sanitization of the sewage sludge. One of the advantages of this system is the non-use of the chemicals usually necessary to treat sludge, such as lime. In addition, the heat recovery from biogas in these systems ensures a high rate of water removal in the sludge, as well as the removal of pathogens, reducing the costs inherent to its handling, transportation and final destination (Chernicharo *et al.*, 2015). The electricity generation from biogas has also been demonstrated to be feasible, particularly for distributed generation, including net metring (Possetti *et al.*, 2019). In addition, the cogeneration of electricity and heat has been considered as on-site utilization of the residual heat produced in the thermodynamic processes of electricity generation. Aiming at the energy efficiency of the process, residual heat can be used also in thermal sludge drying systems.

The key to successful biogas recovery design is the accurate quantification of the biogas available. Each system presents daily and monthly variations in biogas production regardless of the size of the treatment plant. The influent sewage characteristics, reactor tightness to prevent leaks, scum removal devices, operational control and significant rainfall events (where separate sewerage and drainage systems are not effective) all influence the net biogas production, as reported by Possetti *et al.* (2013), Silva (2015), Cabral *et al.* (2017) and Possetti *et al.* (2018). The design and operation of third-generation UASB reactors for sewage treatment should ensure that (1) the reactor is sealed tightly to prevent biogas leakage during operation; (2) scum is removed frequently from the inner part of the GLS separator and (3) the biogas collection system is equipped with a condensate removal device, flame trap and a biogas flare. For those plants in which electricity generation from biogas is economically feasible, gas treatment units for $H_2S$ and siloxane removal and an internal combustion engine should additionally be used. Extended design and construction guidelines for the entire biogas system are discussed in Chernicharo and Bressani-Ribeiro (2019).

Problems related to high pressure (overpressure) or vacuum (underpressure) inside GLS separators can interfere with the correct management of the biogas. Therefore, a minimum pressure of 1,500 Pa (15 cm of water column) should be maintained in the GLS to ensure the correct operation of the system (Chernicharo & Bressani, 2019), for which pressure and vacuum relief valves must be installed in the biogas pipeline. Third-generation UASB reactors have adopted water seals as a mitigating measure related to overpressure and vacuum in GLS separators.

### 2.4.5 Control of diffuse emissions

Diffuse emissions in UASB reactors treating sewage refer to dissolved hydrogen sulphide and methane that can escape the liquid phase (waste gas) (Figure 2.7). $H_2S$ has a characteristic odour that generally masks the presence of other volatile compounds (Bhatia, 1978; Smet & Van Langenhove, 1998). Its formation is inherent to anaerobic reactors treating sewage, which contains sulphate (incoming concentrations rely mainly on the local geology, seawater infiltration in the coastal sewers and presence of industrial contributions). Hydrogen sulphide is not only the primary reason related to odour complaints, but it is also toxic when inhaled and highly corrosive, causing considerable damage to concrete and steel structures under ambient conditions.

The other notable gaseous emission from anaerobic-based plants is methane, which is non-odorous. The quantification of fugitive $CH_4$ emissions has gained significant focus due to its GWP. Point source $CH_4$ emissions are associated with biogas, thus controlled through biogas recovery and usage. Diffuse $CH_4$ emissions from UASB reactors derive from leaks in the biogas collection system, but mainly from the dissolved $CH_4$ fraction in the anaerobic effluent. This latter fraction was typically ignored in the first- and second-generation UASB reactors treating sewage. However, it will desorb to the atmosphere once released into the environment at a rate dependent on mixing (Chernicharo *et al.*, 2015).

Managing diffuse gases involves both the preventive control of the gaseous emissions and the containment/collection of the waste gases for subsequent abatement (Figure 2.7 – see also Chapter

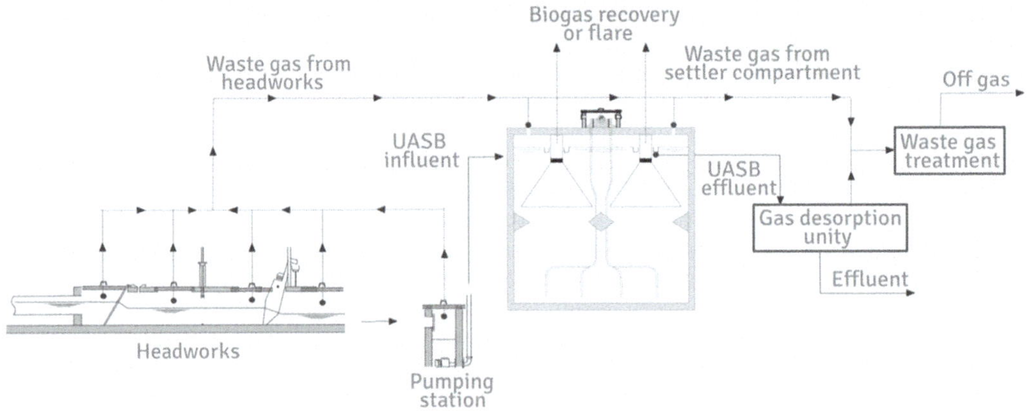


**Figure 2.7** Source and control of diffuse gaseous emissions in UASB-based STPs. *Source*: Chernicharo and Bressani-Ribeiro (2019).

7). Various methods have been tested for stripping or desorption of gases dissolved in the effluent of anaerobic reactors and these methods also produce waste gases that need subsequent abatement (e.g. in biofilters). Currently, third-generation UASB reactors comprises gas desorption units (Chernicharo & Bressani-Ribeiro, 2019). Managing diffuse gaseous emissions should include actions to maximize $CH_4$ capture in the biogas (e.g. proper scum withdrawal) and minimize $CH_4$ diffuse emissions as waste gas. Preventive actions to minimize waste gas emissions from UASB reactors treating sewage are summarized in Table 2.6, considering design and operation phases. It is worth mentioning that odour nuisance can be caused upstream the UASB reactor, jeopardizing controlling efforts associated with the anaerobic process. Therefore, a comprehensive analysis of the treatment plant should be performed.

**Table 2.6** Preventive actions to minimize waste gas emissions from UASB-based treatment plants.

| Possible Problem | Recommended Action |
|---|---|
| *Design* | |
| Accumulation of scum in the sewerage and inside the GLS separator | • Inclusion of systems to prevent the accumulation of scum or to allow the removal of accumulated scum (e.g. points of water jets in pumping stations, sieves and aerated grit chambers upstream of the anaerobic reactors). |
| Sediment build-up in the sewerage and the STP | • Accurate quantification of sewage flow and organic loads for appropriate hydraulic design.<br>• Control cross-section variations in channels of the preliminary treatment units, to avoid organic solids deposition.<br>• Design proper units for receiving sludge from other STPs or septic tanks. |
| Odour dispersing around the STP because of diffuse emissions | • Devices for capturing and conducting emissions in headworks.<br>• Minimizing turbulence during the conduction of anaerobic effluent, preventing hydraulic falls, reducing the volumes of pumping wells.<br>• Submersible incoming pipelines or using turbulent regime and hydraulic falls to force the liberation of dissolved gases in properly covered structures.<br>• Using perimeter vegetation and structures that induce atmospheric turbulence (aromatic trees, such as eucalyptus; 4-m high walls).<br>• Containing, conducting and treating waste gases. |
| Desorption and emission of gases dissolved in the effluent from the anaerobic reactors (odours and $CH_4$) | • Use of modular GLS separators (e.g. internal devices for scum removal and viewing windows for inspection).<br>• Using downstream devices to allow the controlled desorption and treatment of gases dissolved in the effluent. |
| *Operation* | |
| Odour generation upstream of the STP because of organic matter decomposition (e.g. sediments, scum) and release of precursor compounds in the sewer system | • Implementing specific rules (pre-treatment requirements) for receiving non-domestic effluents.<br>• Restricting the discharge of industrial effluents containing high loads of nitrogen and sulphur compounds, among others.<br>• Chemical dosing for preventative and reactive responses to odours. |
| Odour emissions from the headworks | • Frequently removing retained solids from the headworks.<br>• Proper guidelines for receiving sludge from septic tanks.<br>• Chemical dosing for preventive and reactive responses to odours. |
| Odour emissions from sludge streams in the STP | • Restricting the acceptance of non-stabilized sludge.<br>• Chemical dosing for preventive and reactive responses to odours. |
| Odour dispersing around the STP because of diffuse emissions | • Frequently removing scum accumulated in the UASB reactors.<br>• Cleaning and washing units after daily operational interventions. |

*Source:* Adapted from Chernicharo and Bressani-Ribeiro (2019).

## 2.5 FINAL CONSIDERATIONS AND FUTURE PERSPECTIVES

The anaerobic technology for mainstream sewage treatment is fully developed in warm climate regions. Three decades of continuous implementation have led to the development of the so-called third-generation UASB reactors. Therefore, current design guidelines comprehensively address the problems associated with earlier applications, comprising integrated management of sludge, scum and biogas. Extensive applied scientific research and operational practice shed light on control protocols, considering the management of diffuse emissions. The particular understanding of biogas production patterns allows for the design of feasible infrastructure for energy recovery. Moreover, construction processes have also evolved to cope with the peculiarities of anaerobic mainstream sewage treatment (e.g. protective coatings in top slabs).

Notwithstanding, the observed progress using UASB reactors for sewage treatment still presents challenges that prevent their maximum operational performance. Although there is currently available technical knowledge and practical experience for the design, construction and operation, many problems arise from situations beyond the limits of the treatment plant. The issues span from administrative to financial failures, which often prevent the implementation of design controls and the execution of essential operational procedures. The future of the anaerobic technology for mainstream sewage treatment has been at risk of discredit in several places where UASB reactors (early generations) were implemented, compromising the significant advances made to date. Nevertheless, the present societal appeal for developing circular economy models opens the window for effectively recognizing the advantages of direct anaerobic sewage treatment, as it enables the recovery of energy (biogas) and compounds in the sewage (e.g. nitrogen, phosphorus, sulphur and water itself).

## REFERENCES


ANA (Agência Nacional de Água) (2017). Atlas Esgotos. Despoluição de bacias hidrográficas (Atlas sewage. Remediation of watersheds). Agência Nacional de Água (National Water Agency), Brasilia, Brazil, Executive summary. 88pp. Available at atlasesgotos.ana.gov.br.

Baréa L. C. and Alem Sobrinho P. (2006). Behavior of polishing lagoon in metropolitan Curitiba and the possibility of using of duckweed for improvement of effluent quality. *Sanare Revista Técnica Sanepar Curitiba*, **24**(24), 46–60.

Bhatia S. P. (1978). Organosulfur emissions from industrial sources. In: Sulfur in the Environment, Part I, J. O. Nriagu (ed.), John Wiley & Sons, Canada.

Brandt E. M. F., Noyola, A. and McAdam, E. J. (2019). Control of diffuse emissions in UASB reactors treating sewage. In: Anaerobic Reactors for Sewage Treatment: Design, Construction and Operation, C. A. L. Chernicharo and T. Bressani-Ribeiro (eds.), IWA Publishing, London, England. https://doi.org/10.2166/9781780409238_0237

Busato R. (2004). Desempenho de um filtro anaeróbio de fluxo ascendente como tratamento de efluente de reator UASB: estudo de caso da ETE de Imbituva. (Performance of an anaerobic filter for the treatment of UASB reactor effluent: case study). PhD Thesis, Universidade Federal do Paraná, Curitiba/PR, Brazil.

Cabral C. B. G., Chernicharo C. A. L., Platzer C. J., Barjenbruch M. and Filho P. B. (2017). Evaluation of biogas production and energy recovery potential in 5 full-scale WWTPs with UASB reactors. *Chemical Engineering and Chemical Process Technology*, **3**(3), 1043.

Carraro A. (2006). Evaluation of a sewage treatment system with intermittent receipt of water treatment plant sludge following RALF and conventional aerobic biological filtration. *Sanare Revista Técnica Sanepar Curitiba* **24**(24), 46–60.

Chernicharo C. A. L. and Bressani-Ribeiro T. (eds.) (2019). Anaerobic Reactors for Sewage Treatment: Design, Construction, and Operation. IWA Publishing, London. https://doi.org/10.2166/9781780409238

Chernicharo C. A. L., Almeida P. G. S., Lobato L. C. S, Couto T. C. and Borges J. M. (2009). Experience with the design and start up of two full-scale UASB plants in Brazil: enhancements and drawbacks. *Water Science and Technology* **60**(2), 507–515, https://doi.org/10.2166/wst.2009.383

Chernicharo C. A. L., van Lier J. B., Noyola A. and Bressani Ribeiro T. (2015). Anaerobic sewage treatment: state of the art, constraints, and challenges. *Reviews in Environmental Science and Biotechnology*, **14**(4), 649–679, https://doi.org/10.1007/s11157-015-9377-3

Díaz Flórez C. A. (2016). Avaliação de protocolos de descarte e desidratação da escuma acumulada no interior de separadores trifásicos de reatores UASB (Evaluation of protocols for the disposal and dewatering of accumulated scum inside GLS separators of UASB reactors). MSc Thesis, Graduate Programme on Sanitation, Environmental and Water Resources, Federal University of Minas Gerais, Belo Horizonte, Minas Gerais (in Portuguese).

Florencio L., Kato M. T. and Morais J. C. (2001). Domestic sewage treatment in full-scale UASB plant at Mangueira, Recife, Pernambuco. *Water Science and Technology*, **44**(4), 71–77, https://doi.org/10.2166/wst.2001.0182

Franco P. L. P. (2010). Análise da Potencialidade do Reuso Indireto Potável: Estudo de Caso da ETE Atuba Sul, Região Metropolitana de Curitiba (Analysis of the potential indirect drinking reuse: case study of Atuba Sul WWTP, Curitiba's Metropolitan Region). MSc Thesis, Universidade Federal do Paraná, Curitiba/PR, Brazil.

Halalsheh M., Sawajneh Z., Zu'bi M., Zeeman G., van Lier J.B., Fayyad M. and Lettinga G. (2005). Treatment of strong domestic sewage in a 96 $m^3$ UASB reactor operated at ambient temperatures: two-stage versus single-stage reactor. *Bioresource Technology*, **96**, 577–585, https://doi.org/10.1016/j.biortech.2004.06.014

Intergovernmental Panel on Climate Change (2014). Synthesis Report, Contribution of Working Groups I, II and III to the Fifth Assessment Report of the Intergovernmental Panel on Climate Change, Core Writing Team, R. K. Pachauriand and L. A. Meyer (eds.), IPCC, Geneva, Switzerland, pp. 1–151.

Khalil N., Mittal A. K., Raghav A. K. and Rajeev S. (2006). UASB technology for sewage treatment in India: 20 years experience. *Environmental Engineering and Management Journal*, **5**(5), 1059–1069, https://doi.org/10.30638/eemj.2006.082

Khan A. A., Gaur R. Z., Tyiagi V. K., Khursheed A., Lew B., Mehrotra I. and Kazmi A. A. (2011). Sustainable options of post treatment of UASB effluent treating sewage: a review. *Resources, Conservation and Recycling*, **55**, 1232–1251, https://doi.org/10.1016/j.resconrec.2011.05.017

Leitão R. C., Santaellla S. T., van Haandel A. C., Zeeman G. and Lettinga G. (2010). The effect of operational conditions on the hydrodynamic characteristics of the sludge bed in UASB reactors. World Congress on Anaerobic Digestion, International Water Association (ed.), 12th edn, Guadalajara, México.

Lettinga G. (2014). My Anaerobic Sustainability Story, Lettinga Foundation.

Metcalf and Eddy (2014). Wastewater engineering: treatment and reuse. In: G. Tchobanoglous, H. D. Stensel, R. Tsuchihashi, F. Burton, M. Abu-Orf, G. Bowden and W. Pfrang (eds.), 5th edn, The McGraw-Hill Companies, AECOM, New York, USA.

Monroy O., Famá G., Meraz M., Montoya L. and Macarie H. (2000). Anaerobic digestion for wastewater treatment in Mexico: state of the technology. *Water Research*, **34**(6), 1803–1816, https://doi.org/10.1016/S0043-1354(99)00301-2

Mungray A. K. and Patel K. (2011). Coliforms removal in two UASB/ASP based systems. *International Biodeterioration & Biodegradation*, **65**, 23–28, https://doi.org/10.1016/j.ibiod.2010.04.011

Nada T., Ali H. I., Farid M. N., El-Gohary F. A. and Moawad A. (2006). A full-scale application for anaerobic wastewater treatment technology (up-flow anaerobic sludge blanket UASB in Egypt). Conference National Research Centre, 13–15 March, Cairo, Egypt.

Noyola A., Morgan-Sagastume J. M. and López-Hernández J. E. (2006). Treatment of biogas produced in anaerobic reactors for domestic wastewater: odor control and energy/resource recovery. *Reviews in Environmental Sciences and Bio/Technology*, **5**(1), 93–114, https://doi.org/10.1007/s11157-005-2754-6

Noyola A., Padilla-Rivera A., Morgan-Sagastume J. M. L., Guereca L. P. and Hernandez-Padilla F. (2012). Typology of municipal wastewater treatment technologies in Latin America. *CLEAN - Soil, Air, Water*, **40**(9), 926–932, https://doi.org/10.1002/clen.201100707

Oliveira S. C. and Von Sperling M. (2011). Performance evaluation of different wastewater treatment technologies operating in a developing country. *Journal of Water, Sanitation and Hygiene for Development*, **1**(1), 37–56, https://doi.org/10.2166/washdev.2011.022

Pandey N. and Dubey S. K. (2014). Up-flow anaerobic sludge bed (UASB) based sewage treatment plant (STP) at Mirzapur: a review. *International Research Journal of Environmental Sciences*, **3**(8), 67–71.

Park K., Lee H., Phelan S., Liyanaarachchi S., Marleni N., Navaratna D., Jegatheesam V. and Shu L. (2014). Mitigation strategies of hydrogen sulphide emission in sewer networks. *International Biodeterioration & Biodegradation*, **95**(A), 251–261, https://doi.org/10.1016/j.ibiod.2014.02.013

Peña M. R., Rodriguéz J., Mara D. D. and Sepulveda M. (2000). UASBs or anaerobic ponds in warm climates? A preliminary answer from Colombia use of biogas for cogeneration of heat and electricity for local application: performance evaluation of an engine power generator and a sludge thermal dryer. *Water Science and Technology*, **42**(10), 59–65, https://doi.org/10.2166/wst.2000.0609

Peña M. R., Mara D. D. and Avella G. P. (2006). Dispersion and treatment performance analysis of an UASB reactor under different hydraulic loading rates. *Water Research*, **40**(3), 445–452, https://doi.org/10.1016/j.watres.2005.11.021

Possetti G. R. C., Jasinski V. P., Mesquita N. C., Kriguel, K. and Carneiro C. (2013). Medições em tempo real do biogás produzido em reatores UASB alimentados com esgoto doméstico (Real-time measurements of biogas produced in UASB reactors treating sewage). Congresso Brasileiro de Engenharia Sanitária e Ambiental, 27th edn, Goiânia, Goiás, Brazil.

Possetti G. R. C., Rietow J. C., Costa F. J. O. G., Wagner L. G., Lobato L. C. S., Bressani-Ribeiro T., Melo D. F., Reis J. A. and Chernicharo C. A. L. (2018). Contribuição para o aprimoramento de projeto, construção e operação de reatores UASB aplicados ao tratamento de esgoto sanitário – Parte 5: Biogás e emissões fugitivas de metano (Contribution to the improvement of design, construction and operation of UASB reactors treating sewage – Part 5: biogas and fugitive emissions of methane). *Revista DAE*, **66**(214), 73–89.

Possetti G. R. C., Rietow J. C., Cabral C. B. G., Moreira H. C., Platzer C. J., Bressani-Ribeiro T. and Chernicharo C. A. L. (2019). Energy recovery from biogas in UASB reactors treating sewage. In: Anaerobic Reactors for Sewage Treatment: Design, Construction and Operation, C. A. L. Chernicharo and T. Bressani-Ribeiro (eds.), IWA Publishing, London, England. https://doi.org/10.2166/9781780409238_0194

Rosa A. P., Lobato L. C. S., Chernicharo C. A. L., Martins D. C. R. B., Maciel F. M. and Borges J. M. (2012). Improving performance and operational control of UASB reactors via proper sludge and scum discharge routines. *Water Practice & Technology*, **7**(3), 11.

Santos A. (2014). Remoção, tratamento e valoração de escuma proveniente de reatores UASB aplicados ao tratamento de esgoto doméstico (Scum removal, treatment and evaluation from UASB reactors applied to domestic sewage treatment). PhD Thesis, Graduate Programme on Sanitation, Environmental and Water Resources, Federal University of Minas Gerais, Belo Horizonte, Minas Gerais (in Portuguese).

Sato N., Okubo T., Onodera T., Ohashi A. and Harada H. (2006). Prospects for a self-sustainable sewage treatment system: a case study on full-scale UASB system in India's Yamuna River Basin. *Journal of Environmental Management*, **80**, 198–207, https://doi.org/10.1016/j.jenvman.2005.08.025

Silva T. C. F. (2015). Utilização de biogás de estações de tratamento de esgoto para fins energéticos. MSc Thesis, Institute of Energy and Environment, University of São Paulo, São Paulo, Brazil.

Silva F. J. A., Lima M. G. S., Mendonca L. A. R. and Gomes M. J. T. L. (2013). Septic tank combined with anaerobic filter and conventional UASB – results from full scale plants. *Brazilian Journal of Chemical Engineering*, **30**(1), 133–140, https://doi.org/10.1590/S0104-66322013000100015

Smet E. and Van Langenhove H. (1998). Abatement of volatile organic sulfur compounds in odorous emissions from the bioindustry. *Biodegradation*, **9**(3–4), 273–284, https://doi.org/10.1023/A:1008281609966

Tachini M., Filho P. B. and Pinheiro A. (2006). Evaluation of integrated treatment of sanitary sewage and septic tank sludge in a RALF: case study. *Sanare Revista Técnica Sanepar Curitiba*, **24**(24), 46–60.

Uemura S. and Harada H. (2010) Application of UASB technology for sewage treatment with a novel post treatment process. In: Environmental Anaerobic Technology: Applications and New Developments, Chapter 6, H. H. P. Fang (ed.), World Scientific, Imperial College Press, London. ISBN 978-1-84816-542-7

van Haandel A. C. and Lettinga G. (1994). Tratamento anaeróbio de esgoto: um manual para regiões de clima quente (Anaerobic sewage treatment: a manual for hot climate regions). University Federal of Paraíba, Campina Grande, Paraíba, Brazil.

van Lier J. B., Vashi A., van der Lubbe J. and Heffernan B. (2010). Anaerobic sewage treatment using UASB reactors: engineering and operational aspects. *Environmental Anaerobic Technology.*

von Sperling M. and Chernicharo C. A. L. 2005. Biological Wastewater Treatment in Warm Climate Regions. Two volumes. IWA Publishing. Available at http://www.iwapublishing.com/open-access-ebooks/3567.

Walia R., Kumar P. and Mehrotra I. (2011). Performance of UASB based sewage treatment plant in India: polishing by diffusers an alternative. *Water Science and Technology*, **63**(4), 680–688, https://doi.org/10.2166/wst.2011.333

WERF (2010). Energy efficiency in wastewater treatment in North America: a compendium of best practices and case studies of novel approaches. Final report. Water Environment Research Foundation, Alexandria.

doi: 10.2166/9781789063479_0051

# Chapter 3

# Anaerobic membrane bioreactors

D. C. Stuckey*

*Department of Chemical Engineering, Imperial College London, London SW7 2AZ, UK*
**Correspondence: d.stuckey@imperial.ac.uk*

**ABSTRACT**

Anaerobic membrane bioreactors (AnMBRs) are a recent development in wastewater treatment driven by concerns about energy use and sludge disposal. It separates hydraulic retention time from solid retention time enabling short retention times (as low as 3–6 h, but normally 12–24 h), and excellent performance (85–95% chemical oxygen demand (COD) removal, no effluent solids and high bacterial/virus removal). It produces low-sludge yields (0.04–0.12 g sludge/g $COD_{rem}$), and methane in both the gas and dissolved in the effluent (lower temperatures yield higher soluble methane). It can be net energy positive depending on its configuration, for example, using 'dynamic membranes', mechanically shaken membrane or a two-stage granular activated carbon (GAC) bed reactor. However, membranes ranging from ultrafiltration (0.04 μm) to microfiltration (0.4 μm) foul quite quickly, with Soluble Microbial Products (SMPs)/Extra Cellular Polymers (ECPs-which include extracellular polysaccharides) and cells depositing and growing on the surface. At a certain cell density Quorum Sensing (QS) occurs and there is a rapid increase in trans-membrane pressure ('TMP jump'). Fouling can be ameliorated by managed gas sparging, mechanical shaking, addition of flocculants/adsorbents, for example, powdered activated carbon, or quorum quenching. Nevertheless, some fouling is important as it enhances membrane performance. Due to the membrane rejecting cells and many low molecular weight solutes, AnMBRs tolerate shock loads and toxins well, and enable microbial adaptation to occur. However, to improve performance more research is needed to minimize overall energy use, explore enhanced performance with bioaugmentation, enhance rates of solid hydrolysis, optimize its performance in the overall flowsheet (global optimization), use life-cycle analysis to reduce its environmental impact, control sulphate reduction and improve post-treatment of effluents to enable water recycling.

**Keywords:** anaerobic membrane bioreactor (AnMBR), antibiotic resistant genes (ARGs), bioaugmentation, energy use, fouling, global optimization, life-cycle analysis (LCA), microbial ecology, pharmaceutical, personal care products (PPCPs) removal, quorum quenching (QQ).

## 3.1 BACKGROUND

Anaerobic bacteria have long doubling times due to their low-energy yields per gram of substrate (Gibbs free energy yield per gram of substrate), because a lot of catabolic energy yield is lost in the form of gaseous/soluble methane. Hence process engineers have been forced to innovate and develop efficient reactor designs which separate hydraulic retention time (HRT) from the solid retention time

(SRT). In recent years there has been increasing focus on designing anaerobic reactors that not only have very short HRTs, but also reduce the footprint of the process by intensifying or combining unit operations, and produce (consume less) energy. Hence over the last 50 years reactor design has evolved from standard sludge continuous stirred tank reactors (CSTRs) with SRTs of 20–30 days, to anaerobic filters developed by Young and McCarty (1969), operated either upflow, or sometimes downflow, with full of different types of support media. HRTs as low as 20–24 h were possible, and with sewage at ambient temperatures, chemical oxygen demand (COD) removals up to 70% were possible. While this design was effective, the media sometimes get blocked, and solid removal becomes a problem. In addition, wastewaters (WWs) with high levels of sulphate produced hydrogen sulphide and the gas percolating upwards resulted in toxicity to the biomass.

In the late 1970s Lettinga in The Netherlands noticed that anaerobic biomass could granulate under certain conditions, and this led to the development of upflow anaerobic sludge blanket (UASB) reactors using granular biomass. This reactor type is discussed thoroughly in Chapter 2; while it has many advantages, when operated with suspended biomass, it also results in quite high solids in the effluent (~200 mg/L), it is unstable to shock loads and it is not suitable for certain types of feeds. However, this is one of the most successful anaerobic reactor designs. Around the same time as UASBs were being developed, McCarty had noticed that a baffled configuration trapped biomass in the anaerobic reactor, and enabled HRTs as low as 8–12 h, with acceptable COD removal (Bachmann *et al.*, 1985). This design had the major advantage of not requiring power input so that it could be constructed at the lower level than the input flow, for example, below ground, and could be operated in communities without a reliable power source. Hence this type of treatment has been used in developing countries such as South Africa and Indonesia.

Around the same time as UASBs and anaerobic baffled reactors (ABRs) were being developed, Switzenbaum and Jewell (1980) found that with sufficient upflow velocity a bed of sand granules could be expanded under anaerobic conditions, and cells attached themselves to the grains. This was found to be very efficient at treating unscreened sewage at retention times down to 30 min with ~70% COD removal even at low temperatures (13°C). However, this is a sophisticated process requiring close monitoring and control because when the biofilm on the particles grow the density of the particle reduces and the particles rise through the bed. These need to be intermittently removed and the biomass to be removed and recycled back to the bottom of the reactor. However, this demonstrated that very short retention time reactors (0.5 h) are possible to use in anaerobic treatment because the substrate, both soluble and insoluble, is rapidly removed from solution by adsorption onto the biomass before it is catabolized. While this design has been used for higher-strength industrial WWs, it has not been used much with sewage treatment.

At the time of these latest designs, the cost of permeable membranes was quite high; but over time their industrial use, for example, in biotechnology production, has increased, and their unit cost has decreased. These factors led to their adoption in aerobic wastewater treatment (WWT), and early reports reveal good performance and reasonable HRTs (Muller *et al.*, 1995). However, sludge yields were still quite high, and above certain loading rates oxygen mass transfer became a problem in addition to high-energy inputs. In addition, the question of membrane lifetime, at this point being only 2 years, was of considerable financial and practical concern. The anaerobic membrane bioreactor (AnMBR) concept was initially proposed by Grethlein in 1978, who used external cross-flow membrane to treat sewage treatment effluent and achieved high biomass concentrations with 85–95% biochemical oxygen demand (BOD) reduction and 72% nitrate removal. The first commercially available AnMBR was constructed in the early 1980s, by Dorr-Oliver to treat high-strength whey processing WWs. The membrane bioreactor (MBR) process was first introduced in 1969 by Dorr-Oliver Inc.; however, the initial developments could not be translated to widespread industrial applications, owing to the high expenses associated with membrane material and energy. At this point there was little interest in the use of membranes in anaerobic systems, although Bill Ross from South Africa in 1985 had tried using cheaper 'dynamic membrane' systems in treating wine production WWs (personal communication, 1994). He used a nylon cloth with large pores, and over the period of days a 'fouling layer' built up that acted as an effective membrane.

However, as installed membrane costs decreased, the advantages of combining them with anaerobic treatment became apparent, and this has led to the development of AnMBRs which incorporate solid removal and COD reduction in one reactor. This design has evolved from aerobic membrane bioreactors (AeMBRs) which are further developed, but have obvious drawbacks, such as high-energy use and solid yields, and emission of greenhouse gases such as carbon dioxide and nitrous oxide.

This chapter will focus on recent advances in AnMBRs, and on specific areas and questions such as the role of soluble microbial products (SMPs) in membrane fouling, and fouling amelioration using additives such as powdered activated carbon (PAC and polymers/metal salts/biopolymers), removal performance and HRT/SRT, economics and life-cycle analysis (LCA), while drawing on past research from both AeMBRs and AnMBRs.

## 3.2 MEMBRANE TYPE

One of the key questions in the design of an AnMBR is what type of material and pore size the membranes used should be, and whether they should be housed internally in the reactor, or externally in a separate module? In the early days this was important as the capital cost of membranes was a significant fraction of the cost of a reactor. However, with experience from Japan over the years, the lifetime of most polymer membranes has been shown to be at least 10 years. In addition, the installed costs per $m^2$ have decreased substantially in the past decade due to the increasing use of membranes. Hence the discounted costs over 10 years have meant the additional membrane costs are minimal.

Ceramic and polymeric materials can also be used as membranes in an AnMBR. Ceramics such as (1) alumina, (2) silicon carbide, (3) titanium dioxide and (4) zirconia often demonstrate better filtration performance compared with other membrane types, due to their excellent chemical resistance, flexibility for cleaning and fouling resistance. However, the high cost of fabrication often makes them less economically viable; therefore, polymeric membranes, such as (1) polyacrylonitrile, (2) polyethylsulphone, (3) polysulphone, (4) polytetrafluoroethylene, (5) polyvinylidene difluoride (PVDF), are the most used membranes in wastewater treatment plants (WWTPs). About 50% of MBR modules available on the market are based on PVDF. PVDF-based membranes exhibit high-mechanical strength and enhanced flexibility, which makes them a good choice for users (Rahman *et al.*, 2023). Finally, the use of 'dynamic membranes', that is, a large pore nylon cloth that over a short time fouls and builds up a layer of organic detritus and cells which act as an effective filter media is also of interest and is a low-cost option; however, in most cases their performance is not satisfactory (Siddiqui *et al.*, 2021).

The most typical kinds of MBRs are submerged MBRs and side-stream MBRs (Al-Khafaji *et al.*, 2022; de Andrade *et al.*, 2014; Martínez *et al.*, 2021). Each of the modules has different characteristics; for instance, side-stream MBRs provide direct hydrodynamic fouling control but require high energy, and are used primarily in industrial WWT. In contrast, a submerged anaerobic MBR (SAnMBR) operates at reduced water flux with higher permeability and is used for treating municipal WWs on a large scale. According to Al-Khafaji *et al.* (2022), submerged MBRs require a larger membrane area, and are better suited with excellent sewage filterability (Al-Khafaji *et al.*, 2022). Compared with a submerged MBR, a side-stream MBR needs a smaller membrane area and functions well for strong sewage with low filterability. According to this study, there is no obvious selection criterion for the use of either module (submerged or side-stream MBRs) (Al-Khafaji *et al.*, 2022). Instead, engineering judgement should be applied before adopting a specific module. In a study by Rahman *et al.* (2023), a comparison table was introduced between these two modules (Table 3.1).

Pore size is one of the most important properties in the successful operation of membrane-based bioprocesses for the treatment of municipal WWs. Ji *et al.* (2020) compared the characteristics of two AnMBRs, one with a hollow-fibre (HF) membrane of 0.4 μm pore size (AnMBR1), and the other with a membrane of 0.05 μm pore size (AnMBR2) for the treatment of municipal WWs at room temperature (25°C) under various HRTs. Performance was evaluated in terms of organic removal efficiency, biogas production and membrane filtration behaviour during long-term continuous operation. Both AnMBRs

**Table 3.1** Comparison between submerged and side-stream MBR systems.

| MBR Type | Submerged MBR | Side-stream MBR |
|---|---|---|
| Compatibility with wastewater type | Low-strength wastewater with good filterability | Higher strength with poor filterability |
| Membrane flux | Lower membrane flux or lower permeate per unit area of membrane | Higher membrane flux or higher permeate per unit area of membrane |
| Transmembrane pressure | Reduced transmembrane pressure needed | Increased transmembrane pressure is required |
| Power requirement | Lower power per $m^3$ of wastewater treated needed | High power per m3 of wastewater treated needed |
| Susceptibility to variations | Less susceptible to changes in the characteristics of the wastewater and flow irregularities | More susceptible to changes in the characteristics of the wastewater and flow irregularities |
| Requirement of membrane area | Large surface area needed | Less surface area needed |
| Backwashing and cleaning of the membrane | More frequently needs backwashing and cleaning | Less frequently needs backwashing and cleaning |
| Operational flexibility | Less-flexible operation | Control parameters provide for more operational flexibility |
| Expansion of WWTP capacity | Problematic to extend capacity | Simpler to extend |

*Source:* Rahman *et al.* (2023).

resulted in good performance, with COD and BOD removals of ~89 and ~93%, respectively. High-energy recovery was achieved, with the biogas yield ranging between 0.20 and 0.26 L gas/g $COD_{rem}$ and a methane content of ~75%. The difference in membrane filtration behaviour in the two AnMBRs included different permeate fluxes and total filtration resistance ($R_t$). In the AnMBR with a 0.4 μm pore size membrane, an average $R_t$ of $1.08 \times 10^{12}$/m was obtained even when the permeate flux was high at 0.274 m/day, whereas a higher average $R_t$ of $1.51 \times 10^{12}$/m was observed in the AnMBR with 0.05 μm pore size membrane, even when the flux was low at 0.148 m/day. Offline membrane cleaning for AnMBR1 showed that the membrane restoration efficiency was 90.2%; hence larger pore sizes can lead to better overall performance.

Two types of membrane units are usually used: HF and flat sheets (FSs), or plate and frame. A bundle of hundreds to thousands of HFs constitutes an HF membrane module in a pressure tank. The plate and frame units consist of a number of FS membranes, along with support plates (Al-Khafaji *et al.*, 2022). According to Hashisho *et al.* (2016), FS modules are more costly but easier to manage and less susceptible to fouling. Compared to them, HF modules may resist thorough backwashing despite being susceptible to fouling (Hashisho *et al.*, 2016). Tolu *et al.* (2021) conducted a study to compare HF and FS modules for full-scale leachate treatment; in terms of fouling the HF module performed better and prevented clogging for a long period of time. It resulted in lower cleaning frequency and easier maintenance, and in terms of capital and operation expenses, the HF module proved to be the better choice (Tolu *et al.*, 2021).

## 3.3 CAUSES OF MEMBRANE FOULING

Since membranes constitute the largest capital cost of a membrane reactor, their flux determines the required surface area, and hence their capital cost. A multitude of parameters influence membrane flux, and these are shown in Figure 3.1, and are reviewed in Iorhemen *et al.* (2016). Nevertheless, it is important to highlight the fact that membranes do not function well in these environments, except for nanofiltration, unless they are fouled to 'some degree'. This can be seen in membrane performance

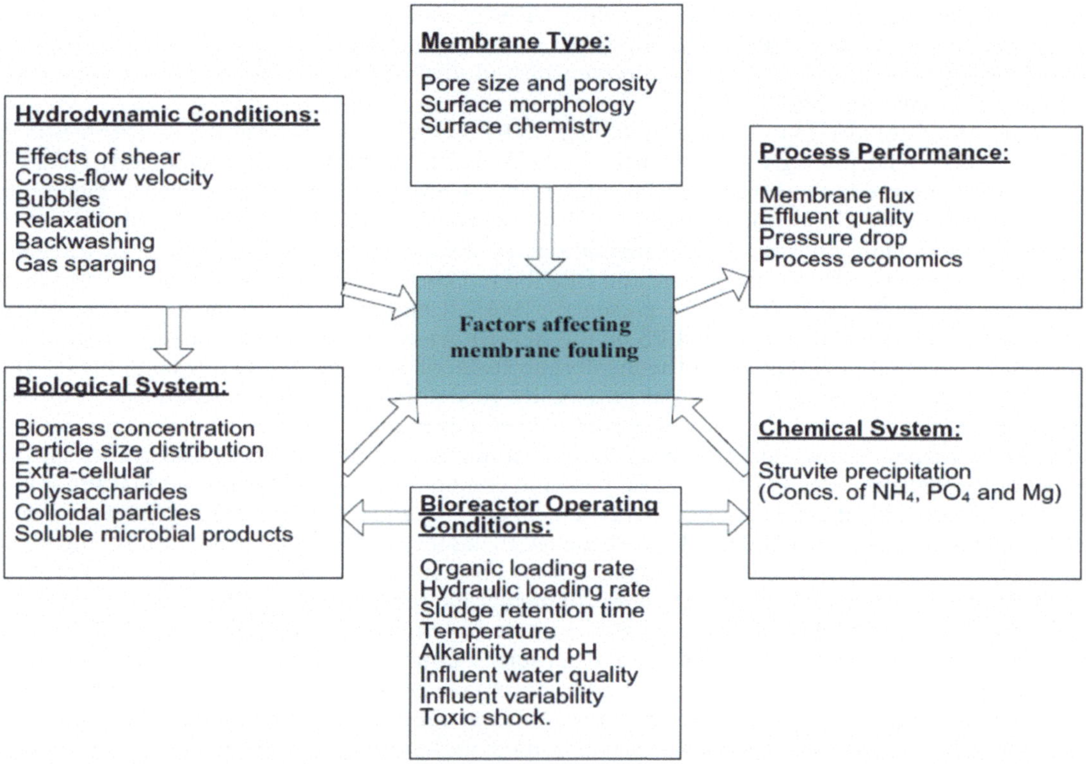


**Figure 3.1** Parameters and interactions that influence process performance in an AnMBR (*Source*: Semmens, 2002).

immediately after a complete clean of the membrane, but even after 5 min a fouling layer has built up and membrane performance increases rapidly until 10–15 min, later membrane performance has attained the 'steady state'. This phenomenon is often not acknowledged under the rubric that 'all membrane fouling is bad' (Vyrides & Stuckey, 2011). Over time, even with membrane gassing or shaking, a fouling layer builds up and the trans-membrane pressure (TMP) rises slowly from only a few kilopascals to 10–15 kPa. At some point after a few days there can be a 'TMP kick' where the TMP/time gradient suddenly changes and rapidly steepens. It seems that this inflection point is caused by quorum sensing (QS) molecules released by the cells present on the surface of the membrane. This triggers the rapid production of extracellular polymer (ECP)/EPS on the membrane surface and a rapid jump in membrane resistance (Oh & Lee, 2018).

### 3.3.1 Difference in fouling between AnMBRs and AeMBRs

Xiong *et al.* (2016) compared membrane fouling in AeMBRs and AnMBRs under similar operating conditions. Although both AeMBRs and AnMBRs achieved more than 90% COD removal, the fouling mechanisms were different. Molecular weight (MW) fingerprint profiles showed that most fragments in anaerobic SMPs were retained by the membrane, and some fragments were present in both SMP and in soluble EPS, suggesting that the physical retention of SMP components contributed to AnMBR membrane fouling. One of the dominant fragments was glycolipoproteins (size: 630–640 kDa), and correlated in abundance in AnMBR-EPS with the extent of anaerobic membrane fouling. In contrast, all detected AeMBR-SMP fragments permeated through the membrane. Aerobic SMP and soluble EPS also showed very different fingerprinting profiles. A large amount of adenosine triphosphate (ATP) was present in the AeMBR-EPS, suggesting that microbial activity arising from certain bacterial populations, such as unclassified *Comamonadaceae* and unclassified *Chitinophagaceae*, may play a role in aerobic membrane fouling.

Xiong *et al.* (2016) also found that despite high COD removal in both AeMBRs and AnMBRs (94 and 90%, respectively), fouling was seen earlier in the AeMBR. The MW distribution of SMPs and EPS in the AnMBR showed that most of the EPS came from the SMP fractions retained by the membrane, implying that the major contributor to AnMBR membrane fouling was physical retention of SMP components by the microfiltration (MF) membrane. In the case of the AeMBR, all SMP fragments completely passed through the membrane and EPS did not correlate with SMP profiles. Furthermore, significantly higher levels of ATP were present in AeMBR biofilms than in those of the AnMBR, suggesting that a larger fraction of the EPS on the AeMBR was a product of microbial activity arising from certain bacterial populations on the membrane surfaces, such as *Comamonas* and unclassified *Chitinophagaceae*. This study shows that potentially very different fouling mechanisms controlled AeMBR and AnMBR biofouling, and highlights the importance of a multifaceted approach in studying membrane fouling and developing control strategies for different MBR systems. In our own laboratory, unpublished results (Kunacheva, personal communication, 2016) showed that the soluble effluent CODs from identically operated AeMBR (0.04 μm HF) and AnMBR (0.45 μm FS) were similar (~96% COD removals from a 500 mg/L synthetic feed), but the soluble COD was substantially higher within the anaerobic reactor (200 vs. 25 mg/L). These observations support the conclusions of Xiong *et al.* (2016), and demonstrate the effectiveness of membrane rejection in AnMBRs.

Bagheria and Mirbagheri (2018) critically reviewed all types of fouling (Figure 3.2) and the approaches used for membrane (aerobic and anaerobic) fouling control in bioreactors treating water and WW. The first antifouling techniques tried to optimize operational conditions, or used chemical agents to control membrane fouling. Despite their positive impacts, these methods did not provide a sustainable solution, and some chemicals used may affect the microorganisms and have environmental drawbacks. Improved understanding of membrane fouling mechanisms and effective interventions has encouraged researchers to look at novel methods that focus on disrupting fouling by suppressing fouling bacteria, such as quorum quenching (QQ). Employing nanomaterials, cell entrapment and

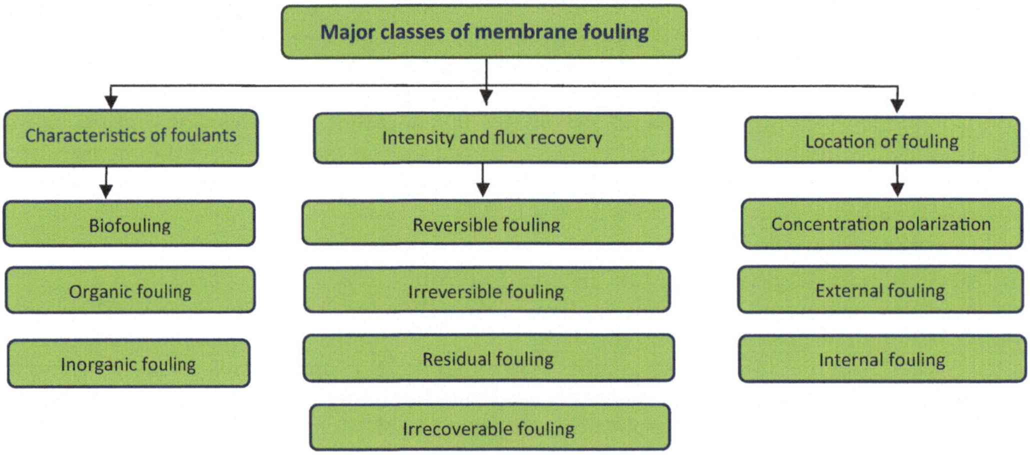


**Figure 3.2** Three main classes for membrane fouling according to published literature (*Source*: Bagheria & Mirbagheri, 2018).

biologically and electrically based methods are the latest techniques to be evaluated. This fact indicated that control of membrane fouling requires employing more than one single approach.

Wang *et al.* (2018) also examined fouling in two types of membrane reactors, and emphasized the recent transition to energy neutral sewage treatment is the key driver for AnMBRs for sewage treatment. However, aerobic municipal WW delivers very little methane compared to the maximum theoretical case study for sewage, and hence energy production must be maximized to enable energy neutrality. Since membrane fouling control is the main energy demand in MBRs the authors tried to determine whether knowledge between fouling characteristics in AeMBRs and AnMBRs could be transferred. They concluded from their study that:

(a) The ratio of proteins to carbohydrates in cell wall EPS (eEPS) is higher in AnMBRs than in AeMBRs. However, the total eEPS appears slightly higher in AeMBRs, and there is a lack of knowledge about the correlation between eEPS concentration and composition, and surface properties such as charge and hydrophobicity in the AnMBR.
(b) There are similar median particle sizes in both AeMBRs and AnMBRs, but fine solids with particle sizes ranging from 1 to 10–15 μm have been widely reported in AnMBRs and these are associated with biomass of high fouling propensity (see later section).
(c) $SMP_{COD}$ concentration in the bulk sludge is 10 times higher in AnMBRs than in AeMBRs, and low temperature and extended SRT enhance SMP production in AnMBRs.
(d) In immersed AnMBRs turbulent gas sparging on membrane performance seems to be limited compared to AeMBRs, and hence fouling is more determined by sludge properties in AnMBRs.
(e) AnMBR fluxes are 33–50% lower than in AeMBRs, whereas permeabilities in AnMBRs are ~50% below than in AeMBRs, and use between 50 and 300% more gas.
(f) The dominant fouling mechanism in AeMBRs and AnMBRs is cake formation, but the effectiveness of backwashing to reduce membrane fouling and permeability recovery after chemical cleaning seems to result in contradictory results and needs further research.

### 3.3.2 Causes of membrane fouling

One of the early insights into membrane fouling when purifying products in other industries, such as biotechnology, was that a concentration gradient builds up on the membrane surface due to a mass balance where the mass flux back into the bulk solution equals the product flow through the membrane. This results in a high concentration of the solute at the membrane surface in the 'boundary layer', and either a 'gel' forming, or in some situations where the solute exceeds its solubility limit, precipitation or crystallization. This phenomenon is referred to as 'concentration polarization (CP)', and in AnMBRs this phenomenon contributes to membrane fouling. In the next section various constituents involved in membrane fouling are discussed in more depth; SMPs, subvisible particles and colloids, inorganic precipitation, and QS.

#### 3.3.2.1 Soluble microbial products

SMPs are primarily from the high-MW fraction which are generated as either SMPs from cell metabolism and lysis (Barker & Stuckey, 1999), or as extracellular polysaccharides (EPSs) which in anaerobic systems constitute only ~20% of the SMPs (Aquino & Stuckey, 2004a). SMPs are critical in membrane fouling but until recently not much has been known about their composition or what parameters influence their production in the reactor. In AnMBRs with high SRTs the production of SMPs from cell lysis (biomass-associated products) is high, and much of the soluble COD in the reactor is SMPs (including ECPs; Vyrides & Stuckey, 2009a). In addition, the way the reactor is operated influences SMP production considerably; Aquino and Stuckey (2003) showed that nutrient deficiencies in a CSTR increases SMP production, whereas the presence of toxic compounds in the waste being treated also enhances their concentration (Aquino & Stuckey, 2004a). Finally, if the reactor is subjected to organic or shock loads then SMP production increases (Aquino & Stuckey, 2004b).

In 2017, Chen *et al.* determined the properties of SMP and EPS, and their role in membrane fouling in an AnMBR treating low-strength WW at room temperature. The properties analysed were specific production, MW distribution and adhesion force, and they showed the primary factor affecting SMP/EPS properties and their diverse membrane fouling performances was the organic loading rate (OLR). An increase in OLR resulted in an increase in the production of specific EPS and macromolecules in the SMP/EPS fractions, in effect exacerbating the flocculation ability of the mixed liquor in the AnMBR and thus facilitating the fast formation of cake layers. Furthermore, the EPS tended to be more viscoelastic and hydrophobic at a higher OLR and because the adhesion forces of the EPS–membrane and EPS–EPS were significantly enhanced as the OLR increased, cake fouling was significantly accelerated. The results indicated that the main cause of fouling was SMP-induced pore blockages, and that membrane resistance increased gradually until an OLR of 0.7 g COD/L/day, but increased rapidly when the OLR was higher than 1.4 g COD/L/day, caused by the EPS-induced fast growth and compact cake layer on the membrane surface.

#### 3.3.2.2 Subvisible particles and colloids

The distribution, composition, and morphological structure of subvisible particles and colloids (0.01–10 μm) were investigated in the supernatant of a lab-scale SAnMBR, and their role in membrane fouling (Zhou *et al.*, 2016). Photometric analysis showed that the supernatant and membrane foulants were dominated by particles and colloids (0.45–10 μm), which accounted for over 90% of the total organics (proteins and polysaccharides). Excitation–emission matrix fluorescence spectra and monosaccharide analysis showed that these particles and colloids were rich in fluorescent proteins, rhamnose, ribose and arabinose, all of which could be related to cellular and extracellular substances. Fluorescence and scanning electron microscopy (SEM) confirmed the presence of bacterial cells in/on the subvisible particles and colloids. The microparticles (5–10 μm) were primarily composed of *Streptobacilli* and/or filamentous bacteria in microcolonies, whereas the submicrometre particles and colloids (1–5 μm and 100 kDa–1 μm) had more free/single cocci and bacilli. The ratio of live/dead cells varied in different size fractions, and the particles (1–10 μm) contained more live cells compared with the colloids (100 kDa–1 μm). Their findings suggest that bacterial cells in/on the particles and colloids could have an important effect on fouling in SAnMBRs as they are pioneering species attaching to membranes to form fouling layers/biofilm. Such insights show that previous foulant-characterization studies tended to overestimate organic fouling, whereas the biofouling induced by these bacteria in/on the particles and colloids was overlooked.

Subvisible particles (0.45–10 mm), with a size between sludge solids and SMP, have recently been identified as a critical foulant in AnMBRs, and recent new insights into the size fractionation and composition of subvisible particles in AnMBRs have enabled fouling to be understood in more depth (Zhou *et al.*, 2019; Yao *et al.*, 2020). The microbial diversity of the subvisible particles in three size fractions (i.e. 5–10, 1–5 and 0.45–1 mm) from bulk and cake solutions in a lab-scale SAnMBR, and their fouling potential, was further explored based on their filtration behaviour and biofilm formation. With decreasing particle size a significant shift in microbial communities was observed for the subvisible particles in both bulk and cake solutions; (a) with notable decreases in filamentous microbes in the order SJA-15, GCA004 and *Anaerolineales* of phylum *Chloroflexi*, and (b) with substantial increases in sulphate-reducing bacteria (i.e. the family *Syntrophobacteraceae*, genus DCE29 of family *Thermodesulfovibrionaceae*, *Desulfovibrio* and *Geobacter*). More importantly, the filamentous microbes associated with microparticles (5–10 mm) led to higher cake fouling resistances, whereas free living cells in the form of colloidal particles (0.45–1 mm) induced severe pore blocking. Moreover, the microparticles had an enhanced capacity to favour biofilm formation ($OD_{595}$, categorized as highly positive), thus potentially aggravating biofouling.

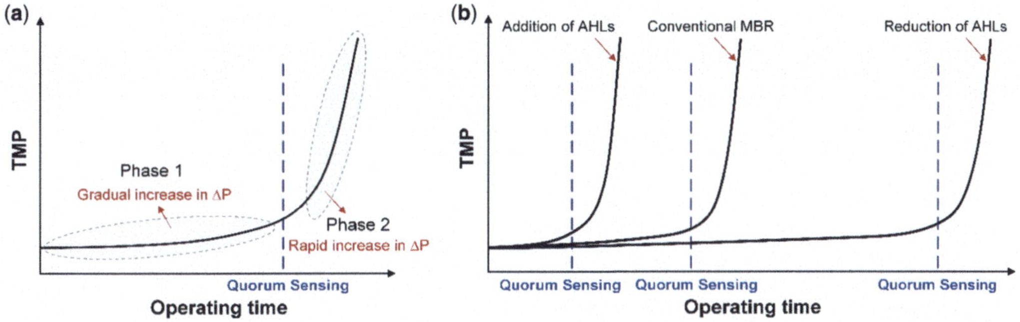


**Figure 3.3** (a) Typical profile of TMP increase during operation of MBRs for WWT, and (b) change in the point of QS and TMP increase as a function of the concentration of AHL signal molecules (*Source*: Oh & Lee, 2018).

### *3.3.2.3 QS molecule release*

It has been understood for a while that, from a fundamental point of view, bacteria regulate their behaviour in groups across species boundaries in areas such as virulence and biofilm formation, in a cell-density-dependent way using signal molecules. This is called 'quorum sensing' (Miller & Bassler, 2001; Waters & Bassler, 2005). In 2002, Lee's group in South Korea suspected that the sudden 'TMP jump' might be closely related to a threshold level of QS between microorganisms in MBR fouling layers (Figure 3.3a).

Bacteria use cell-to-cell communication systems to share information about their population density and to organize group behaviour accordingly. This group behaviour includes virulence factor secretion, symbiosis, competence, bioluminescence, sporulation, antibiotic production and biofilm formation (Lade *et al.*, 2014; Siddiqui *et al.*, 2015; Waters & Bassler, 2005). These processes follow three basic steps: (1) cells in the community produce small diffusible chemical signals (i.e. autoinducers); (2) autoinducers are detected by receptors that exist in the cytoplasm or membranes of cells when the concentrations of autoinducers exceed a threshold, and (3) detection of autoinducers activates not only the expression of specific target genes but also the production of more autoinducers. QS allows bacteria to convert environmental stimuli into specific gene expressions, enabling each individual cell to recognize the number of bacteria in their environment, and thus initiate collective behaviours when their number exceeds a quorum level. In general, QS systems have been divided into three general classes based on the type of autoinducer signal and the apparatus used for its detection: (1) *N*-acyl homoserine lactone (AHL)-type QS in Gram-negative bacteria; (2) peptide-mediated QS in Gram-positive bacteria, and (3) autoinducer-2 QS shared in Gram-negative and Gram-positive bacteria. In addition, several new classes of signal molecules have recently been discovered (Figure 3.4).

Based on this, they postulated that the onset of the second phase corresponding to a quorum level could be shifted either to the right with the reduction of signal molecules (e.g. AHLs) by any means, or to the left with the intentional addition of signal molecules (Figure 3.4). Furthermore, they postulated that membrane biofouling could be alleviated through QS control (i.e. blocking intercellular communication by decomposing AHL signal molecules in the process called quorum quenching; Figure 3.5). Later, Yeon *et al.* proved that the TMP and AHL levels of biofilms increased in parallel in MBRs (Yeon *et al.*, 2009a), and demonstrated that QS-based membrane biofouling control was possible (Wigginton 2009; Yeon *et al.*, 2009b). This use of QQ will be discussed in a later section.

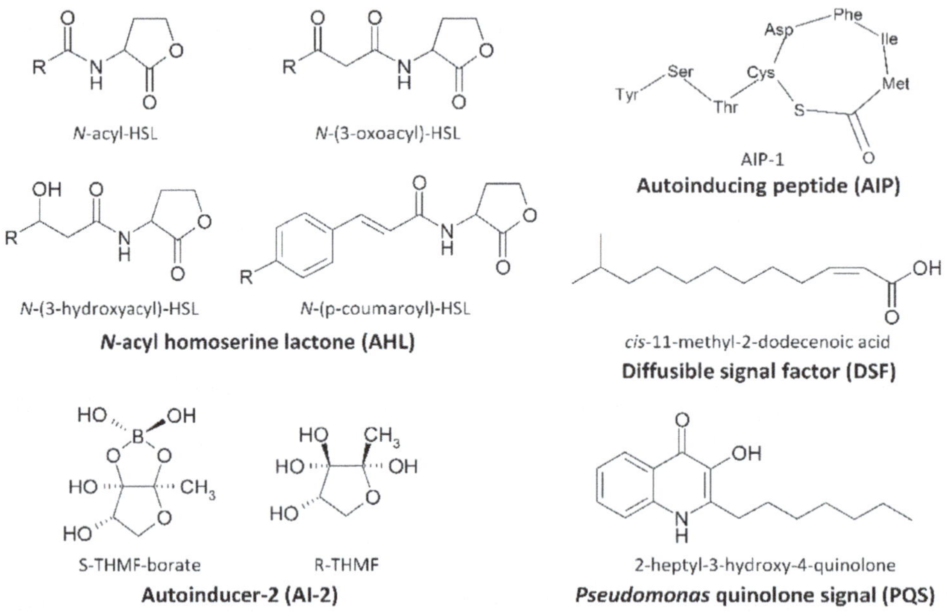


**Figure 3.4** Signal molecules involved in QS (*Source*: Oh & Lee, 2018).

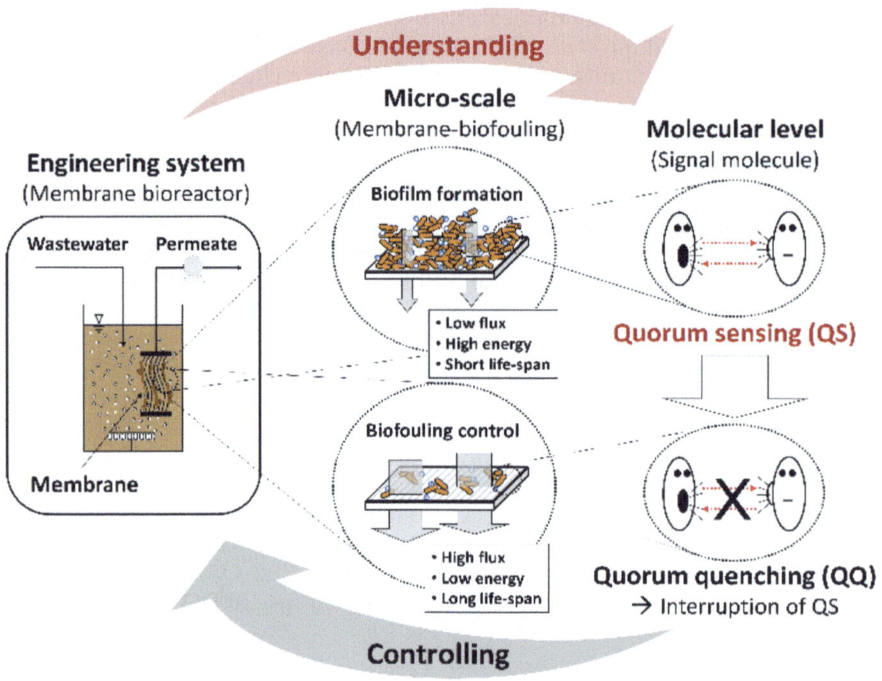


**Figure 3.5** QS–QQ control of biofouling in membrane reactors (*Source*: Oh & Lee, 2018).

#### 3.3.2.4 Inorganic precipitation

Inorganic precipitation (mineral scale) blocks membranes by forming crystals on the surface of the membrane, or within its interstices. The latter fouling is impossible to wash off the surface and hence needs to be cleaned using acids or alkalis. This type of fouling depends on two factors, mainly the concentration and type of ions in the feed (usually low in sewage), and the extent of CP at the surface (dependent on the membrane flux and the extent of physical scouring). In AnMBRs the concentration of inorganic ions ($I \sim 0.1$ molar) means that solubility products ($K_{SP}$) of most precipitates must be adjusted by the Debye–Hückel approximation. Inorganic foulants include cations and anions such as $Ca^{2+}$, $Mg^{2+}$, $Fe^{3+}$, $Al^{3+}$, $SO_4^{2-}$, $PO_4^{3-}$, $CO_3^{2-}$ and $OH^-$, and can precipitate onto/into the membrane due to hydrolysis which leads to pH change, and oxidation. Small amounts of ions, such as $Ca^{2+}$ (<280 mg/L), can help control biofouling due to binding and bridging EPS, whereas concentrations above 800 mg/L significantly increase fouling due to a high inorganic precipitate in the MBR mixed liquor. To remove inorganic precipitation from the membrane surface, chemical cleaning is more effective than physical cleaning (Meng *et al.*, 2017).

Choo and Lee (1996) were among the first researchers to study membrane fouling mechanisms over a long time of a membrane-coupled anaerobic bioreactor designed for the treatment of alcohol-distillery WW. Membrane fouling was primarily due to external fouling, which was closely related to the movement of cell population to the membrane surface and inorganic precipitation on it. The major inorganic foulant was $MgNH_4PO_4 \cdot 6H_20$ (struvite), whose deposition with the cells attached to the membrane surface played an important role in the formation of the strongly attached cake layer limiting membrane permeability. A conceptual resistance-in-series model was used to assess fouling characteristics, and the external fouling was 30 times the internal fouling resistance.

Trzcinski and Stuckey (2016a) examined the treatment of leachate (11.97 g/L, 14.4% soluble) from the organic fraction of municipal solid waste (OFMSW) using an SAnMBR, followed by an AeMBR to polish the effluent. The composition of the inorganic precipitate was investigated in each of the reactors in the process. The flux decreased due to precipitation of calcium as monohydrocalcite ($CaCO_3 \cdot H_2O$) containing traces of metals onto the SAnMBR membrane because of high $CO_2$ partial pressures. Precipitation of calcium in the AeMBR was also observed due to a higher pH. In this case, phosphorus also precipitated with calcium in two different phases: the background layer contained calcium, oxygen, carbon and small amounts of phosphorus (2–6.7%), whereas flakes containing calcium, oxygen and higher amounts of phosphorus (10–17%) were probably hydroxyapatite ($Ca_5(PO_4)_3OH$) demonstrating the different forms of inorganic fouling with different reduction potentials.

## 3.4 AMELIORATION OF MEMBRANE FOULING

A range of techniques are available that can be used to reduce membrane fouling (TMP), and these include gas sparging, mechanical shaking, addition of flocculants and adsorbents, for example granular activated carbon (GAC)/PAC and supressing QS production of cell signals. In addition, membranes can be cleaned intermittently if the TMP rises too high using both physical techniques such as back-flushing or by chemical cleaning using citric acid and hydroxide. Each of these will be discussed in turn.

### 3.4.1 Gas sparging

This is one of the simplest methods of scouring the membrane surface and reducing fouling. The technique is to create a highly turbulent environment at the membrane interface, and using small bubbles is not effective, so large (1–10 mm) bubbles are produced using recycled biogas and a large pore diameter sparger. This technique uses considerable energy, but results in 'slugging bubbles' which have a high shear stress at the leading edge of the bubble. This results in the gas phase being approximately in equilibrium with the liquid phase, and hence minimizes the potential loss of soluble methane in the liquid effluent (Crone *et al.*, 2016). This is because in many cases the anaerobic liquid phase is over-pressured with methane and hence not in equilibrium with the gas phase (Pauss *et al.*, 1990). However, this commonly adopted fouling control method contributes to more than 70% of the overall energy demand in an AnMBR (Pretel *et al.*, 2014), thus offsetting the benefits of bioenergy recovery.

Computational fluid dynamics (CFD) has been used to model liquid flows of one phase (liquid), two phase (liquid/gas) or very complex three phase (liquid/gas/solid) earlier, and is useful to understand the fluid dynamics of fouling at the membrane surface. The use of two-phase flow (gas/liquid) in membrane processes is well established due to its potential to reduce CP and membrane fouling, and therefore enhances membrane flux (Wibisono *et al.*, 2014). Gas/liquid flows create bubble turbulence and instabilities on the surface of membranes in various membrane processes such as MF, ultrafiltration (UF), nanofiltration, reverse osmosis, membrane distillation, electrodialysis and membrane bioreactors. These authors carried out an extensive critical literature review in this area, and data were analysed and normalized based on gas and liquid superficial velocities, gas/liquid ratio and feed types, TMP and membrane module type to identify general characteristics. They wanted to identify key factors in the application of two-phase flows in aqueous separation processes, and optimize this technology in the industry, with the importance of energy saving.

Surprisingly, an extensive study has never been carried out on the complex three-phase (gas, liquid, solid) fluid dynamics existing at the membrane surface in order to optimize membrane scouring and hence reduce fouling. However, three-phase CFD is very complex and is only started to be developed in the last 10 years. Clearly, gas sparging scours the membrane surface and hence alters the thickness and possibly the porosity of the fouling layer, but does this change the effluent COD? However, can ON/OFF gassing be tolerated in an AnMBR to save energy, and even improve effluent quality? Boyle-Gotla *et al.* (2014) developed a new model that allowed for a distributed shear profile from the bulk phase to the membrane surface, with dynamic linking of flux and TMP. The shear profile was calculated using a multi-phase CFD approach, and was applied to a distributed parameter model to simulate membrane fouling profile and flux distribution. This allowed for the simulation of complex flux-step experiments, or situations where non-uniform shear is present. Non-uniform gas distribution decreased critical flux from 12 to 8.5 LMH. This observation emphasizes the importance of local flow conditions on membrane fouling behaviour, and that performance can depend heavily on reactor configuration and hydraulics. Uneven membrane fouling has been noted visually across FS membranes in the past. Furthermore, the incorporation of a flux–pressure feedback loop in the model allows for simulation of flux-step experiments and prediction of critical flux and means model parameters such as cake compressibility and specific resistance can be estimated from dynamic data. A future modification may be inclusion of mechanical shear to simulate HF membrane modules with fibre movement and non-liquid shearing.

In 2017, Meng and co-workers carried out an extensive review of the literature on membrane fouling, although in most examples cited, the focus was more on AeMBRs, and not AnMBRs. They found that biopolymers on fouled membranes are dominated by SMP-polysaccharides and EPS-proteins in the early and late stages of fouling, respectively. The total bacteria and dead/live ratio of bacteria increase significantly after the 'TMP jump'. Such drastic changes of foulants before and after the TMP jump are of great importance for implementing fouling control strategies, that is, TMP-feedback control. In addition, gassing optimization, such as intermittent or cyclic aeration and mechanically assisted gas scouring, has attracted some attention for efficient membrane fouling control with lower-energy consumption. The method of aeration could be further optimized according to the CFD modelling on the fluidization and the scouring behaviour of the particles in MBRs.

Chemical cleaning efficiency seems strongly related to the interaction between the chemicals and foulants. However, some chemical reagents have the potential to decrease the lifetime of membranes, and even lead to the inactivation of microorganisms. Finally, the use of QQ and D-amino acids seems to be efficient in inhibiting bio-cake formation and/or disassembling existing bio-cake in AnMBRs. Enzymatic and bacterial degradation of SMP and EPS could also provide potential applications for enhancing biopolymer elimination. Electrically assisted approaches have been recently incorporated into MBRs with the major aim of improving membrane permeability. The development and use of nanomaterial-based membranes is a promising solution for increasing membrane fluxes and achieving anti-fouling through hydrophilic modifications and in-situ membrane cleaning.

In 2018, Tsibranska *et al.* (2018) examined a two-phase (gas/liquid) reaction process with membrane separation for product removal, favourable shifting of the reaction equilibrium and overcoming eventual inhibitory/toxic effects of the products which also has advantages of being energy and space saving. While these types of systems are not a direct analogue of AnMBRs, they are very similar, and the insights gained in modelling them can help us understand more about membrane fouling in AnMBRs. The effectiveness of such integrated systems is based on fluid dynamics and mass transfer knowledge of flowing matter close to the membrane surface – shear deformation rates and shear stress at the membrane interface, and mass transfer coefficients. A CFD-based approach for assessing the effectiveness of an integrated stirred tank bioreactor with a submerged membrane module was developed. It was related to the hydrodynamic optimization of the selected reactor configuration in two-phase flow, as well as to the concentration profiles and analysis of the reactor conditions in terms of reaction kinetics and mass transfer. A submerged membrane bioreactor (sMBR) is a hybrid vessel with flow characteristics affected both by the turbulence promoter (aeration and/or mixer) and the membrane module. The potential of the CFD for application in an sMBR is most often realized in hydrodynamic studies, especially concerning membrane fouling and aeration (Tsibranska *et al.*, 2018). Recent studies have reported CFD-combined hydrodynamic and mass transfer simulation of membrane modules with different configuration and stirred sMBRs (Haddadi *et al.*, 2018; Vlaev *et al.*, 2020).

Again, surprisingly, despite the concerns about the use of energy in biogas sparging, there has been little research carried out on using intermittent sparging to reduce energy costs. Vyrides and Stuckey (2009b), investigated the performance of an SAnMBR treating saline sewage under fluctuating concentrations of salinity (0–35 g NaCl/L), at 8 and 20 h HRTs, with fluxes ranging from 5 to 8 LMH. The SAnMBR attained a 99% removal of dissolved organic carbon (DOC) with 35 g NaCl/L, whereas removal inside the reactor was significantly lower (40–60% DOC). Even with a sudden drop in salinity, the overall removal recovered quickly, whereas the recovery inside the reactor took place at a slower rate. This highlights the positive effect of the membrane in retaining high-MW organics and biomass in the reactor, so that they can rapidly acclimatize to salinity. The reduction of continuous biogas sparging to intervals of 10 min ON and 5 min OFF resulted in a slight increase in TMP by 2.5 kPa, and also resulted in an increase in effluent DOC removal and inside the SAMBR by 10 and 20%, respectively. Hence, energy input can be reduced by ~33%, and the effluent quality increased.

Fox and Stuckey (2015) developed the idea of 'critical flux' and found that under a sparging rate of 6 L per min (LPM) this was 11.8 LMH; however, membrane 'hysteresis' was found to influence the critical flux, and where the AnMBR had previously been operated at a 2 LPM sparging rate, the critical flux dropped to 7.2 LMH. The existence of a 'critical sparging rate' was also investigated under the condition that 'there exists a sparging rate beyond which any further decrease in sparging rate will cause a dramatic rise in TMP'. For an AnMBR operating at a flux of 7.2 LMH, the critical sparging rate was found to be 4 LPM.

Does the biogas sparging rate influence the reactor effluent quality by controlling the thickness of the fouling layer? Trzcinski and Stuckey (2016b) focused on the treatment of leachate from the OFMSW in an SAnMBR. They found that the permeate COD of SAnMBRs cannot be lowered by decreasing the sparging rate. The main effect of low sparging rate (2 LPM) was an enhanced rejection of the membrane and a significant flux drop while the permeate COD remained constant within the experimental time frame. On the contrary, permeate COD increased due to higher bulk soluble COD (SCOD) when the sparging rate was increased to 10 LPM, in which case the permeate COD increased from 360 to 440 mg/L due to better scouring of the membrane resulting in a thinner fouling layer and better diffusion through the biofilm. The time required to observe the increase in permeate COD depended on the compaction of the cake layer. The increase in sparging rate to 10 LPM allowed the

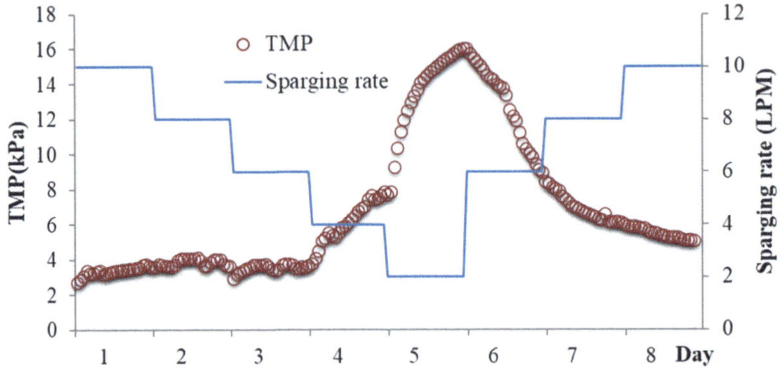


**Figure 3.6** TMP profile with changing sparging rate over time (Zhang *et al.*, 2017a).

reactor TMP to recover to 1 kPa, but not the initial flux. Hence, when the fouling layer is minimal (<1–2 μm) there is an increase in effluent COD, and so there is a correlation between sparging rate and effluent quality. However, at certain thicknesses (low sparging) the quality does not decrease further, but the TMP increases rapidly.

Biogas sparging is an effective way to control membrane fouling in an AnMBR, but is used primarily with a constant gas flow. Zhang *et al.* (2017a) examined whether different gassing strategies could lead to better fouling control, and tried to understand the correlation between sludge properties and the fouling layer by disaggregating them into supracolloidal (1–5 μm), colloidal (0.45–1 μm), and solute (<0.45 mm) fractions. They found that the 'critical constant sparging rate' was 6 LPM at a flux of 26 LMH (see Figure 3.6). However, a sparging strategy of 3 LPM for 30 s alternating with 5 LPM for 30 s (Alter 3/5) reduced membrane fouling by 150% compared with a constant 4 LPM strategy. The viscosity, colloid and floc sizes of the suspended sludge increased with a decreasing sparging rate, whereas the concentration of supracolloidal particles, SMPs, proteins and carbohydrates and EPS decreased with a decreasing sparging rate. Colloids were the major foulants above 2 LPM, whereas high-performance liquid chromatography-size-exclusion chromatography analysis showed that one main SMP fraction was above 1,522 kDa, and the other between 72 and 500 kDa. The high-MW compounds accumulated on the membrane, whereas the later permeated into the effluent. With the Alter 3/5 strategy, SMPs were the primary foulants, and the particle sizes were much smaller than in the supernatant. The foulant SMPs not only originated from the reactor supernatant, but were also produced by biomass in the membrane fouling layer. Chemical analyses reflected the effect of different sparging rates on sludge properties, the production of colloids, SMPs and EPS and the foulant components. The foulant particles were significantly smaller than that in the supernatant, and the SMPs were different between the three strategies. Both SMP production and fouling were low with intermittent 6/0 (30 s OFF) whereas SMP production was high, but SMP fouling was low with constant 4 LPM (Figure 3.7). In contrast, both SMP production and SMP fouling were high with Alter 3/5. The foulant SMPs not only originated from the supernatant, but were also produced by the biomass on the membrane surface.

### 3.4.2 Mechanical shaking and scouring

Lim *et al.* (2023) investigated energy balances when using membrane reciprocation as a fouling control method in a pilot-scale AnMBR treating domestic WW, as compared to using conventional biogas sparging. Both strategies were investigated under different biogas sparging speeds and membrane reciprocation frequencies. Critical flux tests and energy calculations have shown membrane reciprocation to consume 0.59 kWh/m$^3$ at 7 LMH, 62% less energy compared to biogas

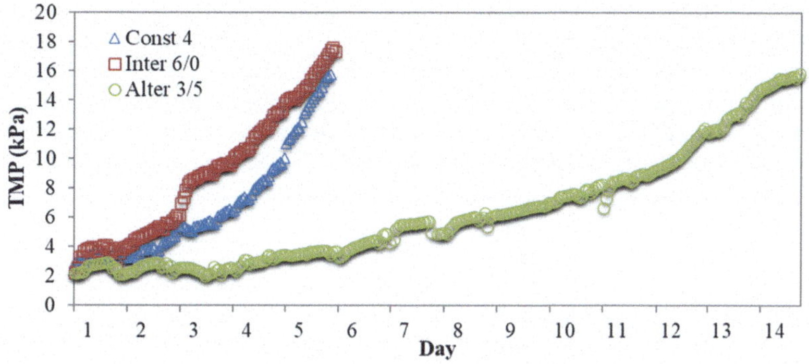


**Figure 3.7** TMP profile under the three biogas sparging strategies: constant 4 LPM (Const 4); intermittent sparging of 6 LPM (1 min) and 0 LPM (30 s) (Inter 6/0); alternating gas rates of 3 and 5 LPM (Alter 3/5) (*Source*: Zhang *et al.*, 2017a).

sparging (1.56 kWh/m$^3$ at 6 LMH). High biogas recovered in the headspace and low-energy demand for membrane reciprocation resulted in a positive energy balance (0.55–0.74 kWh/m$^3$), supporting the prospect of membrane-based technology deploying this fouling control method.

Aslam *et al.* (2017) critically analysed different research studying mechanical cleaning with a scouring agent, and identified a range of R&D needs for this emerging technique for fouling control in AeMBRs, and mechanical cleaning also results in favourable and energy efficient MBR operation. The main action of suspended media is the mechanical scouring of the membrane surface, and its main benefit is that media can impinge onto the membrane surface, whereas hydrodynamic turbulence or air bubbles cannot reach the membrane surface which is protected by the laminar boundary layer. In addition, the use of adsorbents such as GAC as a mechanical scouring agent provides additional benefits of sorption compared to other inert granular materials. PAC can also improve fouling control and membrane flux but steady replacement of aged PAC is important, otherwise fouling could be worse than that without PAC. Consequently, it is very difficult to draw the right conclusions since many parameters listed below are involved in mechanical cleaning success:

(a) hydrodynamics (bubble size, air flow rate, liquid velocity, design),
(b) scouring media (size, shape, density, concentration),
(c) membrane (design, integrity),
(d) bulk composition (sludge concentration, particle size, viscosity),
(e) cake layer (thickness, porosity, composition).

Therefore, the effect of numerous factors impacting mechanical cleaning in various MBRs makes it important to understand the effect of each factor and their interactions. A common methodology needs to be established using standard parameters: shear rate ($\gamma$), fluidization number ($N_f$), momentum transferred ($P_{tr}$) and media volume ($1 - \varepsilon$). Fouling control would be better with larger media size and higher doses where a linear trend is observed. However, the energy cost increases as particle size and dosage increases, and their beneficial effects could also be counter balanced due to breaking of sludge flocs (releasing small particles and organic supernatant) which are currently not represented in models. Thus, there is a trade-off in better fouling reduction with an increase in the shear stress and media size and dosage. Similarly, a better fundamental understanding of how media scrape the membrane surface is needed by determining the media velocity ($V_H$) if this notion makes sense in a fluidized-bed reactor. Hence, to improve the effectiveness of multiphase flow strategies in mechanical cleaning processes for better membrane performance, research should be carried out in the direction of better fundamental understanding of the underlying mechanisms and hydrodynamics of the different mechanical cleaning parameters where media characteristics, membrane module, MBR system geometry and multiphase flow hydrodynamics must be incorporated in a solid/gas/liquid hydrodynamic model. Hence, a universal model able to optimize the hydrodynamics of mechanical cleaning with other parameters in MBRs is still not developed. Since energy efficiency is a major issue in WWT, mechanical cleaning will be key in MBRs in the future.

### 3.4.3 The use of adsorbents/flocculants

The first researchers who examined the effect of adding activated carbon to three, 3 L SAnMBRs were Hu and Stuckey (2007). They focused on COD removal, flux and TMP. The feed was a synthetic substrate (COD: 460 mg/L), with one reactor as a control, one with 1.7 g/L of PAC, and the third with 1.7 g/L of GAC. COD removal was high in all reactors (>90%), in comparison to the control (SAnMBR1), the average COD removal in SAnMBR2 (PAC) increased by 22.4%, whereas SAnMBR3 with GAC was not significantly better. Because PAC (1,300 $m^2/g$) has a significantly greater surface area per mass than GAC (775 $m^2/g$), it is probable that this difference was primarily due to the greater absorbance of fine-colloidal particles and high-MW organics onto the carbon surface. These effects manifested themselves by SAnMBR2 having lower TMPs and higher fluxes than both SAnMBR3 and SAnMBR1. Volatile fatty acids (VFAs) in the effluent from all three SAnMBRs were extremely low (<18 mg/L), even during step changes in HRT, and most of the soluble COD in the effluent was SMPs. Biological Methane Potential (BMP) assays showed that biomass in SAnMBRs was less active than the seed sludge, and it appears that the addition of activated carbon to SAnMBR2 and SAnMBR3 provided a solid support for growth, and hence reduced floc breakage.

Vyrides and Stuckey (2009b) investigated the addition of PAC to an SAnMBR treating saline sewage and found a decrease of 7.0 kPa in the TMP, and an increase in DOC removal in the reactor and effluent by 30 and 5%, respectively. The PAC dramatically decreased the high-MW organics in the reactor over a period of 72 h. SEM images of the membrane and biomass before and after addition of PAC revealed a remarkable reduction in flocs on the membrane surface, and a reduction inside the reactor of SMPs. Finally, energy-dispersive X-ray (EDX) analysis of the membrane pores and biofilm highlighted the absence of organic matter in the inner pores of the membrane.

In 2011, Kim *et al.* (2011) considered the concept of addition of GAC from Hu and Stuckey (2007) to reduce fouling and increase COD removal, and developed a two-stage reactor system consisting of an anaerobic fluidized-bed reactor (AFBR) with 115 g/L of GAC coupled with an anaerobic fluidized membrane bioreactor (AFMBR) with the same GAC concentration, but with a PVDF hollow-fibre membrane (Kolon Inc.) with inside diameter of 1.9 mm, nominal pore size of 0.1 μm and a total membrane surface area of 0.091 $m^2$. The first-stage (AFBR) was operated at 2.0–2.8 h HRT, and was followed by the AFMBR, operating at 2.2 h HRT. COD removals were 88 and 87% in the respective reactors and 99% overall, with a permeate COD of $7 \pm 4$ mg/L. Total energy required for fluidization for both reactors was 0.058 $kWh/m^3$, which could be met by using only 30% of the gaseous methane energy produced, and that of the AFMBR alone was 0.028 $kWh/m^3$, which is significantly less than reported for other SAnMBRs with gas sparging for fouling control.

In addition to adding an adsorbent such as GAC, flocculants can be added to the reactor to increase floc size, and reduce the colloids and even some solutes in solution. Zhang *et al.* (2017b) evaluated eight additives: three PACs, two GACs, one cationic polymer and two metal salts to identify the best additive and dose to minimize membrane fouling in an SAnMBR. Small cross-flow filtration tests showed 400 mg/L PAC SAE2, or 150 mg/L $FeCl_3$, reduced the TMP rise from 0.94 to 0.06 kPa/h, indicating excellent fouling reduction. The best filtration performance correlated with a significant reduction in supernatant supracolloidal particles, colloids and SMPs. Field-emission SEM–EDX showed that PAC SAE2 and $FeCl_3$ reduced the thickness of the fouling layer dramatically, whereas $FeCl_3$ increased sludge floc size and particle size of the colloids, while decreasing the negative charge of colloids, and SMP size. Furthermore, Fe was not found in the supernatant or effluent, but precipitated with the solids, which is beneficial for its long-term use.

Sohn *et al.* (2021) reviewed the field of adsorbent/flocculant addition to AnMBRs to mitigate membrane fouling by altering the feed properties. PAC and GAC have been widely added as an adsorbent to AeMBRs and AnMBRs for membrane fouling control while organic enhancers such as biochar and waste yeast, and inorganic enhancers such as polyaluminium chloride and zeolite have also been used. The study discusses the impacts of different fouling reduction enhancers in the AnMBR as well. In addition, the mechanisms of enhancers mitigating the membrane fouling are summarized for better understanding of the effects of enhancers in AnMBRs. Their main conclusions were as follows: the addition of fouling reduction enhancers, including activated carbon (AC), biochar, zeolite and polyaluminium chloride, could effectively alleviate membrane fouling in AnMBRs; enlarged floc size and decreased soluble organics mainly contributed to the mitigation of fouling; overdosing or large particle size of enhancers could lead to contrary results due to their potential to be a foulant.

### 3.4.4 Quorum quenching

The correlation between QS signal and membrane biofouling was demonstrated in 2009 by Yeon *et al.* An excellent summary of QQ is reported in a review by Oh and Lee (2018). In general, there are three strategies to inhibit a system using QS to consolidate biofilms on membrane surfaces: (a) blocking the signal synthesis, (b) inactivating the quorum signals and (c) interfering with the signal receptor. Inhibition of the best-known QS system (AHL-type) in Gram-negative bacteria is depicted in Figure 3.8.

### 3.4.5 Physicochemical and chemical cleaning

Tomczak *et al.* (2023) focused on the physical cleaning of membranes via biogas sparging, manual membrane cleaning (sponge sweeping), relaxation of permeate flow and ultrasound. Chemically enhanced backwashing is an integral process for mitigating fouling in AnMBRs, and has been assessed

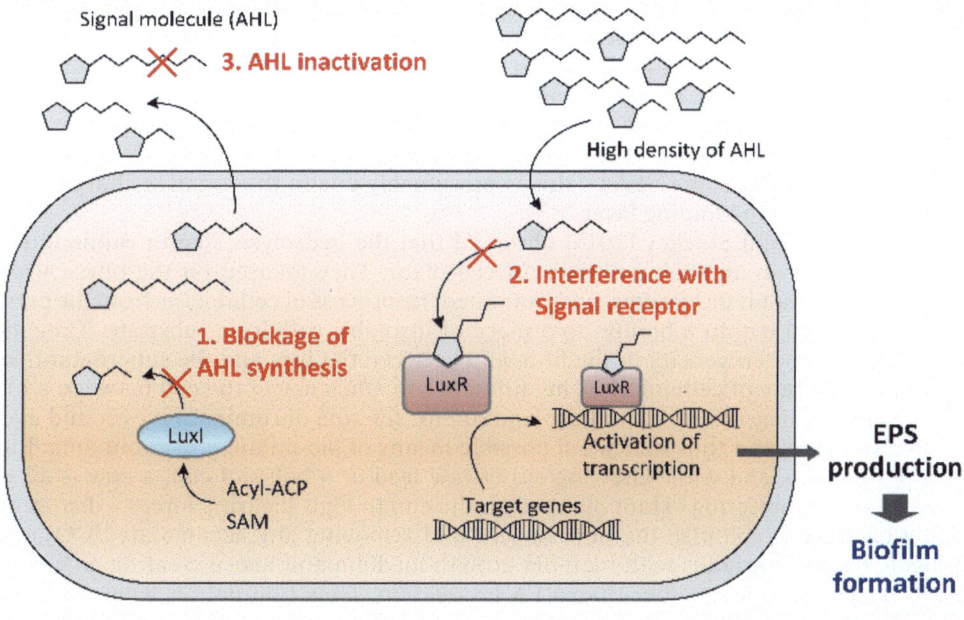


**Figure 3.8** Three strategies to control the AHL-type QS system of Gram-negative bacteria (*Source*: Oh & Lee, 2018).

in many studies (Cheng *et al.*, 2018; Ji *et al.*, 2020; Peña *et al.*, 2019) for both external and SAnMBRs. With backwashing, liquid flows from the permeate side through the membrane removing organic matter on the surface and deposits inside its pores (Chang *et al.*, 2017). In most studies backwashing was enhanced by using sodium hypochlorite (NaClO) and citric acid, and in industrial applications, backwashing is fully automatic (Shi *et al.*, 2014). Chemical membrane cleaning, including soaking, has been successfully used as a fouling mitigation strategy in numerous studies (Vinardell *et al.*, 2021). The most commonly used chemicals are alkalis (NaOH), oxidants (NaClO), chelators (ethylenediamine tetraacetic acid, EDTA), as well as acids (citric and nitric acids). Puspitasari *et al.* (2010) found that the efficiency of these chemicals depended mainly on chemical's properties, membrane characteristics and cleaning operating conditions. Furthermore, the choice of proper cleaning products requires knowledge of the feed, and a mixture of various cleaning agents is often used.

NaClO is the most used oxidant for membrane cleaning and removes both organic and biological foulants via oxidation and disinfection (Terán-Hilares *et al.*, 2022), and enhances the detachment of organic molecules from the membrane by increasing their hydrophilicity. With ceramic membranes, Song *et al.* (2016) showed that permeate flux can be restored using NaOH at $70 \pm 1$°C which hydrolyses colloids into fine particles, and organic matter into small molecules. In addition, NaOH reacts with fats and oils to form water-soluble soap micelles, and can be used to remove silicates and inorganic colloids. However, chemical cleaning with NaOH and NaClO may reduce the integrity of polymeric membranes and shorten their lifespan. It is also understood that membrane cleaning efficiency can be improved by sequentially combining alkaline or oxidant reagents with citric and nitric acids. This strategy of chemical cleaning has been well adopted in studies focused on the applications of AnMBRs for biogas production (Rong *et al.*, 2022). For instance, to restore MF performance, Chen *et al.* (2017) cleaned a polyethylene (PE) membrane with tap water and then soaked it in citric acid and NaClO solutions.

## 3.5 REMOVAL PERFORMANCE

### 3.5.1 COD removal

Most AnMBRs operating on domestic sewage and at an average HRT of ~9 h remove above 90% (Rattier *et al.*, 2022), and with synthetic feeds they can achieve above 90% removal in just 3 h HRT (Hu & Stuckey, 2006). The reason for this is that no cells or colloids are present in the effluent due to membrane rejection. In addition, most MF and UF membranes act as a physical support media for the fouling layer on the surface composed of SMPs, ECP, colloids and sometimes inorganic precipitates. This acts as a 'dynamic' membrane on the surface and hence rejects many solutes from quite high MW to quite low MWs, including some VFAs – this is presumably due to the negative charges present on the dissociated VFAs and the fouling layer.

Interestingly, Mason and Stuckey (2016) observed that the hydrolysis step in ruminants is up to 30 times faster than in an anaerobic digestion (AD) plant. They focused on the physical/chemical environment within cellulolytic biofilms, and examined the process of cellulolysis from the perspective of a bacterium attached within a biofilm to a piece of insoluble cellulosic substrate. They modelled the extent of chemical heterogeneity in the film, and between the film and the supernatant, and how this might control the rate of cellulolysis. The difference in efficiency of the cell between a high- and a low-pH environment may offer a partial explanation for the role of rumination over and above that of surface area increase, and thus indicate a possible means of bio-mimicry of ruminants. Ruminant saliva has a pH of ~8.5, and while chewing either raw feed or a bolus of cud, a cow is (a) creating new surface areas, (b) subjecting extant biofilms on the cud to high shearing forces – disrupting them and shedding bacteria to colonize the new surface and removing any accumulated $CO_2$ and VFAs and (c) infusing the feed or bolus with high-pH growth medium and hence creating an environment favourable to bacterial growth rather than VFA production. Thus rumination seems to introduce a phase of high-bacterial growth to complement the suppressed growth rates that will be seen in biofilms in the rumen where activity shifts from biomass production to VFA production. This may be necessary to avoid washout that could otherwise be caused by the high-flushing rates needed to support the high rate of digestion. Thus the heterogeneity of the environment in a ruminant may well be a contributor to its digestive efficiency in contrast to the typical AD CSTR which strives by stirring to achieve a high degree of homogeneity. Hydrogenotrophic acetogenesis seems to play a potentially important role in moderating the extreme conditions within rumen biofilms. By removing gas and moderating pH may be a crucial component of biofilm maintenance as this leads to the prediction that suppression of methanogenesis could lead to lower rates of cellulose digestion. One key conclusion from this study is that the properties of the supernatant that are easy to measure in practical operations of AD (e.g. pH) are only indirectly related to the environment actually experienced by the bacteria, and that research to improve AD might benefit from a concerted attempt to measure and manage the microenvironment of the microbes rather than their macroenvironment.

### 3.5.2 Bacteria, viruses and antibiotic-resistant gene (ARG) removal

Despite their ability to reject small solutes, there has been little research carried out on one of the advantages of membrane reactors in general, that of reducing pathogens in treated sewage without chemical treatment. Early research on pathogen removal in AeMBRs by Shang *et al.* (2005) used a synthetic feed and a small bacteriophage MS-2 (20–25 nm ~ polio virus size) to demonstrate the ability of (viral) pathogen removal by an MBR. Phage removal can be due to three mechanisms: size rejection by the membrane, biomass adsorption and rejection by the membrane-attached biofilm. Only $0.4 \pm 0.1$ log reduction of MS-2 occurred with a clean membrane without biomass. At 6,000 mg/L mixed liquor suspended solid (MLSS), 200-day SRT and 9-h HRT, suspended biomass resulted in 0.8 log phage removal. With the development of biofilm on the membrane surface the log removal value increased substantially to ~2.5, and this improved by prolonging the time between membrane cleanings. The removal was influenced by system parameters such as the concentration of MLSS, SRT and the Food to Microorganism (F/M) ratio. Operating at lower MLSS, or longer SRT (among 50 and 200 days with the same MLSS) yields faster and larger improvements in phage removal, whereas lowering the F/M ratios diminishes the correlation between TMP and phage removal. Phage removal improves as the biofilm is allowed to develop, but this decreases the effluent flux and/or increases the TMP. Balancing these factors is important if pathogen control relies fully or partially on the MBR. While MS-2 was used to indicate likely pathogen removal by MBR in this study, it is possible that the surface characteristics of viruses and phages vary greatly and hence their adsorption onto the complex biofilm/biomass may also vary.

Lv *et al.* (2006) focused on virus (phage) removal and mechanisms of an aerobic sMBR treating hospital WWs using a large (200 nm) phage T4 as a model. Two membrane modules (0.1 and 0.22 mm pore size) were compared over 75 days running continuously. During stable operation, the sMBR achieved almost complete phage removal for both membrane modules. For the 0.22 mm module, the cake layer, the gel layer and the membrane contributed 6.3, 3.1 and 1.7 log removals, respectively, confirming the importance of the cake/gel layer on the membrane surface. Damage to the cake/gel layer resulted in a decrease in phage removal. For the 0.1 mm membrane, the membrane alone played a major part in phage removal. Inactivation by activated sludge and adsorption by cake/gel layer contributed about 3.6 log to phage removal everyday so that there was no phage accumulation in bulk solution, although what fraction of this removal was due to other hospital constituents in the WW is not known. These results demonstrated that an aerobic sMBR was an efficient system for treating WWs containing viruses.

While these studies on aerobic bioreactors were useful, there was no information available on AnMBRs, so Fox and Stuckey (2014) investigated the removal of phages MS-2 (25 nm) and T4 (200 nm) in an AnMBR with a pore size of 0.4 mm. The membrane reactor without biomass was assessed and its log removal was $0.7 \pm 0.4$ log for the MS-2 phage, and $2.3 \pm 0.2$ log for the T4. When anaerobic biomass was added to the reactor the log removal for both phages increased, and this was thought to be due to a complex relationship with the biofilm on the membrane. Overall, MS-2 rejections ranged from 1.75 up to 5.5 log, with the highest rejections observed at the highest sparging rates after extensive fouling had occurred. For T4, removal in an AnMBR ranged from 5 log to complete removal (>log 7). These results confirm, most importantly, that MBR removal is generic, and similar results can be obtained with either aerobic or anaerobic systems.

### 3.5.3 Antibiotic-resistant genes

These represent an important emerging contaminant due to public health concerns about the spread of infections resistant to common antibiotics, and conventional activated sludge (CAS) processes have shown mixed results regarding ARG removal. ARGs are present in the environment as both intracellular and extracellular genes, and bacteria develop antibiotic resistance in three ways: releasing antibiotic-degrading enzymes, producing the efflux pumps or modifying the antibiotic-binding targets (Aziz *et al.*, 2022). In the first method, certain bacteria release antibiotic-degrading enzymes that can counter the effect of antibiotics. For example, *Enterobacteriaceae* releases $\beta$-lactamase enzymes to counter the effect of penicillin and breaks the $\beta$-lactam ring of penicillin to make it ineffective (Sharma *et al.*, 2016). In the second method, bacteria use an efflux pump through which antibiotics can be pumped out of the cell, whereas in the last method, bacteria modify their structures against that antibiotic and develop antibiotic resistance (Verraes *et al.*, 2013). Once bacteria become resistant, their dissemination starts using two basic mechanisms, that is, vertical gene transfer and horizontal gene transfer (Tian *et al.*, 2016). In vertical gene transfer, resistant genes are passed through bacterial replication or spontaneous mutation. While in horizontal gene transfer, resistant genes are transferred among different bacteria by either conjugation, transduction or transformation (Figure 3.9).

Kappell *et al.* (2018) examined the impact of an AnMBR on ARG removal when treating primary clarifier effluent at 20°C. AnMBR treatment resulted in 3.3–3.6 log reduction of ARGs and the horizontal gene transfer determinate, intI1, copies in the filtrate. Membrane treatment significantly decreased the total biomass as indicated by a decrease in 16S rRNA gene concentration (Figure 3.10). Microbial community analysis showed that the relative abundance of putative pathogens was higher in membrane filtrate compared to primary effluent, although the overall bacterial 16S rRNA gene concentrations were lower in the filtrate. Membrane treatment also substantially reduced microbial diversity in the filtrate compared to anaerobic reactor contents.

Aziz *et al.* (2022) extensively surveyed the data available on ARGs and found that a fouled AnMBR can remove ARGs (sul1, sul2, tet(O), tetW, ermF, erm(B), blaNDM-1, blaCTX-M-15, blaoxa-48, blaoxa-1) and mobile genetic elements (intI1 and tp614), and thus presents a feasible solution in preventing

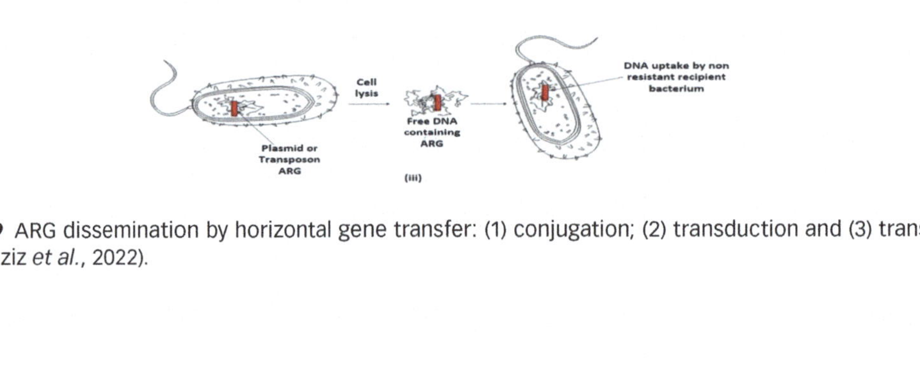


**Figure 3.9** ARG dissemination by horizontal gene transfer: (1) conjugation; (2) transduction and (3) transformation (*Source*: Aziz *et al.*, 2022).

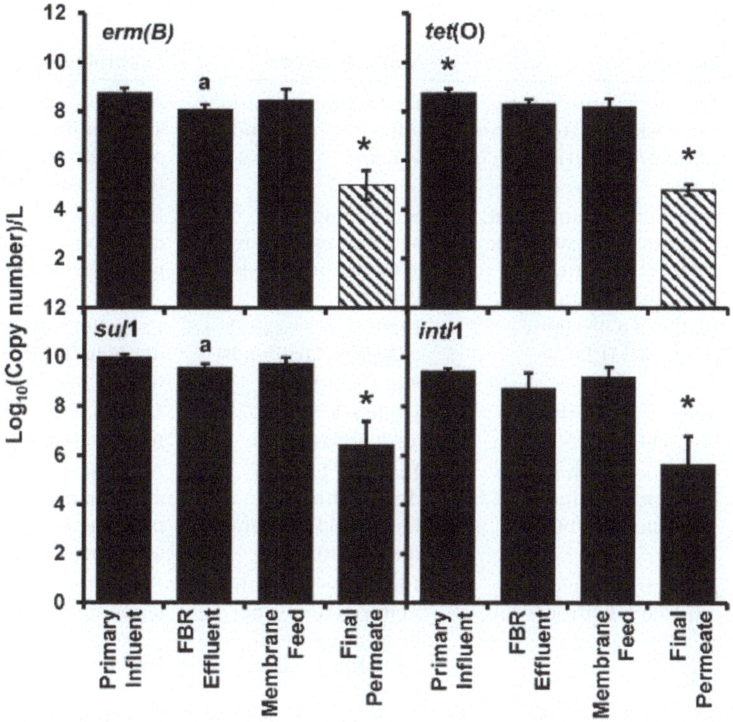


**Figure 3.10** Abundance of ARGs *erm*(B), *tet*(O) and *sul1* and the class 1 associated integrase gene (*intI1*) per litre from the AnMBR. Bars represent the average value, and error bars represent the standard deviation from four sampling events. Striped bars represent reported results equivalent to the detection limit because only one or two samples out of four were quantifiable. The asterisk (*) denotes significant difference from all other sample locations, (a) denotes significant difference compared to primary sample (*Source*: Kappell *et al.*, 2018).

antibiotic resistance. Hence mitigation methods must be employed rather than focusing on complete eradication or removal of the fouling layer. The use of certain biofilm carriers (activated carbon, bioballs and sponge) could control membrane fouling, but more research is still required to evaluate their role while treating real antibiotic WWs. In addition, microbial fuel cells (MFCs) and microbial electrolysis cells (MECs) are also promising technologies for both antibiotics and ARG removal which can also moderate fouling. Hence, future research must investigate the combined effect of AnMBR and MFCs/MECs for eliminating both antibiotics and ARGs and controlling fouling-related issues.

### 3.5.4 Pharmaceutical, personal care products (PPCPs) or organic micropollutants (OMPs)

OMPs in domestic WWs are produced primarily by human and industrial activities, and cover a wide range of contaminants including pharmaceuticals, pesticides, personal care products (PCarePs), industrial chemicals, hormones and other emerging compounds (Lim *et al.*, 2020; Table 3.2). While the OMPs are present in trace amounts in WWs (from ng/L to μg/L), considerable effort has been made to better understand the fate and transport of them in water bodies while considering their significant risks to human health and eco-environmental security. These risks include short-term or long-term toxicity, environmental persistence, antibiotic resistance and endocrine disruption. However, the individual or collective behaviour of OMPs in domestic WWTPs still requires more research since their removal behaviour under anaerobic conditions is still poorly understood.

**Table 3.2** Types of Organic Micropollutants.

| OMPs | Classes | Major Sources | Examples |
|---|---|---|---|
| Pharmaceuticals | Antibiotic, antidiabetic, analgesic, anticonvulsant, stimulant, veterinary drug | Extraction, hospital effluents, farmland waste | Acetaminophen, androstenedione, amoxicillin, cbz, def, idp, keto, penicillin, smx, paracetamol, timolol. salicylic acid etc. |
| Personal care products (PCarePs) | Antiseptic, disinfectant, fragrance and synthetic musk, stimulant, UV filter, insect repellent | Showering, swimming, bathing, shaving, or industrial waste | Benzophenone, caf, diltiazem, chlorophene, triclosan, methylbenzylidene, tonalide, etc. |
| Steroids and hormones | Endocrine disruptive chemicals (EDCs) | Excretion, hospital effluents, farmlands, aquaculture | Estradiol, estrone, diethylstilbestrol, progesterone, testosterone, etc. |
| Pesticides | Insecticide, herbicide, fungicide | Domestic and agricultural uses | Diuron, Mercoprop, MCPA, terbuthylazine, etc. |
| Detergent, surfactant | Cationic, anionic, non-ionic, perfluorinated compounds (PFCs) | Bathing, laundry, dishwashing, household dilutants, dispersants | Alkylphenol ethoxylates, alkylphenols(nonylphenol, and octylphenol), perfluorooctane sulfonates, perfluorooctanoic acid |
| Others | Plasticizer, flame retardant, hydrocarbon, solvent | Leaching out of the material, improper cleaning, run-off from the garden, roadway, etc. | Phthalates, polybrominated compounds, dioxin and furans, polycyclic hydrocarbons, trichlorethylene, benzene, toluene |

*Source:* Lim *et al.* (2020).

Both sorption and biotransformation play important roles in the removal of OMPs during the operation of an AnMBR, although gas stripping can remove some volatile compounds in AnMBRs with gas fouling control (Table 3.3). Sorption of OMPs from the aqueous phase onto biomass should occur due to hydrophobic interactions and electrostatic interactions (cation bridging and exchange). Do and Stuckey (2019) examined the fate of 0.5–1.5 mg CIP/L (ciprofloxacin) in an AnMBR and measured removal by both adsorption and biodegradation. Sorption could be equally well described by both the Langmuir and Freundlich isotherms, but only ~1% of the total mass of CIP was removed by sorption and sludge wasting. They concluded that 50–76% CIP removal was possible when CIP was fed at a concentration of 0.5–1.5 mg/L. At 4.7 mg CIP/L, the removal efficiency was substantially decreased to <20%. Biological degradation was the main mechanism for removing CIP, with a few intermediate compounds detected.

Biotransformation is the process whereby microbes decompose organic pollutants. In an AnMBR, a porous membrane is used to retain the biomass. However, the rejection efficiency of the OMP through MF or UF may not be very high because the size of the OMP molecules is often smaller than the pore size of the membrane, which is in the range from 0.1 to 0.01 μm. Nevertheless, it allows for the retention of colloidal organic carbon which is largely bound to OMPs (Lim *et al.*, 2020). During membrane filtration, fouling caused by the deposition of organic matter on the membrane and a higher concentration of OMPs in WW results in the formation of a denser and more compact structure of a fouling layer on the membrane. This is because the fouling layer formed on the membrane surface can often play a role as a secondary membrane to improve OMP rejection.

Lim *et al.* (2020) conclude that the main pathway for removing OMPs from solution in AnMBRs is sorption which depends on the physicochemical characteristics of both the contaminants and biomass present in bulk and/or grown in suspended carriers. Longer SRTs can increase the removal of OMPs due to higher biomass levels and microbial diversity, and sorption onto biomass can be improved by increasing the HRT. Sorption–desorption between OMPs and biomass should be the rate-limiting step

**Table 3.3** Removal Pathways of OMPs.

| | No. 1 | No. 2 | No. 3 | No. 4 | No. 5 |
|---|---|---|---|---|---|
| Pathway | | | | | |
| Mechanisms | Coagulation/ flocculation Sorption Precipitation | Volatilization | Biological oxidation (biodegradation) | Chemical oxidation | Liquid/liquid separation |
| Variables | Partition coefficient Solubility constant: $K_{d,OMP}$ ($L^3M^{-1}$), S | Henry's Law constant: $k_H$ ($ML^{-4}T^{-2}N^{-1}$) | Biodegradation kinetic constant: $k_{bio,OMP}$ ($T^{-1}$) | Kinetic constant: $k_{O2,OMP}$ ($M^{-1}s^{-1}$) | Molecular weight Charge $MW_{OMP}$ |
| Influencing Parameters | Particle size and charge, concentration and type of organic matter, hydrophobicity of compounds | Gas sparging | Composition of microbial community, redox conditions, SRT, HRT | Cbz, def | Permeate flux, membrane fouling, SRT, HRT |
| Strong influence of the pathway on | PAH, PCB, PBDE, fluoroquinolones | Naphthalene, trichloroethylene | Paracetamol, hormones, ibp | Cbz,def | Smx, cbz, trim, amitriptyline |

*Source:* Lim *et al.* (2020).

to determine the biotransformation rate. Although intrinsic removal of OMPs through the porous membranes often used in an AnMBR is very low, the biofilm formed on the membrane surface can significantly improve their removal as much as dissolved and colloidal matter retention. Similarly, biomass carriers such as carbon-based media used in an AnMBR should contribute to the enhancement of biosorption and biotransformation of OMPs as operational time is increased. Finally, higher temperatures (up to 40°C), biogas production and sparging significantly increased OMP volatilization according to the Arrhenius analysis.

Aziz *et al.* (2022) reviewed the ability of AD to eliminate antibiotics and ARGs from the environment. The study concluded that the AnMBR is the best anaerobic reactor design for removing certain antibiotics (sulphamethoxazole, sulphadiazine, trimethoprim (Tmp), clarithromycin, erythromycin (ERY), CIP, ofloxacin, cefalexin, cephradine). However, inhibition to AD in terms of methane production due to antibiotics exposure has also been observed for sulphamethoxazole (>25 mg/L), tetracycline (TET) (>1 mg/L), ofloxacin (>10 mg/L), CIP (>80 mg/L), sulphamerazine (>90 mg/L), tylosin (>130 mg/L) and ceftiofur (>10 mg/L). They also concluded that microorganisms belonging to phylum *Proteobacteria*, *Firmicutes*, *Chloroflexi*, *Bacteroidetes*, *Thermotogae*, *Euryarchaeota*, *Elusimicrobia*, *Chlorobi*, *Spirochaetes*, *Synergistetes*, and *Actinobacteria* are all important for stable performance of AD in terms of antibiotics removal and bioenergy production. However, the effects of antibiotics (individual and combined) on bioenergy production and microbial compositions have not been extensively investigated for AnMBRs.

Xiao *et al.* (2017) investigated the removal of five selected pharmaceutical compounds in a lab-scale AnMBR treating synthetic sewage. Tmp and sulphamethoxazole (Smx) were removed at efficiencies of 94.2 ± 5.5 and 67.8 ± 13.9%, respectively, whereas in contrast, only marginal removals of carbamazepine (Cbz) (0.3 ± 19.0%) and diclofenac (Dcf) (15.0 ± 7.2%) was observed. Finally, three distinct stages of triclosan (Tcs) removal were observed: 82.3 ± 15.0% within the first 10 days; 54.4 ± 10.4% in the following 30 days and 31.4 ± 12.6% in the last 30 days. Adsorption to the sludge was the major contributing factor for its initial high removal, but clearly it was not further degraded. When PAC was added to the AnMBR, the removal efficiencies of all five pharmaceuticals improved, especially Smx and Tcs, which showed removal efficiencies of $95.5 \pm 4.6$ and $93.2 \pm 6.6\%$, respectively. Higher removals of Cbz ($92.4 \pm 5.3\%$) and Dcf ($82.6 \pm 11.1\%$) were also observed, but only within the initial 5 days of PAC addition, and enhancement in their biotransformation by PAC adsorption was determined to be the major contributing factor for these improved removals. It is possible that adsorption onto PAC changes both the steric structure, and enhances the concentration at the interface, thereby leading to enhanced degradation.

### 3.5.5 Bioaugmentation

Herrero and Stuckey (2015) carried out one of the first reviews on bioaugmentation (BA) (the process of adding selected strains/mixed cultures to WW reactors to improve the catabolism of specific compounds, e.g. refractory organics, or overall COD). Community assembly, ecology and microbial dynamics in bioreactors treating WWs are complex processes. Basic knowledge of the biosystem ecology is required to establish a clear definition of the treatment goal to achieve. Not only should strain (or tailor-made consortium) selection, but also the way of introducing and maintaining the selected microorganisms and/or their activities in the community, and acclimation in scaling up and bioreactor design must be considered. Emerging areas of research should help in developing new bioaugmented bioreactors: protein engineering; *in vitro* compartmentalization; nanomaterials for enhanced reactivity, surface area and/or sequestration characteristics and single-cell analysis and cell–cell interactions under real operating conditions. Studies on BA are difficult due to tracking the changes in microbial ecology (16S RNA sequencing) over time, and correlating ecological changes with reactor performance.

Zhang *et al.* (2017c) also reviewed the literature, and concluded that successful application of BA is quantifying when and how BA can be effectively implemented. The key factors determining success were: main BA time, BA dosage, BA consortium, BA times, BA time interval, interaction between BA microorganisms and indigenous species (accommodation, competition and predation), WW characteristics, operation conditions and so on. Further study is needed into: (1) BA reactor design – two-stage BA reactor was suggested with the first-stage parent reactor supplying BA bacteria. (2) Optimal BA parameters, such as consortium, dosage, time, times and frequency, interval, interaction of BA cells with the indigenous species, WW quality, the reactor configuration and the operation conditions are critical to bioprocesses and should be determined. How to maintain BA effects rather than repeated BA is a key part for improving performance and decreasing operational cost. Even the coupling of BA with immobilization should be expanded in the future. (3) Accurate quantification of the beneficial or adverse effects of BA on bioprocesses is still elusive and a model to assess BA result should be developed. A real-time tracking molecular technique for monitoring of the microbial community structure, especially analysis of the dominant bacteria in BA systems, needs further study. Meanwhile, technology aiming at avoiding biomass loss in the BA process (Menashe & Kurzbaum, 2016) is urgently needed. (4) However, the limiting for full-scale application of BA is ever-changing due to the changes in on-site operational conditions and WW characteristics. Thus, wide employment of full-scale BA needs to be practiced in the future. Meanwhile, selection of the BA consortia, when and how to employ BA for successful adaptation under treatment stress, needs further studies.

Some interesting research has been carried out recently by Fakhria *et al.* (2021) by adding the saprophytic fungus *Trichocladium canadense* to an AnMBR, and studying its impact on membrane biofouling, biogas production, the microbial communities in the reactor and the removal of the common antibiotics ERY, Smx and TET from a synthetic WW reactor. The results showed that BA with 20% *T. canadense* slowed membrane biofouling by 25%, COD removal increased by 16% and a higher removal of ERY and Smx was achieved. The presence of *T. canadense* significantly increased the abundance and diversity of the biofilm archaeal community and the bacterial phylum *Firmicutes*, a known bio-foulant. A *T. canadense* concentration of 20% was optimal among the alternatives for the highest biogas production. The substantial increase in biological performance observed in the study could be attributed to the capability of *T. canadense* to degrade antibiotics and presumably other toxic compounds, thus alleviating stress allowing the microorganisms to perform better. Alleviation of membrane biofouling might be achieved through the biodegradation of EPS by *T. canadense*, as well as by the increase in the methanogenic archaeal community and a decrease in the relative abundance of *Firmicutes*, known to take part in biofouling, by 14% compared with the control with antibiotics.

Aydin *et al.* (2023) examined BA with the green alga *Haematococcus pluvialis* on the anaerobic treatment of pharmaceutical WWs, alleviating membrane biofouling, biogas production and impact on the indigenous microbial communities using AnMBRs. The results revealed that BA with the green alga increased the removal of COD by 12%, delayed membrane fouling by 25% and increased biogas production by 40%. Furthermore, BA with the green alga led to a significant change in relative abundance of archaea and the main methanogenesis pathway shifted from *Methanothermobacter* to *Methanosaeta*, accompanied by their respective syntrophic bacteria.

### 3.5.6 Effect of salinity and sulphate ions on performance

There are a range of other solutes that can influence AnMBR performance in addition to antibiotics, and these include saline WWs (from sewers with seawater intrusion, or using seawater for sewage, e.g. Hong Kong), sewage containing high levels of sulphate (>99 mg $SO_4^{2-}$) again from seawater intrusion or illegal industrial discharge and high levels of heavy metals. Again, the use of a membrane prevents the 'washout' of stressed organisms, and rejects many of the soluble constituents allowing the microbiome to adapt to the stress imposed better than non-membrane systems.

Vyrides and Stuckey (2009a) investigated the role of compatible solutes, extracellular polysaccharides (EPS), and nutrients on anaerobic biomass when stressed with salinity. When 1 mM of osmoregulants glycine betaine, $\alpha$-glutamate and $\beta$-glutamate were added separately to serum bottles containing biomass not adapted to sodium, and fed with glucose and 35 g NaCl/L, all the compatible solutes were found to alleviate sodium inhibition, although glycine betaine was found to be the most effective. The effect of glycine betaine on different anaerobic bacterial groups under salinity stress was monitored using VFAs, and showed that methanogens were more protected than propionate utilizers. Moreover, the addition of 1 mM of glycine betaine to anaerobic biomass not adapted to salinity resulted in significantly higher methane production rates compared with anaerobic biomass that was exposed to 35 g NaCl/L for 4 weeks. Interestingly, under saline batch conditions when the medium was replaced totally, the culture produced less methane than when only new substrate was added due to compatible solutes cycling between the media and the cell. The elimination of macronutrients from the medium was found to have a more pronounced negative effect on biomass under saline compared with non-saline conditions, and because of the synthesis of N-compatible solutes sufficient nutrients should always be present. On the contrary, the absence from the medium of micronutrients did not further reduce biomass activity under salinity. Finally, a higher production of EPS was obtained from biomass exposed to higher salt concentrations, and its composition was found to change under different saline conditions and time. As a result, biomass under saline conditions had a slightly higher mean floc size compared with the biomass that was not subjected to salt.

Vyrides *et al.*, (2010) further investigated fundamental mechanisms that anaerobic biomass uses to cope with salinity, and applied these findings to a continuous SAnMBR. When anaerobic biomass was exposed to 20 and 40 g NaCl/L for 96 h, the main solute generated *de novo* by biomass was trehalose. When trehalose, *N*-acetyl-lysine and potassium were separately introduced into a batch culture a slight decrease in sodium inhibition was observed. In contrast, the addition of 0.1 and 1 mM glycine betaine dramatically improved the adaptation of anaerobic biomass to 35 g NaCl/L, and it continued to enhance the adaptation of biomass to the salt for the next three batch feedings without further addition. No shift in archaeal microbial diversity was found when anaerobic biomass was exposed to 35 g NaCl/L for 360 h in batch mode, and no changes were found when glycine betaine was added. The dominant species identified under these conditions were *Methanosarcina mazeii* and *Methanosaeta* sp. The addition of 5 mM glycine betaine to a continuous SAnMBR at 12 h HRT, and operation in batch mode for 2 days can significantly enhance saline (35 g NaCl/L) synthetic sewage degradation. In addition, the injection of 1 mM glycine betaine into an SAnMBR for 5 subsequent days also significantly enhanced DOC removal from sewage under these conditions. The main compatible solutes generated by anaerobic biomass after 44 days exposure to 35 g NaCl/L in an SAMBR were *N*-acetyl-lysine and glycine betaine. Finally, the addition of 1 mM glycine betaine to the medium was beneficial for anaerobic biomass in batch mode at 20°C under saline and non-saline conditions. The use of compatible solutes to protect against salinity was reviewed by Vyrides and Stuckey (2017).

Vinardell *et al.* (2020) found that sulphate in municipal sewage significantly affects AD and filtration processes (Song *et al.*, 2018). Sulphate-reducing bacteria (SRB) use organics and hydrogen as electron donors to reduce sulphate ( $SO_4^{2-}$) to sulphide ($HS^{-1}$). With sulphate, SRB compete with methanogens for the same substrates reducing substrate availability and methanogenesis. In addition, sulphide production can inhibit methanogenic activity, further reducing methane production. Finally, hydrogen sulphide in biogas also requires the use of corrosive-resistant instrumentation and equipment, and dissolved hydrogen sulphide reduces the lifetime of the membrane. Hence, sulphate concentration in sewage has a direct impact on the economic feasibility of AnMBRs. Shin and Bae (2018) reported that AnMBR pilot plants treating sewage with high-sulphate concentrations obtained poor methane yields (0.08–0.15 L $CH_4$/g COD) when compared to the average methane yield of those treating sewage with low-sulphate concentrations (0.22 L $CH_4$/g COD). Giménez *et al.* (2011), who studied the influence of the COD/ $SO_4^{2-}$–S ratio on AD performance, reported a sharp decrease in methane production as the influent sulphate concentration increased. The methane production nearly ceased when COD/ $SO_4^{2-}$ was below the stoichiometric ratio for sulphate reduction of 2.01 mg COD/mg $SO_4^{2-}$–S (0.67 mg COD/mg $SO_4^{2-}$). Hence SRB outcompete methanogens and nearly all the sulphate is converted to sulphide if enough electron donor COD is available. Furthermore, the presence of dissolved sulphide in the permeate can affect the overall treatment efficiency since sulphide contributes to the effluent COD.

The removal of sulphide from AnMBR effluents has recently been addressed by using coagulation–flocculation (Lim *et al.*, 2019) and membrane distillation (Silva *et al.*, 2020). However, alternative methods such as electrochemical systems are also gaining attention due to their capacity to recover sulphide as sulphur or other oxidized sulphur species. The corrosive nature of sulphide also affects membrane permeability and durability. Sulphide has been reported to damage the internal material when transported through the membrane cell, making it more susceptible to membrane fouling. In this regard, Song *et al.* (2018) observed that membrane fouling increased as the influent sulphate concentration increased. Specifically, the TMP sharply increased from 0.5 to 0.85 bar after the addition of more than 33 mg $SO_4^{2-}$–S/L. They attributed this to the higher release of EPS under high-sulphide concentrations. However, there is a little understanding of the impact of sulphide concentration on the microbial community activity, particle size distribution, EPS composition and membrane performance and durability.

Some studies have evaluated the impact of sulphate concentration on the economic and energetic prospects of the AnMBR process. Ferrer *et al.* (2015) estimated that treating low-sulphate (57 mg COD/mg $SO_4^{2-}$–S/L) municipal sewage is more favourable than treating sulphate-rich (5.7 mg COD/mg $SO_4^{2-}$–S/L) municipal sewage (€0.070 and €0.097/m$^3$, respectively). They also stated that methane recovery from AnMBR effluents is more economically attractive when treating low-sulphate municipal sewage due to the low-methane production at high-sulphate concentrations. These results clearly illustrate that sulphate concentration in sewage has a significant impact on AnMBR performance and profitability and, therefore, it has a key role in decision-making processes. Some research has estimated that COD/ $SO_4^{2-}$–S ratios at or above 30 (10 mg COD/mg $SO_4^{2-}$) could be adequate to sustain good AD performance and high-methane yields (Song *et al.*, 2018). However, further studies into the technical, economic and energy issues relating to sulphate in AD are important. Particularly useful would be to determine a COD/ $SO_4^{2-}$–S ratio above which the AnMBR is best for the treatment of municipal sewage. However, further understanding of the implication of sulphide concentration on AD and membrane performance is needed before carrying out such techno-economic studies.

Nevertheless, there are methods of controlling sulphate in AD. Vu *et al.* (2022) discussed technologies to remove $H_2S$ during AD via input control, process regulation and post-treatment. Post-treatment technologies (trickling filters and scrubbers) are established with >95% removal, but they do not ameliorate $H_2S$ toxicity to methanogens. Input control (substrate pretreatment via chemical addition) reduces sulphur input via sulphur precipitation, and results showed 95% $H_2S$ reduction, stable methane production and low-operational cost.

One interesting idea which has been evaluated for decades is micro-aeration. Further adaptation of micro-aeration relies on a comprehensive design framework and exchange operational experience for eliminating the risk of over-aeration. Pokorna-Krayzelova *et al.* (2017) modified the IWA AD model with sulphur and oxygen (ADM1-S/O) to describe and control sulphate reduction and sulphide oxidation in anaerobic/microaerobic environments. Model validation provided a good fit in terms of $H_2S$ emissions and biogas flow. However, a sulphur balance showed some limitations of the model as it predicted higher $H_2S$ removal (lower $H_2S$ concentrations in the effluent) than observed experimentally. Micro-aeration was achieved by blowing $O_2$ into the liquid phase, and aerobic carbon oxidation and re-reduction of elemental sulphur back to sulphide were considered in the model, whereas the simulated composition of active biomass in the micro-aerobic reactor was not significantly affected by micro-aeration. Sulphur oxidizing bacteria only made up a small fraction of the active biomass (0.05% of active biomass for an influent COD: $SO_4^{2-}$ ratio of 32.3 g/g). $H_2S$ in the biogas increased proportionally with increasing influent S:COD ratio. Maximum $H_2S$ concentrations of 0.32 and 0.14% were observed in the biogas from UASB and Upflow Microaeration Sludge Blanket (UMSB) reactors, respectively, for a S:COD ratio of 3.3 g/g and $O_2$:S ratio of 0.25. The highest $H_2S$ removal efficiency from biogas was obtained for a $O_2$:S ratio of 0.5 kmol $O_2$: kmol $SO_4^{2-}$–S, whereas increasing the $O_2$:S ratio to over 0.5 did not significantly improve $H_2S$ removal.

### 3.5.7 Effect of heavy metals on supplementation and toxicity

The influence of heavy metals on both metabolism and toxicity in AD is critical because many of them are needed to form the co-enzymes required during metabolism, for example, factor $F_{430}$ and coenzyme M, or in overall metabolism such as calcium and magnesium (White & Stuckey, 2000). However, very high concentrations of metals can be toxic, although with the presence of sulphide in solution may precipitate as the sulphide salt which has a very low $K_{SP}$, and hence are probably rendered insoluble and hence non-toxic. In addition, some metals can be complexed or adsorbed, and hence their availability and toxicity is complex. In addition, 'equilibrium chemistry' equations must be modified by the Debye–Hückel equation due to the high ionic strength ($I \sim 0.1$ molar) in most AD systems.

Thanh *et al.* (2016) carried out an extensive review on metal addition and speciation, and found that trace metals were often supplemented in excessive amounts, at times leading to inhibitory effects on digestion due to the lack of understanding of speciation and bioavailability. However, it has been shown many times before that lack of trace metals leads to poor performance in AD (Thanh *et al.*, 2017a). Initially, the bioavailability of trace metals was generally quantified based on the total rather than 'free' or bioavailable metal concentrations; however, it is now clear that chemical speciation is critical when predicting bioavailability. Although trace metal speciation has been studied extensively to determine the fate of trace metals in an anaerobic environment, its relationship with trace metal bioavailability is still not clear. In the past 15 years, interest has increased in how trace metals can be effectively supplemented to anaerobic reactors considering both their speciation and bioavailability. Recent studies have showed that metal sulphides play a significant role in bioavailability as they may act as a store for metals, and as a supply source in anaerobic reactors. On the contrary, trace metal bioavailability in AD is also dependent on natural and synthetic chelating agents, for example, SMPs and EPS, and EDTA and Nitrilotriacetic acid (NTA), respectively. However, whether these chelating agents increase or decrease trace metal bioavailability remains controversial. In addition to the popular chemical sequential extraction, analytical techniques such as Wageningen Donnan membrane technique (WDMT), diffusive gradients in thin films (DGT), X-ray diffraction (XRD), and Fourier Transform Infrared Spectroscopy (FTIR) have been applied to determine the speciation of trace metals (TMs) in various types of environmental samples; however, their application to anaerobic sludge is still in its infancy.

Thanh *et al.* (2016) suggested further research in these areas is important to obtain a better understanding of trace metal speciation and bioavailability in AD:

(1) Metals such as Fe, Ni and Co have been studied extensively; however, other metals (trace nutrients) such as Ca, Mg, Mn, W and Se are also essential to the optimal operation of AD. Hence, more research is required to determine the relationship between their speciation and bioavailability in AD, and the effects of one trace metal on the speciation and bioavailability of another during coprecipitation and adsorption are also important to understand.
(2) Different forms of trace metals, for example, chloride, EDTA complex or vitamins ($B_{12}$ for Co) have been studied for their bioavailability to anaerobic microorganisms. However, trace metals have also been shown to be present in other forms, such as phosphates, sulphides or carbonates; it is not clear whether their bioavailability in AD is the same or not. Research should be carried out to determine in which form trace metals have the most significant effect on AD, and define a more effective metal-dosing strategy for anaerobic reactors.
(3) Although methanogens are thought to be the most sensitive microbial group in AD to changes in trace metal speciation and bioavailability, it is important to determine the effects of these changes on other anaerobic trophic groups such as hydrolytic fermentative bacteria and syntrophic acetogens, and the acetoclastic methanogens and hydrogenotrophic methanogens in standalone as well as competitive environments.
(4) Metal sulphides such as NiS have been shown to be a store of and source for trace metals. To understand which factors influence the bioavailability of metal sulphides, it is necessary to determine their speciation in terms of organic/inorganic sulphides, possible interaction with other metal sulphides, and the degree of crystallinity of the metal–sulphide precipitates using extended X-ray absorption fine structure and XRD.
(5) An in-depth study is required to verify the bioavailability of metals complexed with EDTA as the literature about its stimulation/inhibitory effects is contradictory. Research on the effect of NTA on the bioavailability of other metals, for example, Fe, Co, and Mo, is also required, as identifying other economic metal chelators that can be added to improve trace metal bioavailability. Finally, better characterization and quantification of SMPs, and further development of the EPS extraction procedure are needed to determine how important their roles are in trace metal complexing and binding properties.
(6) It would be useful to harmonize sequential extraction schemes for data comparability, and design new schemes based on the existing ones with shorter extraction times. Also techniques such as X-ray absorption near edge structure, DGT and Donnan membrane technique (DMT) need to be further developed and adapted to better classify metal speciation, and hence predict metal bioavailability in anaerobic reactors.
(7) Development of a simple procedure to study sulphur speciation in AD is necessary, although a combination of acid volatile sulphide extraction and the semi-continuously extracted metals (SEM) technique seem to be promising, because these methods are very expensive to use.
(8) Most studies on trace metal bioavailability were carried out in UASB or CSTR reactors. An AnMBR is one of the most advanced anaerobic reactors, and hence the metal fractions bound to the biomass will not be washed out from the reactor. Hence, it is important to determine how the dynamics of trace metals changes and their requirements in AnMBRs.

Following this fact, Thanh *et al.* (2017b) investigated the effect of changes in process parameters such as pH (7, 6.5 and 6), HRT (6, 4, and 2 h) and SRT (100 and 25 days) on trace metal speciation in an SAnMBR. The metal retention capacity of SAnMBRs reduced when the pH, HRT and SRT were reduced by up to 21.9, 39.1 and 17.1%, respectively, but the speciation of these TMs generally shifted towards highly bioavailable fractions, that is, soluble and exchangeable. The degree of speciation shift depended on the TM affinity for anaerobic sludge and their sensitivity to the changes. TMs with the most and the least significant changes in speciation were Fe and Mn, respectively. Finally, Thanh *et al.* (2017c) examined the effects of a biodegradable chelating agent, ethylenediamine-*N*,*N'*-disuccinic acid (EDDS), on the speciation and bioavailability of iron ($Fe^{2+}$) in AD. $Fe^{2+}$ supplementation at 10 mg/L increased methane yield, but the presence of 8 mg/L sulphide led to the precipitation of $Fe^{2+}$ as FeS which limited its bioavailability. The results confirmed that EDDS could replace common chelating agents with low biodegradability (EDTA and NTA), and improve the bioavailability of $Fe^{2+}$ by forming an Fe–EDDS complex, thereby protecting $Fe^{2+}$ from sulphide precipitation. Experimental findings from sequential extraction using the community bureau of reference method, and quantification of free EDDS and Fe–EDDS complex using ultra-high-performance liquid chromatography, confirmed that 29.82% of $Fe^{2+}$ was present in bioavailable forms, that is, soluble and exchangeable, when EDDS was added at a 1:1 molar ratio to $Fe^{2+}$. As a result, methane production increased by 11.17%, and yield increased by 13.25%. Hence chelating metals can improve their availability in AD, and make them more cost effective.

## 3.6 THE MEASUREMENT OF ANAEROBIC TOXICITY AND EFFECT OF GENERIC TOXICITY

A comprehensive review on anaerobic toxicity was carried out by Chen *et al.* (2014). AD is sensitive to toxicants, and a wide range of compounds can inhibit the process and cause upset or failure. Substantial research has been carried out over the years to identify specific inhibitors/toxicants, and their mechanism of toxicity in AD. These authors presented a detailed and critical summary of research on the inhibition of anaerobic processes by specific organic toxicants (chlorophenols, halogenated aliphatics and long chain fatty acids), inorganic toxicants (ammonia, sulphide and heavy metals) and nanomaterials, focusing on the mechanism of their inhibition/toxicity. These compounds primarily obstruct the metabolism of sensitive obligate hydrogen-producing acetogens and methanogenic portions of the digester population, as well as inhibit methane formation, or can even cause complete failure of methanogenesis. However, because of the difference in anaerobic microorganisms and waste composition, results from previous studies on inhibition of anaerobic processes vary substantially. Chen *et al.* (2014) summarized and highlighted the effect of specific organic toxicants, inorganic toxicants and the emerging nanomaterials on AD. In addition, better understanding the mechanism(s) of inhibition or toxicity of different toxicants in a digester provides insights into overcoming these toxic effects and possible solutions or strategies to properly cope with it, successfully apply AD and significantly improve waste treatment efficiency. On the contrary, measuring the toxicant concentration and monitoring them are an essential precautionary strategy. Currently for anaerobic waste treatment, little research has been carried out on measuring toxicity before the waste is introduced into a digester. Most research has focused on detoxification after inhibition and not on stopping/ameliorating toxicity before it happens. Hence, a rapid response method to determine toxic substances in the feed stream, and toxic byproducts in a digester, needs to be developed to protect the anaerobic microcosm from instability, and hence enable digesters to operate without toxic perturbations.

Previously, the only bioassay to measure potential anaerobic toxicity relatively quickly (1–3 h) was anaerobic toxicity assay (ATA) (Owen *et al.*, 1979). However, while cheap and effective, it was somewhat cumbersome, and even a 1-h warning could be disastrous for a toxic slug entering a digester. Chen *et al.* (2015a) developed a rapid ($\sim$8 min) fluorescence measurement based on resazurin reduction to detect toxicants/inhibitors to anaerobic metabolism. They initially used a pure facultative anaerobic strain, *Enterococcus faecalis* as a model organism; this technique was quick and sensitive at detecting the model toxicant, pentachlorophenol (PCP). The technique revealed significant metabolic changes in *E. faecalis* with a PCP spike ranging from 0.05 to 100 mg/L, and could detect toxicity to *E. faecalis* at only 0.05 mg/L in 8 min. By using this technique in a mixed anaerobic sludge, not only could the effect of 0.05–100 mg/L PCP be determined on anaerobic metabolism within 10 min, but also its rate of biogas production. A kinetic model was then developed to describe the process of resazurin reduced to resorufin, and eventually to dihydroresorufin under anaerobic conditions (Chen *et al.*, 2015b). By modelling the assay results of resazurin (0.05, 0.1, 0.2 and 0.4 mM) reduction by a pure facultative anaerobic strain, *E. faecalis* and fresh mixed anaerobic sludge, with or without 10 mg/L spiked PCP, it was clear that the pseudo-first-order rate constant for the reduction of resazurin to resorufin, $k_1$, was a good measure of 'toxicity' with respect to a control. With lower biomass density and the optimal resazurin addition (0.1 mM), the toxicity of 10 mg/L PCP for *E. faecalis* and fresh anaerobic sludge was detected in 10 min, and the toxicity differences among seven chlorophenols to *E. faecalis* and fresh mixed anaerobic sludge were elucidated within 30 min. The toxicity differences determined by this assay were comparable to toxicity sequences of various chlorophenols reported in the literature. These results suggest that the assay developed in this study can not only quickly detect toxicants for AD, but also can efficiently detect the toxicity differences among a variety of similar toxicants.

These results suggest that a resazurin-based fluorescence measurement can potentially be incorporated into a microfluidic system to develop a biosensor for the real-time monitoring, control and early warning of toxicant/inhibitor loads in the influent to an AD. To rapidly evaluate unknown aqueous effluents, a segmented-flow microfluidic device was developed for real-time toxicity detection/quantification by Ortiz *et al.* (2019). A microfluidic chip integrating a magnetic stirrer for serial dilution was demonstrated to rapidly determine half-maximal effective concentration ($EC_{50}$) values of known and unknown toxicants, where 100 nL samples were suspended into a continuous oil phase with no interfering surfactants. A five logarithmic dilution sequence was evaluated in $<5$ min, enabling feedback for dynamic processes, for example, dilution, ion exchange or absorbent treatments. The assay was refined from traditional laboratory procedures to reduce sample volume and response time, with the advantages of short toxicant exposure ($<30$ s), high-density bacteria ($10^7$ cells/mL) and continuous mixing in an oxygen-free medium. Simple scaling of the stirred chamber volume from 1 to 30 μL enables the number of samples to be increased from 100 to 2,500 dilution droplets, respectively. A detailed analysis of the resazurin kinetics suggests the presence of two or more enzymes with distinct Michaelis constants. Enzyme kinetics and the resazurin reduction rate is dependent on growth phase and $EC_{50}$ inhibition by mercury > zinc > copper > nickel cations. Resazurin concentration of 10 μM is optimal for *E. faecalis* at 0.1 optical density. The adaptable method is transferable to other microorganisms, such as common baker's yeast, *Saccharomyces cerevisiae*, where the resazurin reduction rate is 30% of *E. faecalis* (4 vs. 13 nM/s per 100 nL droplet). Zinc and nickel cations were observed to increase the base resazurin reduction rate of baker's yeast by 25%, whereas copper is found to be more cytotoxic than mercury cations.

This technique can then be 'turned around' to determine the levels of trace nutrients to stimulate AD. Chen *et al.* (2016) investigated the effect of nickel(II), cobalt(II) and their mixture, on the AD of synthetic municipal WWs. Using an anaerobic toxicity assay (ATA), it was revealed that AD was stimulated by the addition of 5 mg/L nickel(II) and cobalt(II), and their mixture in day(s). Using the resazurin reduction assay, they evaluated the stimulatory effect of 5 mg/L nickel(II) and cobalt(II), and their mixture, in just 6 h. By investigating the distribution of these metals in the liquid phase of the anaerobic system and kinetics of resazurin reduction by nickel-spiked anaerobic sludge, the concentration of nickel(II) on AD performance was profiled. Three critical concentrations were determined: stimulation starting (~1 mg/L), stimulation ending (~100 mg/L) and stimulation maximizing (~10 mg/L). Hence, the resazurin reduction assay is a novel and quick protocol for studying the stimulation of anaerobic bioprocesses by bioavailable essential trace metals.

## 3.7 MICROBIAL ECOLOGY OF AnMBRs

Kong *et al.* (2022) followed a large pilot-scale AnMBR and the microbial community development during the treatment of municipal WWs for 217 days operating at HRTs from 24 to 6 h. Characterization and dynamics of the microorganisms revealed that a stable prokaryotic community was gradually achieved.

In the community of methane-producing archaea (or methanogens), the acetotrophic methanogen *Methanosaeta* was significantly enriched at 25°C with an overwhelming relative abundance in the entire community. The abundance of *Methanosaeta* was even higher than the most abundant bacterial phyla *Chloroflexi*, *Firmicutes*, *Proteobacteria* and *Bacteroidetes*, and is quite different from that found in other typical anaerobic systems. This substantial enrichment of methanogens is the key to maintaining stable methane production in the treatment of municipal WWs by AnMBRs. The interspecies cooperation of major functional bacterial groups including protein/carbohydrate/cellulose-degrading (genera *Anaerovorax*, *Aminomonas*, *Levilinea*, *Flexilinea* and *Ruminococcus*), sulphate-reducing (*Desulfovibrio* and *Desulfomicrobium*) and syntrophic (*Syntrophorhabdus*, *Syntrophus*, etc.) bacteria with acetotrophic and hydrogenotrophic archaea enhances the stability of reactor operation and helps to acclimate the entire prokaryotic community to real municipal WWs. This agrees to some extent to other studies (Table 3.4).

Ni *et al.* (2022) also operated two AnMBRs with different membrane pore sizes (0.4 or 0.05 μm) at 25°C with domestic WWs at different HRTs (24 to 4 h). The microbial communities of the two AnMBRs were investigated by 16S rRNA sequencing to determine the effect of HRT.

The predominant archaea was an aceticlastic methanogen *Methanosaeta*, and the composition of hydrogenotrophic methanogens changed with HRT; the population of *Methanobacterium* was higher for longer HRTs, whereas the population of unclassified *Methanoregulaceae* was higher for shorter HRTs. *Anaerolineae*, *Bacteroidia* and *Clostridia* were dominant in both reactors, with a combined relative abundance of over 55%. The relative abundance of *Anaerolineae* was proportional to the biogas

**Table 3.4** The major functional microorganisms in the AD system for the treatment of municipal wastewater and other types of wastewaters using the AnMBR process.

| Wastewater type | Temperature (°C) | Major functional organisms | | Reference |
|---|---|---|---|---|
| | | Bacteria (phylum or genus) | Archaea (genus) | |
| Pharmaceutical | 27 | *Firmicutes, Bacteroidetes, Proteobacteria (phylum)* | *Methanimicrococcus, Methanomassiliicoccus, Methanobrevibacter* | Ng *et al.*, 2016 |
| Petrochemical | 37 | *Proteobacteria, Acetothermia, Patescibacteria (phylum)* | *Methanosaeta, Methanolinea* | Kudisi *et al.*, 2022 |
| Textile | 35 | *Chloroflexi, Proteobacteria, Proteobacteria (phylum)* | *Methanomethylovorans, Methanomassiliicoccus* | Kong *et al.*, 2018 |
| Phenolic | 35 | *Clostridium (genus)* | *Methanobacterium, Methanosaeta* | Munoz Sierra *et al.*, 2018 |
| Municipal | 10–25 | *Clostridium, Arcobacter, Geobacter, Trichoccous, Acinetobacter (genus)* | *Methanosaeta, Methanobacterium, Methanospirillum* | Seib *et al.*, 2016 |
| Municipal | 14–27 | *Bacteroidetes, Proteobacteria, Firmicutes, Chloroflexi (phylum)* | *Methanosaeta* | Kannan *et al.*, 2020 |
| Municipal | 35 | *Chloroflexi, Proteobacteria and Bacteroidetes (phylum)* | *Methanosaeta, Methanolinea* | Mei *et al.*, 2018 |
| Municipal | 10–25 | *Saccharofermentans, Smithella, Acetobacterium, Leptotrichia (genus)* | *Methanosaeta, Methanobacterium, Methanolinea* | Ji *et al.*, 2021 |
| Municipal | 25 | *Flexillinea, Levillinea, Paludibacter, Tangfeifania, Anaerovorax (genus)* | *Methanosaeta* | This study |

*Source:* Kong *et al.* (2022).

production performance. The sum of the relative abundance of *Anaerolineae* and *Clostridia* fluctuated slightly with changes in the HRT in both AnMBRs when the reactor was stable. A co-occurrence analysis revealed the relative abundance of the operational taxonomic units belonging to *Anaerolineae* and *Clostridia* was functionally equivalent during the treatment of domestic sewage. A principal coordination analysis revealed that the changes in the microbiome of the WWT community in each reactor were consistent with the change of HRT. In addition, both the HRT and the stability of the process are important factors for maintaining microbial community structures.

Finally, Tao *et al.* (2020) examined the issues around a long start-up time, and system instability in response to operational shocks from a functional microbiota perspective. Exploring the microbiota–functionality nexus (MFN) could be pivotal to understand the reasons behind these difficulties, and hence improving AD performance. They presented a systematic MFN study based on 138 samples taken from 20 well-profiled lab-scale AD reactors operated for up to 2 years. All the reactors were operated in the same lab within the same period using the same methodology to harvest physico-chemical and molecular data, including key monitoring parameters, quantitative polymerase chain reaction and 16S sequencing results. The results showed a core bacterial microbiota prevailing in all reactor types, including *Bacillus*, *Clostridium*, *Bacteroides*, *Eubacterium*, *Cytophaga*, *Anaerophaga* and *Syntrophomonas*, whereas various methanogens dominated different communities due to different inocula origins, reactor temperatures or salinity levels, and this core bacterial microbiota correlated well with biogas production (Pearson's correlation coefficient of 0.481, $p < 0.0001$). Such a strong correlation was even comparable to that between biogas production and the methanogenic 16S rRNA gene content (Pearson's correlation coefficient of 0.481, $p < 0.0001$). These results indicated that AD performance only modestly correlated with microbial diversity, a key governing factor. AD microbiota was neither functionally redundant nor plastic, and a high variety in communities can exhibit a strong difference in reactor performance. This study demonstrates the importance of a core bacterial microbiota in AD and supports inspiring considerations for design, BA and operational strategies of AD reactors in the future.

## 3.8 ENERGY, GLOBAL OPTIMIZATION AND LCA

Initially, with domestic WWT the primary focus was on removing pollutants and pathogens, and to render the water safe to discharge to the environment. Mainly, the technology used was old and trusted, for example, CAS and CSTRs for anaerobic sludge treatment. However, several decades ago the use of energy in the plant became an issue due to its increasing costs, and this was coupled with the disposal of sewage sludge as land area for disposal was reducing. Hence, the focus shifted to minimizing energy use in WWT and reducing the generation of sludge. Because of these two factors, interest in the use of anaerobic treatment technologies other than CSTRs increased, and novel designs such as anaerobic filters, UASBs, ABRs and AnMBRs were developed. However, the primary focus on these designs was still energy minimization, and hence many papers, even today, focus on making anaerobic treatment technologies energy positive (or at least less energy negative) without considering other issues around this.

Recently, ideas have moved on, and there is a growing realization that WWTPs should not just focus on energy *per se*, but on 'closing the loop', that is, to minimize the disposal of resources into the environment and hence minimize its impact, and to recycle as much as possible having due regard to the environment (Akyol *et al.*, 2020). One of the early authors who advocated this approach was Verstraete (Verstraete *et al.*, 2009) who examined the value of each component of municipal 'used water' (Table 3.5) – this new term started to be used during the reclamation of WWs in Singapore in the early 2000s (Lee and Tan, 2016). Because of its high quality and bacterial sterility, AnMBR effluent can be further post-treated to render drinking quality water (Mai *et al.*, 2018).

Hence, putative solutions need to be assessed by LCA. This is a process of evaluating the effects that a product has on the environment over the entire period of its life, thereby increasing resource-use

**Table 3.5** Potential product recovery from municipal "used water".

| Potential recovery | Per $m^3$ of sewage | Current market prices | Total per $m^3$ sewage (€) |
|---|---|---|---|
| Water | 1 $m^3$ | €0.25/$m^3$ | 0.25 |
| Nitrogen | 0.05 kg | €0.215/kg | 0.01 |
| Methane | 0.14 $m^3$ | €0.338/ $m^3$ $CH_4$ | 0.05 |
| Organic fertilizer | 0.10 kg | €0.20/kg | 0.02 |
| Phosphorus | 0.01 kg | €0.70/kg | 0.01 |
| | | **Total** | **0.35** |

*Source:* Verstraete *et al.* (2009).

efficiency and decreasing liabilities. In addition, since the choices available for different unit operations in the WWT flowsheet are substantial, 'global optimization' is used to derive the optimal flowsheet while minimizing environmental damage. There is a perception in the literature that if a 'unit operation' (e.g. an AnMBR) is 'optimized', then a string of optimized units in a flowsheet is also optimal. For example, if an AnMBR is chosen instead of an UASB then we do not need solid removal or further disinfection, and therefore the complex mathematical technique of 'global optimization' is required to find the best flowsheet. Hence, this section will initially review papers looking solely at energy balances around AnMBRs, and then move on to global optimization of treatment processes, before finally examining the research on LCA of flowsheets.

### 3.8.1 Energy balance

While many papers previously have discussed energy neutral (positive) WWT, the one by McCarty *et al.* (2011) was seminal in emphasizing that energy positive WWT was possible using AD. This topic is covered in-depth in another chapter in this book, but a short summary will be presented here with respect to AnMBRs. Energy is produced during AD in the form of methane gas, both in the gas phase and dissolved in the liquid phase. The lower the temperature of operation, the higher the loss of methane in the effluent, and despite many assumptions, the gas and liquid phases are not always in equilibrium (Crone *et al.*, 2016). With gas sparging this is most likely to occur, whereas with other methods of fouling control the phase concentrations may not be predictable using Henry's law (assumes equilibrium). Net energy is the difference between recoverable energy (and methane recovery, e.g. by solid membranes, still requires capital input), and the energy required for membrane cleaning (gas sparging needs the most) and pumping.

The net energy yield in AnMBRs depends strongly on a lot of factors: type of reactor (gas purged, mechanically mixed, granular with GAC, dynamic membranes, cross-flow with external module), sparging rate and variation (time variant, variable), membrane flux, loading rate, HRT, flocculent or PAC addition, absolute TMP and rate of increase, COD removal and feed type and concentration. Lim *et al.* (2023) compared mechanically reciprocated AnMBR membranes to a gassed system and found that membrane reciprocation consumed 0.59 kWh/$m^3$ at 7 LMH, 62% less energy than biogas sparging (1.56 kWh/$m^3$ at 6 LMH). With high biogas recovery and energy efficient fouling control, a positive net energy balance (0.55–0.74 kWh/$m^3$) was achieved. Using a GAC granular bed two-reactor system, Shin *et al.* (2021) showed with sewage at temperate climates, an energy balance yields net energy positive operation (+0.11 kWh/$m^3$), with energy recovery from produced methane (0.39 kWh electricity/$m^3$ + 0.64 kWh heat/$m^3$) exceeding energy consumption due to GAC fluidization (0.07 kWh electricity/$m^3$) and gas sparging (0.20 kWh electricity/$m^3$ at an optimal flux of 12.2 LMH).

Hu *et al.* (2022) reviewed a number of promising AnMBR configurations including biogas-sparging AnMBRs, particle-fluidized AnMBRs, granular-sludge AnMBRs, anaerobic dynamic membrane bioreactors (AnDMBRs) and anaerobic forward osmosis membrane bioreactors (AnFOMBRs), with the energy balance systematically analysed. The net energy balance was: 0.098, 0.661, 0.715, 0.716 and −0.773 kWh/m$^3$ for biogas sparging, particle-fluidized, granular-sludge, AnDMBRs, AnFOMBRs, respectively. Hence, while gassed and GAC membrane systems were positive, the best system seemed to be the AnDMBR. Therefore, depending on a variety of factors AnMBRs can be net energy positive, and some with low-energy demands can potentially generate significant quantities of energy.

### 3.8.2 Global optimization

More efficient water treatment plants are needed if future global water requirements are going to be met. Plant design, however, is complicated by the need to optimize separate but linked treatment processes, with individual process optima not necessarily leading to the most efficient overall systems. Koleva *et al.* (2016) proposed a superstructure optimization-based methodology for flowsheet synthesis formulated as a mixed integer non-linear programming (MINLP) model. The superstructure encompasses the most used technologies in advanced WWT, which enables the design of 'fit-for-purpose' treatment. The model identifies the optimum configuration of passes and stages in the flow diagram and capital cost estimation. A multi-objective optimization is performed for the minimization of water net cost and minimization of contaminant concentrations using a constraint method. The applicability of the model was verified through a theoretical case study, and the results fall into the lower costs of purification facilities worldwide and indicate the effectiveness and efficiency of the developed approach.

Puchongkawarin *et al.* (2015) also used this approach of superstructure optimization-based methodology with MINLP to develop a decision support system for isolating, among hundreds or even thousands of alternatives, promising resource recovery systems whose development was worth pursuing. Based on this preselection, further simulation and optimization studies can then be undertaken to refine the performance and cost prediction by considering detailed design and operation considerations, as well as process integration. Such decomposition is needed as current computational capabilities and available algorithms do not allow for the optimal design and operation to be solved in a single step due to complex unit configuration, multiple scales, time dependence and uncertainty. They concluded that a major hurdle in applying this methodology is the availability of reliable performance models for the treatment and separation units, as well as reliable (capital and operation) costing data. They advocate the use of state-of-the-art WWT simulators to derive simple response-surface models, which are general enough to be independent of detailed design choices and keep the superstructure optimization model computationally tractable – this approach was demonstrated on a simple case study. Naturally, such simple models carry significant uncertainty and usually only provide a rough approximation of the actual performance of such complex units. A way to refine these models involves performing an iteration between the detailed process simulators and the superstructure optimization problem. Moreover, for those treatment/recovery techniques that are less well established or lack reliable performance models, a scenario-based analysis can be applied, whereby multiple sets of resource recovery systems are determined on account of the forecast performance and cost scenarios. This analysis can be useful for resource allocation, for instance to help determine which technologies are critical and focus further research and development effort.

As for future research directions, a key extension will be to integrate WW and biosolid management within the same recovery system. Another important research activity should be the development and regular update of information databases as new advanced treatment and recovery technologies develop, or as the economic, environmental and socio-cultural contexts evolve. Besides the availability of feasible technologies that can transform WW into a product, and the downstream processing of this product into a saleable item, the circumstances that are required to successfully establish a functioning and sustainable resource recovery system also involves developing a distribution infrastructure and catching investors' interest in developing such technologies.

### 3.8.3 Life-cycle analysis

Corominas *et al.* (2020) reviewed a lot of studies on LCA of municipal WWT, and found that inconsistencies in assumptions and methods made it difficult for researchers and practitioners to compare results across studies. The reported LCAs related to municipal WW management focused on developing systematic guidance for researchers and practitioners to conduct LCA studies to inform planning, design and optimization of WW management and infrastructure (WWTPs; collection and reuse systems; related treatment technologies and policies), and to support the development of new technologies to advance treatment objectives and the sustainability of WW management. The study guides the reader step-by-step through LCA methodology to make informed decisions on: (1) definition of the goal and scope, (2) selection of the functional unit and system boundaries, (3) selection of variables to include, and their sources to obtain inventories, (4) selection of impact assessment methods and (5) selection of an effective approach for data interpretation and communication to decision makers. The paper provided (1) examples of questions which can be addressed (and cannot be addressed) by LCA, (2) background information on the selection of the methodological approach (attributional vs. consequential, process-based vs. input–output), (3) guidance on the selection of the functional unit and on the definition of system boundaries, (4) recommendations to conduct the inventory, (5) guidance on the selection of indicators, impact assessment methodologies and, (6) recommendations for the interpretation of the results to facilitate decision making.

Pretel *et al.* (2015) focused on SAnMBRs and detailed how design and operational decisions influence the technological, environmental and economic sustainability of the system across its life cycle. Specific design and operational decisions evaluated included: SRT, MLSS concentration, sludge recycling ratio ($r$), membrane flux ($J$) and specific gas demand per membrane area. The possibility of methane recovery (both as biogas and soluble methane in reactor effluent) and bioenergy production, nutrient recovery and final destination of the sludge (land application, landfill or incineration) were also evaluated. The implications of these design and operational decisions were characterized by leveraging a quantitative sustainable design framework which integrated steady-state performance modelling across seasonal temperatures (using pilot-scale experimental data and simulating software DESASS), life-cycle cost analysis and LCA. Sensitivity and uncertainty analyses were used to characterize the relative importance of individual design decisions, and to navigate trade-offs across environmental, economic and technological criteria. Based on this analysis, there are design and operational conditions under which SAnMBRs could be net energy positive and contribute to the pursuit of carbon negative WWT.

## 3.9 OPEN RESEARCH QUESTIONS

Based on the current state-of-the-art in AnMBRs it appears there are several fruitful areas of promising research, some which may not always concur with heavily researched areas at present. They are:

(1) *The minimization of energy use* to keep membrane fouling at its 'optimum' level – this can be defined as maximizing effluent quality (solute rejection) with 'acceptable' levels of TMP drop, and hence cost. This would enhance the attractiveness of using AnMBRs. The literature suggests that using sequential reactors containing GAC and membranes considerably reduces energy use (net positive), and these types of reactors need to be developed more. Intermittent gas sparging (ON/OFF or varying rates) seems to save energy and still result in acceptable power use, whereas flocculants/adsorbents also reduce fouling – the question to be answered here is what is their use over time, and the final fate of the additives – all of these questions need to be addressed at pilot scale. In addition, mechanically vibrated membrane units seem to lower power use dramatically, and hence need to be developed more – can they also be used intermittently? Finally, the most intriguing solution is the use of immobilized cells to interrupt QS by quenching the signal and extending the TMP jump considerably – is this technique economic?

(2) *Bioaugmentation*, as opposed to ex-post microbial ecology (of which there seems to be a lot), seems to be an important area to conduct more research. Can we understand more about changing the microbial ecology of reactors to change their function, and how adding different species can alter behaviour. For example, to develop AnMBRs capable of performing well at psychrophilic temperatures with low-energy use; enhance the ability of cultures to degrade some more refractory PPCPs.

(3) *Enhance hydrolysis rates of insoluble feeds* – more research and modelling needs to be conducted on the microbial interactions in flocs in terms of mass transfer, kinetics and thermodynamics. Rumen seem to be able to hydrolyse insoluble lignocellulosics at 30 times the rate of sewage digesters and hence there may be considerable potential in understanding and harnessing this increased potential, for example, by enhancing mixing regimes, and controlling gas ($H_2$, $CO_2$ and $CH_4$) environments through gas membranes.

(4) *Mathematical optimization of AnMBRs* – there is considerable data available on the effect of a variety of system parameters (HRT, SRT, gas scouring volumes, TMP, OLR, cleaning regimes, gas yields) on increased methane yields, net energy and market costs. Operationally useful models need to be developed to maximize energy output, while minimizing costs. In addition, we need to develop global optimization techniques of WW flowsheets for minimizing energy and impacts on the environment.

(5) *Sulphate control with micro-aeration* – high levels of sulphate in influents to anaerobic systems are problematic since they are reduced to sulphide which is partly stripped out in the gas phase. More research needs to be conducted on the micro-aeration of AD, and exploring any inhibitors to sulphate reduction.

(6) *Post-treatment for reuse* – in the future more water will need to be recycled, and AnMBR effluents will need to be treated to high standards. More research is required on removing N and P from effluents to recycle, and remove trace organics (PPCP) using ion exchange for PAC. In addition, we need to focus on viruses in effluents and how to remove them effectively.

## REFERENCES

Akyol C., Foglia A., Ozbayram E. G., Frison N., Katsou E., Eusebi A. L. and Fatone F. (2020). Validated innovative approaches for energy-efficient resource recovery and re-use from municipal wastewater. *Critical Reviews in Environmental Science and Technology*, **50**(9), 869–902, https://doi.org/10.1080/10643389.2019.1634456

Al-Khafaji S. S., Al-Rekabi W. S. and Mawat M. J. (2022). Apply membrane biological reactor (MBR) in industrial wastewater treatment: a mini review. *Eurasian Journal of Science Engineering and Technology*, **7**, 98–106.

Aquino S. F. and Stuckey D. C. (2003). Production of soluble microbial products (SMPs) in anaerobic chemostats under nutrient deficiency. *Journal of Environmental Engineering*, **129**, 1007–1014, https://doi.org/10.1061/(ASCE)0733-9372(2003)129:11(1007)

Aquino S. F. and Stuckey D. C. (2004a). Soluble microbial products formation in anaerobic chemostats in the presence of toxic compounds. *Water Research*, **38**, 255–266, https://doi.org/10.1016/j.watres.2003.09.031

Aquino S. F. and Stuckey D. C. (2004b). The effect of organic and hydraulic shock loads on the production of soluble microbial products in anaerobic digesters. *Water Environment Research*, **76**(6), 2628–2636, https://doi.org/10.1002/j.1554-7531.2004.tb00225.x

Aslam M., Charfi A., Lesage G., Heran M. and Kim J. (2017). Membrane bioreactors for wastewater treatment: A review of mechanical cleaning by scouring agents to control membrane fouling. *Chemical Engineering Journal*, **307**, 897–913, http://dx.doi.org/10.1016/j.cej.2016.08.144

Aydin S., Fakhri H. and Tavsanli N. (2023). Bioaugmentation of the green alga to enhance biogas production in an anaerobic hollow-fiber membrane bioreactor. *Biofouling*, **39**(4), 349–358, https://doi.org/10.1080/08927014.2023.2184689

Aziz A., Sengar A., Basheer F., Farooqi I. H. and Isa M. H. (2022). Anaerobic digestion in the elimination of antibiotics and antibiotic-resistant genes from the environment – a comprehensive review. *Journal of Environmental Chemical Engineering*, **10**(1), 106423, https://doi.org/10.1016/j.jece.2021.106423

Bachmann A., Beard V. L. and McCarty P. L. (1985). Performance characteristics of the anaerobic baffled reactor. *Water Research*, **19**(I), 99–106, https://doi.org/10.1016/0043-1354(85)90330-6

Bagheria M. and Mirbagheri S. A. (2018). Critical review of fouling mitigation strategies in membrane bioreactors treating water and wastewater. *Bioresource Technology*, **258**, 318–334, https://doi.org/10.1016/j.biortech.2018.03.026

Barker D. J. and Stuckey D. C. (1999). A review of soluble microbial products (SMP) in wastewater treatment systems. *Water Research*, **33**(14), 3063–3082, https://doi.org/10.1016/S0043-1354(99)00022-6

Boyle-Gotla A., Jensen P. D., Yap S. D., Pidou M., Wang Y. and Batstone D. J. (2014). Dynamic multidimensional modelling of submerged membrane bioreactor fouling. *Journal of Membrane Science*, **467**, 153–161, https://doi.org/10.1016/j.memsci.2014.05.028

Chang H., Liang H., Qu F., Liu B., Yu H., Du X., Li G. and Snyder S. A. (2017). Hydraulic backwashing for low-pressure membranes in drinking water treatment: a review. *Journal of Membrane Science*, **540**, 362–380, https://doi.org/10.1016/j.memsci.2017.06.077

Chen J. L., Ortiz R., Steele T. W. J. and Stuckey D. C. (2014). Toxicants inhibiting anaerobic digestion: a review. *Biotechnology Advances*, **32**(8), 1523–1534, https://doi.org/10.1016/j.biotechadv.2014.10.005

Chen J. L., Ortiz R., Xiao Y., Steele T. W. J. and Stuckey D. C. (2015a). Rapid fluorescence-based measurement of toxicity in anaerobic digestion. *Water Research*, **75**, 123–130, https://doi.org/10.1016/j.watres.2015.02.035

Chen J. L., Steele T. W. J. and Stuckey D. C. (2015b). Modeling and application of a rapid fluorescence-based assay for biotoxicity in anaerobic digestion. *Environmental Science & Technology*, **49**(22), 13463–13471, https://doi.org/10.1021/acs.est.5b03050

Chen J. L., Steele T. W. J. and Stuckey D. C. (2016). Stimulation and inhibition of anaerobic digestion by nickel and cobalt: a rapid assessment using the resazurin reduction assay. *Environmental Science & Technology*, **50**, 11154–11163, https://doi.org/10.1021/acs.est.6b03522

Chen R., Nie Y., Ji J., Utashiro T., Li Q., Komori D. and Li Y.-Y. (2017). Submerged anaerobic membrane bioreactor (SAnMBR) performance on sewage treatment: removal efficiencies, biogas production and membrane fouling. *Water Science and Technology*, **76**, 1308–1317, https://doi.org/10.2166/wst.2017.240

Cheng H., Hiro Y., Hojo T. and Li Y.-Y. (2018). Upgrading methane fermentation of food waste by using a hollow fiber type anaerobic membrane bioreactor. *Bioresource Technology*, **267**, 386–394, https://doi.org/10.1016/j.biortech.2018.07.045

Choo K. H. and Lee C. H. (1996). Membrane fouling mechanisms in the membrane-coupled anaerobic bioreactor. *Water Research*, **30**(8), 1771–1780, https://doi.org/10.1016/0043-1354(96)00053-X

Corominas L., Byrne D. M., Guest J. S., Hospido A., Roux P., Shaw A. and Short M. D. (2020). The application of life cycle assessment (LCA) to wastewater treatment: a best practice guide and critical review. *Water Research*, **184**, 116058, https://doi.org/10.1016/j.watres.2020.116058

Crone B. C., Garland J. L., Sorial G. A. and Vane L. M. (2016). Significance of dissolved methane in effluents of anaerobically treated low strength wastewater and potential for recovery as an energy product: a review. *Water Research*, **104**, 520–531, https://doi.org/10.1016/j.watres.2016.08.019

de Andrade L. H., dos Santos Mendes F. D., Espindola J. C. and Amaral M. C. S. (2014). Internal versus external submerged membrane bioreactor configurations for dairy wastewater treatment. *Desalination and Water Treatment*, **52**, 2920–2932, https://doi.org/10.1080/19443994.2013.799048

Do M. T. and Stuckey D. C. (2019). Fate and removal of ciprofloxacin in an anaerobic membrane bioreactor (AnMBR). *Bioresource Technology*, **289**, 121683, https://doi.org/10.1016/j.biortech.2019.121683

Fakhria H., Arabacı D. N., Unlu I. D., Yangin-Gomeca C., Oveza S. and Aydin S. (2021). Addition of *Trichocladium canadense* to an anaerobic membrane bioreactor: evaluation of the microbial composition and reactor performance. *Biofouling*, **37**(7), 711–723, https://doi.org/10.1080/08927014.2021.1949002

Ferrer J., Pretel R., Durán F., Giménez J. B., Robles A., Ruano M. V., Serralta J., Ribes J. and Seco A. (2015). Design methodology for submerged anaerobic membrane bioreactors (AnMBR): a case study. *Separation and Purification Technology*, **141**, 378–386, https://doi.org/10.1016/j.seppur.2014.12.018

Fox R. A. and Stuckey D. C. (2014). MS-2 and T4 phage removal in an anaerobic membrane bioreactor (AnMBR): effect of gas sparging rate. *Journal of Chemical Technology & Biotechnology*.

Fox R. A. and Stuckey D. C. (2015). The effect of sparging rate on transmembrane pressure and critical flux in an AnMBR. *Journal of Environmental Management*, **151**, 280–285, https://doi.org/10.1016/j.jenvman.2014.08.011

Giménez J. B., Robles A., Carretero L., Durán F., Ruano M. V., Gatti M. N., Ribes J., Ferrer J. and Seco A. (2011). Experimental study of the anaerobic urban wastewater treatment in a submerged hollow-fibre membrane bioreactor at pilot scale. *Bioresource Technology*, **102**, 8799–8806, https://doi.org/10.1016/j.biortech.2011.07.014

Grethlein H. E. (1978). Anaerobic digestion and membrane separation of domestic wastewater. *Water Pollution Control Federation*, **50**, 754–763.

Haddadi B., Jordan C., Miltner M. and Harasek M. (2018). Membrane modeling using CFD: combined evaluation of mass transfer and geometrical influences in 1D and 3D. *Journal of Membrane Science*, **563**, 199–209, https://doi.org/10.1016/j.memsci.2018.05.040

Hashisho J., El-Fadel M., Al-Hindi M., Salam D. and Alameddine I. (2016). Hollow fiber vs. flat sheet MBR for the treatment of high strength stabilized landfill leachate. *Waste Management*, **55**, 249–256, https://doi.org/10.1016/j.wasman.2015.12.016

Herrero M. and Stuckey D. C. (2015). Bioaugmentation and its application in wastewater treatment: a review. *Chemosphere*, **140**, 119–128, https://doi.org/10.1016/j.chemosphere.2014.10.033

Hu A. Y. and Stuckey D. C. (2006). Treatment of dilute wastewaters using a novel submerged anaerobic membrane bioreactor. *Journal of Environmental Engineering*, **132**(2), 190–198, https://doi.org/10.1061/(ASCE)0733-9372(2006)132:2(190)

Hu A. Y. and Stuckey D. C. (2007). Activated carbon addition to a submerged anaerobic membrane bioreactor (SAMBR): effect on performance, transmembrane pressure and flux. *Journal of Environmental Engineering*, **133**(1), 73–80, https://doi.org/10.1061/(ASCE)0733-9372(2007)133:1(73)

Hu Y., Cai X., Xue Y., Du R., Ji J., Chen R., Sano D. and Li Y.-Y. (2022). Recent developments of anaerobic membrane bioreactors for municipal wastewater treatment and bioenergy recovery: focusing on novel configurations and energy balance analysis. *Journal of Cleaner Production*, **356**, 131856, https://doi.org/10.1016/j.jclepro.2022.131856

Iorhemen O. T., Hamza R. A. and Tay J. H. (2016). Membrane bioreactor (MBR) technology for wastewater treatment and reclamation: membrane fouling. *Membranes*, **6**, 33, https://doi.org/10.3390/membranes6020033

Ji J., Sakuma S., Ni J., Chen Y., Hua Y., Ohtsu A., Chen R., Cheng H., Qin Y., Hojo T., Kubota K. and Li Y.-Y. (2020). Application of two anaerobic membrane bioreactors with different pore size membranes for municipal wastewater treatment. *Science of the Total Environment*, **745**, 140903, https://doi.org/10.1016/j.scitotenv.2020.140903

Kappell A. D., Lee K., Kimbell L. K., Seib M. D., Carey D. E., Choi M. J., Kalayil T., Fujimoto M., Zitomer D. H. and McNamara P. J. (2018). Removal of antibiotic resistance genes in an anaerobic membrane bioreactor treating primary clarifier effluent at 20°C. *Environmental Science: Water Research & Technology*, **4**, 1783–1793, https://doi.org/10.1039/C8EW00270C

Kim J., Kim K., Ye H., Lee E., Shin C., McCarty P. L. and Bae J. (2011). Anaerobic fluidized bed membrane bioreactor for wastewater treatment. *Environmental Science & Technology*, **45**, 576–581, https://doi.org/10.1021/es1027103

Koleva M. N., Liu S., Craig A., Styan C. A. and Papageorgiou L. G. (2016). Multi-objective optimisation approach for the synthesis of water treatment plants. *Computer Aided Chemical Engineering*, **38**, 2379–2384, https://doi.org/10.1016/B978-0-444-63428-3.50401-X

Kong Z., Li L., Wu J., Rong C., Wang T., Chen R., Sano D. and Li Y.-Y. (2022). Unveiling the characterization and development of prokaryotic community during the start-up and long-term operation of a pilot-scale anaerobic membrane bioreactor for the treatment of real municipal wastewater. *Science of the Total Environment*, **813**, 152643, https://doi.org/10.1016/j.scitotenv.2021.152643

Lade H., Paul D. and Kweon J. H. (2014). *N*-acyl homoserine lactone-mediated quorum sensing with special reference to use of quorum quenching bacteria in membrane biofouling control. *BioMed Research International*, **2014**, 1–25, https://doi.org/10.1155/2014/162584

Lee H. and Tan T. P. (2016). Singapore's experience with reclaimed water: NEWater. *International Journal of Water Resources Development*, **32**(4), 611–621, https://doi.org/10.1080/07900627.2015.1120188

Lim K., Evans P. J. and Parameswaran P. (2019). Long-term performance of a pilot-scale gas-sparged anaerobic membrane bioreactor under ambient temperatures for holistic wastewater treatment. *Environmental Science & Technology*, **53**(13), 7173–7174, https://doi.org/10.1021/acs.est.9b02917

Lim M., Patureau D., Heran M., Lesage G. and Kim J. (2020). Removal of organic micropollutants in anaerobic membrane bioreactors in wastewater treatment: critical review. *Environmental Science: Water Research & Technology*, **6**, 1230–1243, https://doi.org/10.1039/C9EW01058K

Lim Z. K., Liu T., Zheng M., Rattier M., Keller J., Yuan Z., Guo J. and Hu S. (2023). Membrane reciprocation as energy-efficient fouling control with high biogas recovery in a pilot-scale anaerobic membrane bioreactor. *Resources, Conservation and Recycling*, **190**, 106849, https://doi.org/10.1016/j.resconrec.2022.106849

Lv W., Zheng X., Yang M. Y., Zhang Y., Liu Y. and Liu J. (2006). Virus removal performance and mechanism of a submerged membrane bioreactor. *Process Biochemistry*, **41**, 299–304, https://doi.org/10.1016/j.procbio.2005.06.005

Mai D. T., Kunacheva C. and Stuckey D. C. (2018). A review of posttreatment technologies for anaerobic effluents for discharge and recycling of wastewater. *Critical Reviews in Environmental Science and Technology*, **48**(2), 167–209, https://doi.org/10.1080/10643389.2018.1443667.

Martínez R., Ruiz M. O., Ramos C., Cámara J. M. and Diez V. (2021). Fouling control of submerged and side-stream membrane bioreactors based on the statistical analysis of mid-term assays. *Journal of Cleaner Production*, **326**, 129336, https://doi.org/10.1016/j.jclepro.2021.129336

Mason P. M. and Stuckey D. C. (2016). Biofilms, bubbles and boundary layers – a new approach to understanding cellulolysis in anaerobic and ruminant digestion. *Water Research*, **104**, 93–100, https://doi.org/10.1016/j.watres.2016.07.063

McCarty P. L., Bae J. and Kim J. (2011). Domestic wastewater treatment as a net energy producer – can this be achieved? *Environmental Science & Technology*, **45**, 7100–7106, https://doi.org/10.1021/es2014264

Menashe O. and Kurzbaum E. (2016). A novel bioaugmentation treatment approach using a confined microbial environment: a case study in a membrane bioreactor wastewater treatment plant. *Environmental Technology*, **37**(12), 1582–1590, https://doi.org/10.1080/09593330.2015.1121293

Meng F. G., Zhang S., Oh Y., Zhou Z. B., Shin H. S. and Chae S. R. (2017). Fouling in membrane bioreactors: an updated review. *Water Research*, **114**, 151–180, https://doi.org/10.1016/j.watres.2017.02.006

Miller M. B. and Bassler B. L. (2001). Quorum sensing in bacteria. *Annual Review of Microbiology*, **55**, 165–199, https://doi.org/10.1146/annurev.micro.55.1.165

Muller E. B., Stouthamer A. H., Van Verseveld H. W. and Eikelboom D. H. (1995). Aerobic domestic waste water treatment in a pilot plant with complete sludge retention by cross-flow filtration. *Water Research*, **29**(4), 1179–1189, https://doi.org/10.1016/0043-1354(94)00267-B

Ni J., Ji J., Li Y.-Y. and Kubota K. (2022). Microbial characteristics in anaerobic membrane bioreactor treating domestic sewage: effects of HRT and process performance. *Journal of Environmental Sciences*, **111**, 392–399, https://doi.org/10.1016/j.jes.2021.04.022

Oh H.-S. and Lee C.-H. (2018). Origin and evolution of quorum quenching technology for biofouling control in MBRs for wastewater treatment. *Journal of Membrane Science*, **554**, 331–345, https://doi.org/10.1016/j.memsci.2018.03.019

Ortiz R., Chen J. L., Stuckey D. C. and Steele T. W. J. (2019). Rapid serial diluting biomicrofluidic provides $EC_{50}$ in minutes. *Micro and Nano Engineering*, **2**, 92–103, https://doi.org/10.1016/j.mne.2019.02.002

Owen W., Stuckey D. C., Healy J. B., Young L. Y. and McCarty P. L. (1979). Bioassay for monitoring biochemical methane potential and anaerobic toxicity. *Water Research*, **13**, 485–492, https://doi.org/10.1016/0043-1354(79)90043-5

Pauss A., Andre G., Perrier M. and Guiot S. R. (1990). Liquid-to-gas mass transfer in anaerobic processes: inevitable transfer limitations of methane and hydrogen in the biomethanation process. *Applied And Environmental Microbiology*, **56**(6), 1636–1644, https://doi.org/10.1128/aem.56.6.1636-1644.1990

Peña M., do Nascimento T., Gouveia J., Escudero J., Gómez A., Letona A., Arrieta J. and Fdz-Polanco F. (2019). Anaerobic submerged membrane bioreactor (AnSMBR) treating municipal wastewater at ambient temperature: operation and potential use for agricultural irrigation. *Bioresource Technology*, **282**, 285–293, https://doi.org/10.1016/j.biortech.2019.03.019

Pokorna-Krayzelova L., Mampaey K. E., Vannecke T. P. W., Bartacek J., Jenicek P., Eveline I. P. and Volcke E. I. P. (2017). Model-based optimization of micro-aeration for biogas desulfurization in UASB reactors. *Biochemical Engineering Journal*, **125**, 171–179, https://doi.org/10.1016/j.bej.2017.06.009

Pretel R., Robles A., Ruano M. V., Seco A. and Ferrer J. (2014). The operating cost of an anaerobic membrane bioreactor (AnMBR) treating sulphate-rich urban wastewater. *Separation and Purification Technology*, **126**, 30–38, https://doi.org/10.1016/j.seppur.2014.02.013

Pretel R., Shoener B. D., Ferrer J. and Guest J. S. (2015). Navigating environmental, economic, and technological trade-offs in the design and operation of submerged anaerobic membrane bioreactors (AnMBRs). *Water Research*, **87**, 531–541, https://doi.org/10.1016/j.watres.2015.07.002

Puchongkawarin C., Gomez-Mont C., Stuckey D. C. and Chachuat B. (2015). Optimization-based methodology for the development of wastewater facilities for energy and nutrient recovery. *Chemosphere*, **140**, 150–158, https://doi.org/10.1016/j.chemosphere.2014.08.061

Puspitasari V., Granville A., Le-Clech P. and Chen V. (2010). Cleaning and ageing effect of sodium hypochlorite on polyvinylidene fluoride (PVDF) membrane. *Separation and Purification Technology*, **72**, 301–308, https://doi.org/10.1016/j.seppur.2010.03.001

Rahman T. U., Roy H., Islam M. R., Tahmid M., Fariha A., Mazumder A., Tasnim N., Pervez M. N., Cai Y., Naddeo V. and Islam M. S. (2023). The advancement in membrane bioreactor (MBR) technology toward sustainable industrial wastewater management. *Membranes*, **13**, 181, https://doi.org/10.3390/membranes13020181

Rattier M., Jimenez J. A., Miller M. W., Dhanasekar A., Willis J., Keller J. and Batstone D. (2022). Long-term comparison of pilot UASB and AnMBR systems treating domestic sewage at ambient temperatures. *Journal of Environmental Chemical Engineering*, **10**, 108489, https://doi.org/10.1016/j.jece.2022.108489

Rong C., Wang T., Luo Z., Hu Y., Kong Z., Qin Y., Hanaoka T., Ito M., Kobayashi M. and Li Y-Y. (2022). Pilot plant demonstration of temperature impacts on the methanogenic performance and membrane fouling control of the anaerobic membrane bioreactor in treating real municipal wastewater. *Bioresource Technology*, **354**(2022), 127167, https://doi.org/10.1016/j.biortech.2022.127167

Semmens M. J. (2002). Personal communication.

Shang C., Wong H. W. and Chen G. (2005). Bacteriophage MS-2 removal by submerged membrane bioreactor. *Water Research*, **39**, 4211–4219, https://doi.org/10.1016/j.watres.2005.08.003

Sharma V. K., Johnson N., Cizmas L., McDonald T. J. and Kim H. (2016). A review of the influence of treatment strategies on antibiotic resistant bacteria and antibiotic resistance genes. *Chemosphere*, **150**, 702–714, https://doi.org/10.1016/j.chemosphere.2015.12.084

Shi X., Tal G., Hankins N. P. and Gitis V. (2014). Fouling and cleaning of ultrafiltration membranes: a review. *Journal of Water Process Engineering*, **1**, 121–138, https://doi.org/10.1016/j.jwpe.2014.04.003

Shin C. and Bae J. (2018). Current status of the pilot-scale anaerobic membrane bioreactor treatments of domestic wastewaters: a critical review. *Bioresource Technology*, **247**, 1038–1046, https://doi.org/10.1016/j.biortech.2017.09.002

Shin C., Tilmans S. H., Chen F., McCarty P. L. and Criddle C. S. (2021). Temperate climate energy-positive anaerobic secondary treatment of domestic wastewater at pilot-scale. *Water Research*, **204**, 117598, https://doi.org/10.1016/j.watres.2021.117598

Siddiqui M. F., Rzechowicz M., Winters H., Zularisam A. W. and Fane A. G. (2015). Quorum sensing based membrane biofouling control for water treatment: a review. *Journal of Water Process Engineering*, **7**, 112–122, https://doi.org/10.1016/j.jwpe.2015.06.003

Siddiqui M. A., Biswal B. K., Saleem M., Guan D., Iqbal A., Wu D., Khanal S. K. and Chen G. (2021). Anaerobic self-forming dynamic membrane bioreactors (AnSFDMBRs) for wastewater treatment – recent advances, process optimization and perspectives. *Bioresource Technology*, **332**, 125101, https://doi.org/10.1016/j.biortech.2021.125101

Silva A. F. R., Riccia B. C., Koch K., Weißbach M. and Amaral M. C. S. (2020). Dissolved hydrogen sulfide removal from anaerobic bioreactor permeate by modified direct contact membrane distillation. *Separation and Purification Technology*, **233**, 1160.

Sohn W., Guo W., Ngo H. H., Deng L., Cheng D. and Zhang X. (2021). A review on membrane fouling control in anaerobic membrane bioreactors by adding performance enhancers. *Journal of Water Process Engineering*, **40**, 101867, https://doi.org/10.1016/j.jwpe.2020.101867

Song X., McDonald J., Price W. E., Khan S. J., Hai F. I., Ngo H. H., Guo W. and Nghiem L. D. (2016). Effects of salinity build-up on the performance of an anaerobic membrane bioreactor regarding basic water quality parameters and removal of trace organic contaminants. *Bioresource Technology*, **216**, 399–405, https://doi.org/10.1016/j.biortech.2016.05.075

Song X., Luo W., McDonald J., Khan S. J., Hai F. I., Guo W., Ngo H. H. and Nghiem L. D. (2018). Effects of sulphur on the performance of an anaerobic membrane bioreactor: biological stability, trace organic contaminant removal, and membrane fouling. *Bioresource Technology*, **250**, 171–177, https://doi.org/10.1016/j.biortech.2017.11.021

Switzenbaum M. S. and Jewell W. J. (1980). Anaerobic attached-film expanded-bed reactor treatment. *Journal Water Pollution Control Federation*, **52**(7), 1953–1965.

Tao Y., Ersahin M. E., Ghasimi D. S. M., Ozgun H., Wang H., Zhang X., Guo M., Yang Y., Stuckey D. C. and van Lier J. B. (2020). Biogas productivity of anaerobic digestion process is governed by a core bacterial microbiota. *Chemical Engineering Journal*, **380**(2020), 122425, https://doi.org/10.1016/j.cej.2019.122425

Terán-Hilares R., Singh I., Tejada Meza K., Andrade C., G. J. and Pacheco Tanaka D. A. (2022). Alternative methods for cleaning membranes in water and wastewater treatment. *Water Environment Research*, **94**, e10708, https://doi.org/10.1002/wer.10708

Thanh P. B., Ketheesan B., Zhou Y. and Stuckey D. C. (2016). Trace metal speciation and bioavailability in anaerobic digestion: a review. *Biotechnology Advances*, **34**, 122–136, https://doi.org/10.1016/j.biotechadv.2015.12.006

Thanh P. M., Ketheesan K., Stuckey D. C. and Zhou Y. (2017a). Effects of trace metal deficiency and supplementation on a submerged anaerobic membrane bioreactor. *Bioresource Technology*, **241**, 161–170, https://doi.org/10.1016/j.biortech.2017.05.098

Thanh P. B., Ketheesan B., Zhou Y. and Stuckey D. C. (2017b). Effect of operating conditions on speciation and bioavailability of trace metals in submerged anaerobic membrane bioreactors. *Bioresource Technology*, **243**, 810–819, https://doi.org/10.1016/j.biortech.2017.07.040

Thanh P. B., Ketheesan B., Zhou Y. and Stuckey DC. . (2017c). Effect of ethylenediamine-*N,N'*-disuccinic acid (EDDS) on the speciation and bioavailability of $Fe^{2+}$ in the presence of sulfide in anaerobic digestion. *Bioresource Technology* **229**, 169–179, https://doi.org/10.1016/j.biortech.2016.12.113

Tian Z., Zhang Y., Yu B. and Yang M. (2016). Changes of resistome, mobilome and potential hosts of antibiotic resistance genes during the transformation of anaerobic digestion from mesophilic to thermophilic. *Water Research*, **98**, 261–269, https://doi.org/10.1016/j.watres.2016.04.031

Tolu H., Iren E. and Altinbas M. (2021). Full scale sanitary landfill leachate treatment by MBR: flat sheet vs. hollow fiber membrane. *Journal of Membrane Science and Research*, **7**, 118–124.

Tomczak W., Gryta M., Grubecki I. and Miłek J. (2023). Biogas production in AnMBRs via treatment of municipal and domestic wastewater: opportunities and fouling mitigation strategies. *Applied Sciences*, **13**, 6466, https://doi.org/10.3390/app13116466

Trzcinski A. P. and Stuckey D. C. (2016a). Inorganic fouling of an anaerobic membrane bioreactor treating leachate from the organic fraction of municipal solid waste (OFMSW) and a polishing aerobic membrane bioreactor. *Bioresource Technology*, **204**, 17–25, https://doi.org/10.1016/j.biortech.2015.12.074

Trzcinski A. P. and Stuckey D. C. (2016b). Effect of sparging rate on permeate quality in a submerged anaerobic membrane bioreactor (SAMBR) treating leachate from the organic fraction of municipal solid waste (OFMSW). *Journal of Environmental Management*, **168**, 67–73, https://doi.org/10.1016/j.jenvman.2015.11.055

Tsibranska I., Vlaev S. and Tylkowski B. (2018). The problem of fouling in submerged membrane bioreactors – model validation and experimental evidence. *Reviews in Physics*, **3**, 1–14, https://doi.org/10.1016/j.revip.2017.11.001

Verraes C., Van Boxstael S., Van Meervenne E., Van Coillie E., Butaye P., Catry B., de Schaetzen M. A., Van Huffel X., Imberechts H., Dierick K., Daube G., Saegerman C., De Block J., Dewulf J. and Herman L. (2013). Antimicrobial resistance in the food chain: a review. *International Journal of Environmental Research and Public Health*, **10**, 2643–2669, https://doi.org/10.3390/ijerph10072643

Verstraete W., Van de Caveye P. and Diamantis V. (2009). Maximum use of resources present in domestic 'used water'. *Bioresource Technology*, **100**, 5537–5545, https://doi.org/10.1016/j.biortech.2009.05.047

Vinardell S., Astals S., Peces M., Cardete M. A., Fernandez I., Mata-Alvarez J. and Dosta J. (2020). Advances in anaerobic membrane bioreactor technology for municipal wastewater treatment: A 2020 updated review. *Renewable and Sustainable Energy Reviews*, **130**, 109936, https://doi.org/10.1016/j.rser.2020.109936

Vinardell S., Astals S., Jaramillo M., Mata-Alvarez J. and Dosta J. (2021). Anaerobic membrane bioreactor performance at different wastewater pre-concentration factors: an experimental and economic study. *Science of the Total Environment*, **750**, 141625, https://doi.org/10.1016/j.scitotenv.2020.141625

Vlaev S. D., Dzhonova-Atanasova D. and Tsibranska I. (2020). CFD evaluation of mass transfer distribution heterogeneity along the membrane-liquid interface in stirred submerged membrane bioreactors. *Chemical Engineering and Processing – Process Intensification*, **147**, 107738, https://doi.org/10.1016/j.cep.2019.107738

Vu H. P., Nguyen L. N., Wang Q., Ngo H. H., Liu Q., Zhang X. and Nghiem L. D. (2022). Hydrogen sulphide management in anaerobic digestion: a critical review on input control, process regulation, and post-treatment. *Bioresource Technology*, **346**, 126634, https://doi.org/10.1016/j.biortech.2021.126634

Vyrides I. and Stuckey D. C. (2009a). Adaptation of anaerobic biomass to saline conditions: role of compatible solutes and extracellular polysaccharides. *Enzyme and Microbial Technology*, **44**, 46–51, https://doi.org/10.1016/j.enzmictec.2008.09.008

Vyrides I. and Stuckey D. C. (2009b). Saline sewage treatment using a submerged anaerobic membrane reactor (SAMBR): effects of activated carbon addition and biogas-sparging time. *Water Research*, **43**, 933–942, https://doi.org/10.1016/j.watres.2008.11.054

Vyrides I., Santos H., Mingote A., Ray M. J. and Stuckey D. C. (2010). Are compatible solutes compatible with biological treatment of saline wastewater? Batch and continuous studies using submerged anaerobic membrane bioreactors (SAMBRs). *Environmental Science & Technology*, **44**, 7437–7442, https://doi.org/10.1021/es903981k

Vyrides I. and Stuckey D. C. (2011). Fouling cake layer in a submerged anaerobic membrane bioreactor treating saline wastewaters: curse or a blessing? *Water Science and Technology*, **63**(12), 2902–2908, https://doi.org/10.2166/wst.2011.461

Vyrides I. and Stuckey D. C. (2017). Compatible solute addition to biological systems treating waste/wastewater to counteract osmotic and other environmental stresses: a review. *Critical Reviews in Biotechnology*, **37**(7), 865–879, https://doi.org/10.1080/07388551.2016.1266460

Wang K. M., Garcia N. M., Soares A., Jefferson B. and McAdam E. J. (2018). Comparison of fouling between aerobic and anaerobic MBR treating municipal wastewater. *H2Open Journal*, **1**(2), 131–159, https://doi.org/10.2166/h2oj.2018.109

Waters C. M. and Bassler B. L. (2005). Quorum sensing: cell-to-cell communication in bacteria. *Annual Review of Cell and Developmental Biology*, **21**, 319–346, https://doi.org/10.1146/annurev.cellbio.21.012704.131001

White C. J. and Stuckey D. C. (2000). The influence of metal ion addition on the anaerobic treatment of high strength, soluble wastewaters. *Environmental Technology*, **21**(11), 1283–1292, https://doi.org/10.1080/09593332108618157

Wibisono Y., Cornelissen E. R., Kemperman A. J. B., van der Meer W. G. J. and Nijmeijer K. (2014). Two-phase flow in membrane processes: a technology with a future. *Journal of Membrane Science*, **453**, 566–602, https://doi.org/10.1016/j.memsci.2013.10.072

Wigginton N. S. (2009). Cutting off communication. *Science*, **326**, 19–19, https://doi.org/10.1126/science.326_19a

Xiao Y., Yaohari H., De Araujo C., Sze C. C. and Stuckey D. C. (2017). Removal of selected pharmaceuticals in an anaerobic membrane bioreactor (AnMBR) with/without powdered activated carbon (PAC). *Chemical Engineering Journal*, **321**, 335–345, https://doi.org/10.1016/j.cej.2017.03.118

Xiong Y., Harb M. and Hong P.-Y. (2016). Characterization of biofoulants illustrates different membrane fouling mechanisms for aerobic and anaerobic membrane bioreactors. *Separation and Purification Technology*, **157**, 192–202, https://doi.org/10.1016/j.seppur.2015.11.024

Yao Y., Zhou Z., Stuckey D. C. and Meng F. (2020). Micro-particles, a neglected but critical cause of different membrane fouling between aerobic and anaerobic membrane bioreactors. *ACS Sustainable Chemistry & Engineering*, **8**, 16680–16690, https://doi.org/10.1021/acssuschemeng.0c06502

Yeon K. M., Cheong W. S., Oh H. S., Lee W. N., Hwang B. K., Lee C. H., Beyenal H. and Lewandowski Z. (2009a). Quorum sensing: a new biofouling control paradigm in a membrane bioreactor for advanced wastewater treatment. *Environmental Science & Technology*, **43**, 380–385, https://doi.org/10.1021/es8019275

Yeon K. M., Lee C. H. and Kim J. (2009b). Magnetic enzyme carrier for effective biofouling control in the membrane bioreactor based on enzymatic quorum quenching. *Environmental Science & Technology*, **43**, 7403–7409, https://doi.org/10.1021/es901323k

Young J. C. and McCarty P. L. (1969). Anaerobic filter for waste treatment. *Journal Water Pollution Control Federation*, **41**(5), R160–R173.

Zhang Q., Tan G. H. (Victor) and Stuckey D. C. (2017a). Optimal biogas sparging strategy, and the correlation between sludge and fouling layer properties in a submerged anaerobic membrane bioreactor (SAnMBR). *Chemical Engineering Journal* **319**, 248–257, https://doi.org/10.1016/j.cej.2017.02.146

Zhang Q., Singh S. and Stuckey D. C. (2017b). Fouling reduction using adsorbents/flocculants in a submerged anaerobic membrane bioreactor. *Bioresource Technology*, **239**, 226–235, https://doi.org/10.1016/j.biortech.2017.05.022

Zhang Q. Q., Yang G.-F., Zhang L., Zhang Z.-Z., Tian G.-M. and Jin R.-C. (2017c). Bioaugmentation as a useful strategy for performance enhancement in biological wastewater treatment undergoing different stresses: application and mechanisms. *Critical Reviews in Environmental Science and Technology*, **47**(19), 1877–1899, https://doi.org/10.1080/10643389.2017.1400851

Zhou Z., Tan Y., Xiao Y. and Stuckey D. C. (2016). Characterization and significance of sub-visible particles and colloids in a submerged anaerobic membrane bioreactor (SAnMBR) environ. *Environmental Science & Technology*, **50**(23), 12750–12758, https://doi.org/10.1021/acs.est.6b03581

Zhou Z., Tao Y., Zhang S., Xiao Y., Meng F. and Stuckey D. C. (2019). Size-dependent microbial diversity of sub-visible particles in a submerged anaerobic membrane bioreactor (SAnMBR): implications for membrane fouling. *Water Research*, **159**, 20–29, https://doi.org/10.1016/j.watres.2019.04.050

doi: 10.2166/9781789063479_0093

# Chapter 4

# Immobilized and granular biomass systems

M. C. Tomei[1]*, M. De Sanctis[2], V. G. Altieri[2], C. Di Iaconi[2] and V. Stazi[1]

[1]Water Research Institute (IRSA), National Research Council (CNR), Roma, Italy
[2]Water Research Institute (IRSA), National Research Council (CNR), Bari, Italy
*Correspondence: concetta.tomei@irsa.cnr.it

**ABSTRACT**

In several geographic areas, domestic wastewater (DWW) is characterized by low temperatures (below 25°C), which enables a stable and effective operation of an anaerobic process, as mainstream treatment option at ambient temperature, quite challenging. Low temperatures reduce the hydrolysis rate of particulate organic matter with consequent reduction of biogas production and energy recovery. In addition, the low growth rate of the anaerobic microorganisms can cause poor effluent quality and risk of biomass washout in case of suspended biomass systems. Increasing biomass concentration in the bioreactor has the beneficial effect of increasing biodegradation kinetics thus allowing the reduction of the reactor volumes and footprint thus making the anaerobic process competitive with the conventional activated sludge process. This operating condition can be achieved with different strategies including biomass granulation and biomass immobilization on inert support media. For both alternatives, a number of studies have been conducted, with different plant configurations, operating the anaerobic process under variable temperature regimes or in isothermal mode in psychrophilic or mesophilic conditions. In this chapter, examples of application to demonstrate the feasibility of immobilized and granular biomass systems for direct anaerobic treatment of DWW are reported and discussed. In the first case, the effect of feeding strategy and the suitability of natural and synthetic filling materials were investigated while, in the second, the effect of temperature and of the hydraulic retention time have been tested in order to maximize the process performance and evaluate the potentialities of the anaerobic process in comparison to the aerobic one.

**Keywords:** anaerobic biofilter, anaerobic sequencing batch biofilm reactor, granular biomass, granulation modelling, immobilized biomass, packing material, sequencing batch reactor, upflow anaerobic filter.

## 4.1 WHY HIGH BIOMASS CONCENTRATION FOR A BETTER PERFORMANCE?

Kinetics of biological processes for chemical oxygen demand (COD) removal is generally expressed, in a simplified form, by the Monod kinetics:

$$\mu = \mu_{\max} X \frac{S}{K_S + S}$$

where $\mu$ is the biomass growth rate; $\mu_{max}$ is the maximum biomass growth rate; $X$ is the biomass concentration, $K_s$ is the half-saturation constant and $S$ is the substrate concentration. Anaerobic biomass is generally characterized by lower $\mu_{max}$ and higher $K_s$ values with respect to the aerobic one, so operating at high biomass concentration, especially for diluted wastewater is mandatory to reach process kinetics comparable to the conventional aerobic processes and operation volumes and footprint suitable for practical application. In suspended biomass systems, increase of biomass concentration is limited by the need of achieving an efficient solid–liquid separation to guarantee the effluent quality in terms of suspended solids content. In addition, anaerobic sludge is characterized by scarce settleability (Akil & Jayanthi, 2012), which makes the solid–liquid separation difficult.

To face the above-mentioned drawbacks, the two main operational strategies to retain high concentrated biomass amounts inside anaerobic bioreactors consist of the attached biomass growth realized through the development of biofilm onto inert support surface, and the operation with high-density granular biomass (Van Lier *et al.*, 2015) where the carrier is the own biomass.

The first option is applied in the so-called immobilized biomass systems where support media such as glass, peat, powdered minerals, and a variety of plastic materials are employed in bioreactors operated in fixed and fluidized-bed configurations under different operating conditions. In the second alternative, the biomass is 'concentrated' in granules obtained with specific start-up procedures whose controlling factors, including biotic and abiotic parameters, are still under investigation.

In this chapter, theoretical and practical aspects of the two systems are presented and discussed. An overview of the technologies will provide a picture of the present state of knowledge and to fill gaps for their upscaling.

## 4.2 IMMOBILIZED BIOMASS SYSTEMS

### 4.2.1 Principle of operation

Anaerobic treatment of diluted wastewater, such as domestic wastewater (DWW), is common in tropical and sub-tropical areas due to the warm environmental condition. The latter allows wastewater reaching centralized wastewater treatment plants (WWTPs) at a temperature usually above 20°C (Chernicharo *et al.*, 2009; Foresti, 2001). At this temperature anaerobic treatment can provide good removal performance of the organic load. Conversely, in temperate climate zones, sewage temperature is below 15°C from autumn to spring, even approaching temperature of few °C (Zhou *et al.*, 2018). In these climatic conditions, given the low concentration of organic matter in DWW, with COD values usually close or below 1 kg/m$^3$, and the specific heat of water (4.186 J/g/°C), the generated methane is far to balance the energy required for warming wastewater. Moreover, operating the process at low temperature strongly reduces the hydrolysis and methanogenesis rates. The consequent limited biogas production, negatively affect the sludge mixing causing a poor contact substrate-biomass, as reported for upflow anaerobic sludge blanket (UASB) reactors in Serrano León *et al.* (2018). In addition, the reduction of the growth rate of microorganisms and the typical poor settling properties of the flocculent anaerobic sludge can expose the digester to the risk of biomass washout (Akil & Jayanthi, 2012; Lettinga *et al.*, 2001). To overcome these drawbacks, a possible technological alternative is the uncoupling of the hydraulic retention time (HRT) and the sludge retention time (SRT) achieved with high biomass concentration in the bioreactor, as discussed in the previous chapters (1–3) of this book.

A strategy to effectively retain the anaerobic biomass consists in the use of a proper inert material as support for biomass growth in attached mode. One of the first studies published in this field was carried out by Young and McCarty in 1969. The packing material can be either fixed or suspended in the reactor, but the fixed biofilm systems are usually more effective in retaining microorganisms. In these systems, the packing material is used to fill the bioreactor and acts a filter, hence limiting the washout of biomass and particulate matter. Anaerobic microorganisms are not only immobilized in the form of biofilms adhered to the support material, but also in the form of occluded biomass in the voids. SRT and HRT are no more 'coupled' and the biomass concentration can be increased.

As already pointed out, the increase in biomass concentration allows an increase of the influent flow rate (i.e., a reduction of the HRT) which would reflect in a higher biogas production.

However, one drawback of fixed-bed biofilm reactors is that biomass growth and detachment are not balanced and it could happen that the voids among media and the pores inside the media can clog because of biofilm overgrowth (Loupasaki & Diamadopoulos, 2012). The reduction of void degree and of the bed porosity negatively affects substrate diffusion and the contact substrate-biomass promoting the apparition of dead volumes or preferential flow circulation in the system, so causing a reduction in the bioreactor performance. Therefore, fixed biofilm reactors need periodic washing operations in order to remove the excess sludge from biomass growth and restore the optimal bed porosity and void degree.

The selection of an adequate carrier material was one of the key points in many of the published studies. Several materials have been investigated for their suitability as packing material, including glass, peat, powdered minerals, natural zeolite, expanded clay, fibrous carriers, ceramic, polyurethane foam cubes, and a variety of plastic materials (Li & Lu, 2017; Loupasaki & Diamadopoulos, 2012). The appropriate texture and porosity of the support material allow microbial populations to grow attached onto the surface as biofilm and colonize the interstitial spaces as suspended biomass (Kaetzl *et al.*, 2018). Moreover, biofilm formation led to a high production of extracellular polymeric substances (EPS) by the microbial community, which also plays a role in the biomass settleability.

### 4.2.2 Overview of the technologies

Immobilized biomass reactors include a number of technologies differentiating for plant design, filling material and operating conditions. This section provides an overview of plant configurations available in the literature by including a summary of relevant case studies for each technology. Performance data are summarized in Table 4.1 that reports the main investigated operating conditions. It is worth noting that the applications are mostly at the laboratory scale and, to a minor extent, at the pilot scale. Upscaling of the proposed technologies is still hindered by their greater complexity and lower stability with respect to more verified technologies as UASB reactors operated with suspended biomass, whose easy management has been also demonstrated at full scale. Another critical aspect is represented by the costs (increased by additional cost of filling material), which, at present, make this solution feasible only for high potentiality WWTPs. In any case, immobilized biomass systems have potentialities of performance improvements, which merit to be investigated as it is highlighted in the subsequent sections analysing the evolution and the results achieved so far for the different bioreactor configurations.

#### *4.2.2.1 Upflow anaerobic fixed-film reactor*

One of the pioneering studies on anaerobic immobilized biomass reactor for sewage treatment was conducted by Noyola *et al.* (1988). The authors studied the potential of a new design upflow anaerobic fixed-film reactor (an advanced scheme for the period) in treating primary settled DWWs under different operating conditions. The reactor consisted of a polyvinyl chloride cylinder (height 1 m, internal diameter 0.19 m, volume 28.4 L, void volume 24 L), packed with 112 circular units of two different types of support material (stationary and rotating, respectively), alternatively placed along a central shaft. The total support surface was 4.84 $m^2$, with a specific surface area of these circular elements of 170 $m^2/m^3$. The rotating units were moved by an electrical motor, placed at the top of the reactor. The reactor was fed upwards with raw DWW coming from a small extended aeration treatment plant. The effluent was discharged from the top; the collected gas flow rate was measured with a gas meter. Several parameters (HRT, temperature and mixing speed) were tested. The investigated temperature was in the range of 16°–29°C to simulate DWW temperatures. The results obtained from Noyola *et al.* (1988) showed that at 29°C, 6 h-HRT, and organic loading rate (OLR) of 1.7 kg COD/$m^3$/d, the COD removal efficiency reached 65% and it was not affected by the mixing. In order to investigate the mixing effect at lower temperatures the bioreactor was operated

**Table 4.1** Summary of the performance of the described technologies and main key operating parameters.

| Technology | Volume (L) - Scale | Packing Media | HRT (h) | T (°C) | OLR (kgCOD/$m^3$/d) | COD Removal (%) | Biogas-Methane Yield (in Gas Phase) | References |
|---|---|---|---|---|---|---|---|---|
| Up-flow rotating-stationary fixed film reactor | 28.4 - Lab | PVC saucer-like | 6 | 16–29 | 1.7 | 65% | 0.04–0.11 $m^3$ biogas/kg $COD_{removed}$ | Noyola *et al.* (1988) |
| UAF | 1.5 - Lab | Plastic material (insulating tubes) | 6–20 | 8–20 | 0.3–1.2 | 46–92 | Not specified | Bodík *et al.* (2002) |
| UAF | 1.5 - Lab | Corrugated plastic rings | 10–17 | 15–17 | 0.99–1.71 | 80 | 0.15 $m^3$ $CH_4$/kg $COD_{removed}$ | Martin *et al.* (2010) |
| UAF | 217 - Pilot | Gravel, ceramsite | 72–96 | 19–23 | 0.1–0.5 | 72–83 | Not specified | Yang *et al.* (2017) |
| Three-stage AWFR | 3 units of 50 L each - Pilot | Wool-felt | 24–72 | 10–31 | 0.075–0.350 | 52–76 | 43.5–52.5% of the COD supplied | Li and Lu (2017) |
| Fixed-bed reactor | 1.8 - Lab | Waste tyre rubber | 8–96 | 30–35 | 0.2–2.8 | 50% | Not reported | Reyes *et al.* (1999) |
| AF | 170 - Pilot | Burnt brickbats | 12 | Not specified | 0.1–2.8 | 91 | $\leq$ 0.35 $m^3$ $CH_4$/kg $COD_{removed}$ | Bodkhe (2008) |
| ASBBR | 0.5 - Lab | Polyurethane foam cubes | 40 | 30 | 0.29 | 86% | Not specified | Ratusznei *et al.* (2000) |
| ASBBR | 1200 - Pilot | Polyurethane foam cubes | 8 | 28–29 | 0.2–1.2 | 60–80% | Not specified | (Sarti *et al.* 2006, 2007) |
| ASBBR | 1.0 - Lab | GAC and natural zeolite | 60 | 35 | 0.4 | 93% | 0.61–0.75 $m^3$ biogas/kg $COD_{removed}$ | Dutta *et al.* (2014) |
| AnSBio | 28 - Lab | Plastic material | 45 | 14–35 | 0.4–0.7 | 82% | 0.12–0.32 $m^3$ $CH_4$/kg $COD_{removed}$ | Pentassuglia *et al.* (2020) |
| UASB/AF | 60–88 Pilot | Sheets of reticulated polyurethane foam | 4 | 13 | 3.1 | 55% | Not specified | Elmitwalli *et al.* (2002b) |
| CABR | 21.6 - Lab | Bamboo hollow spheres | 18–48 | 28 | 0.15–0.4 | 69–79 | Not specified | Feng *et al.* (2008) |
| AnMBR with attached biomass | 2.0 - Lab | Cylindrical ceramic packing medium | Not specified | 35 | 0.72 | 75–85% | 0.23 $m^3$ $CH_4$/kg $COD_{removed}$ | Harb *et al.* (2015) |

*Legend:* AF – anaerobic filter; AFBR – anaerobic fluidized-bed reactors; AnMBR – anaerobic membrane bioreactor; AnSBio – anaerobic sequencing biofilter; ASBBR – anaerobic sequencing batch biofilm reactor; AWFR – anaerobic wool-felt filter reactor; CABR – carrier anaerobic baffled reactor; Lab – laboratory scale; UAF – upflow anaerobic filter; UASB – upflow anaerobic sludge blanket.

at 12-h-HRT and 16°C without mixing, a COD removal efficiency of ~42% was observed, while with mixing (5 rpm) the removal efficiency increased to 60%. This finding suggested that the limiting step in static fixed-film reactors at low temperatures was represented as the substrate diffusion. The mixing also affected the total suspended solids (TSS) concentration in the effluent: when a gentle mixing was applied, the TSS content in the effluent decreased, while a marked increase was observed when a more vigorous mixing (≥15 rpm) was applied. Concerning biogas production, the variability of the operating conditions produced fluctuations in the volume of biogas: at 29°C the production ranged from 0.52 L/d (24-h-HRT) to 4.72 L/d (4-h-HRT); operating at 12-h-HRT, and mixing (5 rpm) similar biogas production have been achieved even at different temperatures (0.99 L/d at 29°C vs 0.93 L/d at 16°C). On the contrary, a marked decrease to 0.66 L/d was observed at 16°C in the operation without mixing. In terms of methane concentration in the biogas, operating at 29°C the $CH_4$ percentage ranged from 65% (at a 4-h-HRT) to 84% (at a 24-h-HRT), while at 16°C and 5 rpm it was 75% on average. The authors concluded that anaerobic treatment of sewage seemed a not sustainable approach as a relevant fraction of $CH_4$ was lost in the effluent (in the dissolved form), and the overall methane yield was far from the theoretical value (i.e., 0.35 $m^3/kgCOD_{removed}$).

#### 4.2.2.2 Upflow anaerobic biofilter

Bodík *et al.* (2002) investigated the effect of the temperature and HRT on the performance of a lab-scale upflow anaerobic biofilter (UAF). The filter (void volume after filling with support media 1.32 L) consisted of an acrylic column (height 50 cm, inside diameter 7 cm), filled with plastic material (pieces of insulating tubes of diameter 2 cm and of length 1.5–2 cm). The bioreactor was fed with raw wastewater coming from a Slovakian municipal WWTP, added with 0.2 g/L of glucose and 0.2 g/L of sodium acetate tri-hydrate as carbon sources given the quite low organic content of the selected wastewater (often ranging from 100 to 250 mg/L of COD). Experiments were conducted for approximately 20 months at three temperatures (8°C, 15°C, 24°C) and three HRTs (6, 10, 20 h). The best results were obtained at 24°C and HRT of 10 and 20 h. Average COD removal was in the range of 90–92% and average BOD and TSS removals were ~95%. At a 6-h-HRT, COD, BOD and TSS removal efficiencies slightly decreased to 77%, 81%, and 81%, respectively. Operating at lower temperature, the UAF performance decreased: poor results were observed at 8°C, with a 10-h-HRT: in these conditions, the average COD and BOD removal efficiencies were lower than 50% (46% and 41%, respectively), while TSS removal efficiency (80%) appeared to be less affected by the operating conditions. At 15°C and 24°C, there were no substantial differences in COD removal by comparing 10-h and 20-h HRT.

Another study on a lab-scale UAF was carried out by Martin *et al.* (2010): the bioreactor was fed with synthetic DWW and operated at psychrophilic temperatures (15–17°C). The anaerobic filter (AF) was composed of an acrylic column (total volume 1.5 L, working volume 1.35 L, height 45 cm, internal diameter 8 cm), packed with corrugated plastic rings (individual diameter 1.5 cm, height 0.5 cm), which acted as a support media. The reactor was operated at different HRTs (from 3.2 to 17.1 h) and organic loading rates (OLRs) ranging from 1 to 4 g COD/L/d. The influent COD and TSS were 705 and 560 mg/L, respectively. During the two months start-up phase the OLR was gradually increased. The UAF showed quite stable performance in terms of COD removal. Indeed, for HRTs ranging from 10 to 17.1 h, COD removal efficiency (~80%) was practically constant and not affected by HRT. For lower HRTs (in the range of 3–7 h), the COD removal decreased to 77% and to 65%. Average methane production was 0.15 NL $CH_4/g\ COD_{removed}$. Data obtained by Martin *et al.* (2010) for COD removal are comparable with those reported by Bodík *et al.* (2002) for an UAF, packed with plastic material and operating at a 10-h-HRT and at 15°C. It is worth noting that COD removal efficiencies of 80% are higher than the ones reported by Kobayashi *et al.* (1983) (i.e., 73–77%) for an UAF packed with PVC tubes and operating at higher HRTs and temperatures.

Yang *et al.* (2017) investigated the performance of UAFs, for different filter media, in treating several wastewaters, from high-strength synthetic DWW to secondary effluent. They employed a bioreactor consisting of a circular column (height 1.2 m, internal diameter 0.48 m, efficient volume

120 L), with a perforated plate at the bottom. The column was equipped with a packing layer (height 0.65 m) at the bottom, and with a clay layer (height 0.20 m) at the top, whose aim was the insulation and the prevention of odour escapes. The wastewater was fed upflow, flowing across the perforated plate, with the biomass growing on the fixed-bed surface. In the first part of the study, two UAFs, operating at 19–23°C and packed with ceramsite (an artificial ceramic sand) or gravel as filter media, were fed with high-strength wastewater (average influent concentrations: COD 1500 mg/L, BOD 900 mg/L, N-$NH_4^+$ 110 mg/L; OLR up to 0.075 kg COD/m$^3$/d) and achieved high COD removal efficiency (which overcame 70% either for ceramsite and for gravel biofilter). Efficiencies decreased for both reactors (28.5% for ceramsite biofilter, 35% for gravel biofilter) once OLR was increased to 0.5 kg COD/m$^3$/d. The performance failure was probably due to volatile acid accumulation and consequent pH decrease, because the dosage of $Na_2CO_3$ (at concentrations ranging from 50 to 200 mg/L) in the influent resulted in a significant improvement of the performance (up to 83% for ceramsite biofilter and to 82.1% for gravel biofilter, respectively). Similar trends were observed also for BOD. In the same study, the authors tested UAFs bioreactors fed with a secondary effluent and operated with different filter media. These included gravel, cinder, ceramsite, concrete block and brick, each of them with peculiar features but characterized by small particle size and high specific surface area (gravel: particle size 2–4 cm; cinder: average density kg/m$^3$, lower strength, larger porosity; ceramsite: average particle size 5 mm, average density 1500 kg/m$^3$, specific surface area 4.99 cm$^2$/g; concrete block: average density 2890 kg/m$^3$, blocky of 2–4 cm; brick: light weight, porous). Five different UAFs were fed with secondary effluent (average influent concentrations: COD 86.2 mg/L, BOD 14.6 mg/L). After the steady-state conditions were attained, different hydraulic loading rates were tested (ranging from 0.077 to 0.270 m$^3$/m$^2$/d). The COD average removal efficiencies of the five biofilters were lower than 60% after 120 days of operation. The highest COD removal efficiency was reported for gravel biofilter (up to 52.8%), followed by cinder biofilter (51%), while the lowest ones were reported for concrete block and brick biofilters (33.2% and 28.6%, respectively). Ceramsite biofilter showed intermediate COD removal efficiencies (44.8%). BOD removal efficiencies were generally higher than those reported for COD and gravel biofilter showed the highest removal (up to 83.9%). Also for BOD, concrete block and brick exhibited the lowest performance (65.3% and 53.5%, respectively), while ceramsite and cinder displayed intermediate results (76.3% and 74.4%, respectively). To sum up, the results from Yang *et al.* (2017) highlighted that UAFs could be a feasible technology to remove the main 'gross parameters' either from high-strength or low-strength wastewater, and can be potentially employed also as a polishing stage.

Another bioreactor operated as UAF was proposed by Li and Lu (2017). They investigated the performance of a pilot-scale three-stage anaerobic wool-felt filter reactor (AWFR). The system consisted of three identical PVC columns (height 2.5 m, internal diameter 170 mm, effective volume 50 L), packed with vertical wool felt carrier as a support media for the biomass growth. The support material is characterized by a high specific surface area (~950 m$^2$/g) and porosity (>95%). The AWFR unit was fed with decentralized DWW, coming from the dormitories and restaurants of a Chinese university. Wastewater was fed upflow, from the bottom of the first AWFR column, while the effluent was discharged from the third AWFR column. In each column, starting from the bottom, there was an influent buffer zone, the wool-felt filler layer, an effluent buffer zone, and a gas chamber, from which biogas was sent to a wet gas meter. The experimental campaign, which lasted one year, was divided into five periods (the start-up phase, plus the four seasons), in order to assess the impact of the seasonal variations on the AWFR treatment performance, either in terms of COD removal efficiency and biogas production. During the study, different influent COD concentrations (fluctuating in the range 157–469 mg/L), as well as different HRT (from 1 to 3 days) and seasonal temperatures (approximately between 10 and 31°C), were tested. From spring to summer, the AWFR showed improving results, despite the HRT shortening (1-d-HRT in summer vs 2-d-HRT in spring): the average COD removal efficiency obtained during summer was 76% (in spring an average COD removal efficiency of 61% was reported), probably due to an enhanced microbial activity. During autumn and winter, the HRT was readjusted at 2 and 3 d, respectively, in order to obtain average COD removal efficiencies at least of 50%: despite a lower organic load during winter, the seasonal temperature decrease affected the AWFR performance significantly, determining a minimum of 52% for COD removal efficiency. These outcomes were better than those reported by Gouveia *et al.* (2015) for a two-stage anaerobic filter used to treat DWW, packed with elastic fibre carriers. A possible explanation relies on the role played by the support material, as the wool felt carriers might ensure an enhanced agglomeration (and consequent increasing concentration) of the biomass in the AWFR system. With reference to the biogas production, the AWFR showed seasonal trends quite similar to those reported for COD removal efficiencies. Notably, a maximum of biogas production (10.7 L/d) was reported during the summer, at a 1-d-HRT: this finding was coherent with the high COD removal efficiency reported during the same period. Conversely, the worst performance was observed during winter, due to the lower degradation rate of organic substrate and the higher solubility of methane at lower temperatures.

#### 4.2.2.3 Anaerobic fixed-bed reactor using waste as filling medium

Given the interdependence among the properties of the support material, the biomass retention and the treatment performance, the selection of filling the material to employ in an anaerobic reactor is a critical issue. A different approach in choosing the filling medium to foster the growth of immobilized biomass included waste materials as a viable option. Reyes *et al.* (1999) assessed the performance of a lab-scale anaerobic fixed-bed reactor, whose peculiarity was the presence of waste tyre rubber as filling material, in treating a low-strength wastewater. The system was composed of five sequential units interconnected, operating alternately (the first, the third and the fifth unit working upflow, while the remaining ones working downflow). Each unit had the same size (total volume 1.8 L, free volume 1.2 L, a bottom space of 0.3 L for sludge accumulation) and a bed composed of pieces of waste tyre rubber (unit volume 1 $cm^3$, bed porosity 66%, specific surface area 5 $m^2/m^3$, total surface area 0.015 $m^2$). The upper part of each unit was connected to a biogas collecting bag. Five different HRTs were tested (4, 2, 1 d, and 12 and 8 h), with each experiment lasting 35 days. The system was fed with piggery waste, diluted to obtain a COD concentration of approximately 1000 mg/L and a suspended solids concentration of about 350 mg/L, similar to that observed for municipal wastewater. Similar results were obtained with HRT of 2 and 4 d, with good removal efficiencies of 70%, 80%, and 90% for COD, BOD, and TSS respectively, a slight decrease in the removal efficiency of the organic matter was observed when HRT was diminished to 1 d, with results $\geq$ 60% and 70% for COD and BOD, respectively. Analysing the liquid phase coming from each treatment unit, an interesting consideration related to the proposed scheme was also made by the authors. Specifically, operating at medium–high HRTs (1, 2 and 4 d), the first unit was able to reach the final removal, while for shorter HRTs (8 and 12 h), at least the first three stages of the scheme were necessary to achieve the final COD removal, with the COD removal distributed among all the stages. This phenomenon was less noticeable for BOD removal and this can be explained by the easier biodegradability of the fraction accounted by the BOD measurement. From the results obtained it was deduced that HRTs of the order of 1 d can achieve good removal efficiencies thus demonstrating the feasibility of the proposed treatment scheme even in comparison with the conventional aerobic plants.

Other experiments in which unconventional materials were employed as support medium in an anaerobic treatment, have been performed by Bodkhe (2008) who investigated the performance of a particular configuration of an anaerobic filter (AF) filled of locally available materials (burnt brickbats) as a support medium. This AF (total empty bed volume 170 L, effective volume 120 L) consisted of an inclined tube settlers (ITS) combined with a fixed-bed fixed film (FFFB) anaerobic reactor. The system was fed with municipal wastewater coming from an Indian municipality (average wastewater concentrations: COD 350–450 mg/L, BOD 200–300 mg/L, TSS 300–450 mg/L). The ITS module, composed of 11 units made of polyvinyl chloride (length 70 cm, diameter 4 cm), was designed to control the amount of suspended solids entering the AF, which might induce clogging phenomena. After this partial pre-treatment, wastewater was sent to the bioreactor, packed with burnt brickbats (average diameter 20 mm, specific surface area 200 $m^2/m^3$, total surface area 8.69 $m^2$). The treated effluent was withdrawn from the top of the media chamber. The experimental campaign was carried out operating at different HRTs (ranging from 3 h to 12 d) and organic loading rates (up to 2.8 kg COD/$m^3$/d and to 1.8 kg BOD/$m^3$/d), and lasted approximately 600 days. Once the steady state (defined as a series of steady determinations of the reduction of the main parameters considered) was reached for a specific HRT, this was shifted to the next lower value by increasing the influent flow rate. In terms of COD removal efficiency, to obtain a 95% removal a 12hHRT was necessary; for longer HRTs, the improvement of the performance was negligible. As regards BOD removal efficiency, the trend was very similar: at a 12hHRT the removal efficiency was 90%, reaching 95% only operating at a 4-d-HRT. The profile of suspended solids showed at an 8 and 12-h-HRT removal efficiency of 90% and 95%, respectively, while for higher HRTs no significant performance enhancement was observed. In terms of biogas production, at a 12-h-HRT the specific yield was 0.35 $m^3$ $CH_4$/kg COD removed, with a percent composition of methane ranging from 60% to 70% (30–35% carbon dioxide, together with nitrogen and hydrogen sulphide in traces).

#### 4.2.2.4 Anaerobic sequencing batch biofilter reactor

The technology of the anaerobic sequencing batch biofilter has been extensively investigated. Based on earlier studies by other authors, focused on the treatment either of domestic or industrial wastewater through anaerobic sequencing batch reactors (ASBR), Ratusznei *et al.* (2000) proposed a new configuration for the period, in which the biomass was immobilized on an inert support material. Afterwards, this configuration will be named as anaerobic sequencing batch biofilm reactor (ASBBR) by Sarti *et al.* (2006). The objective was to achieve high biomass retention: with this approach, the reactor no longer included a settling stage, previously included in a typical work cycle in ASBRs. The system consisted of a cylindrical flask (diameter 15 cm, volume 2.5 L, liquid volume 1 L), packed with polyurethane foam cubes (total mass 30 g, side 5 mm, apparent density 23 kg/$m^3$) as support medium for biomass immobilization and growth. A stirring bar at the bottom of the reactor ensured the mixing. The reactor was fed with 0.5 L/cycle of synthetic wastewater (influent COD concentration: 485 mg/L), and was operated under three daily 8-h cycles, at a temperature of 30°C and at a mixing rate of 500 rpm, with the rate being chosen after some hydrodynamic assays. The organic loading rate was slightly less than 300 g COD/$m^3$/d. The system was monitored during 36 days of operation. The biomass immobilization on polyurethane foam matrix guaranteed a higher solids retention, without the need to achieve the formation of biomass granules, thus resulting in a shorter start-up phase. The average effluent COD concentration ensured by the ASBBR was 68 mg/L, with a low standard deviation which suggested a high process stability. The average COD removal efficiency obtained after the first ten days (considered as the start-up phase) was 86%: this outcome could be considered as a remarkable one, given that a steady state was achieved in such a short time. Even an assessment of dynamic COD profiles was made, and it was highlighted how a minimum COD value in the effluent and the maximum COD removal efficiency could be obtained even after 3 h: these results were quite in accordance with some findings previously reported in the literature (Sarti, 1998).

After the pioneering research of Ratusznei and her co-workers, which turned the anaerobic sequencing batch reactor scheme into the so-called ASBBR one, many researchers continued to study this new approach. Sarti *et al.* (2006, 2007) studied the performance of the ASBBR in treating DWW treatment. The system consisted of a polyethylene cylinder (height 1.5 m, diameter 1.0 m, total volume 1.2 $m^3$, 0.08 $m^3$ at the bottom for sludge accumulation, 0.08 $m^3$ at the top for biogas collection; work volume 1.04 $m^3$: 0.65 $m^3$ bulk liquid volume, 0.39 $m^3$ support media volume), filled with polyurethane foam cubes (50 mm side, apparent density 23 kg/$m^3$, bed porosity 40%), confined into a perforated cylindrical stainless-steel basket. The attached biomass grew up into the inert support material. The system was operated at ambient temperatures (28–29°C) under three daily 8-h cycles (after a 2-h feeding and a 5-h reaction, one experiment involved a 0.5-h discharge and a 0.5-h idle, while a second experiment merged these two stages into a unique stage involving a 1-h discharge), treating

1.95 $m^3/d$ of DWW coming from the University of Sao Paulo campus, with an OLR ranging from 0.2 to 1.2 kg $COD/m^3/d$. Mechanical stirring was ensured by two rotors operating at 40 rpm. The bottom of the reactor allowed either the influent or the effluent to be fed or drawn, respectively. The system was not inoculated: however, the continuous flowing of wastewater through the reactor and the potential of the support material to entrap the biosolids guaranteed the effective adhesion and immobilization of the biomass. The studies performed by Sarti *et al.* (2006, 2007) lasted more than four months. The COD removal efficiency increased along the experimental campaigns (by day 20 it was 53%, approximately by day 120 it was higher than 80%), and despite the great variability of the influent COD (between A and B mg/L), the effluent produced during the last experimental period was characterized by an average COD concentration of $133 \pm 39$ mg/L, showing the operational stability of the system. The suspended solids removal efficiency through the whole period was 63%, with an average concentration of 66 mg/L. Given these results, according to the authors the ASBBR packed with polyurethane foam cubes could be a feasible solution to treat DWW at ambient temperatures in warm regions, such as the tropical and subtropical ones. A microbial characterization of the biomass involved in the process was carried out: from the microscopic observation, the presence of anoxygenic phototrophic bacteria affecting the performance of the system was revealed.

Different support materials have been assessed as carriers in ASBBR systems. Dutta *et al.* (2014) compared the performance of an anaerobic sequential batch reactor without any carrier material and three different ASBBRs (packed in different ways) in treating synthetic municipal wastewater. The investigated systems were four bench-scale reactors, named ASBR-1, ASBBR-2, ASBBR-3, and ASBBR-4. ASBR-1 was the control, without any support medium, while the others were packed with granular activated carbon (GAC), zeolite and both GAC and zeolite, respectively. GAC and zeolite used in this study showed similar features (particle size 16–30 mesh, density 0.8–0.9 $g/cm^3$, specific surface area 500–1000 $m^2/g$). Each reactor (working volume 1 L) contained 100 mL of sludge and 100 mL of support material and was operated under 24-h cycles (feeding 0.2 h, reaction 23 h, settling 0.5 h, decanting 0.25 h), at 35°C and at an HRT of 2.5 d. The chosen OLR was 400 $gCOD/m^3$ d. The reactors were fed with a synthetic wastewater characterized by an average COD concentration of 1000 mg/L, and the performance was studied for a short period of 30 days. Lower effluent COD concentrations were observed for the three ASBBR (77 mg/L for ASSBR-2, and concentrations ranging from 30 to 40 for ASBBR-3 and ASBBR-4), if compared to ASBR-1 (average COD concentration 105 mg/L). Better results from the ASBBRs were observed also in terms of biogas production: in this case, the highest outcomes were shown by ASBBR-4 (278 mL/d) and ASBBR-2 (246 mL/d). A first conclusion deriving from these results was that the use of support material allowed to obtain better performance, helping the microbial retention and minimizing the washout: notably, the addition of zeolite improved the COD removal, while the addition of GAC improved the biogas production. The combined effect of the two different filling materials (zeolite and GAC in ASBBR-4) results in higher COD removal and biogas production. An explanation of this behaviour was provided by scanning electron microscope (SEM) images of the carriers, which showed a rough surface of GAC which was more porous than zeolite, and, for this reason, a higher number of microbes preferentially attached to GAC; on the other hand, the positive charge on zeolite can induce the adhesion of negatively charged microbes on it. The GAC surface aided the microbial adhesion and improved the availability of organic compounds adsorbed onto microorganisms, thus resulting in an easier utilization of these compounds and in their conversion into biogas.

Another recent configuration of the anaerobic SBBR, proposed by Pentassuglia *et al.* (2020), and named AnSBio (anaerobic sequencing biofilter) by the authors, will be discussed later in detail in Section 1.2.3.

#### 4.2.2.5 Anaerobic hybrid reactor

Among the bioreactors operated with attached biomass growth, also some hybrid plant configurations have been investigated with the objective of taking advantage by the combination of the two technologies.

Elmitwalli *et al.* (2002*b*) compared the performance of two different reactors (namely, anaerobic hybrid reactor (AH), and an anaerobic filter (AF)) for the anaerobic pre-treatment of municipal wastewater coming from a Dutch village, operated at 13°C and at a 4-h-HRT for approximately five months. The AH reactor was composed of two different units, a sludge bed in the lower part (where biomass grows in the suspended form) and an anaerobic filter on top of that. This approach was supposed to merge the advantages of UASB and AF reactors, reducing their drawbacks at the same time. The two reactors (AH: height 2.55 m, volume 88 L; AF: height 2.1 m, volume 60 L) were both made of Plexiglas. Both the AF and the upper part of the AH reactor contained peculiar sheets of reticulated polyurethane foam (RPF; knobs at one side, flat on the other side), vertically oriented (AH reactor media height: 0.40 m; AF media height: 1.75 m). RPF sheets are also well-known to be very effective in the entrapment of suspended solids from DWW. Each reactor was fed upwards, and from each reactor the effluent was discharged from an upper section. The biogas produced was delivered to a gas meter. The performance of the two treatment systems were assessed by monitoring total and suspended COD removal efficiencies. The average total COD removal efficiencies were 55% and 34% for AF and AH reactors, respectively. Better results were obtained in terms of suspended COD, where both the reactors leveraged the ability of RPF sheets in entrapping suspended solids, showing suspended COD removal efficiencies of 82% and 53% for AF and AH reactors, respectively. This peculiarity of RPF elements, together with the dragging of the accumulated suspended solids in the bottom of the AH reactor, were the main causes for the higher performance of the AF unit. Besides being better than those reported for AH unit, the performance from AF reactor resulted also to be more stable, with lower fluctuations. According to the authors, these results were also higher than those previously reported in the literature: for instance, Wang (1994) found out only a 42% removal efficiency for suspended COD ensured by a UASB reactor which treated a similar DWW at 12°C and at a 3-h-HRT. Interesting findings were reported by the same authors (Elmitwalli *et al.*, 2002*a*) in a study carried out on a two-step AF/AH system which was operated for a year at 13°C: average suspended COD removal efficiencies of 81%, 57%, and 58% at HRTs of 4, 3, and 2 h were reported, respectively. With regard to biogas production, AF unit produced a greater amount of biogas with higher methane content, compared to the AH reactor (70.7% vs 58.9%). Notably, considering each stage of the anaerobic digestion, methanogenesis and acidification were higher in the AF unit compared with the AH reactor, while hydrolysis was comparable between the two treatments.

Another plant configuration combining two treatment technologies was proposed by Feng *et al.* (2008). The authors investigated the effectiveness of a carrier anaerobic baffled reactor (CABR), whose chambers were filled with hollow-sphere carriers made of bamboo, in treating low-strength DWW. Anaerobic baffled reactors represent a feasible strategy to treat anaerobically low-strength municipal wastewater. These systems are composed of more UASB units, connected through a series of vertical baffles, which force wastewater to follow a defined trajectory from the inlet to the outlet (Bachmann *et al.*, 1985; Cardoso Grangeiro *et al.*, 2019). The CABR used in this study consisted of a rectangular section reactor (length 600 mm, width 120 mm, depth 300 mm, effective volume 17 L), composed of six chambers characterized by an equal volume. These regions were packed with a spherical, hollow carriers made of bamboo (diameter ~1.5 cm). The choice of this material was intentional, as bamboo is a peculiar herb, native to China, mainly consisting of fibres. This material was firstly split into rectangular pieces; then, the pieces were interwoven into spheres and subsequently connected in a frame. The material thus obtained was characterized by a high specific surface area (up to 2100 $m^2/m^3$: this value decreased down to 105 $m^2/m^3$ when bamboo pieces held a 95% porosity). This peculiar support material was able either to foster the biomass retention or to entrap suspended solids from wastewater. Bamboo was chosen as carrier material since, besides being inexpensive, it is also a durable material, hard to degrade in anaerobic conditions. The CABR employed in this study was fed with real, dilute wastewater (average influent composition: COD 305 mg/L, BOD/COD ratio ranging from 0.38 to 0.64, TSS 80 mg/L), coming from dormitories and restaurants from a Chinese university and subjected only to a screening pre-treatment. The reactor was operated under ambient temperature

(28°C) for three months. Once the CABR attained steady-state conditions, different HRTs (ranging from 18 to 48 h) were tested to evaluate the relationship between HRT and COD removal efficiency. Observed COD removal efficiencies varied from 79% (at a 48-h-HRT) to 69% (at 18-h-HRT) that is the decrease in HRT resulted in a consequent COD removal reduction as also observe by Barber and Stuckey (1999). Higher removal efficiencies were reported for suspended solids, whose average value was approximately 82%: a major role could have been played by the peculiar support material, well-known also for its feature to entrap solids. Feng and his co-workers assessed also the effectiveness of CABR in terms of microbiological indicators. While the removal of total bacteria was not affected by the CABR treatment, a significant reduction of fecal coliforms and *Ascaris* eggs was observed. As fecal coliforms are mainly aerobes, their removal (higher than 99%, either at 24-h-HRT or at 48-h-HRT) could be explained by the application of adverse growing conditions. With regard to *Ascaris* eggs, larger than coliforms, their removal (almost 84% at 24-h-HRT, and almost 99% at 48-h-HRT) was attributable to physical mechanisms such as deposition and/or filtration. Nevertheless, the anaerobic process ensured a certain effect in the pathogens' removal, although a further disinfection stage would be necessary in order to ensure a good quality of the final effluent.

The opportunity of growing biomass in the attached form has been investigated also in membrane bioreactors. Harb *et al.* (2015) carried out a comparative study, under mesophilic conditions (35°C), between two different membrane bioreactors (AnMBRs) characterized by as many reactor configurations (namely, an upflow attached-growth anaerobic membrane reactor, UA, and a continuous stirred tank reactor, CSTR), applied to treat synthetic municipal wastewater. The reasons behind the study were based on the assumptions that, according to the literature, the reactor configuration, as well as the nature of the biomass involved in the wastewater treatment (suspended vs attached) affects the overall performance and membrane fouling, with attached-growth reactors having the potential to decrease biofouling effects. The study by Harb and his co-workers was focused on the reactor performance and on the microbial communities in the different reactor configurations. Each configurations had a working volume of 2 L and was operated at a HRT of 26 h and at an OLR of 0.72 g COD/L/d. In the CSTR mode, an internal rotor was operated at 200 rpm, while in the UA configuration the system was filled with cylindrical ceramic packing medium (diameter and length 1.5 cm) and operated in upflow recirculation mode. Two external cross-flow membrane modules (effective membrane surface area 50 $cm^2$ for each module), one polyethersulphone (PES) and one polyvinylidene difluoride (PVDF), were connected to each reactor. Membrane modules were connected in series along the same recirculation line. From the headspace of the reactors, the biogas produced was continuously sampled in gas bags. The synthetic wastewater (average influent COD concentration: 800 mg/L) was composed of a mix of organic and inorganic compounds. Confirming the literature data (Martin Garcia *et al.*, 2013), the study showed that the COD removal efficiencies (90–96%) were comparable for the two configurations, although characterized by a different nature of the biomass. The differences among the two configurations were related to the methane content in the biogas from the two reactors, since the CSTR produced a biogas with a higher methane content compared to the UA (80–90% vs 75–85%) and a slightly higher methane content in the liquid phase was reported for the UA compared to the CSTR. These differences were explained with a different microbial composition among the two configurations. According to this study, methanogens and archaea are more present in attached biomass systems if compared to suspended biomass ones. One of the possible explanations of this phenomenon is related to the longer SRTs of the attached biomass reactors, which would foster the presence of these microbial groups (Rittmann & McCarty, 2012).

### 4.2.3 Case study: anaerobic sequencing biofilter (AnSBio)

The anaerobic sequencing biofilter (AnSBio) is a promising technology born with the idea of combining the advantages related to increasing sludge SRT thanks to the presence of a packed section of the plant (i.e. the bed), and operating the bioreactor in sequential mode. This approach was supposed to support the development of a compact biofilm in the packed section allowing an increase in biomass concentration in the reactor and entrapping particulate organic matter coming from the wastewater. This condition should support particulate organic matter hydrolysis (increasing the contact time between organic matter and microbial extracellular enzymes) and favour a stable methanogenesis even when operating the process at moderate–low temperature.

A study focusing on AnSBio was conducted by Pentassuglia *et al.* (2020) to demonstrate its feasibility in treating DWW. After reaching steady-state conditions, the AnSBio was operated for more than one year at ambient temperature (wastewater temperature range 14–35°C) treating semi-synthetic municipal wastewater. Indeed, sucrose was added to the real DWW to restore its readily biodegradable COD, which quickly disappeared in the storage tank. The average COD of the resulting wastewater was 1.056 ± 137 mg/L. The OLR applied to the plant averaged 0.6 ± 0.1 kg COD/m$^3$ d, and the HRT was kept constant at 45 h during the whole experimental trial. Reactor performances in response to seasonal temperature changes were evaluated in terms of COD and TSS removal efficiencies, as well as biogas yield and composition.

The plant operated with treatment cycles each one including three phases: filling/drawing, recirculation and idle. During filling/drawing phase, a fixed volume of wastewater was supplied in up-flow mode, causing raising of the wastewater level in the reactor and therefore effluent discharge by a dedicated port located in the upper section of the AnSBio. During recirculation, wastewater was forced flowing through the biofilter increasing substrate diffusion in the biofilter and limiting biogas bubbles entrapment by the filling material. The increase in biomass concentration in the bed, potentially due to both suspended solids entrapment and biomass growth, was monitored by a pressure meter on the bottom of the reactor column. A sketch of the AnSBio described in Pentassuglia *et al.* (2020) is reported in Figure 4.1. The reactor column consisted of four sections: 2 L unpacked zone (aimed at wastewater homogenization during filling phase), a 14 L bed (the packed section filled with plastic wheel-shape filling material), a 10 L liquid phase above the bed, and 2 L of headspace.

AnSBio provided a COD and soluble COD (sCOD) removal efficiency averaging 82 ± 5% and 88 ± 5%, respectively. The latter seemed not influenced by the process operating temperature. Differently, COD removal efficiency exhibit a less clear trend. Nevertheless, it never felt below 71% even when the plant operated at an average daily temperature of 14°C. The good performance in COD removal were attributed to the AnSBio design, which effectively decoupled SRT and HRT. Indeed, the implementation of inert porous media has been suggested when dealing with systems operated under psychrophilic conditions, since biofilm formation allows microorganisms to well cope with stress related to low temperature (Zhou *et al.*, 2018). As for example, similar results were obtained by Bodík *et al.* (2002), evaluating the performances of an UAF in the temperature range 8–23°C, which found 84–87% COD removal when operating the plant at 15°C. Lower COD removal was obtained by other authors with different biofilters: Li and Lu (2017) reported only a 52 ± 6% COD removal when operating an anaerobic wool-felt filter reactors (AWFRs) during winter (10°C ± 2) at a 3-d HRT. The COD removal efficiency approached 60% when operating temperature was increased to 21°C (spring). Similarly, López-López and co-workers (2013) obtained only 57% COD removal using volcanic rocks as packing media for an up-flow anaerobic filter when the lowest temperature was tested (20°C). However, despite the good removal performances of AnSBio in COD removal, its concentration in the effluent averaged 188 ± 53 mg/L, thus exceeding the limit set by the Directive 91/271/EEC for wastewater discharge in surface waters, namely 125 mg/L.

Differently from what was observed for COD, the AnSBio was not effective in retaining suspended solids. TSS concentration in the effluent averaged 82 ± 40 mg/L, with few values close to 190 mg/L. However, the authors could not discriminate the share of TSS in the effluent coming from the DWW and those ones representing new biomass detached from the biofilter.

Interestingly, an opposite trend between the biofilter head loss (reflecting biofilter porosity) and operating temperature of the plant is observed. The lower the temperature, the higher the head loss and therefore, the lower the bed porosity. The raising in head losses with declining temperature might be attributed to the progressive reduction of bed porosity resulting from the slower hydrolysis of

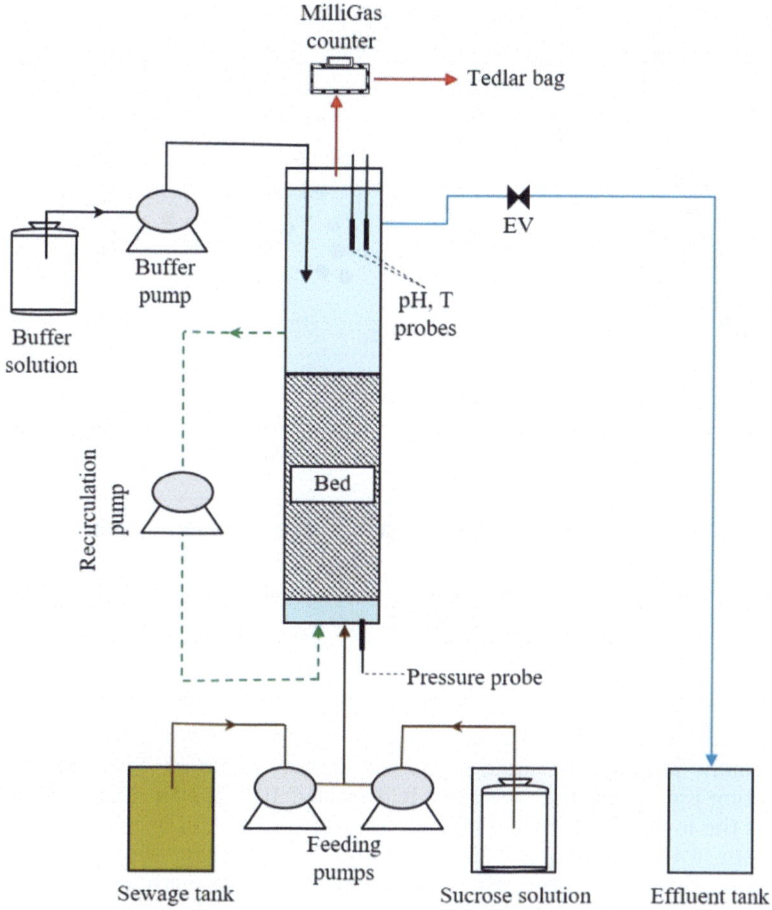


**Figure 4.1** Sketch of the AnSBio utilized in the experimental trial by Pentassuglia *et al.* (2020).

particulate matter, which accumulates under psychrophilic conditions at the lowest temperatures investigated. During warm seasons, microbial population in the AnSBio was supposed to hydrolyse the particulate organic matter entrapped during the cold season due to the increase in the hydrolytic activity (Chernicharo *et al.*, 2015). Another role in the head loss increase during cold seasons was attributed to the production of EPS. Along with structural and trophic functions, a role of EPS in cryoprotection has also been validated (Dev *et al.*, 2019; Pentassuglia *et al.*, 2018), and there are evidences in the literature that enhancement of bed thickness supposedly due to EPS accumulation under psychrophilic conditions results in increased hydraulic shear forces (Maaz *et al.*, 2019). In the AnSBio, the increase in EPS concentration and consequent head loss increase might explain the higher TSS concentration in plant effluent. It is worth noting that, the cyclic increase and decrease of the biofilter head loss, led to a self-regulation of its porosity. Indeed, the AnSBio did not need any washing operation aimed at restoring biofilter porosity during the experimental trial.

As far as biogas production is concerned, Pentassuglia *et al.* (2020) highlighted a clear correlation between biogas yield and operating temperature of the plant (Figure 4.2). This finding is in agreement with other studies reporting that low temperatures led to a reduction in biogas production, even due to

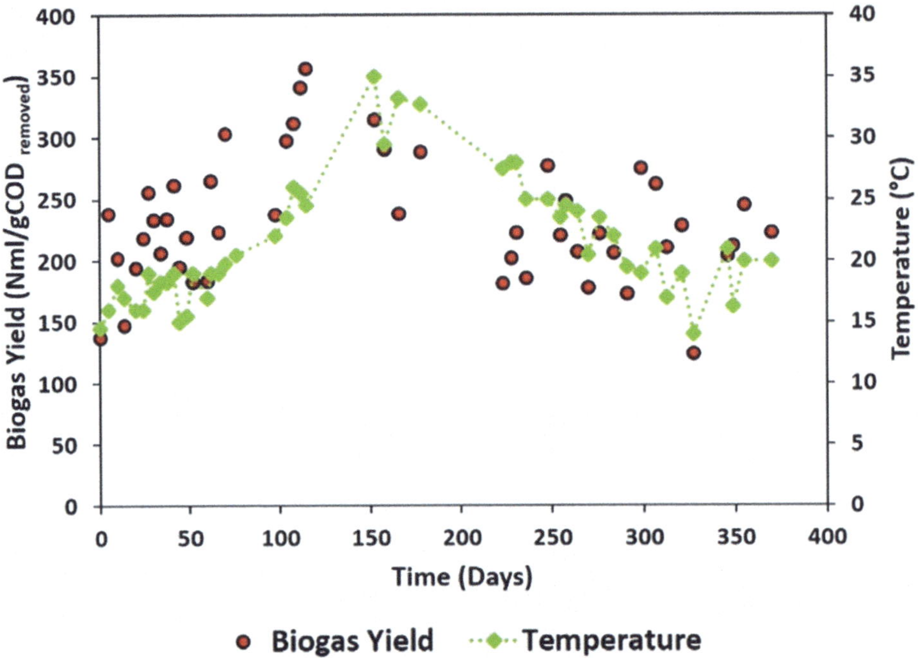


**Figure 4.2** Biogas yield and reactor temperature profiles reported in the study by Pentassuglia *et al.* (2020).

an increase in methane solubility in liquid phase (Bandara *et al.*, 2012; Li & Lu, 2017). According to the biogas production measured, the methane yield ranged 104–267 $NmLCH_4/gCOD_{removed}$ when the AnSBio operate at the lowest and the highest temperature, 14°C and 35°C, respectively. Considering the methane content in the gas phase of the plant and the operating temperature, and according to Henry's law, the authors estimated that about 27% and 11% of the methane produced was lost in the effluent when the plant operated at 14°C and 35°C, respectively. The methane yield recorded at low temperature was far from the maximum theoretical one (0.35 $NmLCH_4/g\ COD_{removed}$). However, this approach might underestimate methane loss. Anaerobic effluents are supersaturated with methane, as was referred for AnMBR and UASB systems operating at low temperatures (Crone *et al.*, 2016; Smith *et al.*, 2013).

The low methane yield during cold seasons would be in line with the hypothesis of EPS accumulation in these periods, which would reduce the amount of COD actually converted to $CO_2$ and $CH_4$. This hypothesis was also supported by the high biogas production during spring, which might be due to the unaccounted fraction of particulate COD, which was converted to methane when wastewater temperature exceeded 25°C, while it remained un-hydrolysed during winter season.

The most relevant outcomes regarding biogas production refers to the methane content in the biogas. The latter ranged 75–90%, with an average value of 84 ± 4%. This value is much higher than that usually reported for mesophilic bioreactors. The high methane fraction was attributed to a relevant presence of hydrogenotrophic methanogens, generating methane by the reduction of $CO_2$ using $H_2$. This metabolic route is thermodynamically favoured under psychrophilic conditions, due to an increase of gas solubility (Lettinga *et al.*, 2001). The relevant abundance of hydrogenotrophic species compared to acetoclastic methanogens at cooler temperatures has been widely described (Bandara *et al.*, 2012; Li & Lu, 2017; McKeown *et al.*, 2009). Indeed, a few studies focusing on anaerobic processes at low temperature reported a methane content in the biogas close (Serrano León *et al.*, 2018) or even higher than 80% (Smith *et al.*, 2015).

### 4.2.4 Discussion of the results and future developments of immobilized biomass systems

As already pointed out in the previous chapters, application of immobilized biomass systems to the anaerobic DWW treatment in moderate and cold climate regions is still limited at lab or pilot scale, and additional research efforts are necessary to move at full scale. A number of studies conducted since the 1980s with plants operating under variable temperature regimes (simulating seasonal wastewater/environmental condition variability) or at constant psychrophilic or mesophilic conditions demonstrated the potentialities of these high-rate bioreactors. Future investigations on optimizing the process performance by varying the operating conditions at higher scale are strongly required. This aspect is of particular relevance in biological systems operating with attached biomass because the scale effect on different factors affecting the bioreactor performance, that is mixing efficiency, bed characteristics in terms of void fraction, biofilm thickness, is very important. Higher scale testing is also necessary to provide a better figure of the cost evaluation because higher cost characterizing these systems, in comparison to simpler ones, that is UASB reactors operating with suspended biomass, is one of the bottleneck for the process feasibility.

Long start-up times characterizing the anaerobic process are more critical for attached biomass operation because process efficiency is determined not only by the specific developed bacterial species, but also by the biofilm thickness which has to be at the optimal size to provide efficient substrate biodegradation and, at the same time, efficient substrate diffusion. This condition requires the optimization of the hydraulic regime in the bioreactor, and specific investigations are required to better highlight the relationship between biofilm thickness and hydraulic conditions.

Complexity of the biological systems and of the technology requires long-term experiments, and availability of representative mathematical models, still lacking in the specialized literature, could help to quickly predict future scenarios saving consistent time amounts.

Finally, a relevant and common drawback to all presented technologies is that, when operating at low temperature, a significant fraction of the methane produced during the process was lost in the effluent due to both, the limited methane production, and the increased methane solubility, thus future research should be devoted to strategies that are able to maximize methane recovery or reuse inside the process. This is analysed in detail in Chapter 7 of this book.

## 4.3 GRANULAR BIOMASS SYSTEMS

### 4.3.1 Granules characterization and principle of operation

Anaerobic granules make their first appearance in a study by Young and McCarty (1969), their presence was reported in the earliest version of an AF, employed to treat low-strength wastewater. They justified the peculiar morphology of these granular agglomerates as the consequence of the action of the gas bubbles which, moving upwards, rose inside the anaerobic filter. In 1976, anaerobic granules were observed in a 6 $m^3$ pilot plant operating at a sugar factory. Thanks to the presence of granules, improved process performance was observed (Hulshoff Pol *et al.*, 2004).

Anaerobic granular sludge is characterized by dense microbial communities that typically include millions of organisms per gram of biomass. None of the individual species in these micro-ecosystems can provide complete biodegradation of the influent pollutants. The degradation of high and low strength wastewaters, indeed, involves the complex interactions among all the resident species (Liu *et al.*, 2002). The size of the granules is a key parameter since, as well as the density, it determines their behaviour in terms of settleability and activity (Trego *et al.*, 2021). Anaerobic granules have typically a spherical shape and their size ranges between 0.1 and 5 mm, with the diameter frequently ranging between 0.5 and 2 mm (Hermansson *et al.*, 2022; Skiadas *et al.*, 2003; Trego *et al.*, 2021). The porosity of anaerobic granules is in the range of 0.64 and 0.9, indicating that they are characterized by highly porous structures (Mu *et al.*, 2006). The settling rate of the granules can vary from 18 to 100 m/h, with an average value of 60 m/h (Hulshoff Pol *et al.*, 2004; Skiadas *et al.*, 2003). The high settling rates indicate that granules are not affected by biomass washout phenomena, thus can operate at high biomass concentration, with a consequent optimal use of the working volume (Stazi & Tomei, 2018).

There are relationships between the geometric parameters of the anaerobic granules, although a bibliographic review provided conflicting results. The porosity of the granules increases proportionally with the size. In turn, porosity also affects phenomena such as mass transfer and gas diffusion (Trego *et al.*, 2021). By conducting an experiment on anaerobic granules of three different sizes (small, medium and large: 0.5–1, 1.5–2 and 3–3.5 mm, respectively), Wu *et al.* (2016) observed that the density of the granules was also directly proportional to the size: medium and small granules were considerably 'lighter' than the larger ones. Similarly, the pore size of the granules also increased with the granule size. These results conflict with the findings by Mu *et al.* (2006), according to which porosity and diameter of the granules are inversely proportional, probably due to the presence of EPS, which could fill the pores of the granules, thus reducing their porosity. Vlyssides *et al.* (2008) proposed a relationship, which would involve granules density and diameter and the ratio between volatile and total suspended solids (VSS/TSS), finding a good correlation for both parameters. The relationship between the diameter of the granules and the specific biogas production was also investigated, taking into account relatively high-strength wastewater with an influent COD concentration of 1000–6000 mg/L. According to this research findings, at the same COD concentration, larger granules ensured a higher specific production of biogas (Wu *et al.*, 2016).

Anaerobic granules are characterized by an organic fraction and an inorganic fraction. The organic component consists of EPS, mainly represented by proteins and polysaccharides, also including lipids, nucleic acids and, to a lesser extent, organic debris, phages, and lysed cells. The EPS composition also influences the physical properties of the granules (Morgan *et al.*, 1990). EPS perform the fundamental role of promoting the formation of stable granules, on the basis of electrostatic interactions between opposite charges, and of encapsulating and protecting the bacteria from the surrounding environment. The EPS percentage is between 0.6% and 20% of VSS (Lim & Kim, 2014; Skiadas *et al.*, 2003). The inorganic component consists of ash, whose fraction can vary in a wide range, from 10% to 90% in dry weight of the granules. This significant variability depends on the wastewater composition, the operating conditions adopted and several experimental parameters (pH, temperature, OLR). Ash is mainly made up of calcium, potassium, iron, phosphorus, magnesium, sodium, and to a lesser extent elements such as nickel and cobalt. A high ash content in the granule, besides increasing its density, can limit the transport of substrates and gas molecules (Lim & Kim, 2014; Skiadas *et al.*, 2003).

### 4.3.2 Granulation modelling

Although cases of anaerobic granules generated by substrates whose initial degradation step is the rate-limiting step and characterized by uniform and non-layered structures are reported (Batstone *et al.*, 2004; Fang *et al.*, 1995), studies carried out by means of microscopic techniques have revealed how anaerobic granules are aggregates composed of several overlapping layers, in which different microbial populations are found. Based on these microscopic observations, MacLeod *et al.* (1990) proposed a multi-layered structural model (Figure 4.3): acidogenic bacteria dominate the outer layer,

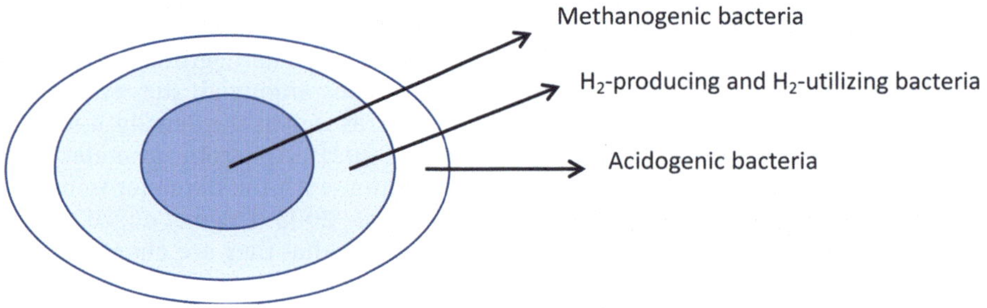


**Figure 4.3** Layered structure of anaerobic granules proposed by MacLeod *et al.* (1990).

methanogenic bacteria predominantly inhabit the inner, $H_2$-producing and $H_2$-utilizing bacteria are mainly localized in the middle layer. Grotenhuis *et al.* (1991) and Fang *et al.* (1995) have also found some granules with uniform structure. In particular, Fang *et al.* (1995) argued that a layered and uniform microstructure would be developed with, respectively, carbohydrates and proteins as substrates.

In the proposed concentric layered structure where each layer is constituted of different bacterial trophic groups, each group is expected to perform its respective role in the organics' degradation, producing biomass and EPS in its vicinity. Among the microflora, the methanogen *Methanosaeta consilii* is believed to play a key role in setting up granulation, because the clumps formed by the growth of these filamentous microorganisms act as nucleation centres where granule development begins. This is followed by subsequent colonization by acetogenic bacteria and hydrogenotrophic methanogens, which often leads to the layered granular biofilm structure mentioned above (Hulshoff Pol *et al.*, 2004).

However, over the years, many other hypotheses have been formulated on the granule formation mode and, consequently, on its structure. This investigation reveals the complexity of a process that is still not fully understood, and which does not seem to offer a unified description.

In general, anaerobic granules form spontaneously by auto-immobilization of bacteria in the absence of a support material. Granulation may start by bacterial adsorption and adhesion to inert matter, to inorganic precipitates, and/or to each other through physical–chemical interactions and syntrophic associations. These substances are the initial precursors (carriers or nuclei) for further bacterial growth. The aggregates are then strengthened by EPS secreted by the bacteria themselves (McHugh *et al.*, 2003). The inorganic fraction of anaerobic granules can be strongly affected by the composition of wastewater and process conditions. The ash content in the granules growing on a complex wastewater is lower if compared with that in the granules growing on a simple wastewater such as acetate, propionate, or butyrate. Moreover, the sizes of the granules in a complex substrate are bigger than those in a simple one. Ash mainly consists of calcium, potassium, and iron. When the ash content in the granules is high, this inert material can act as an inhibitor of the transportation of the substrate, gases, and metabolites between the cells and the bulk solution. In addition, an increase in the ash content results in an increase in the granule's density. On the other hand, there is no relationship between the ash content and the strength of the granule (Lim & Kim, 2014). As mentioned, the phenomenon of granule formation has been deeply investigated, and many models have been proposed to enhance the understanding of the anaerobic granulation mechanism. These models, which include different approaches (structural and thermodynamic), have been reported in several articles (Hulshoff Pol *et al.*, 2004; Liu *et al.*, 2002; 2003; Show, 2006). We will discuss some of them below, but a broader and more exhaustive overview is available in the mentioned literature.

#### 4.3.2.1 Structural models

Structural models mainly focus on the physical, chemical, and microbiological composition of the granules and examine the effects of many factors on sludge granulation.

##### Selection pressure model

In the selection pressure theory (Hulshoff Pol *et al.*, 1983), the granulation process in a UASB reactor is based on the continuous selection of sludge particles. The selection pressure derives by the effects of the hydraulic loading rate and the gas loading rate (function of the sludge loading rate). Under high selection pressure conditions, light and dispersed sludge will be washed out while heavier components can be retained in the reactor. It follows that growth of finely dispersed sludge is minimized, and the bacterial growth is dependent on a limited number of growth nuclei, such as inert organic and inorganic carrier materials or small bacterial aggregates present in the seed sludge. These growth nuclei increase in size until a maximum value, after which parts of the granules detach, producing a new generation of growth nuclei, and so on. On the other hand, under low selection pressure conditions growth will take place mainly as dispersed biomass, which will lead to the formation of a bulking sludge.

### Inert nuclei model

This model was initially proposed by Lettinga *et al.* (1980) and it has been experimentally validated among others by Montalvo *et al.* (2012), who observed the presence of porous media in the bioreactor that fostered the granulation. The sludge granulation starts from inert matter, which acts as a precursor or nucleus on which anaerobic bacteria can attach to form the initial embryonic granule. The mature granules can be further developed through the growth of these attached bacteria under given operation conditions. The inert nuclei model suggests that the presence of nuclei or micro-size bio-carrier for bacterial attachment can be a first step towards anaerobic granulation.

### EPS bonding model

The accumulation of EPS can mediate biological adhesion and aggregation processes and play a crucial role in maintaining the structural integrity of the microbial matrix. The metabolic blocking of exopolysaccharides synthesis was found to prevent microbial adhesion. EPS content in granules was hypothesized to change the surface negative charge of the bacteria, and thereby to bridge two neighbouring bacterial cells physically one to each other as well as with other inert particulate matters, and to settle out as floc aggregates (Cammarota & Sant'Anna, 1998; Schmidt & Ahring, 1994).

### The Capetown model

According to Sam-Soon *et al.* (1987), granulation depends on *Methanobacterium* strain AZ, an organism that utilizes $H_2$ as its sole energy source and can produce all its amino acids, except for cysteine. When this microorganism is under high $H_2$ partial pressure conditions (i.e., excess substrate), cell growth and amino acid production will be stimulated. However, cell synthesis will be limited by the rate of cysteine supply. Moreover, in the presence of ammonium a high production of other amino acids will occur, which *Methanobacterium* strain AZ secretes as EPS binding *Methanobacterium* strain AZ and other bacteria together to form granules. This hypothesis was suggested in the case of a UASB reactor treating a substrate mainly consisting of sugars, with negligible nitrogen content and with adequate nutrients and trace elements for growth.

According to this hypothesis, the conditions that favour granulation are:

- high $H_2$ partial pressure;
- plug flow or semi-plug reactor (in order to achieve phase separation) with a nearly neutral pH;
- non-limiting source of nitrogen, in the form of ammonium;
- limited amount of cysteine.

On the contrary, the situations which hinder granulation are:

- systems where the substrate does not yield $H_2$ in the fermentation process (e.g., acetate) or can be degraded only under low $H_2$ partial pressure conditions (e.g., propionate and lipids);
- completely mixed systems, because of the dilution of the high $H_2$ partial pressure.

### 'Spaghetti' theory

This theory was proposed by Wiegant (1987) for sludge granulation in UASB reactors treating acidified wastewater, solutions of acetate or mixtures of volatile fatty acids (VFA) with predominant *Methanothrix* bacteria. Wiegant divides the granule formation in two phases:

(1) formation of precursors,
(2) actual growth of the granules from these precursors.

Initially, *Methanothrix* bacteria form very small aggregates, due to the turbulence generated by the gas production, or attach to finely dispersed matter. The concentration of SS should not be too high, otherwise the increase in size of the aggregates will be too slow. Selection for aggregates is done by increasing the upflow velocity. Once the precursors are formed, their growth will generate granules of spherical shape by effect of the hydraulic shear forces and the upflowing biogas. In this phase, the granules have a filamentous appearance, like a ball of spaghetti. Afterwards, rod-type granules are formed from these filamentous granules at a high biomass retention time, due to the increase in the density of the bacterial growth.

#### *4.3.2.2 Thermodynamic theories*

Some authors have analysed the granulation mechanism in terms of the energy involved in the adhesion itself, due to the physico-chemical interactions between cell walls or between cell walls and alien surfaces. Aspects like hydrophobicity and electrophoretic mobility have also been considered. When a bacterium approaches another bacterium, their interaction includes repulsive electrostatic force, attractive van der Waals force, and repulsive hydration interaction. Below is a summary of the granulation models based on the thermodynamic theory.

##### Secondary minimum adhesion model

This model is based on the DLVO (Derjaguin, Landau, Vervey, and Overbeek) theory for colloidal particles, which shows that reversible adhesion takes place in the secondary minimum of the DLVO free energy curve. The Gibbs energy of the reversible adhesion is relatively small, and there is always a separation distance between the two adhering bacteria. Thus, the reversible adhesion can change to irreversible adhesion at the primary minimum by overcoming the energy barrier or by protruding fibrils or fimbriae, which bridge the gap between bacteria (Rouxhet & Mozes, 1990). From this model, it appears that anaerobic granulation would start from the self-immobilization of bacteria through reversible and followed by irreversible microbial interaction.

##### Hydrophobic interaction and local dehydration model

Wilschut and Hoekstra (1984) proposed this model suggesting that, under normal physiological conditions, the strong repulsive hydration interaction was the main force to keep the cells apart. The model shows that when bacterial surfaces are strongly hydrophobic, irreversible adhesion will occur. Hydrophobicity and hydrophilicity are usually used to describe a molecule or a structure having the feature of being rejected from an aqueous medium (i.e., hydrophobicity), or being positively attracted (i.e., hydrophilicity). Hydration interaction becomes significant at surface separations of 2–5 nm or less, depending on the nature of bacterial surfaces. According to the surface thermodynamics theory, increasing the hydrophobicity of cell surfaces would cause a corresponding decrease in the excess Gibbs energy of the surface, which in turn promotes cell-to-cell interaction and further serves as driving force for cell self-separation from the liquid phase. Local dehydration of the short-distance-apart surfaces has been identified as the prerequisite for bacterial adhesion.

##### Surface tension model

In a bulk solution with a low surface tension, hydrophilic bacteria adhere to one another, while in solutions with a high surface tension hydrophobic species are likely to adhere to one another. In general, most acidogenic bacteria are hydrophilic, and methanogens appear to be hydrophobic. The surface tension model, proposed by Thaveesri *et al.* (1995), considers the liquid surface tension ($\gamma$) in UASB reactors. Depending on this factor, bacterial cells may grow in rather loose associations, in multilayered granules ($\gamma < 50$ mN/m) or in mixed conglomerates ($\gamma > 56$ mN/m). Thus, the adhesion of hydrophilic cells is enhanced at low liquid surface tension, while the opposite is true for hydrophobic cells.

##### Proton translocation – dehydration theory

Tay *et al.* (2000) proposed a theory based on the proton translocation activity at bacterial membrane surfaces. In this theory, the sludge granulation process was considered to proceed in four steps:

(1) Dehydration of bacterial surfaces: during the start-up period, the organic wastewater is fed into an anaerobic reactor where the seed sludge has been inoculated. The fermentative bacteria

secrete extracellular enzymes into the medium to catalyse the hydrolysis/acidification process. The organic compounds are degraded into VFA coupling with the electron transport. At the same time, the proton pumps on the membranes of these bacteria are activated. Proton translocation can establish a proton gradient across the bacterial cell surface and subsequently cause the surface protonation. The energized bacterial surfaces result in the breaking of hydrogen bonds between negatively charged groups and water molecules as well as partial neutralization of the negative charges on their surfaces. This may induce the dehydration of the bacterial surfaces.

(2) Embryonic granule formation: acidogens, acetogens and methanogens may adhere to each other forming embryonic granules, as a consequence of the upflow hydraulic stress, of this weakened hydration repulsion and of the hydrophobic nature of the cells. In addition, due to the transfer of metabolites between cells, a further dehydration of the bacterial surfaces takes place leading to a strengthening of these initial granules. In this stage, the new physiological environment starts to induce the excretion of EPS to the embryonic granule surfaces.

(3) Granule maturation: the original bacterial colonies continue to grow while other dispersed bacteria may adhere to the embryonic granules. The transfer of intermediates determines the distribution of micro-colonies within the granule, eventually leading to well-structured bacterial aggregates as mature granules. Granule maturation resists and blocks the unrestricted multiplication of bacterial cells because of space restriction. This space restriction and the continuous supply of substrates facilitate the large production of EPS, which causes the hydration of granule surfaces and protects granules against the shear stress and attachment to gas bubbles, with subsequent biomass loss by flotation as EPS is highly hydrophilic and biogas bubbles are highly hydrophobic.

(4) Post-maturation: the proton translocating activity keeps the bacterial surfaces at a relatively hydrophobic state and is mainly responsible in maintaining the structure of the mature granules. On the other hand, the EPS layer outside of a granule causes the hydration of the granule surface, protecting the granule against attachment to gas bubbles and shear stress in the UASB reactor.

### 4.3.3 Overview of the technologies operating with granular biomass

The spatial arrangement of bacteria within an anaerobic granule offers a number of advantages, such as more efficient microbial proliferation, access to resources that cannot be reached by isolated cells, internal physicochemical gradients within the aggregates, defense against antagonists that eliminate isolated cells, optimization of population survival by differentiation into distinct cell types and manipulation of biomass in a single phase (McHugh *et al.*, 2003).

Granular sludge-based bioreactor technologies are recognized worldwide as cost-effective and efficient for the anaerobic treatment of industrial and municipal wastewater (Ahmad *et al.*, 2020), a short description of the technologies suitable for DWW treatment is reported in the following sections.

#### *4.3.3.1 UASB and EGSB bioreactors*

The high-rate bioreactors described in Chapter 1 appears as the most favourable technologies for the anaerobic treatment of DWW. Among them, the UASB and expanded granular sludge bed (EGSB) reactors can operate with high active concentrated granular biomass. This aspect is particularly important because the lower metabolic capacity of anaerobic bacteria than aerobic ones makes the anaerobic process less efficient as compared to a conventional activated sludge system, especially for low-strength wastewaters, thus requiring longer SRTs (Van Haandel *et al.*, 2006). High concentrated granular biomass, indeed, ensures efficient biodegradation kinetics and improves the solid–liquid separation thanks to the excellent sedimentation properties of the granules. Therefore, it is possible to reduce the HRTs of the reactors without lowering the SRT, making this technology competitive with the traditional aerobic process, and to lower the SS concentration in the treated wastewater leaving the reactor.

The UASB bioreactor is the system most employed for the anaerobic treatment of DWW (Chernicharo *et al.*, 2015), a detailed presentation of the principle of operation and technological features of the UASB bioreactors is reported in Chapter 2. Therefore, we provide here only a short reminder. UASB bioreactors are fed in up flow mode. The organic matter contained in DWW is converted into $CH_4$

and $CO_2$ during the passage of the influent through the granular biomass retained into the reactor by settling. At the top of the reactor, a gas–solid–liquid separator allows to collect the produced biogas (Gomec, 2010). The critical aspects, which deserve further investigations to fully exploit the potential of such promising technology are:

- the washout of suspended solids and undigested residual organic matter and nutrients over the discharge requirements, causing the need of effluent post treatment and the related energy demand (Crone *et al.*, 2016),
- the presence of dissolved methane ($dCH_4$) in the treated effluent, which can be released into the atmosphere if not properly recovered.

The emission of this dangerous greenhouse gas represents not only a performance decrease in terms of energy recovery, but also a strong negative environmental impact. Losses of $dCH_4$ through anaerobic effluents can amount between 45 and 88% of the total $CH_4$ produced, depending on load, reactor type and temperature. Consequently, strategies for the recovery or reuse of $dCH_4$ within the same treatment process by biological oxidation are required for making anaerobic treatment an attractive option for energy-neutral and climate-friendly DWW management (Cookney *et al.*, 2016; Lee *et al.*, 2018).

Anaerobic granules have been most commonly developed in UASB and EGSB reactors. Both high-rate systems have been presented in Chapter 1. A brief description of some other anaerobic granules-based bioreactors is reported below, while a more detailed characterization can be found elsewhere (Liu *et al.*, 2002; Liu & Tay, 2004; McHugh *et al.*, 2003).

#### 4.3.3.2 Internal circulation reactor

The internal circulation (IC) reactor, also considered a high-rate bioreactor, consists of two interconnected UASB compartments. In the first highly loaded compartment most of the organic matter is converted into biogas. The biogas is collected by a lower level phase separator and is used to generate a gas lift able to carry upward wastewater and sludge, via a riser pump, to a two-phase (gas/liquid) separator located at the top of the reactor where the biogas is separated from the wastewater/sludge liquor and leaves the system. The mixed liquor is guided to the bottom of the reactor, where it mixes with the influent resulting in the internal circulation flow. The upflow of the biogas drives the internal circulation. The residual biodegradable COD of the effluent coming from the first compartment is removed in the second one. Any biomass lost from the first compartment is retained in the upper section, allowing the application of high organic and volumetric loading rates. The high organic loading rates and low HRTs generate high turbulence within the reactor, causing higher average granular shear rates than those observed in UASB reactors, thus resulting in the development of sludge aggregates larger than granules developed in UASB reactors. Comparison tests showed that the granular sludge generated in IC reactors has higher methanogenic activity than anaerobic sludge of UASB reactors, and that the potential sludge activity is more efficiently used within IC reactors. Therefore, among new generation anaerobic reactor systems, IC reactor looks a viable competitor of the well-established UASB reactor (McHugh *et al.*, 2003).

#### 4.3.3.3 Anaerobic sequencing batch reactor

The major characteristics making an anaerobic sequencing batch reactor (ASBR) different from an UASB reactor are the following: a feed distribution system is not required; there is no three-phase separator; an upflow hydraulic pattern is absent; the operation is in discontinuous mode (Liu & Tay, 2004). Kennedy and Lentz (2000) suggested that, at low organic loading rates, the performances of continuous UASB and ASBR are quite similar. However, continuous UASB reactors perform better than the ASBRs at high organic loading rates. Additionally, ASBR at full scale may require a buffer tank to store the incoming DWW before the feeding phase. Moreover, it is recommended to operate with two or more ASBR units in parallel. A case study involving a UASB reactor working on sequential mode will be displayed in the next section.

#### 4.3.3.4 Anaerobic migrating blanket reactor

The anaerobic migrating blanket reactor (AMBR) is a continuously fed compartmentalized system that does not require a gas–solid separator and systems for feed distribution. The anaerobic granules here developed are darker in colour, smaller, and denser than granules formed in an UASB reactor, which operates under similar conditions. The AMBR has some advantages over the UASB reactor, such as low biomass migration rates, less chance of short-circuiting, efficient removal of poorly biodegradable compounds, and the possibility to be operated in step feed mode for high-strength wastewaters during shock loads. However, the internal structure of this reactor is more complex than that of UASB. For example, the AMBR requires multipoint mechanical mixing to improve feed distribution and prevent clogging by sludge (Angenent & Sung, 2001).

#### 4.3.3.5 Anaerobic hybrid reactor

As extensively discussed in Section 4.2.2.5 AH reactor combines the operating principle of UASB and AF reactors, promoting the advantages of both systems and minimizing their constraints. The bottom section of the reactor is designed as a UASB and contains a granular sludge bed, while the top section has a randomly packed matrix which aids in biomass retention and provides a surface for the attachment of microorganisms. The use of packing media only in the top portion of the reactor minimizes the channelling problems which can occur with granular biomass generally associated with fully packed upflow AFs. Most of the biodegradable COD is converted into biogas in the UASB section and any remaining COD is moved through the AF section where it is degraded by the microorganisms present in the biofilm. In addition, to avoid granular sludge washout, the upper AF section serves as an effluent polishing step. Anaerobic hybrid reactors are reported to offer considerable advantages with respect to buffering against shock loading, with the AF section contributing to reactor stability and reliability (McHugh *et al.*, 2003).

### 4.3.4 Granulation in UASB reactors

#### 4.3.4.1 Start up

The granulation process, the pre-requisite of a UASB system start-up, is usually very time-consuming partially because of the slow growth of methanogenic bacteria and, subsequently, the slow formation of embryonic granules. Therefore, reducing the time required for granulation has always been one of the most important goals of research in UASB technology. Many types of seed sludge can be used for UASB inoculation, such as digested sludge, digested manure, septic tank sludge, pond sediment, and raw waste activated sludge. The best option would be to inoculate with granules already formed, for example coming from other reactors or preserved in case of UASB reactors, which operate seasonally or intermittently. In this latter case, the start-up of new installations can even be completed within a few days (Xu & Tay, 2001).

Reactor start-up is a very important economic process step, because during this period the productivity of the wastewater supplier must be adapted to the capacity of the treatment plant. In addition, inadequate start-up causes poor subsequent treatment and often requires expensive system maintenance and effluent post-treatment. Start-up is often considered the most unstable and difficult phase in UASB reactors. It can require long times, even several months. Therefore, the reduction of the start-up time is one of the key parameters to increase the competitiveness of UASB reactors (Show *et al.*, 2004).

The formation and growth of anaerobic granule is not only important for the start-up of the reactor. Granules, to be formed and acclimatized for the specific wastewater, have to be preserved as long as the reactor is in operation. For this reason, it is important to list and describe the parameters and operating conditions which influence this process.

#### 4.3.4.2 Factors affecting granulation in UASB reactors

Anaerobic sludge granulation is quite complex and is affected by many physicochemical parameters and operating conditions. As a consequence, the information on the major factors influencing this process is fundamental to pursue possible strategies for fast production of anaerobic granules.

### Temperature

Composition of granules in UASB reactors strongly depends on the operational temperature as different species achieve optimum growth rates at different temperatures. Sudden temperature changes could result in granule disintegration in the reactor (Abbasi & Abbasi, 2012). Methanogenic bacteria, the core microbial component of UASB granules, grow slowly and their generation time ranges from 3 days at 35°C to more than 50 days at 10°C. This implies that UASB reactors should be operated at a temperature range of 30–35°C. However, DWW treatment in UASB reactors is not beneficial at such temperatures, since in most cases additional energy is required to heat the wastewater. Heating wastewater increases difficulties associated with operation and management of the system, and the advantages of UASB system, such as simple in construction and operation and no additional electricity requirement would be lost (Liu *et al.*, 2002). Thus, operation at temperatures of 30–35°C is carried out only in regions with a hot climate, where the observed wastewater temperature is often in this range.

### pH

Anaerobic reaction involves a complex consortium of microbial species which can be divided into three categories: bacteria responsible for hydrolysis; acid-producing bacteria and $CH_4$-producing bacteria. In general, acid-producing bacteria can tolerate a low pH with an optimal range of 5.0–6.0, but most $CH_4$-producing bacteria prefer a very narrow pH range of 6.7–7.4. Thus, it can be assumed that a stable pH close to neutrality is desirable to obtain good-quality granulated sludge (Liu *et al.*, 2002).

The acidogenic population is significantly less sensitive to pH fluctuations compared to methanogens. For this reason, in extreme pH conditions, acid formation prevails over methanogenesis, resulting in accumulation of VFA inside the reactor. The differential growth of fermentative bacteria and methanogens may cause pH changes if the VFA concentration exceeds the buffering capacity of the reactor content. These conditions generally lead to granule disintegration causing the death of methanogens (Tiwari *et al.*, 2006).

### Organic loading rate

From a microbiological point of view, the OLR describes the degree of starvation of the microorganisms in a biological system. A low OLR means that the microorganisms in the reactor are starved, while a high OLR would lead a fast microbial growth. There is evidence that anaerobic granulation is accomplished by gradually raising the OLR during the start-up (Liu *et al.*, 2002). Ghangrekar *et al.* (2005) have suggested that OLR in the range of 2.0–4.5 kg-COD/m$^3$/d is suitable for developing good granular sludge, but the optimum range of OLR and HRT is related to other parameters, such as the strength and composition of substrates, temperature and the concentration of nutrients, macro-metals, trace metals, and anions like sulphate.

### Substrate characteristics

Characteristics of organic substrate may influence the formation, composition, and structure of anaerobic granules. Based on the free energy of oxidation of organics, the substrate can be classified into high-energy and low-energy feeds. During the UASB start-up period, high-energy carbohydrate feed can sustain the acidogens and facilitate the formation of EPS (Liu *et al.*, 2002). The presence of a certain species in a location within the granule is governed by the concentration of a specific substrate (e.g. acetate for methanogen and sucrose for acetogen) at that location. The concentration of a substrate in a location within the granule is in turn governed by the intra-granular diffusion and rate of formation (acetate by acetogen) or consumption (acetate by methanogen). Thus, the layers in a granule may form due to the interplay of the rate of intra-granular diffusion of substrate and the reaction rates of different steps in the degradation (Tiwari *et al.*, 2006).

### Upflow velocity and gas production

The upflow velocity of influent and the superficial velocity of biogas have an impact on granules. At upflow velocities above 1 m/h, the granules may disintegrate due to shear stress, and the resulting fragments may wash out of the reactor. Vigorous gas evolution at high OLR may cause the bacterial cells to shear-off from granule surface, eroding the granules. The upflow velocity is generally kept between 1 and 6 m/h and values up to 10 m/h can be employed in the EGSB reactors (Abbasi & Abbasi, 2012). In this way, it is possible to achieve a good expansion of the sludge bed, ensuring the effective contact between biomass and wastewater, but avoiding, at the same time, granules disintegration and biomass dragging in the effluent (Stazi *et al.*, 2022).

#### Cations and heavy metals

The presence of positive divalent and trivalent ions, such as $Ca^{2+}$, $Mg^{2+}$, $Fe^{2+}$, and $Fe^{3+}$, could bind to negatively charged cells to form a microbial nuclei. Therefore, they could be a key factor in granulation phenomenon (Liu *et al.*, 2002). In addition, multivalent cations condense the diffused electric double layer and facilitate flocculation due to van der Waals forces (Liu *et al.*, 2003). The predominant binding groups for metals on the surface of bacteria are carboxyl and amino groups in proteins. Heavy metals compete with other ions in the solution for these binding sites on the cell surface. The relative toxicities of some metals depend on pH, VFA concentration, HRT, type and form of metal ions, and strength and affinity of the binding groups present on the surfaces of prevalent microorganisms. The presence of inert solids in the granules offers some abiotic surfaces to interact with the metal ions and, in turn, increases the toxicity resistances of granules (Tiwari *et al.*, 2006).

#### Synthetic and natural polymers

One of the important factors for the development of granules from non-granular sludge is the presence of nuclei or biocarrier for microbial attachment growth. Synthetic and natural polymers have been widely used in coagulation/flocculation processes. These polymers can promote particle agglomeration (Liu *et al.*, 2002). In anaerobic reactors, polymers are mainly used either to immobilize the anaerobic sludge or to reinforce the strength of the already existing granules by coating the granule surface with a thin layer of polymer (Liu *et al.*, 2003). Adsorption of polymers on the surface of the dispersed bacteria and neutralization of their surface charges is one of the principal mechanisms to promote anaerobic granulation (Kalogo *et al.*, 2001).

### 4.3.5 Case study: granular UASB operated in sequential mode

This section reports the recent experimental study conducted by Stazi *et al.* (2022) on the feasibility of an UASB, working as sequencing batch reactor (SBR), for the treatment of a synthetic DWW at COD concentration of ≈500 mg/L. Tests were conducted at different temperatures (35°C, 25°C and 15°C), representative of three climatic zones, and different HRTs. The SBRs are characterized by high flexibility given by the modularity. For this reason, they can be proposed for DWW treatment plants (WWTPs) characterized by marked variability of the influent load (i.e., seasonal variation in touristic areas). Moreover, they allow a detailed study of the kinetic trends, according to the different temperatures tested, thanks to the sequential operation mode: working on time base enables a direct detection and analysis of the process kinetics. Another aspect investigated, and still few explored in the literature at experimental level, deals with the quantification, through a complete mass balance, of $dCH_4$ in the effluent at different operating conditions.

The experimental set-up of the laboratory SBR-UASB reactor is shown in Figure 4.4. The jacketed glass reactor, with a working volume of 0.9 L, was equipped with three peristaltic pumps for wastewater feeding (0.5 L each cycle), internal liquid recirculation during the reaction phase, and treated effluent discharge (0.5 L each cycle), and with two 0.5 L gas bags to equalize the pressure and maintain anaerobic conditions inside the reactor. The different temperature conditions were controlled by a circulation thermo-cryostat connected to the reactor jacket. Each operating cycle consisted of four phases: feeding, reaction, biomass sedimentation, and effluent discharge. The reaction phase was operated with liquid recirculation from the top to the bottom of the reactor to expand the sludge bed, thus ensuring good contact between the biomass and the wastewater. The up-flow velocity of the

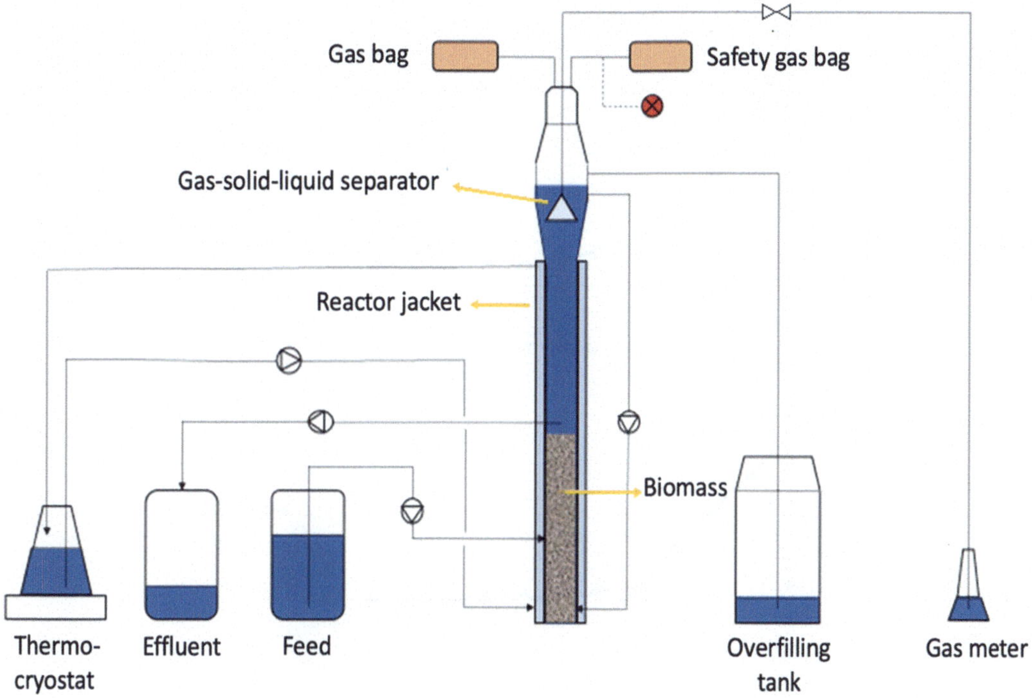


**Figure 4.4** Experimental set-up of the laboratory SBR-UASB reactor tested by Stazi *et al.* (2022).

recirculated liquid was maintained at 8 m/h. This value has been defined, according to the literature (Abbasi & Abbasi, 2012), as the best compromise to achieve an effective expansion of the sludge bed, avoiding, at the same time, biomass dragging. At each temperature, different HRTs have been tested starting from 22 h and gradually decreasing this value.

The UASB reactor exhibited excellent removal efficiencies of the organic matter at 35°C and 25°C for all HRTs tested (22, 14, 9 h). At 15°C there was a reduction of the COD biodegradation kinetics and tests have been stopped at HRT of 12 h because at this value the effluent COD exceeded the limits of 125 mg/L, imposed by the Italian regulations for discharging into receiving water bodies (DLgs 152/2006). The typical COD removal rates obtained, expressed as COD/CODt0, where t0 is the starting time of the reaction phase, are reported in Figure 4.5. At all three temperature conditions it is observed as a higher maximum COD removal rate (given by the slope of the COD concentration curve), as the HRT decreases. This phenomenon may be due to the effect of biomass acclimatization as the tests have been conducted in the sequence of decreasing HRT. The trends shown in Figure 4.3.3 reveal that COD removal kinetics are not significantly different at 35°C and 25°C, while at 15°C the effect of acclimatization is less evident. The results confirmed that the temperature greatly affects the UASB performance, hindering the diffusion of this technology in European countries, where temperatures drop under 15°C in winter.

The UASB reactor revealed similar and good performance at 25°C and 35°C, without substantial differences by varying the HRT: in particular COD removal and specific biogas production were, respectively, 84–94% and 0.14–0.27 $m^3$-biogas/kg-$COD_{removed}$.

In their study, Stazi *et al.* (2022) analysed several parameters. A summary of the values obtained is shown in Table 4.2 for the lowest HRTs, tested at each temperature, which allowed to reduce COD concentration as required by the Italian regulation. Higher HRTs, indeed, do not make the technology

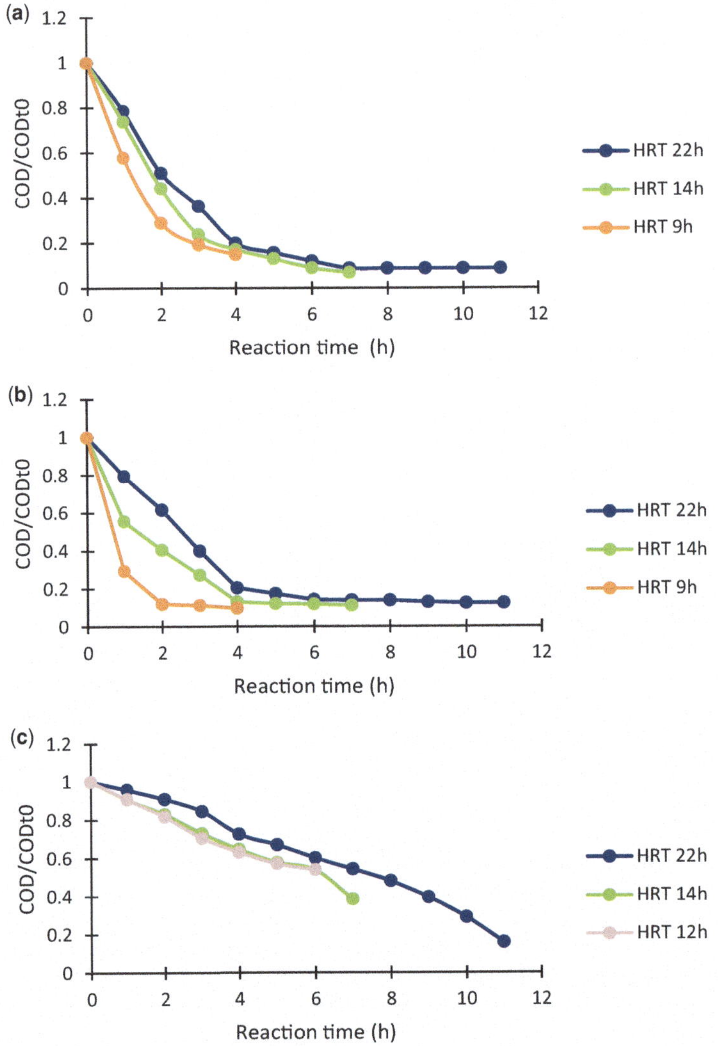


**Figure 4.5** COD removal kinetics as a function of HRT and at different operating temperatures: (a) 35°C; (b) 25°C; (c) 15°C (modified from Stazi *et al.*, 2022).

competitive with the traditional aerobic treatment. Except for the COD removal percentage, and its consequent concentration in the treated wastewater which is in the range 26–44 mg/L at 25°C and 35°C and 93–110 mg/L at 15°C, it can be observed that all the other parameters are comparable. In particular, VFA concentration always stays below 6 $mg_{acetic}$/L and total solids (TS) concentration between 12 and 17 mg/L. In addition, N and P concentrations remain unchanged with respect to the fed wastewater, exceeding the effluent standards, set at 10–15 and 1–2 mg/L, respectively, but without altering the system stability. The best option for these nutrients is their recovery for fertilizing purposes. It is noteworthy how less organic matter removal does not correspond to less biogas production as might be expected. This phenomenon can be due to the entrapment of gas bubbles inside the granular biomass, making it difficult to find, for such systems, a correlation between COD consumption and

**Table 4.2** Summary of SBR-UASB experimental results obtained by Stazi *et al.* (2022).

| | T=35°C HRT=9 h | T=25°C HRT=9 h | T=15°C HRT=14 h |
|---|---|---|---|
| Influent COD (mg/L) | 544–560 | 488–540 | 472–508 |
| Effluent COD (mg/L) | 29–39 | 26–44 | 93–110 |
| COD removal (%) | 89–91 | 85–92 | 64–68 |
| Biogas production ($m^3/kgCOD_{removed}$) | 0.14–0.21 | 0.15–0.25 | 0.11–0.34 |
| Effluent VFA ($mg_{acetic}/L$) | 0 | 2.1–2.2 | 4.2–5.4 |
| Effluent N-$NH_3$ (mg/L) | 59–60 | 51–54 | 36–46 |
| Effluent P-$PO_4$ (mg/L) | 6.8–7.4 | 7.2–7.6 | 6.4–6.8 |
| Effluent TS (mg/L) | 13–17 | 10–17 | 12–17 |

biogas production on the basis of the theoretical value expected by the process chemical reactions. This technological aspect could be improved by increasing the recirculation rate, within the limits of granules disintegration, thus allowing a more effective fluidization of the sludge bed.

The shortest HRT of 9 h reached at 25°C and 35°C, comparable to that of conventional aerobic treatment systems, was deeply investigated with additional measurement of the $CH_4$ fraction in the biogas produced and concentration of d$CH_4$ in the treated effluent. The biogas production and its $CH_4$ fraction were, respectively, 41 N mL and 74–77% at 35°C, and 32 N mL and 53% at 25°C, while d$CH_4$ in the effluent was 5.45–8.72 mg/L at 35°C and 7.13–7.54 mg/L at 25°C. It followed that, with respect to the total $CH_4$ produced, the fraction that is lost for energy purposes, as it is dissolved in the liquid phase, amounts to 36–38% at 25°C and 20–29% at 35°C. The values obtained for $CH_4$ losses with the effluent reveal that higher losses are achieved in these systems at lower temperatures as $CH_4$ solubility increases with decreasing temperatures. Taking into account the measured specific biogas productions and the percent fraction of $CH_4$ in the biogas, specific $CH_4$ production rates were in the range of 0.07–0.14 $m^3CH_4/kgCOD_{removed}$ at 25°C and 0.1–0.18 $m^3CH_4/kgCOD_{removed}$ at 35°C. The measured $CH_4$ yields resulted lower than the theoretical value, calculated according to the stoichiometric anaerobic conversion of glucose into $CH_4$ and $CO_2$ (0.35 $Nm^3CH_4/kgCOD_{removed}$). Possible reasons, in addition to the wastewater composition, are the partial entrapment of gas bubbles in the granular bed and $CH_4$ oversaturation in the liquid phase, with respect to the equilibrium concentration calculated according to Henry's law. This latter phenomenon may also be due to the accumulation of $CH_4$ in the liquid phase of the reactor as an effect of the sequential mode.

The case study here reported the problem of d$CH_4$ in anaerobic effluents of high-rate reactors for its consequent GHG emissions in the atmosphere and energy loss. In the literature, several methods have been proposed to recover or remove $CH_4$ from anaerobic effluents, such as biological oxidation in down-flow hanging sponge reactors, air stripping, mechanical or gas mixing, and membrane-based recovery. Limitations and perspectives for d$CH_4$ recovery or reuse and new methods as post-treatment of anaerobic effluents are widely discussed in Stazi and Tomei (2021).

### 4.3.6 Discussion of the results and future developments of granular biomass systems

Anaerobic granulation is a complex process in which a number of thermodynamic, physical, chemical, and microbiological parameters are involved. Although some mechanisms and models for anaerobic granulation have been proposed, there is still lack of a comprehensive understanding of these processes.

Granular sludge bioreactors are recognized as cost-effective and efficient technologies for the anaerobic treatment of both industrial and municipal wastewaters. High-rate reactor designs, such as UASB, EGSB, and the IC, have allowed for the application of granular sludge technologies to the treatment of a wide range of wastes, from high-strength to very low strength as DWWs, and even in a large range of temperatures. Among them, the system much employed worldwide is represented by UASB bioreactors. One of the biggest challenges regarding UASB technology, operated with granular biomass, is to shorten the start-up time of the reactor by speeding up granule formation. This process is affected by many physicochemical parameters and operating conditions and various external additives have shown promising results in this direction.

The experimental UASB-SBR case study, described by Stazi *et al.* (2022) and here reported, focused on the main aspects which need to be improved for exploiting all the potentialities of UASB bioreactors in DWW treatment: effluent quality, energy and nutrients recovery, and reduction of $CH_4$ losses. The study showed how this technology can successfully treat DWW. Moreover, biogas production and potential nutrients recovery confirm the relevance of anaerobic digestion as sustainable treatment of low-strength wastewaters. The results of the proposed study can be used to provide practical information and suggest lines of investigations on UASB reactors, in the treatment of low-strength wastewaters, especially at temperatures typical of European latitudes, and at HRTs comparable with traditional aerobic WWTPs.

As already discussed for immobilized biomass systems, further investigations are required to formulate detailed process models able to give an exhaustive representation of the complexity of the biological granular systems both at micro (single granule) and macro (bioreactor) scale as well as the effects of the hydraulic regime on the granule maintenance.

In conclusion, we can say that the increased need of sustainability in wastewater treatment systems reveals the importance of capture of the full energy and the water and nutrient resource potential contained in wastewater, thus underlining the importance to improve the performance of anaerobic systems working with granular biomass, such as UASB reactors, in order to realize their full-scale application in a wider spectrum of climate regions.

## REFERENCES

Abbasi T. and Abbasi S. A. (2012). Formation and impact of granules in fostering clean energy production and wastewater treatment in upflow anaerobic sludge blanket (UASB) reactors. *Renewable and Sustainable Energy Reviews*, **16**(3), 1696–1708, https://doi.org/10.1016/j.rser.2011.11.017

Ahmad I., Chelliapan S., Othman N., Abdullah N. and Arumugam N. (2020). Start-up of modified anaerobic baffled reactor for the treatment of landfill leachate by using granular sludge. *Journal of Environmental Treatment Techniques*, **8**(3), 988–992.

Akil K. and Jayanthi S. (2012). Anaerobic sequencing batch reactors and its influencing factors: an overview. *Journal of Environmental Science and Engineering*, **54**(2), 317–322.

Angenent L. T. and Sung S. (2001). Development of anaerobic migrating blanket reactor (AMBR), a novel anaerobic treatment system. *Water Research*, **35**, 1739–47, https://doi.org/10.1016/S0043-1354(00)00447-4

Bachmann A., Beard V. L. and McCarty P. L. (1985). Performance characteristics of the anaerobic baffled reactor. *Water Research*, **19**(1), 99–106, https://doi.org/10.1016/0043-1354(85)90330-6

Bandara W. M. K. R. T. W., Kindaichi T., Satoh H., Sasakawa M., Nakahara Y., Takahashi M. and Okabe S. (2012). Anaerobic treatment of municipal wastewater at ambient temperature: analysis of archaeal community structure and recovery of dissolved methane. *Water Research*, **46**(17), 5756–5764, https://doi.org/10.1016/j.watres.2012.07.061

Barber W. P. and Stuckey D. C. (1999). The use of the anaerobic baffled reactor (ABR) for wastewater treatment: a review. *Water Research*, **33**(7), 1559–1578, https://doi.org/10.1016/S0043-1354(98)00371-6

Barros P., Ruiz I. and Soto M. (2008). Performance of an anaerobic digester-constructed wetland system for a small community. *Ecological Engineering*, **33**(2), 142–149, https://doi.org/10.1016/j.ecoleng.2008.02.015

Batstone D. J., Keller J. and Blackall L. L. (2004). The influence of substrate kinetics on the microbial community structure in granular anaerobic biomass. *Water Research*, **38**, 1390–1404, https://doi.org/10.1016/j.watres.2003.12.003

Bodík I., Herdová B. and Drtil M. (2002). The use of upflow anaerobic filter and AnSBR for wastewater treatment at ambient temperature. *Water Research*, **36**(4), 1084–1088, https://doi.org/10.1016/S0043-1354(01)00308-6

Bodkhe S. (2008). Development of an improved anaerobic filter for municipal wastewater treatment. *Bioresource Technology*, **99**(1), 222–226, https://doi.org/10.1016/j.biortech.2006.11.026

Cammarota M. C. and Sant'Anna G. L. (1998). Metabolic blocking of exopolysaccharides synthesis: effects on microbial adhesion and biofilm accumulation. *Biotechnology Letters*, **20**, 1–4, https://doi.org/10.1023/A:1005394325549

Cardoso Grangeiro L., Coelho de Almeida S. G., Sampaio de Mello B., Fuess L. T., Sarti A. and Dussan K. J. (2019). New trends in biogas production and utilization. In: Sustainable Bioenergy, M. Rai and A. P. Ingle (eds.), Elsevier, Amsterdam (Netherlands), pp. 199–223.

Chernicharo C. A. L., Almeida P. G. S., Lobato L. C. S., Couto T. C., Borges J. M. and Lacerda Y. S. (2009). Experience with the design and start-up of two full-scale UASB plants in Brazil: enhancements and drawbacks. *Water Science and Technology*, **60**(2), 507–515, https://doi.org/10.2166/wst.2009.383

Chernicharo C. A. L., Van Lier J. B., Noyola A. and Bressani Ribeiro T. (2015). Anaerobic sewage treatment: state of the art, constraints and challenges. *Reviews in Environmental Science and Bio/Technology*, **14**, 649–679, https://doi.org/10.1007/s11157-015-9377-3

Cookney J., Mcleod A., Mathioudakis V., Ncube P., Soares A., Jefferson B. and McAdam E. J. (2016). Dissolved methane recovery from anaerobic effluents using hollow fibre membrane contactors. *Journal of Membrane Science*, **502**, 141–150, https://doi.org/10.1016/j.memsci.2015.12.037

Crone B. C., Garland J. L., Sorial G. A. and Vane L. M. (2016). Significance of dissolved methane in effluents of anaerobically treated low strength wastewater and potential for recovery as an energy product: a review. *Water Research*, **104**, 520–531, https://doi.org/10.1016/j.watres.2016.08.019

Decree Italian Law 152/2006. Regulations on Environmental Matters. Available online: https://www.gazzettaufficiale.it/atto/serie_generale/caricaDettaglioAtto/originario?atto.dataPubblicazioneGazzetta=2006-04-14&atto.codiceRedazionale=006G0171

Dev S., Saha S., Kurade M. B., Salama E. S., El-Dalatony M. M., Ha G. S., Chang S. W. and Jeon B. H. (2019). Perspective on anaerobic digestion for biomethanation in cold environments. *Renewable and Sustainable Energy Reviews*, **103**, 85–95, https://doi.org/10.1016/j.rser.2018.12.034

Dutta K., Tsai C. Y., Chen W. H. and Lin J. G. (2014). Effect of carriers on the performance of anaerobic sequencing batch biofilm reactor treating synthetic municipal wastewater. *International Biodeterioration & Biodegradation*, **95**, 84–88, https://doi.org/10.1016/j.ibiod.2014.04.021

Elmitwalli T. A., Oahn K. L., Zeeman G. and Lettinga G. (2002a). Treatment of domestic sewage in a two-step anaerobic filter/anaerobic hybrid system at low temperature. *Water Research*, **36**(9), 2225–2232, https://doi.org/10.1016/S0043-1354(01)00438-9

Elmitwalli T. A., Sklyar V., Zeeman G. and Lettinga G. (2002b). Low temperature pre-treatment of domestic sewage in an anaerobic hybrid or an anaerobic filter reactor. *Bioresource Technology*, **82**(3), 233–239, https://doi.org/10.1016/S0960-8524(01)00191-2

Fang H. H. P., Chui H. K. and Li Y. Y. (1995). Effect of degradation kinetics on the microstructure of anaerobic biogranules. *Water Science and Technology*, **32**, 165–172, https://doi.org/10.2166/wst.1995.0289

Feng H., Hu L., Mahmood Q., Qiu C., Fang C. and Shen D. (2008). Anaerobic domestic wastewater treatment with bamboo carrier anaerobic baffled reactor. *International Biodeterioration & Biodegradation*, **62**(3), 232–238, https://doi.org/10.1016/j.ibiod.2008.01.009

Foresti E. (2001). Perspectives on anaerobic treatment in developing countries. *Water Science and Technology*, **44**(8), 141–148, https://doi.org/10.2166/wst.2001.0485

Ghangrekar M. M., Asolekar S. R. and Joshi S. G. (2005). Characteristics of sludge developed under different loading conditions during UASB reactor start-up and granulation. *Water Research*, **39**, 1123–33, https://doi.org/10.1016/j.watres.2004.12.018

Gomec C. Y. (2010). High-rate anaerobic treatment of domestic wastewater at ambient operating temperatures: a review on benefits and drawbacks. *Journal of Environmental Science and Health, Part A*, **45**(10), 1169–1184, https://doi.org/10.1080/10934529.2010.493774

Gouveia J., Plaza F., Garralon G., Fdz-Polanco F. and Peña M. (2015). Long-term operation of a pilot scale anaerobic membrane bioreactor (AnMBR) for the treatment of municipal wastewater under psychrophilic conditions. *Bioresource Technology*, **185**, 225–233, https://doi.org/10.1016/j.biortech.2015.03.002

Grotenhuis J. T., Smit M., Plugge C. M., Xu Y. S., Van Lammeren A. A., Stams A. J. and Zehnder A. J. (1991). Bacteriological composition and structure of granular sludge adapted to different substrates. *Applied and Environmental Microbiology*, **57**(7), 1942–1949, https://doi.org/10.1128/aem.57.7.1942-1949.1991

Harb M., Xiong Y., Guest J., Amy G. and Hong P. Y. (2015). Differences in microbial communities and performance between suspended and attached growth anaerobic membrane bioreactors treating synthetic municipal wastewater. *Environmental Science: Water Research & Technology*, **1**(6), 800–813, https://doi.org/10.1039/C5EW00162E

Hermansson A., Jacobsson S., de Jonge N., Nielsen J. L. and Morgan-Sagastume F. (2022). Impact of the restraint of biofilm volume and thickness on the performance and microbial composition in anaerobic moving-bed biofilm reactors (AnMBBRs). *Journal of Environmental Chemical Engineering*, **10**(3), 107741, https://doi.org/10.1016/j.jece.2022.107741

Hulshoff Pol L. W., De Zeeuw W. J., Velzeboer C. T. M. and Lettinga G. (1983). Granulation in UASB reactors. *Water Science and Technology*, **15**(8/9), 291–304, https://doi.org/10.2166/wst.1983.0172

Hulshoff Pol L. W., de Castro Lopes S. I., Lettinga G. and Lens P. N. L. (2004). Anaerobic sludge granulation. *Water Research*, **38**, 1376–89, https://doi.org/10.1016/j.watres.2003.12.002

Kaetzl K., Lübken M., Gehring T. and Wichern M. (2018). Efficient low-cost anaerobic treatment of wastewater using biochar and woodchip filters. *Water*, **10**(7), 818, https://doi.org/10.3390/w10070818

Kalogo Y., Seka A. M. and Verstraete W. (2001). Enhancing the startup of a UASB reactor treating domestic wastewater by adding a water extract of *Moringa oleifera* seeds. *Applied Microbiology and Biotechnology*, **55**, 644–651, https://doi.org/10.1007/s002530000544

Kennedy K. J. and Lentz E. M. (2000). Treatment of landfill leachate using sequencing batch and continuous upflow anaerobic sludge blanket (UASB) reactors. *Water Research*, **34**, 3640–3656, https://doi.org/10.1016/S0043-1354(00)00114-7

Khan A. A., Gaur R. Z., Mehrotra I., Diamantis V., Lew B. and Kazmi A. A. (2014). Performance assessment of different STPs based on UASB followed by aerobic post treatment systems. *Journal of Environmental Health Science and Engineering*, **12**, 1–, https://doi.org/10.1186/2052-336X-12-1

Kobayashi H. A., Stenstrom M. K. and Mah R. A. (1983). Treatment of low strength domestic wastewater using the anaerobic filter. *Water Research*, **17**(8), 903–909, https://doi.org/10.1016/0043-1354(83)90164-1

Lee H. S., Tang Y., Rittmann B. E. and Zhao H. P. (2018). Anaerobic oxidation of methane coupled to denitrification: fundamentals, challenges, and potential. *Critical Reviews in Environmental Science and Technology*, **48**(19–21), 1067–1093, https://doi.org/10.1080/10643389.2018.1503927

Lettinga G., van Velsen A. F. M., Hobma S. W., de Zeeuw W. and Klapwijk A. (1980). Use of the upflow sludge blanket (USB) reactor concept for biological wastewater treatment especially for anaerobic treatment. *Biotechnology and Bioengineering*, **22**, 699–734, https://doi.org/10.1002/bit.260220402

Lettinga G., Rebac S. and Zeeman G. (2001). Challenge of psychrophilic anaerobic wastewater treatment. *Trends in Biotechnology*, **19**(9), 363–370, https://doi.org/10.1016/S0167-7799(01)01701-2

Li J. and Lu X. (2017). Effect of seasonal temperature on the performance and on the microbial community of a novel AWFR for decentralized domestic wastewater pretreatment. *Applied Sciences*, **7**(6), 605, https://doi.org/10.3390/app7060605

Lim S. J. and Kim T. H. (2014). Applicability and trends of anaerobic granular sludge treatment processes. *Biomass and Bioenergy*, **60**, 189–202, https://doi.org/10.1016/j.biombioe.2013.11.011

Liu Y. and Tay J. H. (2004). State of the art of biogranulation technology for wastewater treatment. *Biotechnology Advances*, **22**(7), 533–563, https://doi.org/10.1016/j.biotechadv.2004.05.001

Liu Y., Xu H. L., Show K. Y. and Tay J. H. (2002). Anaerobic granulation technology for wastewater treatment. *World Journal of Microbiology and Biotechnology*, **18**, 99–113, https://doi.org/10.1023/A:1014459006210

Liu Y., Xu H. L., Yang S. F. and Tay J. H. (2003). Mechanisms and models for anaerobic granulation in upflow anaerobic sludge blanket reactor. *Water Research*, **37**(3), 661–673, https://doi.org/10.1016/S0043-1354(02)00351-2

López-López A., Albarrán-Rivas M. G., Hernández-Mena L. and León-Becerril E. (2013). An assessment of an anaerobic filter packed with a low-cost material for treating domestic wastewater. *Environmental Technology*, **34**(9), 1151–1159, https://doi.org/10.1080/09593330.2012.743588

Loupasaki E. and Diamadopoulos E. (2012). Attached growth systems for wastewater treatment in small and rural communities: a review. *Journal of Chemical Technology & Biotechnology*, **88**(2), 190–204, https://doi.org/10.1002/jctb.3967

Maaz M., Yasin M., Aslam M., Kumar G., Atabani A. E., Idrees M., Anjum F., Jamil F., Ahmad R., Khan A. L., Lesage G., Heran M. and Kim J. (2019). Anaerobic membrane bioreactors for wastewater treatment: novel configurations, fouling control and energy considerations. *Bioresource Technology*, **283**, 358–372, https://doi.org/10.1016/j.biortech.2019.03.061

MacLeod F. A., Guiot S. R. and Costerton J. W. (1990). Layered structure of bacterial aggregates produced in an upflow anaerobic sludge bed and filter reactor. *Applied and Environmental Microbiology*, **56**, 1598–1607, https://doi.org/10.1128/aem.56.6.1598-1607.1990

Martin M. A., De la Rubia M. A., Martín A., Borja R., Montalvo S. and Sánchez E. (2010). Kinetic evaluation of the psychrophylic anaerobic digestion of synthetic domestic sewage using an upflow filter. *Bioresource Technology*, **101**(1), 131–137, https://doi.org/10.1016/j.biortech.2009.08.010

Martin Garcia I., Mokosch M., Soares A., Pidou M. and Jefferson B. (2013). Impact on reactor configuration on the performance of anaerobic MBRs: treatment of settled sewage in temperate climates. *Water Research*, **47**(14), 4853–4860, https://doi.org/10.1016/j.watres.2013.05.008

McHugh S., O'reilly C., Mahony T., Colleran E. and O'flaherty V. (2003). Anaerobic granular sludge bioreactor technology. *Reviews in Environmental Science and Biotechnology*, **2**, 225–245, https://doi.org/10.1023/B:RESB.0000040465.45300.97

McKeown R. M., Scully C., Enright A. M., Chinalia F. A., Lee C., Mahony T., Collins G. and O'Flaherty V. (2009). Psychrophilic methanogenic community development during long-term cultivation of anaerobic granular biofilms. *The ISME Journal*, **3**(11), 1231–1242, https://doi.org/10.1038/ismej.2009.67

Montalvo S., Guerrero L., Borja R., Sánchez E., Milán Z., Cortés I. and De La La Rubia M. A. (2012). Application of natural zeolites in anaerobic digestion processes: a review. *Applied Clay Science*, **58**, 125–133, https://doi.org/10.1016/j.clay.2012.01.013

Morgan J. W., Forster C. F. and Evison L. J. W. R. (1990). A comparative study of the nature of biopolymers extracted from anaerobic and activated sludges. *Water Research*, **24**(6), 743–750, https://doi.org/10.1016/0043-1354(90)90030-A

Mu Y., Yu H. Q. and Wang G. (2006). Permeabilities of anaerobic $CH_4$-producing granules. *Water Research*, **40**(9), 1811–1815, https://doi.org/10.1016/j.watres.2006.03.006

Noyola A. (2004). Anaerobic digestion applied to municipal wastewater treatment: facts and limitations of an adapted technology for Latin America. In: WEFTEC 2004 (10), Water Environment Federation, Alexandria, VA (USA), pp. 1–12.

Noyola A., Capdeville B. and Roques H. (1988). Anaerobic treatment of domestic sewage with a rotating stationary fixed-film reactor. *Water Research*, **22**(12), 1585–1592, https://doi.org/10.1016/0043-1354(88)90172-8

Pentassuglia S., Agostino V. and Tommasi T. (2018). EAB–Electroactive biofilm: a biotechnological resource. In: Encyclopedia of Interfacial Chemistry, K. Wandelt (ed.), Elsevier, Amsterdam (Netherlands), pp. 110–123.

Pentassuglia S., De Sanctis M., Altieri V. G., Sgherza D. and Di Iaconi C. (2020). Anaerobic sequencing biofilter (AnSBio): a novel approach for high-quality biogas production from low-strength wastewaters at ambient temperature. *Bioresource Technology Reports*, **12**, 100554, https://doi.org/10.1016/j.biteb.2020.100554

Ratusznei S. M., Rodrigues J. A. D., Camargo E. F., Zaiat M. and Borzani W. (2000). Feasibility of a stirred anaerobic sequencing batch reactor containing immobilized biomass for wastewater treatment. *Bioresource Technology*, **75**(2), 127–132, https://doi.org/10.1016/S0960-8524(00)00048-1

Reyes O., Sánchez E., Rovirosa N., Borja R., Cruz M., Colmenarejo M. F., Escobedo R., Ruiz M., Rodriguez X. and Correa O. (1999). Low-strength wastewater treatment by a multistage anaerobic filter packed with waste tyre rubber. *Bioresource Technology*, **70**(1), 55–60, https://doi.org/10.1016/S0960-8524(99)00010-3

Rittmann B. E. and McCarty P. L. (2012). Environmental Biotechnology: Principles and Applications. McGraw-Hill Series in Water Resources and Environmental Engineering. Tata McGraw-Hill Education Private Limited.

Rouxhet P. G. and Mozes N. (1990). Physical chemistry of the interaction between attached microorganisms and their support. *Water Science and Technology*, **22**, 1–16, https://doi.org/10.2166/wst.1990.0131

Sam-Soon P., Loewenthal R. E., Dold P. L. and Marais G. (1987). Hypothesis for pelletisation in the upflow anaerobic sludge bed reactor. *Water SA*, **13**(2), 69–80.

Sarti A. (1998). Performance evaluation of horizontal-flow anaerobic immobilized biomass (HAIB) reactor in the treatment of synthetic substrate simulating domestic sewage. Doctoral dissertation, MSc thesis, EESC/USP.

Sarti A., Pozzi E., Chinalia F. A., Zaiat M. and Foresti E. (2006). The performance of an anaerobic sequencing batch biofilm reactor treating domestic sewage colonized by anoxygenic phototrophic bacteria. *Chemosphere*, **62**(9), 1437–1443, https://doi.org/10.1016/j.chemosphere.2005.06.007

Sarti A., Garcia M. L., Zaiat M. and Foresti E. (2007). Domestic sewage treatment in a pilot-scale anaerobic sequencing batch biofilm reactor (ASBBR). *Resources, Conservation and Recycling*, **51**(1), 237–247, https://doi.org/10.1016/j.resconrec.2006.09.008

Schmidt J. E. and Ahring B. K. (1994). Extracellular polymers in granular sludge from different upflow anaerobic sludge blanket (UASB) reactors. *Applied Microbiology and Biotechnology*, **42**, 457–462.

Serrano León E., Vargas-Machuca J. A. P., Corona E. L., Arbib Z., Rogalla F. and Boizán M. F. (2018). Anaerobic digestion of municipal sewage under psychrophilic conditions. *Journal of Cleaner Production*, **198**, 931–939, https://doi.org/10.1016/j.jclepro.2018.07.060

Show K. Y. (2006). Mechanisms and models for anaerobic granulation. *Waste Management Series*, **6**, 1–33, https://doi.org/10.1016/S0713-2743(06)80103-0

Show K. Y., Wang Y., Foong S. F. and Tay J. H. (2004). Accelerated start-up and enhanced granulation in upflow anaerobic sludge blanket reactors. *Water Research*, **38**(9), 2293–2304, https://doi.org/10.1016/j.watres.2004.01.039

Skiadas I. V., Gavala H. N., Schmidt J. E. and Ahring B. K. (2003). Anaerobic granular sludge and biofilm reactors. In: Biomethanation II, B. K. Ahring, I. Angelidaki, J. Dolfing, L. Euegaard, H. N. Gavala, F. Haagensen, A. S. Mogensen, G. Lyberatos, P. F. Pind, J. E. Schmidt, I. V. Skiadas and K. Stamatelatou (eds.), Springer Berlin, Heidelberg (Germany), pp. 35–67.

Smith A. L., Skerlos S. J. and Raskin L. (2013). Psychrophilic anaerobic membrane bioreactor treatment of domestic wastewater. *Water Research*, **47**(4), 1655–1665, https://doi.org/10.1016/j.watres.2012.12.028

Smith A. L., Skerlos S. J. and Raskin L. (2015). Anaerobic membrane bioreactor treatment of domestic wastewater at psychrophilic temperatures ranging from 15°C to 3°C. *Environmental Science: Water Research & Technology*, **1**(1), 56–64, https://doi.org/10.1039/C4EW00070F

Stazi V. and Tomei M. C. (2018). Enhancing anaerobic treatment of domestic wastewater: state of the art, innovative technologies and future perspectives. *Science of the Total Environment*, **635**, 78–91, https://doi.org/10.1016/j.scitotenv.2018.04.071

Stazi V. and Tomei M. C. (2021). Dissolved methane in anaerobic effluents: a review on sustainable strategies for optimization of energy recovery or internal process reuse. *Journal of Cleaner Production*, **317**, 128359, https://doi.org/10.1016/j.jclepro.2021.128359

Stazi V., Annesini M. C. and Tomei M. C. (2022). Anaerobic domestic wastewater treatment in a sequencing granular UASB bioreactor: feasibility study of the temperature effect on the process performance. *Journal of Environmental Chemical Engineering*, **10**(5), 108512, https://doi.org/10.1016/j.jece.2022.108512

Tay J. H., Xu H. L. and Teo K. C. (2000). Molecular mechanism of granulation. I: $H^+$ translocation-dehydration theory. *Journal of Environmental Engineering*, **126**, 403–410, https://doi.org/10.1061/(ASCE)0733-9372(2000)126:5(403)

Thaveesri J., Daffonchio D., Lessens B., Vandermeren P. and Verstraete W. (1995). Granulation and sludge bed stability in upflow anaerobic sludge bed reactors in relation to surface thermodynamics. *Applied and Environmental Microbiology*, **61**, 3681–3686, https://doi.org/10.1128/aem.61.10.3681-3686.1995

Tiwari M. K., Guha S., Harendranath C. S. and Tripathi S. (2006). Influence of extrinsic factors on granulation in UASB reactor. *Applied Microbiology and Biotechnology*, **71**, 145–154, https://doi.org/10.1007/s00253-006-0397-3

Trego A. C., Mills S. and Collins G. (2021). Granular biofilms: function, application, and new trends as model microbial communities. *Critical Reviews in Environmental Science and Technology*, **51**(15), 1702–1725, https://doi.org/10.1080/10643389.2020.1769433

Van Haandel A., Kato M. T., Cavalcanti P. F. and Florencio L. (2006). Anaerobic reactor design concepts for the treatment of domestic wastewater. *Environmental Science and Bio/Technology*, **5**(1), 21–38, https://doi.org/10.1007/s11157-005-4888-y

Van Lier J. B., Van der Zee F. P., Frijters C. T. M. J. and Ersahin M. E. (2015). Celebrating 40 years anaerobic sludge bed reactors for industrial wastewater treatment. *Reviews in Environmental Science and Bio/Technology*, **14**, 681–702, https://doi.org/10.1007/s11157-015-9375-5

Vlyssides A., Barampouti E. M. and Mai S. (2008). Determination of granule size distribution in a UASB reactor. *Journal of Environmental Management*, **86**(4), 660–664, https://doi.org/10.1016/j.jenvman.2006.12.018

Wang K. (1994). Integrated anaerobic and aerobic treatment of sewage. PhD thesis, Agricultural University, Wageningen, The Netherlands.

WEF (Water Environment Federation) (2007). Design of Municipal Wastewater Treatment Plants MOP 11. 5th edn, WEF Press, Water Environment Federation, Alexandria, VA (USA).

Wiegant W. M. (1987). The 'spaghetti theory' on anaerobic sludge formation, or the inevitability of granulation. In: Granular Anaerobic Sludge: Microbiology and Technology, G. Lettinga, A. J. B. Zehnder, J. T. C. Grotenhuis and L. W. Hulshoff Pol (eds.), Pudoc. Wageningen, The Netherlands, pp. 146–52.

Wilschut J. and Hoekstra D. (1984). Membrane fusion: from liposome to biological membrane. *Trend in Biochemical Science*, **9**, 479–483, https://doi.org/10.1016/0968-0004(84)90316-5

Wu J., Afridi Z. U. R., Cao Z. P., Zhang Z. L., Poncin S., Li H. Z., Zuo J. E. and Wang K. J. (2016). Size effect of anaerobic granular sludge on biogas production: a micro scale study. *Bioresource Technology*, **202**, 165–171, https://doi.org/10.1016/j.biortech.2015.12.006

Xu H. L. and Tay J. H. (2001). Preserved granular sludge for inoculation of new UASB reactors. *Journal of Environmental Science and Health, Part A*, **36**(9), 1747–1756, https://doi.org/10.1081/ESE-100106256

Yang X. J., Tong Y. M., Song Y., Qi W. K., Li Y. Y. and Guo Y. L. (2017) Domestic sewage and secondary effluent treatment using vertical submerged biological filter. *IOP Conference Series: Earth and Environmental Science*, **82**(1), 012067.

Young J. C. and McCarty P. L. (1969). The anaerobic filter for waste treatment. *Journal (Water Pollution Control Federation)*, **41**(5), R160–R173.

Zhou H., Li X., Xu G. and Yu H. (2018). Overview of strategies for enhanced treatment of municipal/domestic wastewater at low temperature. *Science of the Total Environment*, **643**, 225–237, https://doi.org/10.1016/j.scitotenv.2018.06.100

doi: 10.2166/9781789063479_0127

# Chapter 5

# Post-treatment of anaerobically digested sewage for nutrient removal

A. V. Haandel[1]*, S. L. Santos[2], T. Allegue[3] and J. M. Garrido[4]

[1]Department of Civil Engineering, Universidad Federal of Campina Grande, Campina Grande- PB, Brazil
[2]Department of Civil Engineering, Universidad Federal of Rio Grande do Norte, Natal – RN, Brazil
[3]Department of Chemical Engineering, Khalifa University, Abu Dhabi, United Arab Emirates
[4]Department of Chemical Engineering, Universidade de Santiago de Compostela, CRETUS, Santiago de Compostela, Galicia, Spain
*Correspondence: adrianusvh@gmail.com

**ABSTRACT**

Although modern anaerobic sewage treatment systems and in particular the upflow anaerobic sludge blanket reactor (UASB) have demonstrated remarkable performance in tropical and subtropical regions, the effluent quality is generally not compatible with discharge standards in many countries, so post-treatment becomes necessary. In practice, the most widely applied post-treatment options are polishing ponds (PPs) and other aerobic treatment bioreactors. PPs are treatment units treating anaerobically pre-treated wastewater, which distinguishes them from conventional waste treatment ponds, wherein normally raw wastewater is treated. Anaerobic pre-treatment has a very profound influence on the configuration and operation of PPs, as will be shown in this chapter. Other aerobic post-treatments can be realized with conventional activated sludge systems, or more complex technological solutions including membrane bioreactors (MBRs) or trickling filters. The anaerobic–aerobic treatment has important advantages compared to the purely aerobic treatment: the volume and sludge production is much smaller, there is a possibility that the anaerobic–aerobic system operates with energetic self-sufficiency and the concentration of excess sludge is much higher facilitating its processing. An important problem of anaerobic–aerobic treatment is the difficulty of producing a final effluent with low nutrient concentration. Anaerobic sewage treatment reduces the content of biodegradable organic matter in the effluent, limiting in practice the removal of phosphorus to the use of chemical precipitation systems. For the case of nitrogen, the alternatives only go through new processes, which complement the poor capacity of conventional denitrification due to the limited concentration of biodegradable organic matter. This limits in practice, the nitrogen elimination capacity of activated sludge systems or trickling filters treating effluents from UASB reactors. A notable breakthrough could occur if reliable partial nitrification anammox processes for the water line were developed, as these are still under development, or systems in which the dissolved methane present in the anaerobic effluents was used to denitrify, as it has already been observed in some MBRs.

**Keywords:** aerobic post-treatment, anaerobic sewage digestion model for pH change, nutrient removal, organic material removal, pathogen removal, per capita area in polishing ponds, polishing ponds, transfer ponds.

## 5.1 INTRODUCTION

Anaerobic sewage treatment was applied in the first stage, to give adequate treatment to urban wastewater, in warm regions of the planet, through a system with low energy demand and scarce sludge production. This system was able to provide a cost-effective response to the elimination of oxygen-consuming organic substances that were previously discharged into the aquatic environment with untreated sewage. On the other hand, there is a growing concern about developing energy self-sufficient processes, even in other colder regions of the world as the US and Europe, for treating wastewater (Delgado Vela *et al.*, 2015; EEA, 2022), where energy recovery through the methanation of organic matter can be the way to increase sustainability of the WWTPs.

Although efficient, anaerobic sewage treatment does not produce an effluent quality that is compatible with most environmental standards, and post-treatment is required to reduce the concentrations of four components: organic material, nitrogen, phosphorus, and pathogens. Poor nutrient removal is one of the main drawbacks of the anaerobic sewage technology, so its application in nutrient-sensitive areas will require the development of post-treatment processes that effectively remove nitrogen and phosphorus compounds, avoiding those that might imply eliminating the energy benefits of anaerobic treatments. The elimination of a large fraction of the organic matter in anaerobic treatment system limits, in practice, the use of conventional biological nutrient elimination, using enhanced biological phosphorus removal (EBPR) or conventional nitrification–denitrification processes due to the low $BOD_5/P$ and $BOD_5/TN$ of anaerobic effluents.

The effluents from anaerobic upflow anaerobic sludge blanket (UASB) treating sewage contain COD values between 100 and 200 mg/L; $BOD_5$ 60–120 mg/L; $NH_4^+$-N 30–50 mg/L, and $PO_4^{3-}$-P 10–17 mg/L (Foresti *et al.*, 2006). Additionally, the presence of other compounds as dissolved methane and hydrogen sulphide, varies depending on temperature and sewage composition. In practice two post-treatment biological systems have been mostly applied for anaerobically treated sewage: (1) polishing ponds (PPs) and (2) other aerobic post-treatment systems.

Treatment through PP is an evolution of the waste stabilization ponds (WSP). WSP are the oldest communal sewage treatment units (Parker *et al.*, 1950). The proposed Australian system had the limited objective of removing organic material. To achieve this objective, the system was composed of a series of three ponds: an anaerobic pond (AP), for the removal of organic material by anaerobic digestion; followed by a facultative pond (FP), with mixed anaerobic and aerobic conditions for additional organic removal; and, one or more maturation ponds (MP) for further organic material removal in a predominantly aerobic environment. For the FP, the main design criterion is that there must be an equilibrium between the production of dissolved oxygen (DO), due to the photosynthesis of algae present in the pond, and DO consumption, due to the oxidation of organic material by bacteria. This means that the FP must be at least partially in an aerobic environment. In MPs, DO production is normally larger than DO consumption, so that MPs are predominantly aerobic.

Marais and Shaw (1961) established that WSPs were also efficient in removing pathogens: helminth eggs by sedimentation, and thermotolerant coliforms (TTC) decay with time. Marais (1974) established that TTC decay was a first-order process and optimized the MPs, showing the advantage of subdivision. Thus, a system was developed by combining the Australian system for organic material removal and Marais' system for pathogen removal, but for both purposes, a long retention time was required. In the 1960s and 1970s, it became clear that nutrient removal was necessary to avoid eutrophication. Although nutrient removal was attempted in WSPs, Pano and Middlebrooks (1982), Bastos *et al.* (2018), and Zimmo *et al.* (2003) have shown that nitrogen removal is at most partial and phosphorus removal is poor (Gomez *et al.* 2000). Thus, even though much research has been conducted (Ho *et al.*, 2017), much less progress has been made to improve the four main problems of WSPs: (1) large surface areas required, (2) the liberation of biogas from the AP releases odorous compounds and greenhouse gases into the atmosphere, (3) long outfall to combat odour problems for the contributing population and (4) poor effluent quality since efficient nutrient removal is impossible. Secondary problems due to the use of AP are (5) that the sewage suspended solids accumulate, so that operation must be interrupted to remove them every few years, and (6) due to the odour problem the WSP system must be constructed far away from the urban region, which means that a long and costly outfall is necessary.

Most of the disadvantages of anaerobic ponds are eliminated when they are replaced by an efficient anaerobic treatment unit like UASB (Lettinga *et al.*, 1980). The biogas is captured and can be used productively or flared off so that methane is converted into less offensive carbon dioxide. Also, the sewage suspended solids accumulate in the anaerobic reactor and can be discharged and transformed into a semi-solid or a solid, with the possibility of being used as an organic fertilizer. In addition, since at least in principle there is no biogas emission to the atmosphere, the odour problem does not exist and construction near or even within the urban region to be served is feasible. Today, more than 900 units are operating in Brazil alone (Van Haandel & Van der Lubbe, 2019; Van Lier *et al.*, 2020).

As anaerobic pre-treatment has a profound impact on all aspects of pond design and operation, it is convenient to use the term WSP for raw sewage treatment (Mara, 1997), and PP for post-treatment of digested sewage (Cavalcanti, 2003). For instance, the need for an aerobic condition in the FP integrated in WSPs, leads to a large area, because the oxidation of the AP effluent requires much oxygen, and the photosynthesis rate is relatively low because of the low transparency (presence of colloids) of the AP effluent. By contrast, after efficient anaerobic treatment photosynthesis develops at a higher rate in the clear effluent, whereas the oxygen demand is relatively low, so the oxygen production rate will always be greater than the oxygen consumption rate. This in turn means that there is an equivalent consumption of carbon dioxide to oxygen production, and this will trigger a pH increase in the PP, a feature that is not normally observed in WSPs. The pH increase makes nutrient removal feasible: nitrogen removal by desorption of ammonium gas and phosphorus by precipitation of phosphate.

The pH increase in the PP has a strong influence on the processes that develop, especially concerning the nutrients. In Figure 5.1, the fractions of gaseous ($NH_3$), and saline ammonia ($NH_4^+$) have been plotted as a function of pH. The $NH_3$ fraction increases from a very small value at pH = 7 (the likely pH of anaerobic effluent) to a fraction of 50% of the total ammonium concentration at pH = 9.3. The mechanism of nitrogen removal in PP is the desorption of gaseous ammonia from the liquid phase. The desorption/absorption processes can be described in terms of Fick's law, which says that the rate of the processes is proportional to the inductive force of those processes. In the case of volatile components ($CO_2$ and $NH_3$), the inductive force is the difference between the saturation concentration and the actual concentration.

In Figure 5.2, the fractions of $H_2PO_4^-$, $HPO_4^{2-}$, and $PO_4^{3-}$ are plotted as functions of pH. The removal mechanism of phosphate in PP is due to $PO_4^{3-}$ precipitation with cations that may be present

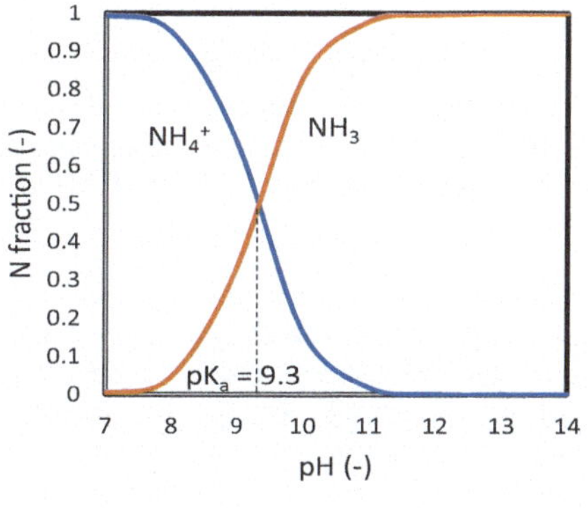


**Figure 5.1** $NH_3$ and $NH_4^+$ fractions as a function of pH.

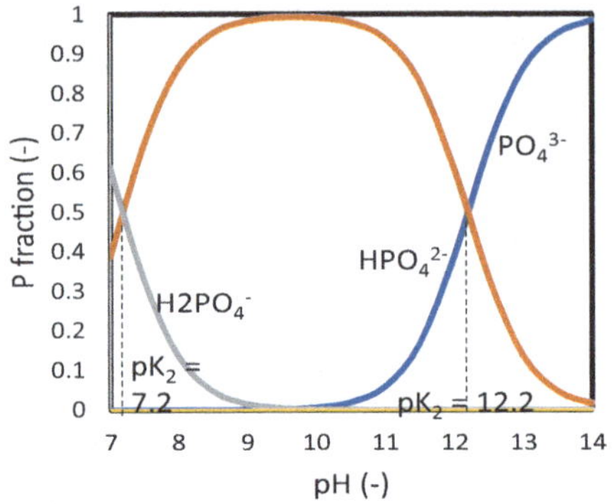


**Figure 5.2** $H_2PO_4^-$, $HPO_4^{2-}$ and $PO_4^{3-}$ fractions as a function of pH ($H_3PO_4$ is not shown).

in the wastewater (like $Ca^{2+}$) or may be added during or after the treatment in the PP (like $Al^{3+}$). Figure 5.2 shows that the pH must be high before a considerable $PO_4^{3-}$ fraction is formed and precipitation can take place. When Figures 5.1 and 5.2 are compared it is concluded that nitrogen will be removed before phosphate removal will occur.

In anaerobic–aerobic treatment systems, an aerobic post-treatment system like activated sludge or biofilm bioreactor is preceded by efficient anaerobic treatment for example in a UASB reactor. In principle, this anaerobic–aerobic set-up can produce a final effluent equivalent to purely aerobic systems except, as stated, for the possibility of removing nutrients efficiently: the removal of a large part of the biodegradable material in the anaerobic pre-treatment is an impediment for removal of nitrogen through nitrification and heterotrophic denitrification and of phosphorus through luxury uptake.

Important advantages of anaerobic pre-treatment are that all factors of investment and operational costs are strongly reduced: the anaerobic–aerobic system can be constructed with a much smaller volume than purely aerobic systems, and the costs for aeration energy and destination of sludge production are also significantly reduced. In many anaerobic–aerobic treatment systems, the UASB reactor is not only used for digestion of influent organic material but also for the stabilization of aerobic excess sludge. In this chapter, it will be shown that this procedure often leads to very serious operational problems and for that reason is not recommendable without a suitable adaptation which will be presented in the following sections.

Chemical precipitation of phosphorus is the only alternative available to reduce its concentration in wastewater with low organic matter content, such as anaerobic bioreactors effluents. For chemical precipitation, calcium, aluminium, or iron salts are used, which form an insoluble metal phosphate salt that can be removed by settling. Hydrated aluminium sulphate ($Al_2(SO_4)_3{\cdot}14H_2O$) is the most common aluminium salt used to remove phosphate. One mole, that is 594 g salt, reacts with 2 moles of phosphate containing 61.9 g of phosphorus to form 244 g of $AlPO_4$.

$$PO_4^{3-} + Al^{3+} \rightarrow AlPO_4 \downarrow$$

Iron salts are commonly used in the precipitation of phosphate from wastewater. Both ferrous ($Fe^{2+}$) and ferric ($Fe^{3+}$) ions can be used in the form of chloride or sulphate. If ferrous salt is used, this should be oxidized to the ferric ion ($Fe^{3+}$), using oxygen, this should be taken in the oxygen balance. Stoichiometrically, 55.85 g of Fe ion will react with 30.95 g of P to form 150.8 g of $FePO_4$.

$$PO_4^{3-} + Fe^{3+} \rightarrow FePO_4 \downarrow$$

Unluckily, the metal ions should be present in excess relative to the phosphorus ion concentration to achieve a low level of dissolved phosphorus compounds, increasing the expected cost of chemical precipitation. The EPA Design Manual (1976) suggests that 20% excess of aluminium ions is necessary if an 80% reduction of phosphorus is desired, while 100% excess is recommended for 95% phosphorus removal. The exact dosage is determined by on-site testing and varies with the specific characteristics of the sewage.

Lime is used as an alkalinity and calcium source to promote chemical P precipitation. Although it is possible to calculate an approximate lime dose for phosphorus removal, this is generally not necessary. In contrast to iron and aluminium salts, the lime dose is largely determined by other reactions that take place when the pH of wastewater is raised. Only in waters of very low bicarbonate alkalinity would the phosphate precipitation reaction consume a large fraction of the added lime. Lime, which previously was one of the main chemicals used for phosphorus removal is nowadays less frequently used because of increased production of sludge as well as the operational and maintenance problems associated with its use. The presence of Ca and Mg in the treated sewage is the main cause of the phosphate removal described for PPs, as will be discussed much more thoroughly in this chapter, due to its precipitation determined by the pH increase generated by photosynthesis.

For the case of nitrogen, traditionally the denitrification process in biological systems is carried out by heterotrophic bacteria, which reduce nitrate to molecular nitrogen by oxidizing organic material. Mulder (1995) discovered anammox bacteria that are capable of autotrophic denitrification by oxidizing ammonia with nitrite anaerobically. If the concentration of organic material is low, such as in anaerobically pre-treated wastewater, heterotrophic denitrification may not be possible, because of the imbalance between the oxidant (nitrate) and the reductant (organic material in this process). In that case, autotrophic denitrification is very advantageous not only because no organic material is required, but also for the reduction of the oxygen demand. However, the removal of nitrogen in the water line through this process is still a challenge.

In this chapter are presented in more detail strategies to improve the removal of nutrients for anaerobically treated sewage with poor organic matter content, using PPs and aerobic biological systems.

## 5.2 POST-TREATMENT IN PPS

### 5.2.1 Main processes that develop in PPs

In a PP, the concentration of the many constituents of the digested sewage changes over time, because of biological, chemical, and physical processes taking place in the unit. The most important biological processes are: (1) photosynthesis, (2) oxidation of organic material by bacteria that use oxygen, and (3) fermentation of organic material by anaerobic digestion. These three processes directly affect the removal of organic material. Figure 5.3 shows a schematic representation of a pond and the interaction between the different biological processes. During photosynthesis the algae use solar energy to reduce carbon dioxide, transforming it into cellular organic matter. In a simplified way this process can be represented by Equation (5.1a) where the organic matter is expressed as chemical oxygen demand (COD):

$$CO_2 + H_2O \rightarrow COD + O_2 \tag{5.1a}$$

It is important to note that, necessarily, the generation of organic material (in terms of COD mass) accompanies the generation of oxygen, that is, the mass of oxygen generated by photosynthesis is exactly the stoichiometric amount required to oxidize the organic material that is also generated in the same process. However, this oxygen is generally not used for oxidation of cellular material of algae, but rather to oxidize the faecal organic material present in the sewage. This bacterial oxidation can be represented by the following schematic equation:

$$COD + O_2 \rightarrow CO_2 + H_2O \tag{5.1b}$$

When analysing Figure 5.3, it is observed that the processes of photosynthesis and bacterial oxidation are complementary in the sense that the products of one process are the reactants of the other. This shows that, in principle, there can be no net removal of the organic material through oxidation if there is no source of oxygen other than photosynthesis, such as atmospheric oxygen uptake. However, the combined action of photosynthesis and oxidation may result in the removal of organic matter from the liquid phase by other mechanisms. The net result of the development of the two processes is the transformation of faecal organic matter into the mass of microalgae and bacteria. This mass of microorganisms can form flocs through spontaneous flocculation and these flocs can settle down and become part of the bottom sludge of the pond, where they can be digested or accumulated as non-biodegradable compounds (Figure 5.1).

The relative rates of the two biological processes, directly affect important parameters such as DO, $CO_2$, biochemical oxygen demand ($BOD_5$), COD, suspended solids (SS), and pH, while others are indirectly affected, such as alkalinity and nutrient concentrations. The relative rates of the two processes are determined mainly by the conditions of transparency, solar irradiation, temperature, depth of the pond available for photosynthesis, and concentration of the biodegradable organic material for bacterial oxidation.

In the case of PP, the concentration of organic matter from the digested sewage is low and the ponds will have typical characteristics of maturing ponds in WSP, especially if they are shallow (larger area favourable to photosynthesis). Under these conditions, it is possible to achieve high oxygen concentration, and there may even be oxygen desorption from the supersaturated liquid phase into the atmosphere. The predominance of photosynthesis over oxidation has another consequence: $CO_2$ is consumed, which increases pH. In this respect, the PP is different from WSP where there is an equilibrium between oxygen production by algae and consumption by bacteria.

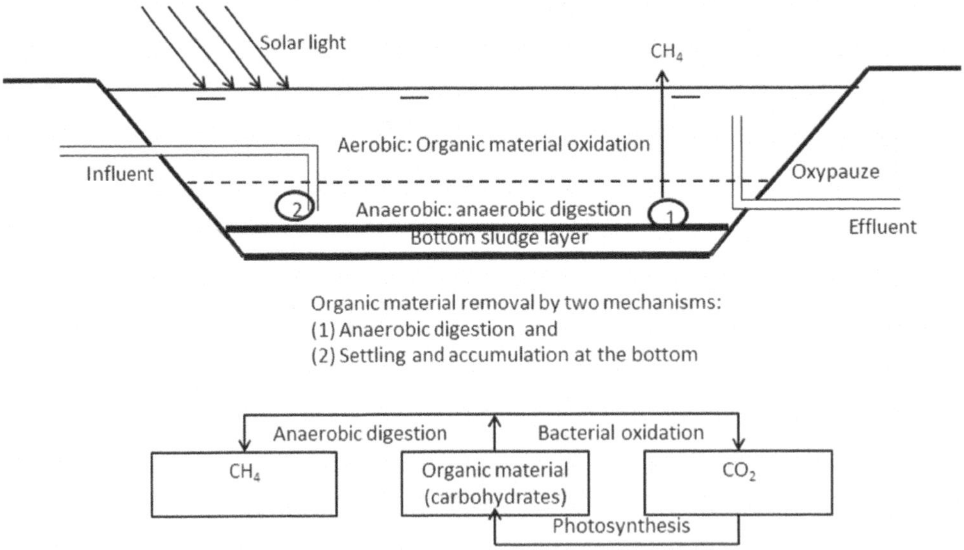


**Figure 5.3** Schematic representation of a PP and of the biological processes that develop in it.

The third important biological process is anaerobic digestion, which develops at the bottom of the PP, where the oxygen from photosynthesis does not reach or, if it arrives, is immediately consumed. The process can be represented schematically as

$$COD \rightarrow CH_4 + CO_2 \quad (5.1c)$$

Strictly speaking, anaerobic digestion also does not destroy the organic material, merely turns it into a gas (methane) that is released, thus effecting the physical removal of the organic material from the liquid phase and therefore the effluent quality.

PPs are affected by five components of the anaerobic effluent: (1) the organic material concentration (BOD and COD), (2) pathogens, especially TTC, (3) pH, (4) nitrogen, and (5) phosphorus.

Concerning organic material removal in PP, Figure 5.3 shows that if photosynthesis has a higher rate than organic material oxidation, there is an increase of algae in the pond. However, on the contrary Aguiar *et al.* (2021) have shown that the BOD in PP is reduced to low values (order of 20–40 mg/L) after a short retention time of only 5 days. The final COD concentration after 5 days is much higher (150–250 mg/L). The explanation is that because the efficient pretreatment in the UASB reactor, it is much easier to remove the residual organic material in the PP. The BOD and COD tend to pass through a minimum at 5 days, whereafter their values tend to increase due to algae production.

Marais (1974), reporting on the pioneering work on kinetic modelling of decay (Marais & Shaw, 1961) has shown that this process can be described as a first-order reaction with the rate of reduction of TTC bacteria proportional to the existing number, that is, it occurs according to Chick's law (Chick, 1908):

$$r_d = (dN/dt) = K_b N_{ttc} \quad (5.2)$$

where $r_d$ = decay (death) rate of the TTC; $N_{ttc}$ = concentration of the TTC (UFC/100 mL); $t$ = decay time (d); $K_b$ = decay constant for TTC (per d).

In chemical engineering, particularly in reactor design calculations, first-order processes have been thoroughly studied, because they are the type that occurs very frequently in practice. It is possible to use the theoretical models developed for the calculation of reactors, to describe the process of decay of TTC in ponds. The most important aspect of pond design is the influence on bacterial decay is the required retention time for a particular TTC removal efficiency. The differential equation (5.2) can be solved for flow through ponds and sequential batch ponds. For a pond operating at constant flow and volume, the solutions are

(1) Sequential batch ponds: $N_e/N_i = \exp(-k_b R_h)$ (5.2a)

(2) Flow through ponds: $N_e/N_i = 1/(1 + k_b R_h)$ (5.2b)

(3) For a series of equal ponds ($M$), the expression becomes: $N_e/N_i = 1/(1 + k_b R_h/M)M$ (5.2c)

It can be noted that all expressions have the same dimensionless group $k_b R_h$.

where $N_e$ = TTC concentration in the effluent; $N_i$ = TTC concentration in the influent; $R_h$ = retention time in a pond system; $R_h/M$ = retention time in pond '$M$' of a series; $M$ = number of ponds in series.

Figure 5.4 shows a graphical representation of Equations (5.2a) to (5.2c): the removal efficiency of TTC (log scale) is plotted in function of the product $k_b R_h$, which is a dimensionless number. The value of the decay constant can be determined experimentally when the decay of TTC is observed as a function of time in batches of PPs at different temperatures and depths: Batista *et al.* (2021) proposed the following equation:

$$k_b = (1.6/H)1.07^{(t-25)} d^{-1} \quad (5.2d)$$

where: $H$ = pond depth in m; $T$ = temperature in °C.

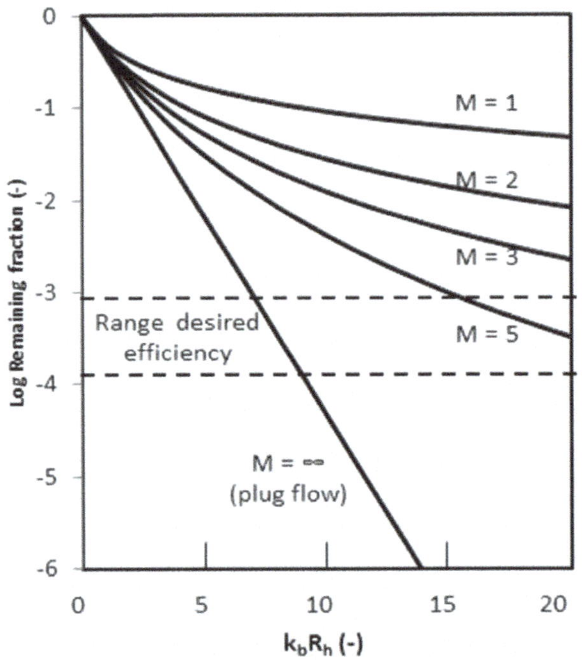


**Figure 5.4** Relationship between the reminiscent fraction of TTC and the dimensionless product $k_b \times R_h$, for a number of lagoons, *M*, of 1, 2, 5, and infinite.

Digested sewage may have a concentration of $10^6$–$10^7$ UFC × 100 per mL (i.e., 10% of the concentration in raw sewage), whereas the WHO standard specifies a maximum of $10^3$ UFC × 100 per mL. Hence to comply with the recommendations of WHO for unrestricted irrigation, the reduction of the number of TTC must be of the order of $10^3$–$10^4$ UFC × 100 per mL, that is, the desired removal efficiency must be of the order of 99.9 to 99.99%.

### 5.2.2 Hydrodynamic regime in PPs

The most common form of operation of PPs is a continuous flow regime (CFPP), due to greater operational simplicity. However, Albuquerque *et al.* (2021) showed that it is possible to operate them in sequential batch mode regime (SBPP) and that this alternative can have important advantages, such as a faster removal of nutrients and pathogens, resulting in a reduction of the required footprint. The main difference between CFPP and SBPP operation mode is the pH trend. In the CFPP pond, the pH has the maximum value during the entire retention time (which is also the discharge pH) and the high pH reduces the photosynthetic activity. On the contrary, in the SBPP, the pH is initially neutral (UASB effluent), increasing during the retention time, reaches its maximum value just before the effluent discharge. In Table 5.1, the differences between the various operating parameters in SBPP and CFPP are summarized.

While sequencing, batch PPs undoubtedly have significant advantages over continuous flow PP, a problem is that the sewage flow is continuous and the SBPP discontinuous. One solution for this problem is to introduce a transfer pond that continuously receives the UASB effluent, from where it can be distributed in batches to PPs as shown in Figure 5.5. The transfer pond would have at least the same volume as the PPs. The advantage of having a transfer pond is that besides transferring digested effluent to the PPs, it can realize several other functions, the most important being:

**Table 5.1** Differences between SBPP and CFPP.

| | SBPP | CFPP |
|---|---|---|
| Feed | Batch-wise | Continuous flow |
| DO concentration | Initially zero<br>Final maximum | Constant, lower than maximum of SBPP |
| pH | Initially 7, Final maximum. | Constant lower than maximum of SBPP |
| BOD & COD removal | Slow COD and BOD increase due to algae flocculation | Rapid increase of COD and BOD |
| N removal | Minimum time for removal | Maximum time for removal |
| P removal | Minimum time for removal | Maximum time for removal |
| Thermotolerant bacteria removal | Minimum time for removal | Maximum time for removal |

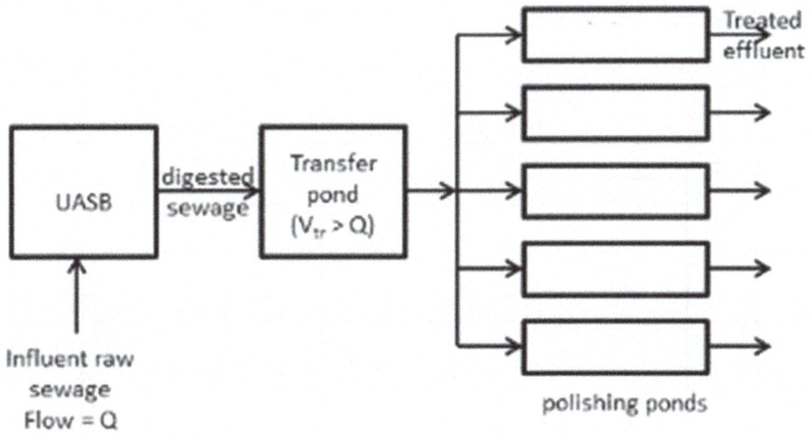


**Figure 5.5** Schematic representation of the flow sheet and operation of PPs fed with sequential batches from the transfer pond.

- The transfer pond also functions as a settling tank for sludge solids eventually discharged from the UASB reactor. Part of the settled solids could then be returned to the UASB reactor, and the remainder be processed to produce a sludge cake or dry solids.
- If the transfer pond is a settler, there is no need to have a settler on top of the UASB reactor. If this structure is modified, the operation of UASB reactors becomes very easy. The principal operational problem of UASB reactors is the scum layer that forms at the surface of the interface under the separator elements. If the separator is not more necessary, the scum problem no longer exists.
- If the retention time of the transfer pond is high enough, the process of photosynthesis can begin and establish a population of algae in the discharge to the PPs, so photosynthesis can start immediately, and algae are transferred to the PPs at the beginning of their operation. If sufficient oxygen is produced by photosynthesis in the transfer pond it can oxidize sulphides that may have been formed in the UASB reactor, thus suppressing the bad odours that could otherwise arise.
- If the retention time in the transfer pond is sufficient, helminth eggs from the UASB reactor may be retained and not discharged into the PPs, thus ensuring that these pathogens will not be present in the final effluent. At any rate, a large fraction of the helminth eggs will be retained in the transfer pond and will end up in the excess sludge.

- The concentration of carbon dioxide dissolved in the UASB reactor effluent will be reduced by the desorption mechanism, reducing the acidity and facilitating the increase of pH by photosynthesis, thus enabling the removal of ammonia nitrogen from the liquid phase in the subsequent PP by desorption.

### 5.2.3 Experimental research

Even in efficient anaerobic treatment systems, the concentration of several constituents of the effluent is high and post-treatment is required. The most important constituents that normally need correction are (1) organic material, (2) nitrogen, (3) phosphorus, and (4) pathogens. An experimental investigation was carried out at a pilot scale to establish if the removal of these constituents is feasible in a sequencing batch PP and principally what are the required depths and retention times.

Having established that SBPPs are more efficient than FTPPs, experiments were carried out at a pilot scale to evaluate the feasibility of removing pathogens, residual BOD and nutrients in SBPP. Figure 5.6 is a flow sheet of the employed system and a photo of a series of four pilot scale ponds.

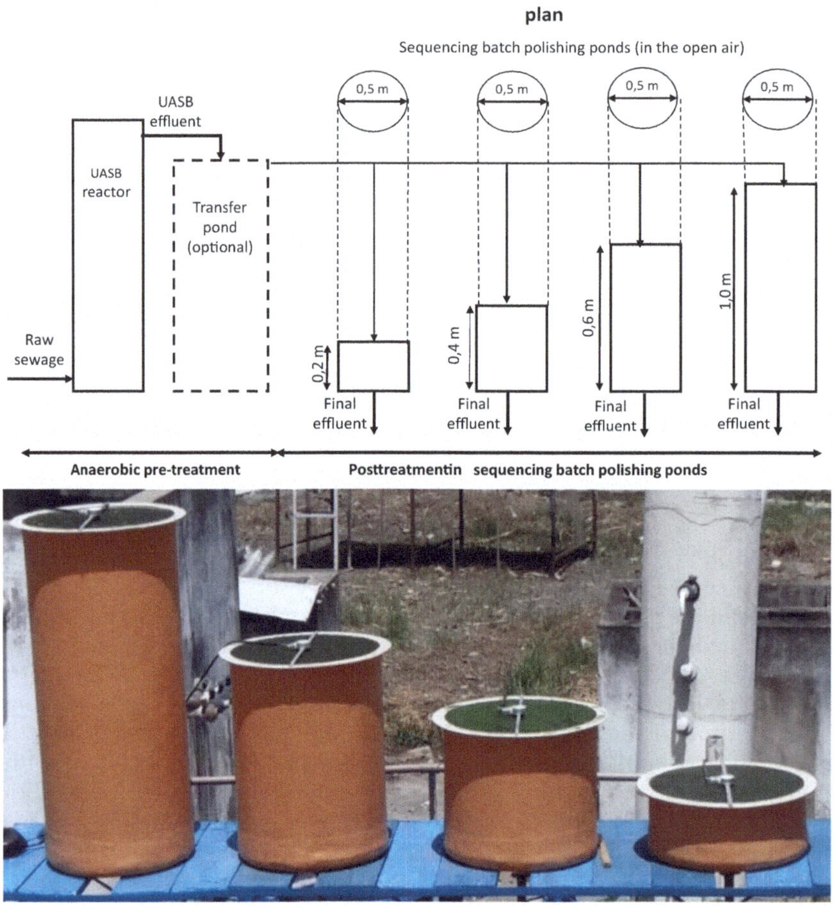


**Figure 5.6** Flow sheet and photograph of a system of SBPPs with depths of 0.2, 0.4, 0.6, and 1.0 m with mild stirring of the pond contents.

The experimental investigation was carried out in Campina Grande-Brazil (7 degrees south), using municipal raw sewage as influent of a UASB reactor, operated at a retention time of 6 h. Shallow ponds were used, because it was shown that these operated at a high rate. These pond models were operated with very gentle superficial stirring with a shallow metal bar attached to a small motor (6 rpm) in order to re-suspend any algae, floated by bubbles of dissolved oxygen, evolving from the ponds when these were supersaturated with DO. At the same time, the agitation served to even out stratification in the liquid phase. Later experiments showed that this agitation had a negligible effect on the performance of the ponds.

The experiments were carried out on four glass fibre cylinders with a diameter of 0.5 m and depths of 0.2, 0.4, 0.6, and 1.0 m. The experiments were carried out in the open air over a period of nine months: during summer (6 months) and winter periods (3 months). In Campina Grande there is abundant sunshine and sewage temperature is always 25°C, but in 'winter' there is more rainfall.

Tests for DO, pH, alkalinity, COD, nitrogen and phosphorus were carried out, according to standard methods and procedures. Distilled water was used to compensate losses due to evaporation and sampling. SBPP operation was continued until pH reached a value of about 9.7 or when the maximum retention time of 30 days was exceeded. For measurements of DO, temperature, and pH, a multi-parameter instrument was used (Hanna, model HI 98196).

Figure 5.7 shows experimental results of the investigation. For comparison, only retention times of 30 days were investigated, as this is the normal retention time in WSP.

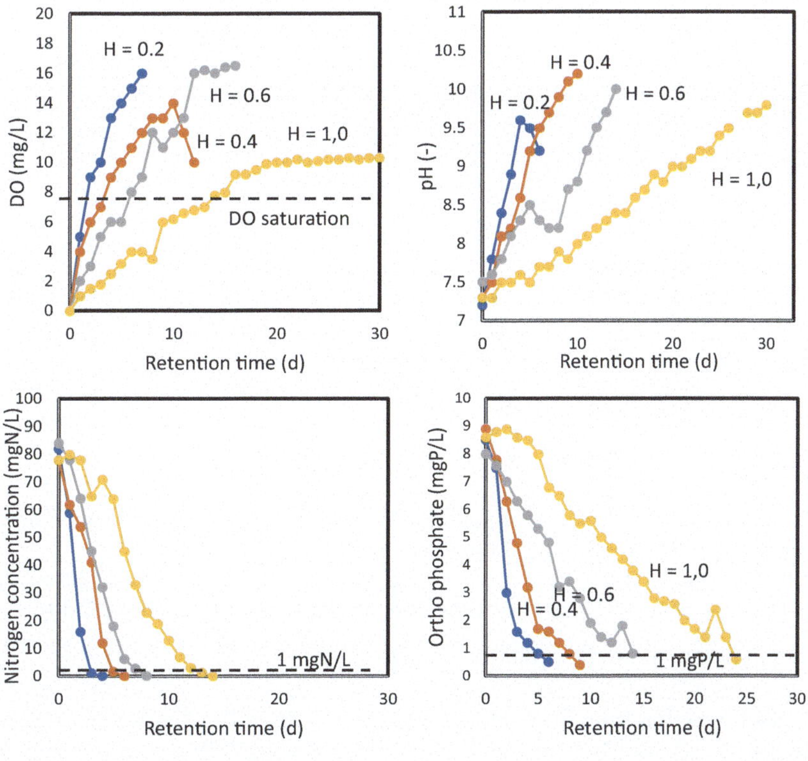


**Figure 5.7** Average values of DO concentration, pH, ammonium nitrogen and phosphate concentrations as functions of time in SBPP with different depths (0.2–1.0 m).

Average values of the DO concentration, the pH value, ammonium nitrogen and phosphate were represented over time for the four investigated depths and led to the following observations:

The DO concentration increased in all SBPP from the first day onwards, but the rate of increase heavily depended on the depth of the pond. Very high DO concentration $> 20$ mg/L tended to become unstable, possibly due to a decreasing oxygen production at high pH values.

pH increased with time, but at very different rates, also depending on the pond depth. Very high pH values (pH $>11$) could not be sustained in the ponds, probably due to reduced photosynthetic $CO_2$ consumption.

Nitrogen removal was feasible in shallow ponds, but the required retention time was longer during rainy periods, when there was less sunshine.

The pond depth has a very strong influence on the time required for total nitrogen removal, varying from 4 to 14 days at an initial concentration of about 80 mg/L. A pH of 8.5 was sufficient to remove nitrogen.

Efficient phosphate removal was possible in all ponds, but the retention time varied strongly with depth (6–24 days) for depths of 0.2–1.0 m. A pH of 9.5–9.7 is required for efficient P removal.

It is important to note that the profiles of Figure 5.5 are all influenced by the climate so that under different circumstances the results will be different. Ideally for any project an experimental investigation should be carried out.

With the aid of the data in Figure 5.6, models were developed that can foresee the values of pH, nitrogen, and phosphorus vs time. The following sections show the development of the models for pH, nitrogen, and phosphorus removal.

### 5.2.4 Model for pH change and nitrogen removal in sequencing batch PPs

The pH value and its stability in sewage treatment plants are determined almost entirely by the carbonic system. If the state of the carbonic system is to be evaluated, the definitions of alkalinity and acidity are important. Loewenthal and Marais (1976) have proposed:

$$Alk = 2[CO_3^{=}] + [HCO_3^{-}] + [OH^{-}] - [H^{+}] \tag{5.3a}$$

$$Ac = 2[CO_2] + [HCO_3^{-}] + [H^{+}] - [OH^{-}] \tag{5.3b}$$

The pH variation in treatment systems is due to variations in alkalinity and acidity, which in turn are affected by the physical, chemical, and biological processes occurring in these systems. To analyse the pH variation, the Deffeyes diagram (Deffeyes, 1956), improved by Loewenthal and Marais (1976), is very useful. It shows that there is a linear relationship between alkalinity and acidity for any pH. By using the dissociation equations for the carbonic system, Equations (1a) and (1b) can be used to relate three variables: alkalinity, acidity, and pH:

$$Alk = [Ac + 10^{(pH\text{-}pK1)} - 10 - pH] * [1 + 2.10^{(pH\text{-}pK2)}] / [1 + 2.10^{(pK1\text{-}pH)}] \tag{5.4a}$$

where $k_1$ = first ionization constant of carbonic system (Harned & Scholes, 1943); $k_2$ = second ionization constant of carbonic system (Harned & Davis, 1943).

If the pH is in the neutral range ($5 < pH < 9$) the influence of the carbonic system is decisive and the influence of the ions $H^+$ e $OH^-$ can be neglected, so Equation (2) is simplified to

$$Alk \approx Ac[1 + 2 \times 10^{(pH\text{-}pK2)}] / [1 + 2 \times 10^{(pK1\text{-}pH)}](5 < pH < 9) \tag{5.4b}$$

Figure 5.8 is a Deffeyes diagram, drawn from Equation (5.3). In the diagram, two values of the three variables: (i) alkalinity, (ii) acidity or (iii) pH define the ionic equilibrium point, which characterizes the condition of the liquid phase. Figure 5.8 shows how the ionic equilibrium point moves, when adding to or removing materials from the liquid phase. Thus, according to Equation (5.3a), the

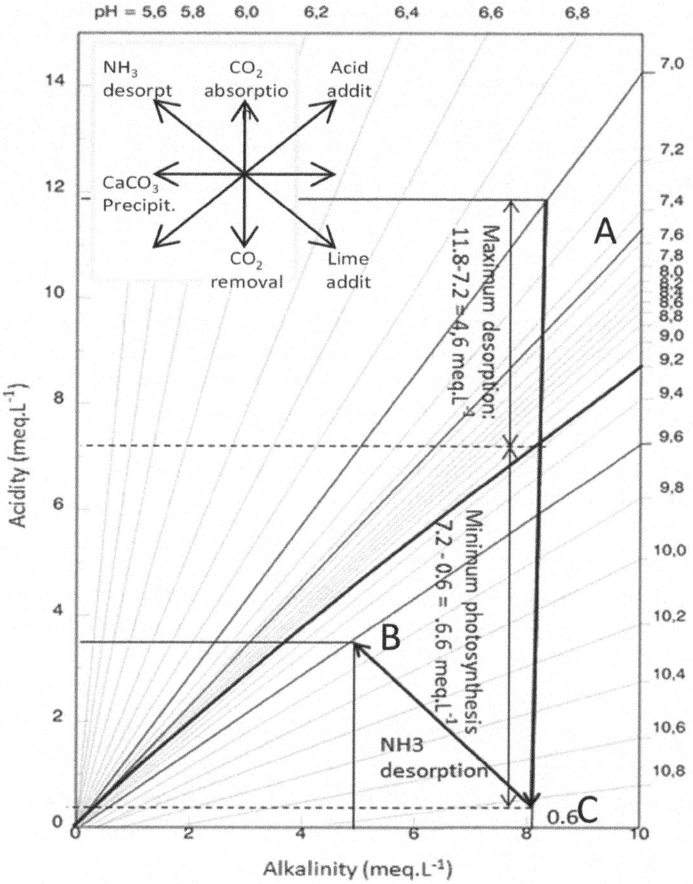


**Figure 5.8** Deffeyes diagram (temperature of 25°C: $pK_1 = 6.33$, $pK_2 = 10.33$, $pK_w = 14$).

alkalinity does not change upon $CO_2$ removal, but the acidity decreases by 2 eq/mol. For example, if the effluent from a UASB reactor has a pH = 7 and an alkalinity of 8 meq/L, (point A), the acidity will be about 12 meq/L.

Once established, the processes occurring in the treatment system and knowing what are the effects of these processes on the alkalinity and acidity, it is possible to establish the variation of alkalinity and acidity and determine the pH in the treatment system and in the effluent. For example, when it is known that in the treatment system, the alkalinity decreases by 3 meq/L (with a final alkalinity of 5 meq/L) and if a final pH = 9.6 is desired, then B is the final ionic equilibrium point, with an acidity of 3.4 meq/L. For this to take place the acidity has to decrease by 0.6 from 12 to 11.4 meq/L. Such a decrease results from desorption or biological consumption of carbon dioxide.

In PPs, the main biological, chemical, and physical processes that can alter alkalinity and/or acidity and therefore the pH are: (1) biological or physical removal of $CO_2$ and (2) physical removal of $NH_3$ by desorption, which is equivalent to the addition of a strong acid. Although at high pH the liquid phase may be super saturated with calcium carbonate, precipitation of $CaCO_3$ does not take place in ponds. With Figure 5.7, it is possible to predict the effect of varying the concentration of carbon dioxide and ammonia on the value of alkalinity, acidity, and pH as shown in Table 5.2.

**Table 5.2** Effect of $CO_2$ and $NH_3$ removal on alkalinity and acidity change and on pH variation in treatment systems. (OM = organic matter).

| Process | Reaction | ΔAlk (eq/mol) | ΔAc (eq/mol) | ΔpH |
|---|---|---|---|---|
| Physical $CO_2$ removal | No | 0 | −2 | Increase |
| Biological $CO_2$ removal | $CO_2 + H_2O \rightarrow OM + O_2$ | 0 | −2 | Increase |
| $NH_3$ desorption | $NH_4^+ \rightarrow NH_3 + H^+$ | −1 | 1 | Decrease |

In PPs, it is important to know the time it takes for the pH to reach a certain value which depends on the kinetics of the relevant processes. To establish the retention time required to obtain a certain pH change in a SBPP it is necessary to formulate equations expressing the rates of the three pH affecting processes. The desorption rate of volatile components in water like $CO_2$ and $NH_3$ can be described by Fick's equation (Equation (5.4)), which states that the desorption rate of a volatile compound in a liquid is proportional to the degree of supersaturation existing between the current concentration of the compound and the saturation concentration:

$$r_d = k_d(C_l - C_s) \tag{5.5}$$

where: $r_d$ = desorption of the volatile compound; $k_d$ = desorption constant; $C_s$ = saturation concentration of the volatile compound; $C_l$ = concentration of the volatile compound in the liquid phase.

Thus, the rates of transfer for $NH_3$ and $CO_2$ can be expressed as

$$r_{dN} = k_{dN} \cdot [NH_3], \tag{5.5a}$$

$$r_{dC} = k_{dC} \cdot ([CO_2] - [CO_2]_s) \tag{5.5b}$$

When the differential equations (5.4a) and (5.4b) are integrated expressions found for $CO_2$ and $NH_3$ as functions of time so that the values of the constants can be determined experimentally:

$$\mathrm{Ln}\{([CO_2] - [CO_2]_s) / ([CO_2]_o - [CO_2]_s)\} = -k_{dC}t \tag{5.6a}$$

$$\mathrm{Ln}([NH_3] / [NH_3]_o) = -k_{dN}t \tag{5.6b}$$

Morais *et al.* (2023) determined experimentally the values of the transfer constants as functions of temperature ($T$) and pond depth ($H$):

$$k_{dC} = (0.56/H)1.056(T - 20) \tag{5.7}$$

$$k_{dN} = (0.20/H)1.095(T - 20) \tag{5.8}$$

For a specific depth and temperature (which define the values of $k_{dC}$ and $k_{dN}$), the only process that is not yet defined in Table 5.2 is the rate of $CO_2$ consumption or oxygen production (OPR), but this can be measured as the rate of change in the DO concentration (Equation (5.9)). The rate of $CO_2$ consumption can be conveniently determined by the oxygen production rate in the pond by considering that if 1 mol of oxygen is produced 1 mol of $CO_2$ is consumed. The rate of oxygen production can be expressed as

$$\mathrm{OPR} = -(dCO_2/dt)_c - (dDO/dt)_p \tag{5.9}$$

where:

$\mathrm{OPR} = (dDO/dt)_p$ = DO production rate in the pond (mol DO/L/d)
= photosynthesis rate – DO consumption rate
= biological $CO_2$ consumption rate (mol $CO_2$/L/d)

Equation (5.9), together with Equations (5.5)–(5.8), are applied for the calculation of pH variation and ammonia removal in SBPP. Since these equations cannot be solved analytically, they must be solved numerically. The differential equations must be written in terms of finite time elements that will cause finite changes in the concentrations of $CO_2$ and $NH_3$, as in Equations (5.10) and (5.11):

$$\Delta[CO_2]/\Delta t = -\ k_{dC}([CO_2]-[CO_2]_s) - OPR \quad (5.10)$$

$$\Delta[NH_3]/\Delta t = [NH_3] - k_{dN}[NH_3]_o \quad (5.11)$$

These variations, in turn, result in alkalinity and acidity changes as well as ammonia concentration variations that can be expressed as

$$\Delta Alk = -\ k_{dN}[NH_3]\Delta t \quad (5.12)$$

$$\Delta Ac = -2\{k_{dC}([CO_2]-[CO_2]_s) - 2OPR + k_{dN}[NH_3]\}\Delta t \quad (5.13)$$

$$\Delta N_{tot} = N_{tot} -\ k_{dN}[NH_3]\Delta t \quad (5.14)$$

#### 5.2.4.1 pH variation in SBPP

The pH variation can be calculated from the alkalinity and acidity changes. Considering that $H^+$ e $OH^-$ concentrations are small and do not influence the alkalinity and acidity one has

$$Alk = Ac(1 + 2k_2/[H^+])/(1+[H^+]/k_1) \quad (5.15)$$

This is an implicit quadratic expression and can be solved for $[H^+]$, resulting in

$$[H^+] = \{k_1(Ac/Alk - 1)/2 + (k_1(Ac/Alk - 1)/2)2 - 4k_1k_2)0.5\}/2$$

Since pH = −log $[H^+]$ one has

$$pH = -\log\{(0.5(k_1(Ac1/Alk1 - 1)/2 + ((k_1(Ac1/Alk1/2)2 - 4k_1k_2)1/2)\} \quad (5.16)$$

#### 5.2.4.2 Nitrogen removal in SBPP

To calculate nitrogen removal in SBPP, it is considered that the ammonia ion ($NH_4^+$) and gaseous ammonia ($NH_3$) are in equilibrium in the liquid phase:

$$NH_4^+ \leftrightarrow NH_3 + H^+ \quad (5.16a)$$

$$K_a = [H^+]\cdot[NH_3]/[NH_4{}^+] \quad (5.17)$$

Equation (5.17) shows that the removal of ammonium is equivalent to the addition of a strong acid, that is, per mmol of desorbed $NH_3$, there is a decrease of 1 meq of alkalinity and an increase of 1 meq of acidity.

According to Emerson *et al.* (1975), the ionization constant is given by Equation (5.19):

$$pK_a = 0.09018 + 2729.92/T\ (T \text{ is in Kelvin}) \quad (5.19)$$

Thus, at a temperature of 298 K and a fixed solids concentration of 1 g/L (i.e., an activity coefficient of about 0.9), the dissociation constant is $pK_a = 9.1$. Therefore, for pH values greater than 9.1, the non-ionized form, $NH_3$, is predominant, according to Equation (9.13). In contrast, for neutral pH values the non-dissociated ammonia fraction is small (1% for pH = 7.1) and accordingly, the desorption rate $NH_3$, will be low.

Table 5.3 shows a summary of the equations to evaluate the effects of the above-mentioned processes on alkalinity and acidity and consequently on pH trends in PPs vs time. Now the basis for a model is developed, able to evaluate the variations in pH, alkalinity, acidity, and ammonium concentration as

**Table 5.3** Equations to calculate the rates of processes and concentration trends of the key operating variables in SBPP.

| Variable | Equation | No. |
|---|---|---|
| DO production rate | $\mathrm{OPR} = \Delta\mathrm{OD} / \Delta t = -(\Delta\mathrm{CO_2} / \Delta t)_f$ | (5.8) |
| $NH_3$ desorption rate | $r_{dN} = (\Delta[\mathrm{NH_3}] / \Delta t)_d = k_{dN}[\mathrm{NH_3}]$ | (5.4a) |
| $CO_2$ desorption rate | $r_{dC} = (\Delta[\mathrm{CO_2}] / \Delta t)_d = k_{dC}([\mathrm{CO_2}] - [\mathrm{CO_2}]_s)$ | (5.4b) |
| Alkalinity change: | $\Delta\mathrm{Alc} / \Delta t = -k_{dn}[\mathrm{NH_3}]$ | (5.11) |
| Acidity change: | $\Delta\mathrm{Ac} / \Delta t = -2\{k_{dc}([\mathrm{CO_2}] - [\mathrm{CO_2}]_s) - 2\mathrm{OPR} + k_{dn}[\mathrm{NH_3}]\}$ | (5.12) |
| pH change: | $\mathrm{pH} = -\log\{(0.5(k_1(\mathrm{Ac} / \mathrm{Alc} - 1) / 2 + ((k_1(\mathrm{Ac} / \mathrm{Alc} / 2)^2 - 4k_1k_2)^{1/2})\}$ | (5.15) |
| $CO_2$ conc. change | $\Delta[\mathrm{CO_2}] / \Delta t = -k_{dc}([\mathrm{CO_2}] - [\mathrm{CO_2}]_s) - \mathrm{OPR}$ | (5.9) |
| $NH_3$ conc. change | $\Delta[\mathrm{NH_3}] / \Delta t = -k_{dn}[\mathrm{NH_3}]$ | (5.13) |
| Change of total N | $\Delta N_{tot} / \Delta t = N_{tot} - k_{dn}[\mathrm{NH_3}]$ | (5.16) |

*Legend:* $K_{dN}$, $k_{dc}$ = desorption constant for $NH_3$ and $CO_2$, respectively; indices s and o stand for saturation and time = 0.

function of time in ponds operated in batch mode. The model is a set of differential equations defined by Equations (5.8)–(5.18).

The model illustrated in Table 5.3 can be used to calculate the values of the alkalinity, acidity, pH, carbon dioxide, and total nitrogen as a function of time in SBPPs. The simulated curves thus generated are compared to the experimental values in Figure 5.9a and b, where experimental values of pH and nitrogen (reported in Figure 5.6) and the simulated curves calculated from Table 5.3 are presented for the four investigated depths. There is a good correlation between the experimental and simulated values for all depths. The correlation for pH sometimes is not good for high pH values. This finding can probably be attributed to the fact that at very high pH values the photosynthesis rate decreases, so that the rate of pH increase is reduced and may even become negative, and this occurrence is not included in the model of Table 5.3.

The pH curves show that there are three phases of pH change: (1) an initial rapid change due to $CO_2$ desorption (2) a slower intermediate rate of pH change due to acidity production by ammonium desorption and (3) a quick final change due to completion of ammonia removal. Similarly, ammonia removal has also three phases: (1) a slow initial phase because the pH and ammonia desorption are still low, (2) a higher rate of ammonia removal when the $NH_3$ concentration is high, and (3) a low final rate of ammonia removal when the $NH_3$ concentration is low at the end of a batch, because the total N concentration is low even if the pH is high.

### 5.2.5 Model for phosphorus removal in sequential batch PPs

To model the P concentration in the PP is difficult because several compounds interact. In fact, there are different phosphorus compounds that may precipitate, and several cations in the liquid phase. Therefore, instead of trying to develop a model with a theoretical solution, an empiric approach has been applied.

In Figure 5.10, the experimental data of Figure 5.7 were used to plot experimental values of the logarithm of the total phosphate concentration, $\log(C_t)$, in the SBPP with depths varying from 0.2 to 1.0 m as a function of pH. The result of the experimental investigation is that for any pH and pond depth the data plot is, in good approximation, a straight line with an average slope of 0.43. The experimental results also indicate that the pH for a final P concentration of $C_t = 1$ mg/L or $\log(Ct) = 0$ is, in good approximation, 9.7 for all depths so that the average of the data leads to the following linear equation:

$$\mathrm{Log}(C_t) = -0.43 \cdot \mathrm{pH} + 4.29 \tag{5.20}$$

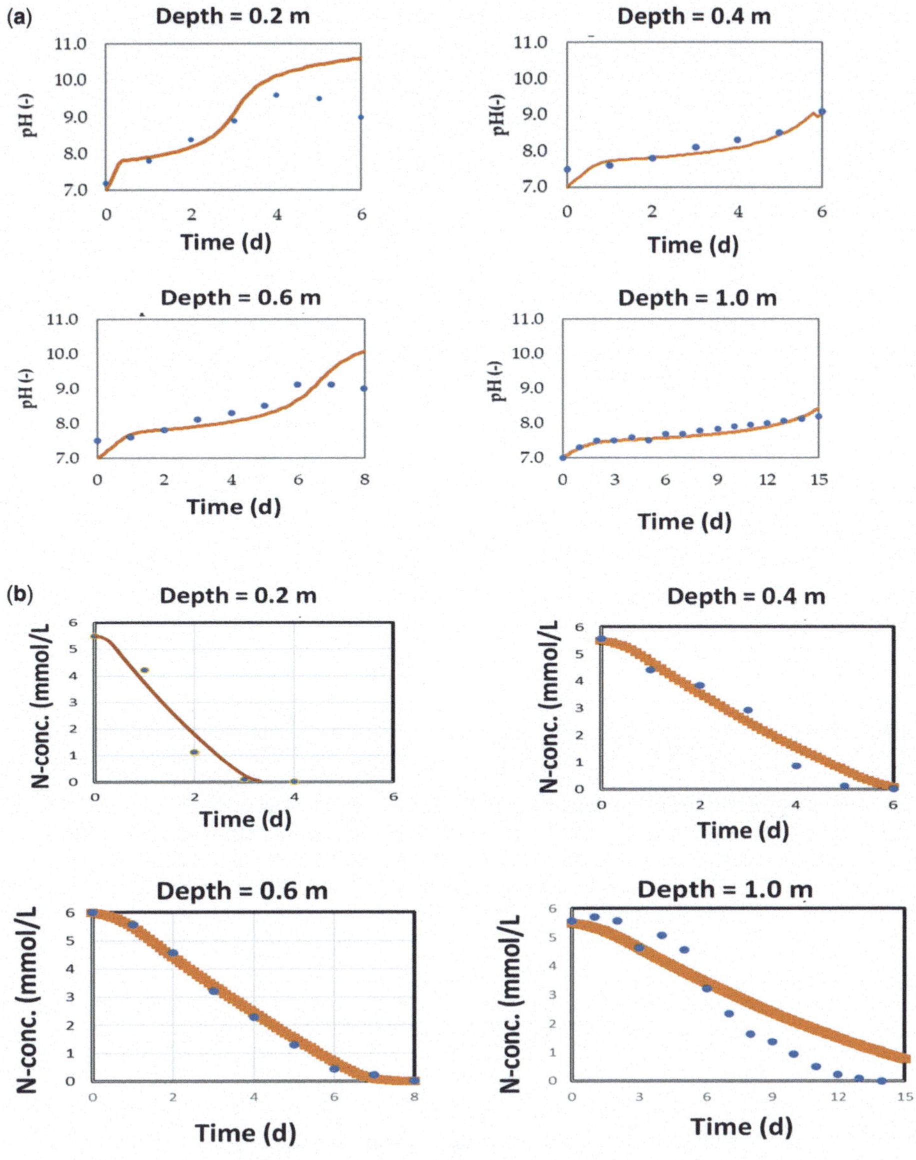


**Figure 5.9** (a) Experimental data and simulated curves for pH vs time in SBPPs with depths varying from 0.2 to 1 m. (b) Experimental data and simulated curves for the nitrogen concentration and pH vs time in SBPPs with depths varying from 0.2 to 1.0 m.

Now the required pH for any final P concentration can be calculated. For example, the pH necessary for a final P concentration of 2 mg/L can be calculated as follows (See Fig 5.11):

For an effluent P concentration of 2 mg/L: $\log C_t = \log 2 = 0.32$;

The straight line with a slope of 0.43 and passing $C_t = 1$ or $\log C_t = 0$ has a pH value of 9.2 for $C_t = 2$ or $\log C_t = 0.32$.

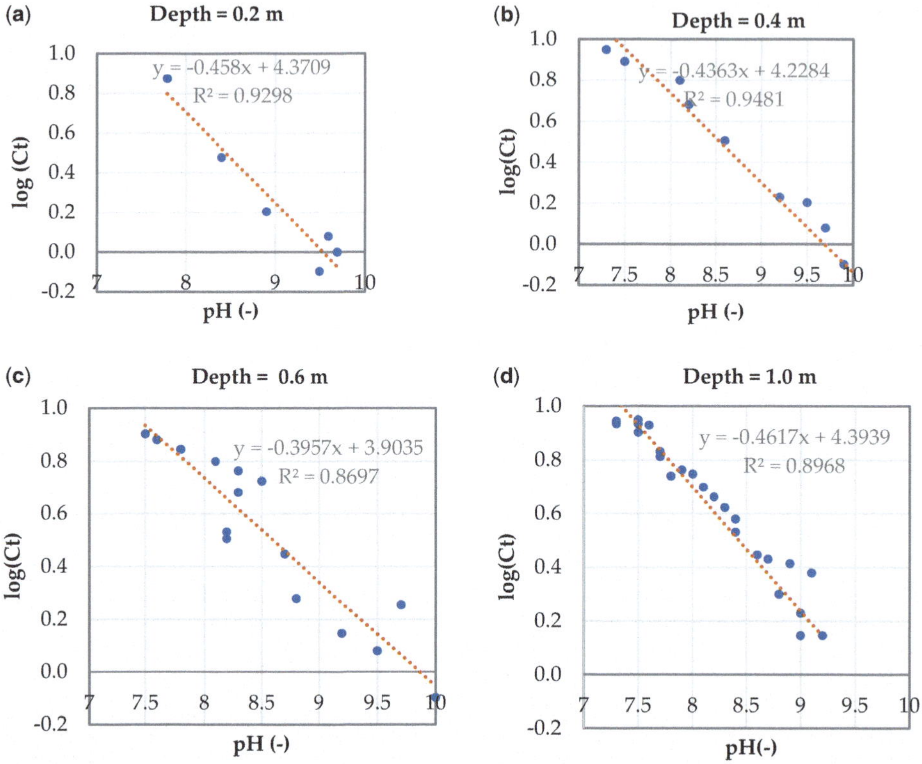


**Figure 5.10** (a) Experimental profiles of the logarithm of the total phosphorus concentration as a function of pH for a pond depth of 0.2 m. (b) Experimental profiles of the logarithm of the total phosphorus concentration as a function of pH for a pond depth of 0.4 m. (c) Experimental profiles of the logarithm of the total phosphorus concentration as a function of pH for a pond depth of 0.6 m. (d) Experimental profiles of the logarithm of the total phosphorus concentration as a function of pH for a pond depth of 1.0 m.

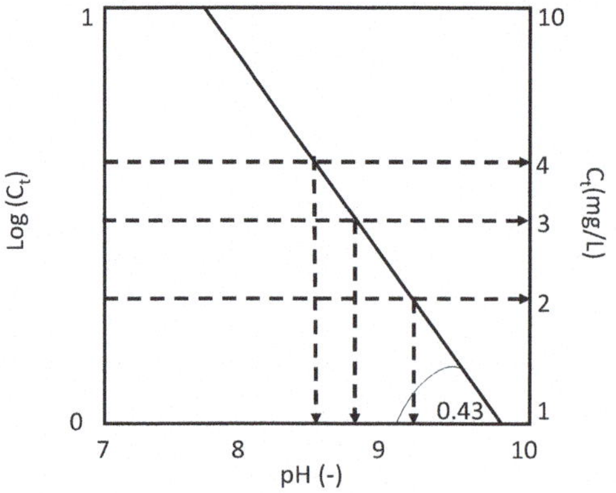


**Figure 5.11** Logarithm of residual phosphorus concentration in SBPP as a function of the pH.

Figure 5.9 is a graphical representation of Equation (5.19), from which the residual phosphate concentration can be read off for any pH, or, conversely where the required pH can be read off for any desired residual phosphate concentration.

The retention time necessary to acquire the required pH for a particular residual phosphorus concentration can be determined by the pH-nitrogen model presented in Table 5.3.

It is important to note that Figures 5.9 and 5.10 are representatives only of the investigated wastewater and have no general applicability because in wastewater the concentrations of cations like $Ca^{2+}$ are variable and this directly influences the solubility of phosphate for any particular pH.

### 5.2.6 Per capita area required for UASB–SBPP system

In Sections 5.4 and 5.5, the required retention times for the removal of the nutrients N and P are calculated. However, the variable of greatest interest for the design optimization is not the retention time, but the area per capita. This variable is related to depth. The area required for the construction of PPs can be calculated by considering that the retention time is the ratio between the volume of the pond and the applied daily flow so that the area per capita can be expressed as

$$A_{ie} = (R_h \times Q_{ie})/H \tag{5.21}$$

where: $A_{ie}$ = area required for SBPP ($m^2$ per inhabitant equivalent (ie)); $R_h$ = hydraulic retention time (days); $Q_{ie}$ = daily contribution per inhabitant flow ($m^3$/ie/d); $H$ = pond depth in m.

Using the data in Figure 5.6, Table 5.4 shows the observed HRT for a final effluent with a concentration of <1 mgN/L and <1 mgP/L in ponds with different depths (columns 2 and 3). Columns 4 and 5 show the values of the area per capita for the residual concentrations of 1 mgN/L and 1 mgP/L, calculated with Equation (5.20) for a contribution per capita of 0.1 $m^3$/d. In Figure 5.12 the values of the per capita areas have been plotted as a function of the pond depth. It can be observed that in shallow ponds it is feasible to obtain a very low nutrient concentration in SBPP, but also that the required area is much smaller than the area of conventional water sewage plant (WSP), where the area per capita is of the order of 3 $m^2$/ie, but where the removal of nutrients is not possible. The figure can be easily adapted for other per capita contributions.

Data in Figure 5.12 show that

(1) In the range of 0.2–1.0 m there is little influence of the pond depth on the pond area. The optimal pond depth is about 0.5 m, which is much smaller than the depth of MP in WSP where the depth is usually in the range of 1.0–1.2 m. The shallow depth of the pond is an important factor in the reduction of construction costs.

(2) The area of shallow ponds is much smaller than 3 $m^2$ per inhabitant equivalent, normally used for the design of conventional WSP, where nutrient removal does not occur.

(3) The area required for P removal is about double of the value for N removal.

**Table 5.4** Average values of the retention time and required area per ie for achieving concentrations <1 mg/L of nitrogen and phosphorus in the final effluent.

| | Retention Time (d) | | Per Capita Area ($m^2$/ie) | |
|---|---|---|---|---|
| **Pond Depth** | **for1 mgN/L** | **for1 mgP/L** | **for1 mgN/L** | **for1 mgP/L** |
| $L_1 = 0.2$ m | 3 | 5 | 1.52 | 2.65 |
| $L_2 = 0.4$ m | 4.5 | 8.5 | 1.11 | 2.12 |
| $L_3 = 0.6$ m | 7 | 13 | 1.15 | 2.19 |
| $L_4 = 1.0$ m | 13 | 24 | 1.28 | 2.40 |

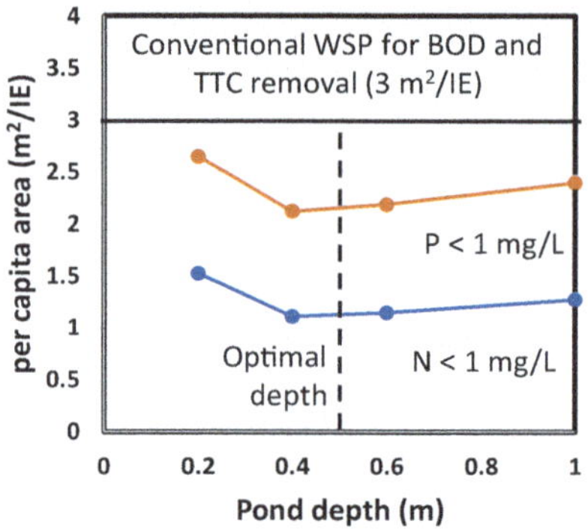


**Figure 5.12** Per capita area for nutrient and pathogen removal in SBPP to produce effluent with less than 1 mgN/L and 1 mgP/L as a function of the pond depth.

### 5.2.7 Production of high-quality water

It is important to note that nutrient removal should be the goal of post-treatment only when the final effluent is not used in agriculture. For industrial reuse or discharge of final effluent into surface water, nutrient removal is required. When wastewater is treated for industrial reuse, it is also necessary to remove the algae growing during the post-treatment in SBPP. One possibility is clarification (coagulation–flocculation–sedimentation–filtration) of SBPP effluent. In this case, the use of a coagulant (usually poly aluminium chloride or aluminium sulphate) will also remove the phosphate, so, in that case, an additional removal unit is not required. Figure 5.13 shows the flowchart of a treatment system that can produce water for industry. The time required for each treatment step is also indicated. As in Figure 5.12, it was established that the retention time (and hence the pond area) for P removal is about double of the retention time for N removal. In cases where clarification is applied it is possible to reduce the retention time in the PP, however, a mechanism must be used to remove the algae.

The same figure also shows that it is possible to treat the effluent after clarification to remove dissolved salts by applying reverse osmosis, possibly preceded by micro or ultrafiltration. This demineralized water may be required in special industrial applications such as water for refrigeration as well as a source for public water supply. Clearly, the additional processes of clarification and reverse

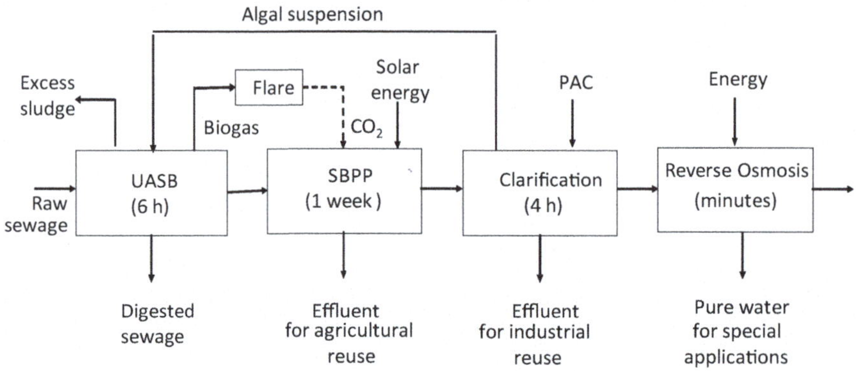


**Figure 5.13** Flow sheet to produce water for industry for general use or to produce demi water for special applications.

osmosis will lead to an increase in the treatment costs and will normally be applied only if there is a specific demand for high-quality water. This is often the case in regions with scarce water resources.

## 5.3 POST-TREATMENT WITH ACTIVATED SLUDGE SYSTEMS AND MBRS

An important disadvantage of anaerobic pre-treatment is that conventional nutrient removal of nitrogen by nitrification and heterotrophic denitrification and phosphorus removal by luxury uptake is limited due to the low concentration of biodegradable organic material in the effluent of the anaerobic pre-treatment system. This low concentration affects both conventional denitrification after nitrification in the anoxic section and the feasibility of luxury P uptake (Van Haandel & Van der Lubbe, 2012). Two alternative biological processes may increase the removal of nitrogen in anaerobic–aerobic systems:

Autotrophic denitrification: the anammox system.
Denitrification coupled with methane oxidation.

This section presents the potential of various post-treatment technologies that can be used to improve the quality of anaerobically treated sewage, with particular emphasis on those that have promoted either the anammox or denitrification processes coupled with methane oxidation for improving nitrogen removal.

### 5.3.1 The anaerobic–aerobic treatment system

In this section, the feasibility of producing an effluent with activated sludge quality, after applying an anaerobic pre-treatment will be evaluated. Equation (5.24) shows that anaerobic treatment results in the division of influent organic material into three parts: (1) a part that is composed of soluble non-biodegradable as well as degradable material, but not digested in the reactor; (2) a part of suspended solids composed of biological sludge and non-biodegradable and particulate organic material from the influent; (3) a part transformed into methane. The objective of the anaerobic treatment is the maximization of the fraction transformed into methane and the minimization of the organic material in the effluent and the discharged sludge. In the aerobic treatment, the organic material is also divided into three fractions: (1) a fraction in the effluent, (2) a fraction in the produced sludge, and (3) an oxidized fraction of inorganic products. Marais and Ekama (1976) have presented a model for activated sludge behaviour in purely aerobic systems. Van Haandel and van der Lubbe (2012) have shown that the same basic theory also applies to the aerobic sections of anaerobic–aerobic systems.

Figure 5.14 shows the interaction of aerobic and anaerobic treatment and the possible advantages that can be obtained in comparison with conventional aerobic treatment. The advantages can be listed:

(a) Drastic reduction in investment cost by reducing the volume of the treatment system.
Research carried out in Campina Grande by Santos (2017) led to the development of empiric expressions for the three fractions or organic material in anaerobic treatment systems: (COD in the effluent, COD in the produced sludge and digested COD):

$$f_s = f_{ust} + f_{bst} = f_{us35} 1.037^{(35-T)} + 0.20(1.011)^{(T-35)} \mathrm{Exp}^{(-0.04(Rs-3.8\times 1.06(35-T)))} \quad (5.22)$$

$$f_x = f_{up35} 1.015^{(35-T)} + 0.16 \times 1.014^{(35-T)} \exp^{(-0.04(Rs-3.8\times 1.06(35-T)))} \quad (5.23)$$

and

$$f_d = 1 - f_s - f_x \quad (5.24)$$

where: $f_s$ = soluble COD fraction discharged in the effluent; $f_x$ = COD fraction discharged as volatile solids in sludge; $f_d$ = digested COD fraction; $R_s$ = sludge age in the UASB reactor; $T$ = temperature in °C; $f_{us}$ = COD fraction in the influent that is non-biodegradable and soluble; $f_{up}$ = COD fraction in the influent that is non-biodegradable and particulate.

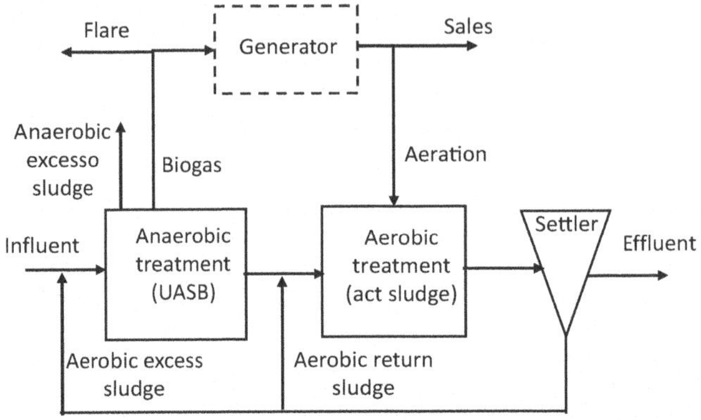


**Figure 5.14** Basic scheme of the anaerobic–aerobic system composed of a UASB reactor and an activated sludge system.

The equations show that the division of the three fractions depends on the composition of the organic material in the influent (the non-biodegradable soluble and particulate COD fractions, $f_{us}$ and $f_{up}$) and two operational variables: sludge age ($R_s$) and temperature ($T$). In anaerobic treatment, what matters is maximizing the digested fraction ($f_d$). Figure 5.15 shows the digested fraction as a function of sludge age at different temperatures. For all temperatures, the digested fraction of organic material is maximum when the sludge age is 100 days or more. Therefore, there is no merit in operating a UASB reactor with a sludge age of more than 100 days: only the volume of the UASB reactor increases, without this resulting in greater efficiency because the removal of the biodegradable organic material is already essentially complete.

It is not known *a priori* how long the liquid will remain in the UASB reactor to have a sludge age of 100 days. However, the Brazilian standard, NBR 12.209/2011, establishes the residence time for the maximum efficiency of the UASB reactor as a function of temperature. Figure 5.16 graphically shows the residence time for maximum efficiency in the UASB reactor as a function of sewage temperature.

As for the volume of the post-treatment system, it is necessary to optimize the volume of the aerobic reactor and the final settler. To optimize the post-treatment, it is sought to establish the smallest volume of the units of the aeration tank and final settler as a function of the sludge concentration. Van

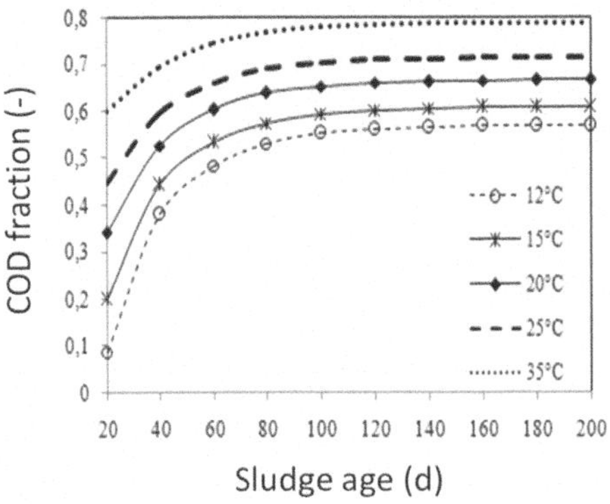


**Figure 5.15** COD fraction converted into biogas as a function of the anaerobic sludge age for different temperatures (fus25 = 0.10 and fup25 = 0.08).

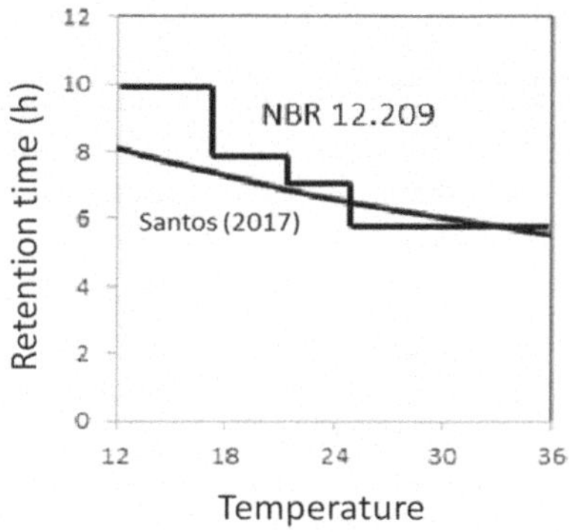


**Figure 5.16** Retention time for maximum efficiency in a UASB reactor as a function of temperature conforming NBR 12.209/2011.

Haandel and van der Lubbe (2012) proposed the following two equations to calculate the volume of the reactor and the settler:

$$v_r = \left[(1 - f_{us} - f_{up})(1 + f_{bh}R_s)YR_s / (1 + b_hR_s) + f_{up}Rs/f_{cv}\right]Q_aS_{ta}/(S_{ta}f_v) \tag{5.25}$$

$$v_d = S_f \cdot H / v_o \cdot \exp^{(kXt)} \tag{5.26}$$

where: $v_r$ = volume of the aerobic reactor per unit of applied sewage flow; $v_d$ = volume of the final settler per unit of applied sewage flow; $S_{ta}$ = total COD in the influent; $f_{us}$ = non-biodegradable and soluble COD in the influent; $f_{up}$ = non-biodegradable and particulate COD in the influent; $f_v$ = volatile sludge fraction; $k$ = sludge compressibility (Vesilind 1968; Leitão, 2004); $v_o$ = limiting sludge settling velocity (Vesilind 1968; Leitão, 2004); $R_s$ = sludge age; $T$ = temperature; $S_f$ = safety factor; $H$ = reactor height; $X_t$ = total sludge concentration.

Figure 5.17 shows the total retention time of anaerobic–aerobic treatment systems as a function of the anaerobic sludge age. The shortest retention time depends on several factors, among which the most important are: (a) concentration and composition of organic material in the influent; (b) sludge settleability; (c) sludge age; (d) temperature; (e) safety factor in the settler. In Figure 5.17, the retention

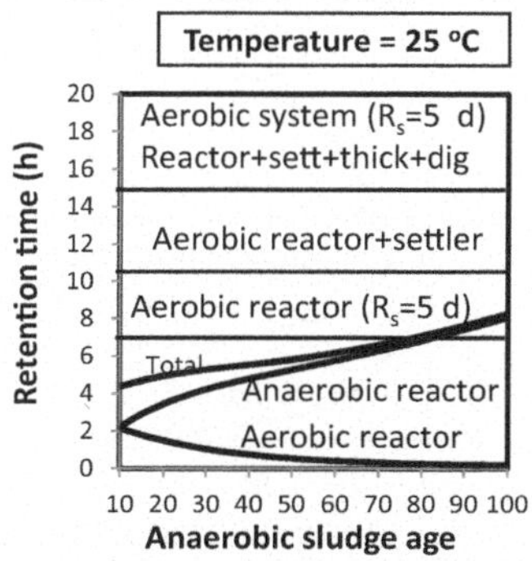


**Figure 5.17** Hydraulic retention time in the anaerobic reactor and the complementary aerobic treatment, as well as the sum of the two, as a function of the age of the anaerobic sludge. The residence time in the aerobic system is indicated (Rsae = 5 days).

time of a purely aerobic system is also indicated for an aerobic sludge age of 5 days, which is the minimum for good effluent quality. It can be observed that the residence time of the anaerobic–aerobic system (total) is much shorter than for the complete aerobic system (reactor + settler + thickener + sludge digester). For an anaerobic sludge age of 100 days, the residence time of the anaerobic–aerobic system is around half the value for an aerobic system (aerobic sludge age of 5 days). Therefore, there is room for a very considerable investment cost reduction.

(b) Reduction of energy consumption for aeration: in principle, the system can be independent on the external energy by generating electricity from methane.

Energy consumption is affected by the two components of the anaerobic–aerobic system. In the pre-treatment, methane is generated that can be used directly to generate electricity. In post-treatment, the energy demand for aeration is reduced compared to the conventional aerobic system, because there is relatively little residual biodegradable material in the UASB effluent. The simplified model of the activated sludge system shows that energy consumption increases with aerobic sludge age because the demand for endogenous respiration increases. Depending on the age of the sludge, there may or may not be energy demand for nitrification.

Figure 5.18 shows the potential for energy production in the anaerobic system as well as the energy demand for aeration in the aerobic post-treatment (both in W/ie) for the case with and without nitrification at 25°C. To estimate the energy production potential, a biogas leakage of 30% and a methane-to-electric energy conversion efficiency of 40% were adopted, which means a production of a power of 0.2 kW per $kgCH_4/d$ was applied to the generator. To estimate the energy demand, an oxygen transfer efficiency of the aerators of 2 $kgO_2/kWh$ was adopted. It is observed that the potential for energy production is much greater than the consumption for aeration so that, in principle, there is a production potential that generates a surplus. In any case, the energy consumption in the anaerobic–aerobic system will be much lower than in the purely aerobic system, even in cases where the energy production potential is not used.

(c) Considerable reduction in sludge production, reducing the cost of its final disposal.

Equations (5.22) and (5.23) allow the comparison between sludge production in the anaerobic–aerobic system and the purely aerobic system. Figure 5.19 shows the sludge production in the two systems as a function of the anaerobic sludge age for a temperature of 25°C. The graphs show that there is indeed a large reduction in sludge production in the anaerobic–aerobic system, especially when the anaerobic sludge age is high, which are normally applied in practice.

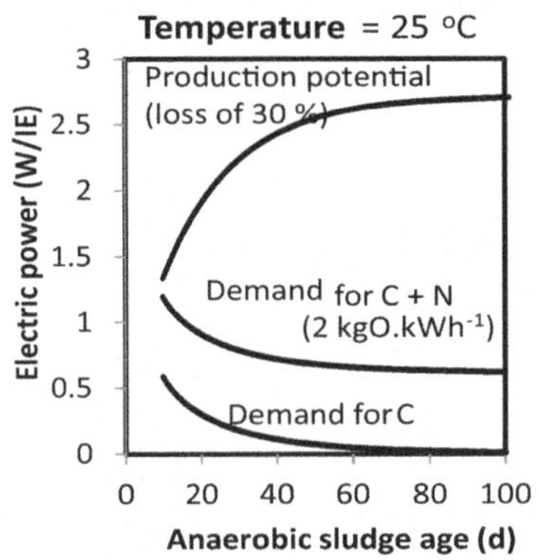


**Figure 5.18** Per capita energy consumption for aeration and electric energy generation potential as a function of the anaerobic age of the sludge at 25°C. The aerobic sludge age applied is 5 days and cases with (C + N) and without (C) nitrification are considered.

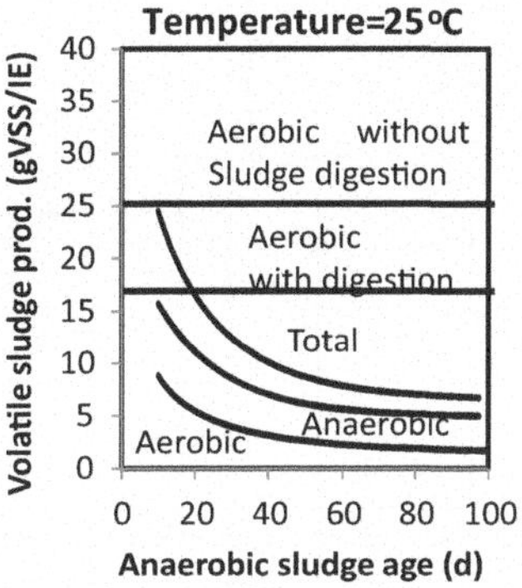


**Figure 5.19** Anaerobic sludge production and excess aerobic sludge as a function of the anaerobic sludge age and excess aerobic sludge production at 25°C (aerobic sludge age = 5 days).

(d) The concentration of the digested sludge is high, facilitating its final disposal.The average sludge concentration in the UASB reactor is 25–30 gTSS(total solids concentration)/L, while in the aerobic system, it is normally not more than 5–10 g/L Therefore, the mixture of the two sludges from anaerobic–aerobic systems will have a concentration much higher than sludge generated in purely aerobic systems, which can greatly facilitate pumping and dehydration.

(e) The excess aerobic sludge can be stabilized in the anaerobic pre-treatment system.In anaerobic–aerobic systems, the possibility exists of using the anaerobic pre-treatment reactor not only for the anaerobic treatment of raw sewage but also for the stabilization of excess sludge generated in the aerobic post-treatment. The excess sludge generated in the anaerobic pre-treatment is already stabilized and does not require specific treatment before its dehydration. To assess the feasibility of using the anaerobic reactor for stabilization of post-treatment active sludge, it is necessary to verify that:
   (1) The introduction of excess aerobic sludge into the UASB reactor does not affect its performance as an anaerobic pre-treatment unit for the sewage to be treated.
   (2) Stabilization of the aerobic sludge in the anaerobic reactor is feasible.
   (3) The aerobic sludge digestion efficiency is equal to the value obtained in a conventional activated sludge digester.
   (4) The non-digestible part of the stabilized sludge mixes with the anaerobic sludge remains in the pre-treatment reactor and is discharged together with the excess anaerobic sludge.

de Silva Filho and van Haandel (2014) showed experimentally that the first three assumptions materialize, but the fourth condition does not: after aerobic sludge digestion, the non-digestible part of the aerobic excess sludge is expelled from the anaerobic reactor, because of its weak settleability. The consequential accumulation of these non-biodegradable solids in the aerobic reactor continues until the sludge concentration becomes so high, that the settler cannot separate the solids from the liquid phase and the system starts discharging sludge along with the effluent.

It is concluded that, although it is possible to digest the aerobic excess sludge in the anaerobic pre-treatment reactor, operational difficulties tend to arise that make this operation not feasible, as in practice has been experienced in many anaerobic–aerobic systems: non-biodegradable residual of aerobic sludge digestion will form an accumulate in the reactors and eventually will be released through the single outlet which is discharged in the final settler as indicated in Figure 5.20. Therefore, practice and theory show that of the five possible advantages of anaerobic–aerobic treatment four can

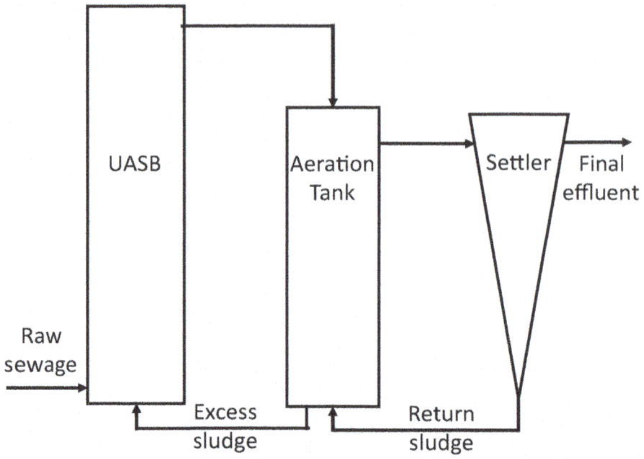


**Figure 5.20** Layout of the activated sludge UASB reactor configuration without intermediate settler for stabilization of excess aerobic sludge in the UASB.

be materialized, but the fifth (the stabilization of aerobic excess sludge in the anaerobic reactor) leads to very serious operational problems.

A solution to this problem is to separate the non-biodegradable solids in the anaerobic effluent before discharge into the aerobic reactor, for example by introducing an intermediate settler as indicated in Figure 5.21. In that case, the accumulation of these solids in the treatment system and the associate operational problems can be avoided.

### 5.3.2 Application of the anammox process

As it was stated in the introduction of this chapter, phosphorus removal could be carried out by chemical precipitation of the orthophosphate ion with calcium, aluminium, or iron salts. On the other hand, conventional nitrogen removal is at most partial in anaerobic–aerobic treatment systems.

Autotrophic nitrogen removal, through the combined action of anammox and ammonium oxidizing microorganisms, is undoubtedly one of the most promising nitrogen removal processes for anaerobically treated effluents. In a first stage, the oxidation of around 50% of the ammonium to

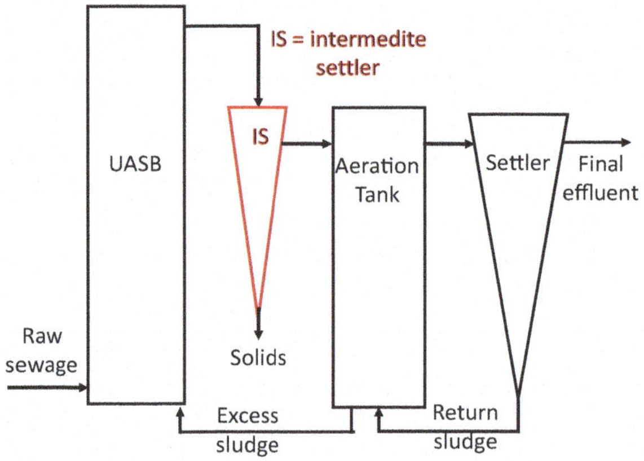


**Figure 5.21** Layout of the activated sludge UASB reactor configuration with an intermediate settler for stabilization of excess aerobic sludge in the UASB.

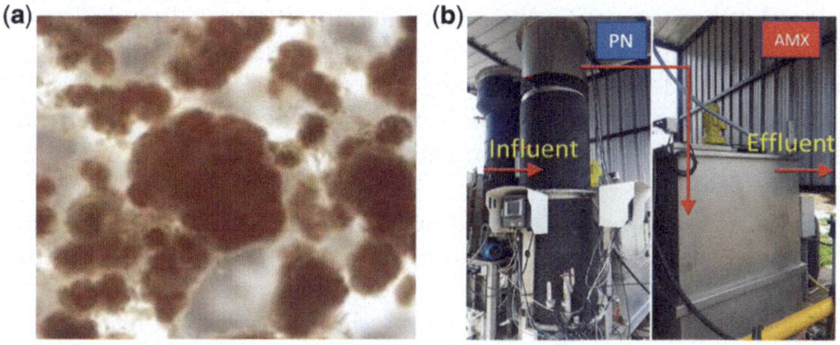


**Figure 5.22** Left (A), anammox granules, anammox biomass is characterized by its intense reddish colour; right (B) picture of the two-stage PN/A prototype operated by Pedrouso *et al.* (2023).

nitrite is sought by means of nitrite-oxidizing autotrophic microorganisms (Equation (5.27)). In a second stage, the ammonium that has not reacted is used as an electron donor to denitrify nitrite, using anammox autotrophic microorganisms (Equation (5.28) and Figure 5.22a).

$$1.32\,NH_4^+ + 1.98\,O_2 \rightarrow 1.32\,NO_2^- + 1.4\,H_2O + 2.64\,H^+ \qquad \text{ammonia oxidation} \tag{5.27}$$

$$1\,NH_4^+ + 1.32\,NO_2^- + 0.13\,H^+ \rightarrow 1.02\,N_2 + 0.26\,NO_3^- + 2.03\,H_2O \qquad \text{anammox} \tag{5.28}$$

$$1\,NH_4^+ + 0.85\,O_2 \rightarrow 0.45\,N_2 + 0.11\,NO_3^- + 1.08\,H^+ + 1.44\,H_2O \qquad \text{overall reaction} \tag{5.29}$$

Because anammox sludge is autotrophic, the $CO_2$ used in sludge production must be partially reduced (oxidation number from 8 to 4), which is possible by partial nitrite to nitrate oxidation. According to Equation (5.29), the overall nitrogen efficiency is limited to a maximum of 89%, due to the production of nitrate. The real efficiency of the process could be affected by the presence of some residual ammonia concentration in the effluent.

The process has multiple advantages, organic matter is not required to denitrify, oxygen consumption is significantly reduced, and a small amount of biological sludge is generated. This process has been successfully applied for the removal of nitrogen of anaerobic digestion centrates in the sludge line of urban STPs, with many installations built after the first full-scale system installed in the Rotterdam STP in 2002, and also for the treatment of certain industrial streams with high nitrogen concentration at temperatures of 30–35°C. At these high temperatures, the nitrite production rate is greater than the nitrate production rate, which is a key for the success of anammox.

Its application in the water line is especially interesting, as it would make feasible the energy sustainability of STPs treating organic matter and nitrogen. Two different reactor configurations could be applied for the autotrophic removal of nitrogen: a one-stage process, in which both partial ammonia oxidation to nitrite and anammox are carried out in the same bioreactor; and a two-stage process, in which partial ammonia oxidation into nitrite is carried out in a first reactor, and the anammox reaction in a second bioreactor coupled in series. From these two configurations, the use of a two-stage process is recommended for treating anaerobically treated sewage, since it gives better results in the water line at temperatures below 30°C (Pedrouso *et al.*, 2021; Wu *et al.*, 2020).

Nevertheless, most of the attempts carried out in the water line have generally been unsuccessful. The reasons are multiple:

(1) Anammox bacteria do not support high residual COD concentrations due to the competitive growth of heterotrophs. It is recommended that the COD/TN ratio of the wastewater be lower than 2–3 g/g, fixing a TN to be treated of 50 mgN/L. This means that the COD in the anaerobically treated sewage should be lower than 100–150 mg/L.

(2) At temperatures lower than 30°C, the anammox activity is reduced. Thus, the reactor volume and biomass concentration of the anammox process should be increased. Additionally, the temperature dependencies on the growth rate for ammonia oxidizers and anammox microorganisms are quite different, which can cause decompensation in the populations, and volumetric activities, of said microorganisms at lower temperatures (Pedrouso *et al.*, 2021; Wu *et al.*, 2020).

(3) Stable suppression of nitrite-oxidizing bacteria is a key aspect in bioreactors treating sewage, with low ammonia concentration, around 50 mgN/L and temperatures of 15–25°C, in comparison with the 500–1,500 mg/L observed in centrates at temperatures of 30–35°C. This complicates the required inhibition of nitrite-oxidizing bacteria by free ammonia (FA), which interferes negatively with the overall process by oxidizing nitrite to nitrate (Wang *et al.*, 2022). Nitrite oxidizers are inhibited at threshold FA concentrations of 0.08–0.82 mg$NH_3$-N/L. Additionally, nitrite oxidizers can also be inhibited by free nitrous acid (FNA) at concentrations higher than 0.02 mg$NO_2^-$-N/L (Pedrouso *et al.*, 2017, 2023; Zheng *et al.*, 2023). The use of a strategy to inhibit nitrite oxidizers by either FA or FNA, or even both, is conditioned by sewage alkalinity, pH, and TN in the wastewater stream. For hard waters, rich in ions such as magnesium and calcium, in which a pH higher than 7 can be easily maintained in the bioreactor, the alternative of free ammonia is more viable, since the presence of FA is favoured at higher pH. On the contrary, for soft waters, where the pH of the bioreactor can reach easily acidic values, decreasing FA concentration, the alternative of FNA can be the solution to inhibit nitrite oxidizers.

For all these reasons, nitrogen treatment, through the anammox process, is still a challenge for the treatment of the effluents from anaerobic systems in STPs. An exception would be hot climate zones with water temperatures above 25°C and the segregated treatment of black waters with much higher nitrogen content. For instance, autotrophic nitrogen removal was responsible for a large part of the nitrogen removal in the STP of Changi (Singapore), using a step-fed activated sludge secondary system, with water temperatures between 28 and 32°C (Wang *et al.*, 2022). Additionally, Zheng *et al.* (2023) treated urban sewage in a pilot system, using a high-rate activated sludge system and a two-stage hybrid pilot plant with both suspended biomass and biofilms, operating at temperatures between 20 and 32°C. The treated effluent contained around 10 mg/L of TN, mainly ammonia. Most of the nitrogen was removed autotrophically in the post-treatment system. It should be highlighted, that the observed COD removal in the first high-rate activated sludge system, around 60%, was quite similar to that observed in many anaerobic UASB reactors. Pedrouso *et al.* (2023) also treated sewage in a pilot system, composed of a first activated sludge reactor, and a two-stage partial nitrification–anammox system, observing a nitrogen removal efficiency of 80%, with temperatures between 23 and 26°C (Figure 5.22a). These studies clearly suggest that there is also an opportunity to treat nitrogen from anaerobic reactor effluents at such temperatures.

Another possibility is the segregated treatment of domestic black water and grey water. Black water contains more than 90% of the COD, and more than 80% of the TP, and TN daily load generated in a municipality. COD, TN, and TP concentrations usually are between 1,000–7,000, 100–1,000, and 25–200 mg/L, respectively (Pedrouso *et al.*, 2020). Thus, the segregated treatment of black water, using a first anaerobic treatment stage, makes it much easier for the application of an autotrophic TN removal post-treatment, due to the higher FA and FNA concentrations observed in the bioreactors. Pedrouso *et al.* (2020) were able to remove above 80% of the TN from anaerobically treated black waters, with around 120 mgN/L, using a one-stage autotrophic process.

### 5.3.3 Post-treatment with membrane reactors

MBRs are perhaps one of the technologies with the greatest potential for wastewater treatment, including the polishing of anaerobically treated sewage. Although there are various configurations of biological membrane systems, this chapter refers to the use of aerobic MBRs using submerged membrane modules. MBR is one of the modifications of the activated sludge process, where the secondary clarifier has been replaced by microfiltration (MF) or ultrafiltration (UF) membranes, with a pore-size between 0.01 and 0.1 μm for UF membranes, and 0.1–0.4 μm for MF membranes. Through sieving, the membranes separate those components larger than the pore size, such as bacteria (>1 μm), helminth eggs (20–80 μm), protozoa (<50 μm), and for the case of UF even colloidal substances, making possible to obtain a high-quality effluent free of suspended solids and microbial indicators, facilitating the reuse of the reclaimed water. Thus, the use of MBRs may probably be the best technology available for water reuse. In any case, more information regarding MBRs and water reuse is provided in Chapters 3 and 8 of this book, respectively.

At present, anaerobic sewage treatment presents two main drawbacks: (i) negligible nitrogen removal, and (ii) the presence of dissolved methane in their effluents, which may be released into the atmosphere, notably increasing the greenhouse gas emissions of the process. Methane is a strong greenhouse gas with a global potential 28 times higher than carbon dioxide. Cakir and Stemstrom (2005) compared greenhouse emissions from both aerobic and anaerobic sewage treatment, concluding that aerobic technologies emit less greenhouse gases, especially for the treatment of diluted sewage, due to the dissolved methane emissions associated with anaerobic sewage treatment. There are various strategies to reduce the impact of dissolved methane emissions, including desorption-based techniques to recover part of the dissolved methane, and biological methane oxidation reactors, in which methane may be used to denitrify (Chapter 7 of this book; Stazi and Tomei, 2021).

An innovative MBR process developed at the University of Santiago de Compostela that aims to respond to the problems associated with anaerobic sewage treatment, the SIAM® treatment technology (US patent 9,725,345B2), obtains the removal of dissolved methane and total nitrogen, so killing two birds with one stone. This technology is based on the use of a hybrid membrane bioreactor (hybrid MBR) to treat the effluent of an UASB system fed with sewage.

The first development of the hybrid MBR of the SIAM® technology included two compartments: an anoxic compartment that accounted for 70% of the volume of the hybrid MBR; and an aerobic membrane filtration compartment, where ammonium was mostly nitrified to nitrate, and the remaining dissolved methane not eliminated in the anoxic stage was partly oxidized or desorbed to the atmosphere. A hollow-fibre ultrafiltration membrane was used to separate the permeate from the mixing liquor, preventing microorganisms from being washed out.

In the first development of the SIAM®, plastic carrier particles were placed in the anoxic compartment to promote the growth of a biofilm with methanotrophic or anammox microorganisms, which are maintained, unlike biomass in suspension, always in the anoxic conditions to favour growth.

The elimination of nitrogen associated with the biological oxidation of methane can follow two pathways: one aerobic, known as aerobic methane oxidation coupled to denitrification (Amo-D); and another anaerobic, known as nitrite/nitrate-dependent anaerobic methane oxidation (N-damo).

In the Amo-D, aerobic methanotrophs are able to convert methane into methane oxidation products that are employed as a carbon source by conventional heterotrophic denitrifiers (Equation (5.30)) in a subsequent reaction (Thalasso *et al.*, 1997).

$$CH_4 + 1.1O_2 + 0.72NO_3^- + 0.72H^+ \rightarrow 0.36N_2 + CO_2 + 2.36H_2O \quad (5.30)$$

N-damo process is carried using either nitrite by N-damo bacteria (*Candidatus Methylomirabilis oxyfera*) or nitrate by N-damo archaea (*Candidatus Methanoperedens nitroreducens*) (Haroon *et al.*, 2013; Equations (5.31) and (5.32), respectively). Unfortunately, these newly discovered N-damo microorganisms are characterized by extremely slow growth rates of approximately 1–2 weeks (Ettwig *et al.*, 2009; Haroon *et al.*, 2013).

$$\text{N-damo bacteria: } CH_4 + 8/3NO_2^- + 8/3H^+ \rightarrow 4/3N_2 + CO_2 + 10/3H_2O \quad (5.31)$$

$$\text{N - damo archaea: } CH_4 + 4NO_3^- \rightarrow 4NO_2^- + 1CO_2 + 2H_2O \qquad (5.32)$$

The methane removal rate in the first two-compartment pilot units reached up to 195 ± 17 mgCH$_4$/L/d, and the nitrogen removal rate, referred to the anoxic compartment, up to 144 ± 20 mgNO$_x$-N/L/d (Silva-Teira *et al.*, 2017). The dissolved methane removal percentage reached 60–80%, while nitrogen removal was 15 mgN/L. From this concentration, and considering the difference between the COD fed to the MBR and that of the permeate, it was estimated that only 4 mgN/L was due to conventional heterotrophic elimination, estimating that aerobic and anaerobic methanotrophs were responsible for the remaining 11 mgN/L.

In a second development, a notable advance of the SIAM® process was the inclusion of an additional aerobic compartment, between the anoxic and membrane filtration compartments (Figure 5.23). This compartment also contained plastic supports for biofilm growth. This development boosted the nitrogen capacity of the hybrid MBR process. Nitrogen removal increased from 15 to 40 mg N/L, for sulphate depleted wastewater, to 60–70 mgN/L for sewage with around 50 mgSO$_4^{2-}$-S/L (Allegue *et al.*, 2023). Those results were obtained in a small 176 L SIAM® pilot plant (120 L UASB + 56 L Hybrid MBR) treating synthetic sewage, maintaining HRT in the hybrid MBR around 0.41 days. Similar results were obtained in a SIAM® prototype with a 3.87 m$^3$ UASB and a 2.06 m$^3$ hybrid MBR, located in the STP of Cartagena (Murcia, Spain) in the frame of the European Union project Siamec (Domínguez *et al.*, 2019; Figure 5.24) maintaining an HRT, referred to the MBR post-treatment system, of around 0.43 days. The TN removal efficiencies were around 75–80%, diminishing TN concentration from 75–90 to some 20 mg/L.

The presence of sulphate in raw sewage can alter the performance of the UASB system, as this ion is reduced in the anaerobic system to sulphur-reduced compounds including elemental sulphur and hydrogen sulphide. Sulphate content in raw sewage may have either an anthropic or natural origin or be caused by seawater intrusion in sewerage systems, especially in coastal towns. Sulphide may be utilized by autotrophic sulphur-oxidizing bacteria (SOB) as an electron donor for denitrification, which could enhance the nitrogen removal capabilities of UASB post-treatment systems. SOB can use reduced sulphur compounds, such as sulphide, as electron donor to transform nitrite/nitrate into N$_2$ (Equations (5.33) and (5.34)).

$$3HS^- + 8NO_2^- + 5H^+ \rightarrow 3SO_4^{2-} + 4N_2 + 4H_2O \qquad (5.33)$$

$$5HS^- + 8NO_3^- + 3H^+ \rightarrow 5SO_4^{2-} + 4N_2 + 4H_2O \qquad (5.34)$$

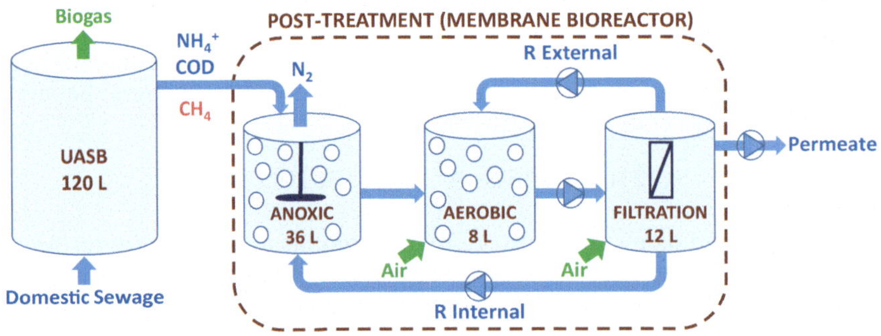


**Figure 5.23** Schematic of the second development of the SIAM® process, with a first anaerobic UASB stage for removing COD; and a hybrid MBR post-treatment for polishing the effluents of the first anaerobic stage. Plastic carrier particles were introduced in both the anoxic and aerobic compartments of the hybrid-MBR system.

**Figure 5.24** Image of the SIAM® prototype (using the second development) operated in the sewage treatment plant of the city of Cartagena (Spain), in the frame of the European Union Life Siamec project.

In principle, the presence of either dissolved methane or sulphur-reduced compounds in the anaerobically treated sewage increases the TN removal capacity of the post-treatment system, as was previously observed for the two SIAM® plants. However, hydrogen sulphide generated in the UASB stage may also inhibit the activity of microorganisms involved in the TN and dissolved methane removal, as was reported for nitrifying (Bejarano Ortiz *et al.*, 2013), anammox (Carvajal-Arroyo *et al.*, 2013) and methane-oxidizing microorganisms (Xu *et al.*, 2020). The inhibition of nitrification was observed in the SIAM® prototype, around day 480 (Figure 5.25) due to an electrical failure in the plant. This caused a rapid loss of nitrification, and no nitrate ions entered the anoxic compartment to oxidize the reduced sulphur products. This led to a vicious circle in which nitrification was not

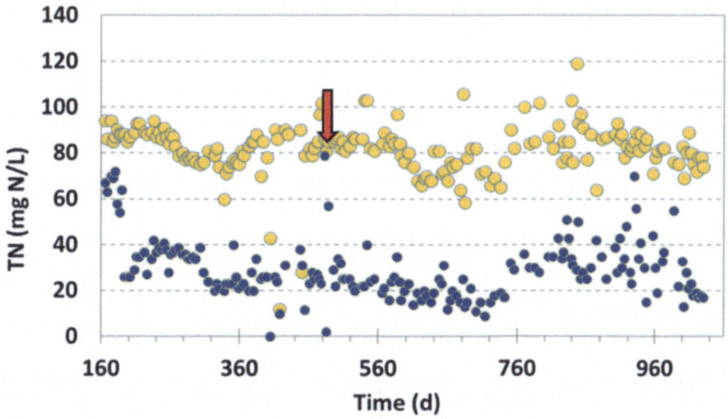


**Figure 5.25** Evolution of the TN in the hybrid MBR stage of the SIAM® prototype operated in the STP of the city of Cartagena (Spain), in the frame of the EU Life Siamec project: (●) TN in the anaerobically treated sewage; (●) TN measured in the permeate of the MBR system. The red arrow indicates a nitrification inhibition event due to the accumulation of reduced sulphur compounds, which occurred around operating day 480.

achieved due to inhibition, and the inhibitor was not eliminated due to the absence of nitrate and nitrite in the anoxic compartment. In the prototype plant, this was solved by reducing the flow of residual water fed for a few days.

One of the advantages of using the SIAM® process, as occurred with some other MBR processes, is that it facilitates water reuse of the reclaimed water. Microbial (*E. coli*; *Legionella* spp.; Nematodes eggs) and chemical indicators (TSS, $BOD_5$, turbidity) in the permeate of the SIAM® prototype treating sewage were below the limits for water reuse in agriculture set by the EU regulation 2020/741 for reclaimed water quality class A. On the other hand, the observed overall sludge production is similar to that referred for anaerobic UASB systems, around 0.1 gTSS/gCOD, reducing the cost associated with sludge management.

Concerning the microbial population developed, the biofilms of the anoxic and aerobic compartments were enriched in anammox microorganisms. N-damo microorganisms (detected by q-PCR, Illumina, and FISH) were only observed in aerobic and anoxic biofilms of the prototype treating urban sewage in Cartagena, around 0.02% by Illumina. The presence of aerobic methanotrophs was always observed in the hybrid MBR, both in suspension and the biofilms, with values between 1.5 and 8%.

Despite the low N-damo content, denitrification associated with anaerobic methane oxidation was always detected in batch activity tests (Silva-Teira *et al.*, 2017). Xie *et al.* (2023) also detected some kind of N-damo activity without any known N-damo, in a moving-bed biofilm reactor (MBBR) under hypoxic conditions, with only 0.02 mg/L of dissolved oxygen (DO). The presence of aerobic methane oxidizers was remarkable, despite the low DO. They suggested the possibility that some unknown strains may harbour uncharacterized N-damo capacity that contributed to nitrogen removal.

The use of a hybrid MBR process with several anoxic and aerobic chambers, as proposed in the SIAM® system, basically is in line with what has been observed for anammox systems in the water line, using hybrid bioreactors (Zheng *et al.*, 2023). The use of carrier particles promotes the growth of sensitive microorganisms, avoiding the strong dissolved oxygen variations to which the biomass in suspension is subjected.

To interpret the obtained results in a more analytical way, a mathematical model to describe the biological processes in the hybrid MBR was developed taking account of the kinetic and stoichiometric parameters of existing microbial populations: heterotrophs, sulphate reducing bacteria, sulphide oxidizing bacteria, aerobic and anaerobic methanotrophs (including N-damo bacteria and archaea), ammonium and nitrite oxidizing bacteria and anammox. The model considered only the existence of suspended biomass, neglecting the presence of biofilms, and the mass balances in the hybrid MBR considered three tanks connected in series, representing the anoxic, aerobic and membrane filtration compartments of this system. The mathematical model developed may be considered a way of describing the system under study, the hybrid MBR, through mathematical equations, allowing the simulation, once it has been correctly adjusted, to verify the behaviour of the system.

The model developed is similar to some of the existing activated sludge models (ASM1 and ASM2) and contained 22 components, including sulphur compounds, dissolved methane and oxygen. The microbial reactions were modelled by 27 kinetic processes (Liñares-Lamas, 2018; Mauricio-Iglesias *et al.*, 2016) and contained 97 kinetic and stoichiometric parameters. Some of the parameters of the model were calibrated using the experimental results obtained during 248 days of operation, and others fixed considering the typical values referred to the ASM models. The model was verified considering that the experimental results obtained in a later period fit correctly with the predictions obtained.

The use of the model allows the study of the impact of various operating conditions different from those tested, allowing the identification of which variables can positively or negatively affect the biological elimination of total nitrogen or dissolved methane in the hybrid MBR.

Heat maps were used to represent the different operating conditions on the removal of total nitrogen and dissolved methane. Figure 5.26 shows the total nitrogen and dissolved methane removal predicted by the model, as a function of TSS in the aerobic compartment and HRT of the hybrid MBR.

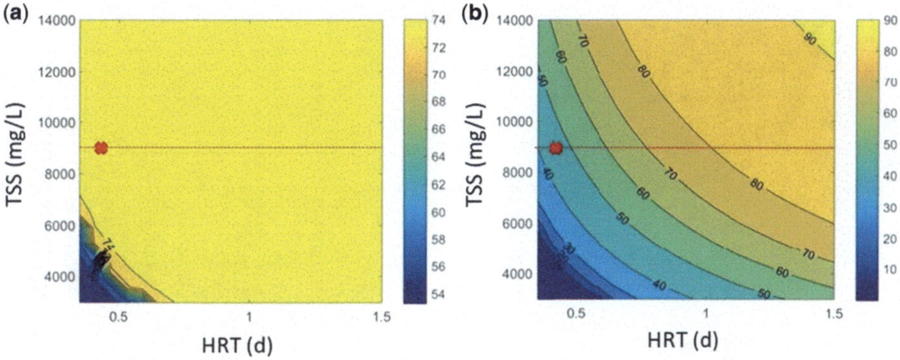


**Figure 5.26** (a) Percentage of total nitrogen removal and, (b) percentage of dissolved methane removal predicted in the hybrid MBR process as a function of the HRT and TSS concentration in the aerobic compartment (Liñares-Lamas, 2018). The red dot indicated the experimental results obtained in the hybrid MBR of a SIAM® prototype.

The red dot indicates the observed experimental result observed maintaining HRT, DO and TSS at 0.4 day, 1 mg/L and 9 gTSS/L, respectively.

At short HRT the model predicts that nitrogen removal depends notably on the TSS concentration of the post-treatment bioreactor (Figure 5.26a), especially at low HRT. On the other hand, it might be necessary to operate with high HRT, greater than 0.5 days to obtain high nitrogen efficiency, when the post-treatment bioreactor operates below 4000 mgTSS/L, a concentration typically observed in activated sludge bioreactors.

Figure 5.26b shows the percentage of biological methane removal in the system carried out by aerobic methanotrophic organisms as a function of solids concentration and HRT. Since these microorganisms have a slow growth, below a certain value of solids they were not able to compete with the rest of the microorganisms and end up disappearing from the system. Therefore, methane removal increases with increasing solids concentration and with increasing HRT. Therefore, the use of an MBR system that allows maintaining high sludge concentrations may be a very good option for the removal of dissolved methane and nitrogen in anaerobically treated sewage.

Figure 5.26 shows the dissolved methane and nitrogen removals based on dissolved oxygen concentrations observed in the aerobic compartment of the hybrid MBR and the HRT applied. This parameter, DO, is one of the operating variables that is usually controlled, in fact, in aerobic bioreactors. The removal of nitrogen is practically constant as a function of the DO of the aerobic compartment (Figure 5.27a) since the volume of the membrane chamber is sufficient to carry out nitrification, in fact the removal range depicted in the heat map varies only between 74.2 and 75.4%. However, for the removal of methane (Figure 5.27b) an optimum value of oxygen is observed. Below this, oxygen limits the activity of aerobic methane oxidizers, while with a higher value, methane desorption due to the stripping to the atmosphere, increases. It is observed that the DO optimum values are between 0.4 and 1 mg $O_2$/L, and that methane removal would be strongly encouraged if the volume of the MBR was increased. The model predicts that methane removal could be increased from 45%, obtained in the experimental system, to more than 80%, if the HRT were increased from 0.4 to 1 day, maintaining the same concentration of dissolved oxygen. This would obviously increase the volume of the post-treatment system, increasing the capital expenses of the sewage treatment plant.

The model indicates that the elimination of dissolved methane depends strongly on the concentration of organic matter in the anaerobically treated sewage fed to the MBR, with a COD of around 300 mg/L, improving the removal significantly if the COD efficiency of the UASB increased. In fact, the removal of dissolved methane determined in a hybrid MBR, fed with an

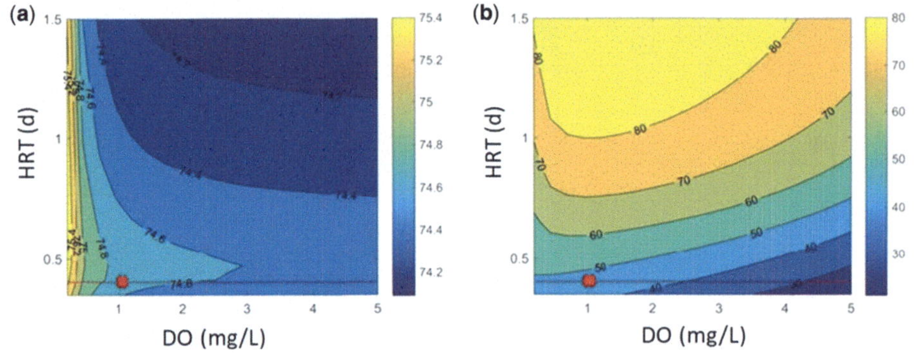


**Figure 5.27** (a) Percentage of total nitrogen removal, and (b) percentage of dissolved methane removal predicted in the hybrid MBR process as a function of the HRT and DO concentration in the aerobic compartment (Liñares-Lamas, 2018). The red dot indicated the experimental results obtained in the hybrid MBR of a SIAM® prototype.

anaerobically treated wastewater containing less than 100 mgCOD/L, were higher, around 70–80% (Alvarino *et al.*, 2019; Silva-Teira *et al.*, 2017). This is due to the lower presence of conventional heterotrophs in the sludge when COD fed diminishes. On the other hand, the model indicates that the development of N-damo microorganisms is problematic under the operating conditions used. This agrees with the experimental results, where these microorganisms were present in a very low proportion in the biofilms, although some anaerobic methane oxidation activity was detected (Silva-Teira *et al.*, 2017).

The model indicates that most of the methane removal takes place by aerobic methanotrophic bacteria, which were detected in the sludge samples. In any case, the role of biofilms or the development of unknown anaerobic methanotrophic communities as referred in the literature (Xie *et al.*, 2023) could open new ways to improve the simultaneous removal of dissolved methane and total nitrogen from anaerobically treated sewage.

## 5.4 POST-TREATMENT WITH BIOFILM REACTORS

### 5.4.1 Trickling filters

Trickling filters are a type of non-submerged biofilm three-phase bioreactors, in which a plastic or rock packing material contained in a tower, is used to promote biofilm growth. Sewage is sprayed over the top of the trickling filter, and trickles to the bottom, wetting the biofilm. The treated sewage is collected at the bottom of the tower. Part of the biodegradable organic matter and ammonium contained in the sewage are oxidized by the biofilm, thanks to the oxygen transferred through natural or forced draft aeration.

Rocks with a nominal size of around 5–20 cm, random plastic packing media, and plastic bundles are the typical packing materials used in TFs. The height of the trickling filter is conditioned, among others, by the weight of the packing material, limiting the height of the trickling filters to 1–3 m or 4–12 m if rock or plastic packing materials are used, respectively.

Although trickling filters with rock are still in use, there is a tendency to replace them with plastic packing materials, due to the higher specific surface area, better air circulation, and the absence of ponding, which is caused by solids accumulation in the rock material (Bressani-Ribeiro *et al.*, 2017). This not only improves the efficiency of the filter but also reduces the nuisance problems or the proliferation of insects associated with the use of rock in TFs.

It is recommended to pretreat the sewage fed by primary sedimentation in conventional TF facilities. Trickling filters used as the post-treatment stage of UASB effluents are widely used in countries such as Brazil. Several studies have demonstrated that the combined UASB + TF systems used for sewage treatment have a similar overall BOD removal efficiency than the conventional primary settling + TF sewage treatment (Bressani-Ribeiro *et al.*, 2017).

The size of the TF required can be estimated in terms of the hydraulic loading rate (HLR; $m^3/m^2/d$) and the organic loading rate (OLR; $kgBOD/m^3/d$) selected. TF are classified in five different categories depending on the HLR, OLR and the nature of the packing material used (Metcalf & Eddy 2014): low rate with rocks (1–4 $m^3/m^2/d$ and 0.07–0.22 $kgBOD/m^3/d$); intermediate rate with rocks (4–10 $m^3/m^2/d$ and 0.24–0.48 kg $BOD/m^3/d$); high rate with rocks (10–40 $m^3/m^2/d$ and 0.4–2.4 $kgBOD/m^3/d$); high rate with plastic (10–75 $m^3/m^2/d$ and 0.6–3.2 $kgBOD/m^3/d$); and roughing TFs with plastic packing (40–200 $m^3/m^2/d$ and >1.5 $kgBOD/m^3/d$). According to Bressani-Ribeiro *et al.* (2017), most of the UASB + TF full-scale plants in Brazil were designed as high-rate TFs with rocks.

Nitrogen or phosphorus removal with the above indicated TFs is negligible, and only certain ammonia removal could be expected, depending on the operational conditions, especially the applied loading rate and temperature.

### 5.4.2 Down-flow hanging sponge

Down-flow hanging sponge (DHS) technology can be considered an improvement of the trickling filter process, where the packing material is made up of polystyrene sponges. This technology, developed in Japan, has been successfully applied to the post-treatment of effluents from UASB systems, in urban STPs. This effluent trickles down through the sponges by gravity. In DHS systems the growth of a biofilm, covering the polystyrene particles used, eliminates in contact with air a large part of the organic matter and TKN fed. The sponge used as carrier in DHS is characterized by its large porosity (around 98.5%), which favours the retention of water, increasing the contact time between the residual water and the attached microorganisms, regarding other packing materials used in trickling filters, enhancing the pollutants removal.

Okubo *et al.* (2015) operated a full-scale UASB + DHS with polyurethane foam sponge as DHS support, for 5 years, without any significant sponge damage that could endanger the process.

DHS operated at short HRT, around 1.5 h referred to the sponge volume in full-scale facilities (Okubo *et al.*, 2015; Mazhar *et al.*, 2021). DHS towers contained around 24.7–33% carriers. Sludge retention time is around 90–120 days (Tandukar *et al.*, 2007; Tawfik *et al.*, 2006). DHS has low energy requirements regarding aerobic activated sludge systems, caused natural draft aeration. Additionally, excess of biomass growing onto the carrier is easily detached by the water that is percolating in the system, avoiding the clogging of the filter media. Excess biomass could easily be removed by secondary sedimentation, and this stage could even be dispensed in those locations where the TSS discharge limits are not rigorous (Bressani-Ribeiro *et al.*, 2017). Biomass yield observed in DHS is very low, around 0.04 gTSS/gCOD as stated by Okubo *et al.* (2015).

In DHS, most of the COD is removed in the upper part of the filter. Organic matter removal efficiency of UASB + DHS systems are comparable with those of conventional activated sludge reactors (Tandukar *et al.*, 2007), with removal BOD and COD efficiencies of around 90% for pilot plants (Tandukar *et al.* 2007; Tawfik *et al.* 2006). For two full-scale facilities located in India, Agra (Aamir Mazhar *et al.*, 2021) and Karnaoba (Okubo *et al.*, 2015), the COD efficiencies were 83 and 89%, respectively. The lower COD efficiency of the Agra facility might be caused by the absence of secondary sedimentation of the DHS effluent. TSS of the final effluent was higher, 42.7 ± 10 versus the 19 ± 8 mgTSS/L observed in the Karnaoba facility.

COD removal is hardly affected by temperature in UASB + DHS systems, with a slight improvement in any case at temperatures above 20°C (Tandukar *et al.*, 2007). A large fraction of the ammonium fed is oxidized to nitrate in the lower part by aerobic nitrifiers. Around 50–70% of the fed ammonia is oxidized to nitrate in DHS (Okubo *et al.*, 2015). The existence of anaerobic and anoxic environments inside the sponges of the lower part facilitates the elimination of part of the fed TKN due to denitrification processes. DHS systems may eliminate a fraction of the fed TN, around 6–12 mg/L, mainly due to biological nitrification–denitrification processes, as the contribution of assimilation to nitrogen removal is almost negligible (Okubo *et al.* 2015; Onodera *et al.* 2014; Tawfik *et al.* 2006).

## 5.5 CONCLUSIONS

### 5.5.1 Combination of anaerobic pre-treatment and PPs

Anaerobic pre-treatment in combination with PP is highly advantageous when compared with conventional WSP:

- A large reduction of the pond can be achieved.
- Nutrient removal becomes feasible in PP, but not in WSP.
- There is a possibility of productive use of biogas and sludge production.
- Decreasing of effluent quality and quantity by evaporation is reduced.

The preferable hydrodynamic regime for PPs is sequential batch (SBPP) rather that continuous flow (CFPP).

- SBPP can remove nutrient more rapidly than CFPP, but it requires a transfer pond to feed the SBPP.
- Transfer ponds can have important secondary advantages: settling of anaerobic solids and odour reduction.

The increase of pH in PPs is due to $CO_2$ removal by photosynthesis and desorption. Ammonium desorption tends to reduce pH.

The principal mechanism for ammonium removal is desorption, although some nitrogen may be removed by incorporation in the produced biosolids.

The principal mechanism for phosphorus removal in PPs is precipitation with naturally present or added cations, although some phosphorus may be removed by incorporation in the produced biosolids. High-quality effluent may be produced by clarification of PP effluent and can further be improved to pure water by advanced treatment like inverse osmosis.

The construction costs for SBPP are much smaller than for WSP for three reasons:

- A per capita reduction of the pond area of 2/3 (67%) can be achieved if only nitrogen is to be removed from sewage. If nitrogen and phosphate are to be removed the reduction is 1/2 (50%).
- The depth for SBPP (0.5 m) is much smaller than for WSP (1.2 m), so that much less excavation is required, and the slopes are much smaller. The SBPP do not need to have all the same bottom level, again reducing excavation costs.
- The reduction of odour from SBPP plants allows construction near the contributing population reducing the costs for the outfall.

### 5.5.2 Anaerobic pre-treatment and other aerobic activated sludge post-treatment systems

Aerobic post-treatment, using activated sludge systems (after anaerobic pre-treatment) removes organic material and suspended solids with the same efficiency as purely aerobic sewage treatment, but has great advantages:

- Strong reduction in the volume of the treatment system, lowering the investment costs.
- Reduction of energy demand with the possibility of being self-sufficient through using the energy from the methane generated in the pre-treatment reactor.
- Reduction of sludge production, thus reducing operating cost.
- Increase in excess sludge concentration, facilitating dehydration and final disposal.
- Nitrogen removal is difficult and is limited by the low COD contained in the anaerobically treated effluent.

### 5.5.3 Post-treatment with membrane and biofilm reactors

The use of MBRs, as post-treatment systems, allows obtaining a high-quality effluent, facilitating the reuse of water. Additionally, the higher microbial concentrations and the use of a membrane that prevents the washing of slow-growing microorganisms, facilitate the use of dissolved methane as an additional carbon source for denitrification.

The integrated SIAM® process, which consists of a UASB system coupled in series to a hybrid MBR, is effective in removing total nitrogen and dissolved methane. The development of second-generation SIAM® systems, in which the hybrid MBR consists of three chambers, allows the removal of total nitrogen to be significantly improved, going from a TN removal of 15–20 mg/L previously obtained, to more than 40 mg/L, and maintaining the removal of dissolved methane.

The use of trickling filters, using DHS systems, significantly improves the results previously observed for traditional trickling filters with stones or conventional plastic fillers. TN removal is limited to some 10 mg/L and cannot be easily controlled.

## REFERENCES


Aguiar M. R. L., Santos S. L. and van Haandel A. (2021). Polishing ponds operating in batch mode for post-treatment of UASB reactors: organic material removal. Revista DAE: Edição especial, São Paulo, n. 229, março 2021.

Albuquerque M. S., Santos S. L. and van Haandel A. (2021). Influence of the hydrodynamic regime on the performance of polishing ponds. Revista DAE: Edição especial, São Paulo, n. 229, março 2021.

Allegue T., Arias A., Liñares A., Omil F. and Garrido J. M. (2023). Impact of dissolved sulfide on a hybrid membrane bioreactor treating the effluent of a mainstream up-flow anaerobic sludge blanket. *Environmental Science: Water Research and Technology*, **9**, 2733–2744, https://doi.org/10.1039/D3EW00404J

Alvarino T., Allegue T., Fernandez-Gonzalez N., Suarez S., Lema J. M., Garrido J. M. and Omil F. (2019). Minimization of dissolved methane, nitrogen and organic micropollutants emissions of effluents from a methanogenic reactor by using a preanoxic MBR post-treatment system. *Science of the Total Environment*, **671**, 165–174, https://doi.org/10.1016/j.scitotenv.2019.03.169

Bastos R. K. X., Rios E. N. and Sánchez I. A. (2018). Further contributions to the understanding of nitrogen removal in waste stabilization ponds. *Water Science and Technology*, **77**(11), 2635–2641, https://doi.org/10.2166/wst.2018.218

Batista M. M., Lambais G. R., Santos S. L. and van Haandel A. (2021). Bacterial decay in polishing ponds treating UASB reactor effluents. Revista DAE: Edição especial, São Paulo, n. 229, março 2021.

Bejarano Ortiz D. I., Thalasso F., de Cuervo López F. M. and Texier A. C. (2013). Inhibitory effect of sulfide on the nitrifying respiratory process. *Journal of Chemical Technology and Biotechnology*, **88**(7), 1344–1349, https://doi.org/10.1002/jctb.3982

Bressani-Ribeiro T., Brandt E. M. F., Gutierrez K. G., Díaz C. A., Garcia G. B. and Chernicharo C. A. L. (2017). Potential of resource recovery in UASB/trickling filter systems treating domestic sewage in developing countries. *Water Science and Technology*, **75**(7), 1659–1666, https://doi.org/10.2166/wst.2017.038

Cakir F. Y. and Stenstrom M. K. (2005). Greenhouse gas production: A comparison between aerobic and anaerobic wastewater treatment technology. *Water Research*, **39**, 4197–4203, https://doi.org/10.1016/j.watres.2005.07.042

Carvajal-Arroyo J. M., Sun W., Sierra-Alvarez R. and Field J. A. (2013). Inhibition of anaerobic ammonium oxidizing (anammox) enrichment cultures by substrates, metabolites, and common wastewater constituents. *Chemosphere*, **91**(1), 22–27, https://doi.org/10.1016/j.chemosphere.2012.11.025

Cavalcanti P. F. F. (2003). Integrated Application of the UASB Reactor and Ponds for Domestic Sewage Treatment in Tropical Regions. Doctoral thesis, University of Wageningen, The Netherlands.

Chick H. (1908). An investigation of the laws of disinfection. *Journal of Hygiene*, **8**(1), 92–158, https://doi.org/10.1017/S0022172400006987

Deffeyes K. S. (1956). Carbonate equilibria: a graphic and algebraic approach. *Limnology and Oceanography*, **10**, 412–426, https://doi.org/10.4319/lo.1965.10.3.0412

Delgado Vela J., Stadler L. B., Martin K. J., Raskin L., Bott C. B. and Love N. G. (2015). Prospects for biological nitrogen removal from anaerobic effluents during mainstream wastewater treatment. *Environmental Science and Technology Letters*, **2**(9), 234–244, https://doi.org/10.1021/acs.estlett.5b00191

de Silva Filho J. A. and van Haandel A. C. (2014). Simultaneous sludge stabilization of aerobic post treatment and sewage digestion in an anaerobic pre-treatment reactor. *Revista Dae*, **194**, 86–102.

Domínguez J., Rodríguez L., Silva-Teira A., Serna T., Sánchez E., Simón P., Moya N., Arias A., Lema J. M. and Garrido J. M. (2019). Removal of dissolved methane and nitrogen from anaerobically treated sewage using the SIAM process. (proceedings) Anaerobic Digestion Conference (AD16), Delft, The Netherlands.

Emerson K., Russo R. C., Lund R. E. and Thurston T. T. (1975). Aqueous ammonium calculations: effect of pH and temperature. *Journal of the Fisheries Research Board of Canada*, **23**, 2379–2383, https://doi.org/10.1139/f75-274

Environmental European Agency (2022). Beyond water quality. Report 05/2022. Publications Office of the European Union, Luxembourg, https://www.eea.europa.eu/publications/beyond-water-quality-sewage-treatment

Ettwig K. F., Van Alen T., Van De Pas-Schoonen K. T., Jetten M. S. M. and Strous M. (2009). Enrichment and molecular detection of denitrifying methanotrophic bacteria of the NC10 phylum. *Applied and Environmental Microbiology*, **75**, 3656–3662, https://doi.org/10.1128/AEM.00067-09

Foresti E., Zaiat M. and Vallero M. (2006). Anaerobic processes as the core technology for sustainable domestic wastewater treatment: Consolidated applications, new trends, perspectives, and challenges. *Reviews in Environmental Science and Bio/Technology*, **5**(1), 3–19, 2744, https://doi.org/10.1007/s11157-005-4630-9

Gomez E., Paing J., Casellas C. and Picot B. (2000). Characterisation of phosphorus in sediments from waste stabilization ponds. *Water Science and Technology*, **42**, 257–264, https://doi.org/10.2166/wst.2000.0656

Harned H. S. and Davis R. (1943). The ionization constant of carbonic acid in water from 0 to 50°. *Journal of the American Chemical, Society*, **65**(10), 2030–2037, https://doi.org/10.1021/ja01250a059

Harned H. S. and Scholes S. R. (1943). The ionization constant of HCO3-from 0 to 50 deg C. *Journal of the American Chemical, Society*, 1706.

Haroon M. F., Hu S., Shi Y., Imelfort M., Keller J., Hugenholtz P., Yuan Z. and Tyson G. W. (2013). Anaerobic oxidation of methane coupled to nitrate reduction in a novel archaeal lineage. *Nature*, **500**, 567–570, https://doi.org/10.1038/nature12375

Ho L. T., Van Echelpoel W. and Goedhals P. (2017). Design of waste stabilization pond systems: a review. *Water Research*, **123**, 236–248, https://doi.org/10.1016/j.watres.2017.06.071

Leitão R. C. (2004). Robustness of UASB reactors treating sewage under tropical conditions. PhD thesis, Wageningen University, Netherlands.

Lettinga G., van Velsen A. F. M., Hobma S., de Zeeuw W. and Klapwijk A. (1980). Use of the upflow sludge blanket (USB) reactor concept for biological wastewater treatment, especially for anaerobic treatment. *Biotechnology and Bioengineering*, **22**(4), 699–734, https://doi.org/10.1002/bit.260220402

Liñares-Lamas A. (2018). Modeling of the sulfur cycle in post-treatment of anaerobic reactor effluents. Master thesis, Universidade de Santiago de Compostela, Spain (in Galician).

Loewenthal R. E. and Marais G. V. R. (1976). Carbonate Chemistry of Aquatic Systems: Theory and Application. Ann Arbor Sci. Publishers, Michigan, USA.

Mara D. D. 1997. Design Manual for Waste Stabilization Ponds in India. Lagoon Technology International, Leeds, 125 p.

Marais G. V. R. (1974). Fecal bacteria kinetics in stabilization ponds. *Journal of the Environmental Engineering Division ASCE*, **100**(1), 119–139, https://doi.org/10.1061/JEEGAV.000013

Marais G. V. R. and Ekama G. A. (1976). The activated sludge process part I: Steady state behaviour, Water SA, 163–200.

Marais G. V. R. and Shaw V. A. (1961). A rational theory for design of waste stabilization ponds in South Africa. *Transactions of the South African Institute of Civil Engineers*, **3**, 205–227.

Mauricio-Iglesias M., Garrido J. M. and Lema J. M. (2016). Operation of an innovative WWTP with environmental objectives. A model-based analysis. *IFAC-PapersOnLine*, **49**(7), 539–543, https://doi.org/10.1016/j.ifacol.2016.07.398

Mazhar M. A., Khan N. A., Khan A. H., Ahmed S., Siddiqui A. A., Husain A., Rahisuddin Tirth V., Islam S., Shukla N. K., Changani F., Yousefi M., Hassaballa A.E. and Radwan N. (2021). Upgrading combined anaerobic-aerobic UASB-FPU to UASB-DHS system: Cost comparison and performance perspective for developing countries. *Journal of Cleaner Production*, **284**, 124723, https://doi.org/10.1016/j.jclepro.2020.124723

Metcalf E. and Eddy M. (2014). Wastewater Engineering: Treatment, Disposal and Reuse. 5th edn. McGraw-Hill, New York, USA.

Morais C. A., dos Santos S. L. and van Haandel A. C. (2023). A predictive model for ammonium removal in polishing ponds operated in sequential batch mode. *Water Science and Technology*, **87**(1), 13–26, https://doi.org/10.2166/wst.2022.418

Mulder A., van de Graaf A. A., Robertson L. A. and Kuenen J. G. (1995). Anaerobic ammonium oxidation discovered in a denitrifying fluidized bed reactor. *FEMS Microbiology Ecology*, **16**(3), 177–184, https://doi.org/10.1111/j.1574-6941.1995.tb00281.x

Okubo T., Onodera T., Uemura S., Yamaguchi T., Ohashi A. and Harada H. (2015). On-site evaluation of the performance of a full-scale down-flow hanging sponge reactor as a post-treatment process of an up-flow anaerobic sludge blanket reactor for treating sewage in India. *Bioresource Technology*, **194**, 156–164, https://doi.org/10.1016/j.biortech.2015.07.015

Onodera T., Tandukar M., Sugiyana D., Uemura S., Ohashi A. and Harada H. (2014). Development of a sixth-generation down-flow hanging sponge (DHS) reactor using rigid sponge media for post-treatment of UASB treating municipal sewage. *Bioresource Technology*, **152**, 93–100, https://doi.org/10.1016/j.biortech.2013.10.106

Parker C. D., Jones H. L. and Taylor W. S. (1950). Purification of sewage in lagoons. *Sewage and Industrial Waste*, **22**(6), 760–775, https://doi.org/10.2307/25031317

Pano A. E. and Middlebrooks J. (1982). Ammonia Nitrogen Removal in Facultative Wastewater Stabilization Ponds. *Journal Water Pollution Control Federation*, **54**(4), https://doi.org/10.2307/25041312

Pedrouso A., Morales N., Rodelas B., Correa-Galeote D., Val del Rio A., Campos J. L., Vazquez-Padin J. and Mosquera-Corral, A. (2023). Rapid start-up and stable maintenance of the mainstream nitritation process based on the accumulation of free nitrous acid in a pilot-scale two-stage nitritation-anammox system. *Separation and Purification Technology*, **317**, 123851, https://doi.org/10.1016/j.seppur.2023.123851

Pedrouso A., Val del Río Á, Morales N., Vázquez-Padín J. R., Campos J. L., Méndez R. and Mosquera-Corral A. (2017). Nitrite oxidizing bacteria suppression based on in-situ free nitrous acid production at mainstream conditions. *Separation and Purification Technology*, **186**, 55–62, https://doi.org/10.1016/j.seppur.2017.05.043

Pedrouso A., Tocco G., Val del Río A., Carucci A., Morales N., Campos J. L., Milia S. and Mosquera-Corral A. (2020). Digested blackwater treatment in a partial nitritation-anammox reactor under repeated starvation and reactivation periods. *Journal of Cleaner Production*, **244**, 1–9, https://doi.org/10.1016/j.jclepro.2019.118733

Pedrouso A., Val del Rio A., Morales N., Vazquez-Padin J. R., Campos J. L. and Mosquera-Corral A. (2021). Mainstream anammox reactor performance treating municipal wastewater and batch study of temperature, pH, and organic matter concentration cross-effects. *Process Safety and Environmental Protection*, **145**, 195–202, https://doi.org/10.1016/j.psep.2020.07.052

Santos S. L., Simões J. P. O, Paiva F. V. and Van Haandel A. (2017). Design optimization of a simple single family anaerobic sewage treatment system. *Water Practice and Technology*, **12**, 55–71, https://doi.org/10.2166/wpt.2017.010

Silva-Teira A., Sánchez A., Buntner D., Rodríguez-Hernández L. and Garrido J. M. (2017). Removal of dissolved methane and nitrogen from anaerobically treated effluents at low temperature by MBR post-treatment. *Chemical Engineering Journal*, **326**, 970–979, https://doi.org/10.1016/j.cej.2017.06.047

Stazi V. and Tomei M. C. (2021). Dissolved methane in anaerobic effluents: A review on sustainable strategies for optimization of energy recovery or internal process reuse. *Journal of Cleaner Production*, **317**, 128359, https://doi.org/10.1016/j.jclepro.2021.128359

Tandukar M., Ohashi A. and Harada H. (2007). Performance comparison of a pilot-scale UASB and DHS system and activated sludge process for the treatment of municipal wastewater. *Water Research*, **41**(12), 2697–2705, https://doi.org/10.1016/j.watres.2007.02.027

Tawfik A., Ohashi A. and Harada H. (2006). Sewage treatment in a combined up-flow anaerobic sludge blanket (UASB)-down-flow hanging sponge (DHS) system. *Biochemical Engineering Journal*, **29**(3), 210–219, https://doi.org/10.1016/j.bej.2005.11.018

Thalasso F., Vallecillo A., García-Encina P. and Fdz-Polanco F. (1997). The use of methane as a sole carbon source for wastewater denitrification. *Water Research*, **31**, 55–60, https://doi.org/10.1016/S0043-1354(96)00228-X

Van Haandel A. C. and van der Lubbe J. (2012). Handbook Biological Wastewater Treatment: Design and Optimization of Activate Sludge Systems. IWA Publishing, London, UK.

Van Haandel A. C. and van der Lubbe J. (2019). Anaerobic Sewage Digestion: Theory and Applications. IWA Publishing, London, UK.

Van Lier J. B., Mahmoud N. and Zeeman G. (2020). Anaerobic Wastewater Treatment. IWA Publishing, London, UK, https://doi.org/10.2166/9781789060362_0701

Vesilind P. A. (1968). Theoretical considerations: design of prototype thickeners from batch settling tests. *Water and Sewage Works*, **115**, 302–307.

Wang Z., Zheng M., Duan H., Yuan Z. and Hu S. (2022). A 20-year journey of partial nitritation and anammox (PN/A): from sidestream toward mainstream. *Environmental Science & Technology*, **56**(12), 7522–7531, https://doi.org/10.1021/acs.est.1c06107

Wu P., Zhang X., Wang X., Wang C., Faustin F. and Liu W. (2020). Characterization of the start-up of single and two-stage anammox processes with real low-strength wastewater treatment. *Chemosphere*, **245**, 125572, https://doi.org/10.1016/j.chemosphere.2019.125572

Xie T., Liu X., Xu Y., Bryson S., Zhao L., Huang K., Huang S., Li X., Yang Q., Dong H. and Winkler M. H. (2023). Coupling methanotrophic denitrification to anammox in a moving bed biofilm reactor for nitrogen removal under hypoxic conditions. *Science of the Total Environment*, **856**, 158795, https://doi.org/10.1016/j.scitotenv.2022.158795

Xu M., Zhou H., Yang X., Angelidaki I. and Zhang Y. (2020). Sulfide restrains the growth of *Methylocapsa acidiphila* converting renewable biogas to single cell protein. *Water Research*, **184**, 116138, https://doi.org/10.1016/j.watres.2020.116138

Zheng M., Li H., Duan H., Liu T., Wang Z., Zhao J., Hu Z., Watts S., Meng J., Liu P., Rattier M., Larsen E., Guo L., Dwyer J., Van Den Akker B., Lloyd J., Hu S. and Yuan Z. (2023). One-year stable pilot-scale operation demonstrates high flexibility of mainstream anammox application. *Water Research X*, **19**, 100166, https://doi.org/10.1016/j.wroa.2023.100166

Zimmo O. R., van der Steen N. P. and Gijzen H. J. (2003). Comparison of ammonia volatilization rates in algae and duckweed-based waste stabilization ponds treating domestic wastewater. *Water Research*, **37**(19), 4587–4594, https://doi.org/10.1016/j.watres.2003.08.013

doi: 10.2166/9781789063479_0167

# Chapter 6

# Maximizing sustainability by energy recovery in anaerobic treatment systems

A. V. Haandel[1]*, S. L. Santos[2], T. Allegue[3] and J. M. Garrido[4]

[1]Department of Civil Engineering, Universidad Federal of Campina Grande, Campina Grande, PB, Brazil
[2]Department of Civil Engineering, Universidad Federal of Rio Grande do Norte, Natal, RN, Brazil
[3]Department of Chemical Engineering, Khalifa University, Abu Dhabi, United Arab Emirates
[4]Department of Chemical Engineering, Universidade de Santiago de Compostela, CRETUS, Santiago de Compostela, Galicia, Spain
*Correspondence: adrianusvh@gmail.com

**ABSTRACT**

Traditionally, wastewater treatment systems require considerable energy input, but this has changed since modern anaerobic treatment systems were developed in the second half of the last century. The production of biogas opens the possibility to produce electric power, which may make these systems independent of external energy and may even become net energy producers. A second source of energy production in wastewater treatment systems is sludge production. After drying the sludge its combustion can generate energy, though less than the energy potential from biogas, not only due to the reduced sludge mass, but also because solid fuel has a lower conversion efficiency than gaseous fuel. It has been demonstrated that in an efficient anaerobic treatment system operating under favourable conditions (warm climate, low-sulphate concentration in the influent) ~1/6 of the influent organic material is converted into sludge whereas 4/6 is converted into biogas and 1/6 leaves the anaerobic treatment system with the treated effluent. Normally, the production of electric power from the biogas is more than sufficient to run the treatment system and an external source energy is not required. However, this does not mean that energy production is always a sound economic solution. Economic feasibility depends on the price of electric power and generally requires the operation of a large treatment system. In this chapter, equations are derived to estimate the composition of biogas and sludge in anaerobic wastewater treatment systems and specifically the influence of sulphate in the influent is discussed. Moreover, the processes for collection and treatment of biogas and sludge are presented. The potential for useful energy generation from anaerobic treatment systems is analysed and several examples are presented in order to highlight the feasibility of the energy recovery under different operating conditions.

**Keywords:** anaerobic wastewater treatment, biogas production and composition, electric power from biogas and sludge, influence of sulphate in the influent, volatile solids production.

## 6.1 INTRODUCTION

All biological wastewater treatment systems seek to reduce the concentration of organic material since its presence is highly undesirable. When raw wastewater is discharged on surface water, it leads to consumption of dissolved oxygen, and, depending on the ratio between surface water flow and organic

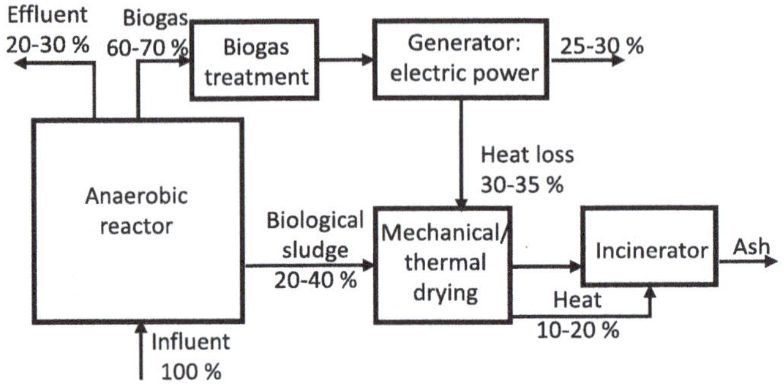


**Figure 6.1** Distribution of organic material of influent wastewater in anaerobic treatment systems. Numbers indicate the estimated fractions under normal operational conditions.

load, it can cause death of microorganisms. Conventional aerobic wastewater treatment systems are energy consumers, but anaerobic treatment systems offer the possibility of recovering the energy content of the influent organic matter, which can be profitably employed in wastewater treatment plants or sold as electric energy. As shown in Figure 6.1 the organic material of wastewater after the anaerobic treatment is distributed in three fractions: (1) a gaseous fraction composed of biogas, (2) a solid fraction composed of biological sludge and (3) a fraction dissolved in the effluent. In this distribution, the largest fraction is biogas, in which methane is the main component and can be used for electric energy generation. Sludge combustion can provide useful energy production after separation from the liquid phase and drying. The residual organic material in the effluent is composed of a biodegradable and a non-biodegradable fraction. The biodegradable components can be removed in a post-treatment unit as activated sludge or polishing ponds (see Chapter 5) whereas the non-biodegradable fraction will only be removed if the production of high-quality water is an objective of the treatment system. In that case advanced processes such as reverse osmosis can be used to produce water without impurities.

In this chapter, an estimate is made of the composition of biogas and sludge as well as the potential for useful energy generation from these in anaerobic treatment systems. Moreover, the processes for collection and treatment of biogas and sludge are presented.

## 6.2 ENERGY RECOVERY IN ANAEROBIC TREATMENT SYSTEMS

The production of biogas is the most characteristic feature of the anaerobic treatment. Biogas is composed mainly of methane and carbon dioxide, which are decomposable products of organic material and have a limited solubility in water. There are other components, whose magnitude depends on the operating conditions in digesters and the characteristics of wastewater. In this chapter, it will be shown that for sewage treatment contributions of nitrogen and water vapour in biogas may exceed that of carbon dioxide. If sulphate is present in wastewater hydrogen sulphide will also be formed, which generally has a low concentration, but is nonetheless important because it can cause several problems in anaerobic digesters as it will be reported in detail in Section 6.4.7.

The destination of biogas depends on the produced quantity and the quality. If high amounts are available, it is possible to use biogas in a productive way, generating electricity and heat in combined heat and power units or refining and pressurizing it for subsequent use, for example for automotive purposes or injection in the gas grid. The technical feasibility of using biogas for power generation is not in question: there are many manufacturers who produce generators specifically designed for the use of biogas as fuel, with power generation ranging from 30 kW to over 5 MW. The transformation efficiency of these units is in the range of 35–40%. For very large biogas flows (production $>5\ MW_{el}$) there is the possibility of using gas turbines, which have a high initial cost, but with an efficiency of $\geq 50\%$.

A more complicated application is the upgrading of biogas for producing pure methane. In this case, the purified biogas must be compressed to a pressure high enough for grid injection (16 bar) or to a high pressure of 200–250 atm to be used in cars which are adapted for use of compressed natural gas. The upgrading of biogas into methane has also been demonstrated, but the feasibility of this alternative depends on the price of conventional fuel, which is too low for widespread application.

Raw biogas treatment processes are well established and the economic feasibility depends mainly on the benefit of the productive use, that is, the cost of fuel that otherwise would have to be purchased. In small treatment systems, it is usually not economically convenient using the biogas for energy production and it is preferable to burn it with a flare. It is worth noting that the combustion is important, and that omission of this step can result in the generation of greenhouse gases (GHG) by the anaerobic treatment higher than the one derived by the aerobic. The methane is a potent GHG with a carbon equivalence of 25:1, that is 1 mol of methane gas is 25 times more harmful than carbon dioxide.

***Example 6.1***

Estimate the per capita production of gases contributing to the greenhouse effect in aerobic and anaerobic systems for a per capita contribution of 100 g chemical oxygen demand (COD)/day and making the following assumptions:

(1) Aerobic: oxygen consumption: 0.72 g $O_2$/g COD; aeration efficiency: 1 kWh/kg $O_2$; fuel consumption for power generation: 0.2 kg $CH_4$/day/kW of electric power.
(2) Anaerobic: anaerobic digestion efficiency: 70%
(3) Average elemental composition of sewage is $C_2H_2O$, close to the value of $C_2H_{19/9}$ $ON_{1/9}$ determined by Henze and Harremoes (2001).

***Solution***

Assuming that carbon dioxide is produced in equimolar ratio with the consumption of oxygen in the aerobic system, it follows that the oxygen consumption of 72 g $O_2$/inh/day or 72/32 = 2.25 mol/inh/day generates 2.25 mol/inh/day of $CO_2$. On the contrary, the energy consumed in oxygen transfer (1 kWh/kg $O_2$) will be ~72 Wh/day or 3 W. For power generation of 1 W, consumption of 5 g $CH_4$/day is required and this generates 5/16 = 0.31 mol/day $CO_2$ (other fuels have roughly the same contribution). For a power generation of 3 W, the production of $CO_2$ is therefore 0.93 mol/inh/day leading to a total of 2.25 + 0.93 = 3.2 mol $CO_2$/inh/day in aerobic systems.

In anaerobic systems the per capita production is 0.7 × 100/4 = 17.5 g $CH_4$/day or 17.5/16 = 1.1 mol $CH_4$/day. If this amount is not burned, fatally it escapes into the atmosphere as a contributor to the GHG effect. Not considering the production of $CO_2$ in anaerobic digestion (its major fraction does not pass into the atmosphere but remains dissolved), methane production is equivalent to 21 × 1.1 = 23 mol $CO_2$/inh/day, a factor 23/3.2 = 7 times greater than the calculated value for the aerobic generation system! By burning the methane production of biogas, the $CO_2$ production will be 1.1 mol/inh/day, ~1/3 of the production in aerobic systems.

If methane in the biogas is burnt, there is still the dissolved $CH_4$, with amounts to ~20 mg $CH_4$/L generating from the digestion of 80 mg COD/L. If the dissolved $CH_4$ is not metabolized, for example by methanotrophic bacteria it will eventually be released into the atmosphere. If a per capita contribution of 120 L/day is assumed, the dissolved methane would be 20/16 × 120 = 150 mmol $CH_4$/inh equivalent to 3.1 mol $CO_2$/inh, equal to the production in aerobic systems.

It can be concluded that even if all the biogas produced is burnt, the $CO_2$ production of the burnt biogas and the release of dissolved methane leads to at least the same contribution of GHGs by anaerobic and aerobic treatment systems. Therefore, the sustainability of anaerobic systems for sewage treatment is actually questionable, due to the unavoidable release of GHGs into the atmosphere. In the above analysis, anaerobic digestion of aerobic sludge is not considered, but in reality, ~30% of the organic influent load to aerobic systems is converted into methane and will increase the output of GHGs, so that in the end the two systems produce practically equivalent amounts of GHGs. If methane can be productively used, anaerobic systems become more attractive in terms of sustainability, because they reduce the use of fossil fuels.

## 6.3 SOLUBILITY OF GASES

Biogas is a mixture of various gases and vapours that desorb from the liquid phase in an anaerobic digester because of the limited solubility of these products in water. To estimate the flow of biogas, it is necessary to know the production and solubility of its components. Calculation requires a first assumption of equilibrium conditions between the gaseous phase (biogas) and liquid phase in a bioreactor whose composition is assumed to be equal to one of the effluents, and that there is a uniform composition in both phases. In good approximation, this will normally be the case, even though the solid phase in the reactor can be highly stratified (Cozma *et al.*, 2013).

The solubility of gases in the liquid phase can be expressed with Henry's law, stating that if a liquid and a gaseous phase are in equilibrium, there is proportionality between the saturation (maximum) concentration of a gas component in the liquid phase and its concentration in the gas phase:

$$C_s = k_d C_g \tag{6.1}$$

where $C_s$ = saturation concentration of the dissolved gas (mg/L or mmol/L); $C_g$ = gas concentration in the gas phase (mg/L or mmol/L) and $k_d$ = distribution coefficient (dimensionless).

The distribution coefficient indicates the solubility of a gas: the higher the coefficient, the higher the solubility. The value of this coefficient has been determined and tabulated for the most important volatile compounds. Table 6.1 shows the values of the distribution coefficients for different temperatures of gases which can be biogas components. Concentrations may be expressed in any unit (g/L, mol/L, etc.).

Equation (6.1) can also be written in a different form:

$$C_s = k_h P_p \tag{6.2}$$

where $k_h$ = Henry's constant; $P_p$ = partial pressure of the component in the biogas (g/(m$^3$ Pa) or g/J).

The relationship between the distribution coefficient and the Henry constant is:

$$k_h = k_d W_m / RT \; (g/J) \tag{6.3}$$

where $W_m$ = molar weight of the gas; $R$ = universal gas constant = 8.3 J/kmol; $T$ = temperature (K).

**Table 6.1** Distribution coefficients of several biogas components.

| Gas | Formula | Molecular Weight (g/mol) | Temperature (°C) | | |
|---|---|---|---|---|---|
| | | | $k_{d20}$ | $k_{d25}$ | $k_{d30}$ |
| Methane | $CH_4$ | 16 | 0.034 | 0.032 | 0.031 |
| Carbon dioxide | $CO_2$ | 44 | 0.94 | 0.83 | 0.74 |
| Nitrogen | $N_2$ | 28 | 0.017 | 0.016 | 0.015 |
| Oxygen | $O_2$ | 32 | 0.034 | 0.032 | 0.030 |
| Ammonium | $NH_3$ | 17 | 0.76 | 0.71 | 0.67 |
| Hydrogen sulphide | $H_2S$ | 34 | 2.87 | 2.55 | 2.27 |

## 6.4 BIOGAS FLOW

To estimate the flow of biogas generated in anaerobic treatment, firstly it is necessary to evaluate the volumetric flow of its components. In Figure 6.2 the gases normally present in biogas are presented. Along with methane, which is the main component, significant contributions of carbon dioxide, water and nitrogen are expected. In addition, there are other compounds which are less bulky, but whose presence is important because of their nature: $O_2$, $NH_3$ and $H_2S$ (Van Haandel & van der Lubbe, 2019).

Regarding hydrogen sulphide, the contribution of this component depends on the concentration of sulphate in the water supply generating the sewage. Usually, this concentration will be small but there are regions where water resources with a high-sulphate concentration are the only source for public water supply. In these cases, sulphide production may still not be high, but due to sulphate reduction, there is a consumption of organic matter and methane production may be strongly reduced. Nitrogen, oxygen and ammonium are dissolved in influent whereas methane, carbon dioxide and hydrogen sulphide gas are generated in the anaerobic treatment unit. Being a vapour and not a gas at ambient temperature, the partial pressure of $H_2O$ is constant for a given temperature and is independent of the pressure of the gas phase.

### 6.4.1 Production potential of methane

Methane production depends directly on the applied organic load and digestion efficiency. Stoichiometric production is calculated as 0.25 kg of $CH_4$/kg digested COD. When it is assumed that ~10% of influent COD is transformed into volatile solids, the production of methane can be calculated from the applied organic load if the removal efficiency of the organic material is known. The removal efficiency depends on a variety of factors: the nature of the organic material, environmental conditions, operating conditions and reactor characteristics in terms of design and construction. The daily mass of methane produced may be expressed as

$$M_{CH_4} = (MS_{ta} - MS_{te} - MS_{xv}) / 4 \tag{6.4}$$

As a first estimate of the methane production, it can be assumed that all produced methane desorbs from the liquid phase. Knowing that the molar volume of a gas at atmospheric pressure is given by

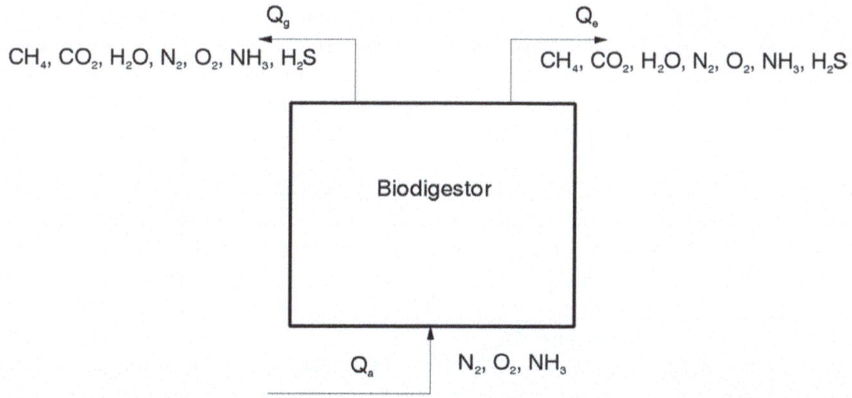


**Figure 6.2** Schematic representation of the different components of biogas and the equilibrium between the gas and liquid phases.

$22.4 \times (273 + T)/273$ ($T$ in °C) and 1 mol of methane has a mass of 16 g, the produced methane volume is calculated as

$$\begin{aligned} V_{CH_4} &= 22.4 \times (273+T)/273 \times M_{CH_4}/16 \\ &= 22.4(273+T)/273(MS_{ti} - MS_{te} - MS_{xv})/64 \end{aligned} \tag{6.5}$$

where $V_{CH_4}$ = first estimate of methane volume in the biogas (L/day); $M_{CH_4}$ = mass of the produced methane (g/day); $MS_{ti}$ = mass of applied COD (g/day); $MS_{te}$ = mass of COD in the effluent (g/day); $MS_{xv}$ = COD mass converted into volatile sludge (g/day).

***Example 6.2***

Determine the potential for per capita methane production from sewage by anaerobic digestion, making the following assumptions: (1) per capita organic matter production: 100 g COD/inh/day, (2) the organic material removal efficiency: 80%, (3) conversion into volatile solids: 10%.

***Solution***

From Equation (6.4):

$$\begin{aligned} M_{CH_4} &= (MS_{ti} - MS_{te} - MS_{xv})/4 \\ &= (100 - 20 - 10)/4 = 17.5\,\text{g}\,CH_4\text{/inh/day.} \end{aligned}$$

From Equation (6.5) at a temperature of 25°C:

$$\begin{aligned} V_{CH_4} &= 22.4 \times (273+T)/273 \times M_{CH_4}/16 \\ &= 22.4 \times (273+25)/273 \times 17.5/16 \\ &= 26.7\,L_{CH_4}\text{/inh/day.} \end{aligned}$$

However, the estimated volume of the produced methane is not precise since part of the methane produced will remain dissolved in the liquid phase. The solubility of methane is low (as shown in Table 6.1) being the distribution coefficient $k_d = 0.03$. To make a first estimate of dissolved methane concentration can be assumed that biogas is composed by only methane. As 1 mol of methane (or any other gas) has a volume of 24.4 L at a temperature of 25°C and atmospheric pressure, its concentration is 1/24.4 mmol/L = 41 mmol/L. Now, applying Equation (6.1) for the case of having only methane in the biogas is: $C_s = 41 \times 0.03 = 1.2$ mmol/L = 20 mg/L. Thus, it is estimated that the dissolved methane is 20 mg $CH_4$/L. As is known that for producing 1 mg $CH_4$, about a mass of 4 mg of COD must be digested, it is estimated that a maximum of 80 mg COD/L will be found as dissolved methane after anaerobic digestion. This concentration is not negligible in the case of municipal sewage, which normally has a concentration of digestible COD in the range of 300–500 mg/L. In Example 6.2, for a per capita contribution of 120 L/day, methane loss in the effluent would be 0.020 g $CH_4$/L × 120 L/inh/day = 2.4 g $CH_4$/inh/day or 2.4/17.5 = 14%. So the best estimate of $CH_4$ production in the biogas would be $M_{CH_4} = 17.5 - 2.4 = 15.1$ g/inh/day, whereas the per capita volume is $V_{CH_4} = 22.4 \times (273 + 27)/273 \times 15\ 1/16 = 23.1$ $CH_4$/L/day. It is important to notice that as the temperature becomes lower, the digested COD becomes smaller and the dissolved methane concentration becomes higher. As a result, the dissolved fraction increases and may be as high as 50% in case with a low influent COD and a low temperature.

***Example 6.3***
Estimate the fraction of the produced methane that remains dissolved in a upflow anaerobic sludge blanket (UASB) reactor operating at a temperature of 15°C and a sludge age of 75 days, treating sewage with a COD of 500 mg/L in the influent.

***Solution***
The fractions of the influent COD which are discharged in the effluent ($f_s$) and transformed into sludge ($f_x$) can be estimated (from Equations (5.1) and (5.2)) as $f_s = 0.22$ and $f_x = 0.11$ (see also dos Santos, 2013). Hence, the digested fraction is $f_d = 1 - f_s - f_x = 0.67$. Therefore, the digested COD is $0.67 \times 500 = 330$ mg/L, equivalent to 330/4 = 82 mg $CH_4$/L. On the contrary, at 15°C the solubility of methane in water at 15°C is ~23 mg/L ($k_d = 0.036$), so that a fraction of 23/82 = 28% is retained in the liquid phase and 72% is released into the biogas. Figure 6.3 shows the methane fraction in the gas phase for different temperatures and influent COD concentration as a function of the sludge age.

To make a more accurate estimate of the biogas volume and the partial pressures of methane and other components in the biogas, it is necessary to apply an iterative calculation, taking into account also the contributions of other gases. Firstly, the contribution of each component will be individually evaluated and then the iterative calculation procedure will be applied.

### 6.4.2 Carbon dioxide

The presence of $CO_2$ in the biogas is due to its production in the anaerobic digestion process. It can be assumed that the $CO_2$ production in molar terms is equivalent to the methane production, considering an average carbon oxidation number close to zero for organic matter, accordingly with the elemental composition of sewage used for Example 6.1, but its presence in the gas phase is much smaller because of its greater solubility. The partial $CO_2$ pressure can be estimated from the equilibrium of the carbonic system:

$$CO_2 + H_2O \leftrightarrow H^+ + HCO_3^- \qquad (6.6a)$$

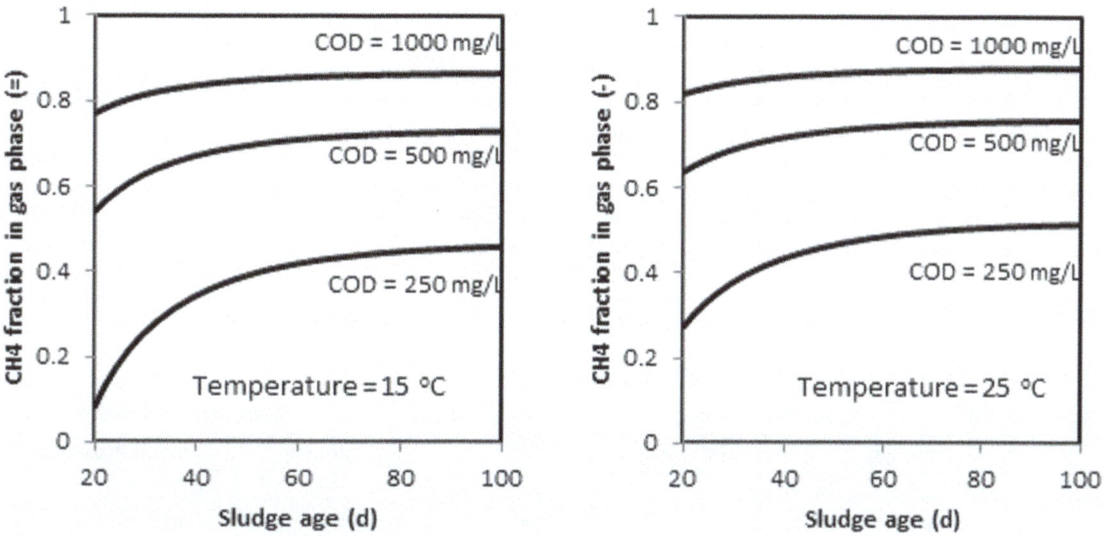


**Figure 6.3** Fraction of the methane production that desorbs from the liquid phase as a function of the sludge age for different temperatures and influent COD concentrations.

or

$$[HCO_3^-][H^+]/[CO_2] = k_1 \tag{6.6b}$$

or

$$[HCO_3^-][H^+]/\left(k_d[CO_2]_g\right) = k_1 \tag{6.6c}$$

By applying Equation (6.3) this can also be written as

$$p_{CO_2} = [HCO_3^-]10^{(pk_1+pK_H-pH)} \tag{6.7a}$$

where $pX = -\log[X]$; $k_1$ = first dissociation constant of the carbonic system $= 4.6 \times 10^{-7}$; $k_H$ = Henry constant for $CO_2$ = 0.034 mol/L/atm at 25°C; that is, $pK_H = 1.47$ (Loewenthal & Marais 1976).

Since at pH ≈ 7 the alkalinity is practically equal to the concentration of bicarbonate, Equation (6.7a) can also be written as

$$p_{CO_2} = \text{Alc.} \times 10^{(pk_1+pK_H-pH)} \tag{6.7b}$$

It is concluded that in the case of sewage treatment the $CO_2$ partial pressure will be low: for instance for an alkalinity of 350 ppm or 7 meq/L (normal for domestic wastewater) and a pH = 7.0 a partial pressure of only 0.04 at 25°C is calculated. The $CO_2$ pressure can be calculated provided there are data of alkalinity and pH, as well as for temperature. The presence of $CO_2$ in the biogas normally is not problematic. In contrast dissolved $CO_2$ in the liquid phase results in aggressive water, which can lead to serious concrete corrosion problems.

### 6.4.3 Water vapour

The pressure of water vapour is directly related to the temperature. At 25°C, the water saturation pressure is 23.7 mm of mercury (3.1 Pa), or at a pressure of 760 mm or 1.013 kPa partial pressure is 0.031. Table 6.2 shows some values of water vapour pressure for different temperatures. The presence of water vapour in the biogas can cause problems, particularly when the ambient temperature greatly varies. During periods of low ambient temperature, water may condensate in biogas piping, because the biogas is generated in the relatively warm liquid phase. The biogas will have water vapour at the pressure of the liquid phase in the reactor, but out of the treatment system the biogas temperature decreases and water vapour will tend to become supersaturated and condensate. Apart from causing blockages, the condensed water tends to cause corrosion problems, especially if hydrogen sulphide is also present in the biogas. For those reasons, there will usually be a condensate retention device in the biogas transport system, which may be the hydraulic seal of the UASB reactor, if located at the lowest point of biogas piping.

**Table 6.2** Values of the saturation pressure of water vapour at different temperatures.

| Temperature (°C) | Saturation Pressure | | Temperature (°C) | Saturation Pressure | |
|---|---|---|---|---|---|
| | mm Hg | Partial Pressure | | mm Hg | Partial Pressure |
| 0 | 4.6 | 0.006 | 20 | 17.5 | 0.023 |
| 5 | 6.5 | 0.009 | **25** | **23.7** | **0.031** |
| 10 | 9.2 | 0.012 | 30 | 31.7 | 0.042 |
| 15 | 12.8 | 0.017 | 35 | 41 | 0.054 |

### 6.4.4 Nitrogen

In many cases, it may be assumed that wastewaters are saturated with atmospheric nitrogen (which has a partial pressure of 0.8 atm). In that case, the saturation concentration is 14.8 mg/L at 25°C (Table 6.1). If it is assumed that nitrogen is not produced or lost during anaerobic treatment, the influent nitrogen will leave the system in the biogas and in the effluent, thus from the mass balance:

$$Q_l C_a = Q_l C_s + Q_g C_g \tag{6.8a}$$

or

$$C_a = C_s + (Q_g / Q_l) C_g \tag{6.8b}$$

Hence

$$C_g = C_a / (k_d + (Q_g / Q_l)) \tag{6.8c}$$

and

$$C_s = k_d C_a / (1 + (Q_g / Q_l)) \tag{6.8d}$$

where $C_a$ = nitrogen concentration in the influent; $C_g$ = nitrogen concentration in the gas phase; $Q_g/Q_l$ = ratio between the biogas and wastewater flow.

In order to calculate nitrogen in the effluent and in the biogas, it is necessary to know the concentration in the influent and the ratio of the flows of biogas and wastewater.

***Example 6.4***

For the conditions of Example 6.2 and having a per capita sewage contribution of 120 L/day, estimate the partial pressure of nitrogen on the biogas.

***Solution***

The concentration of nitrogen in the influent is estimated from the distribution constant ($k_d = 0.016$, Table 6.1). At a partial pressure of 0.8 bar, the concentration of nitrogen in the air is $41 \times 0.8 = 34$ mmol/L. Now the saturation concentration of nitrogen is calculated as

$$C_s = 0.016 \times 34 = 0.54 \text{ mmol/L and } C_a = 14.8 \text{ mg N/L}$$

Since $Q_g/Q_l = 26.7/120 = 0.22$ (Example 6.2), the nitrogen concentration in the biogas is

$$C_g = C_a / (k_d + (Q_g/Q_a))$$
$$= 14.8/(0.016 + 0.22) = 63 \text{ mg/L or } 63/28 = 2.2 \text{ mmol/L.}$$

Hence

$$C_s = k_d C_g = 0.016 \times 63 = 1.0 \text{ mg/L.}$$

The calculations indicate that nitrogen is efficiently removed by stripping in the anaerobic digester and almost all transferred to the gas phase, due to its very low solubility.

At a temperature of 25°C and a pressure of 1 bar the concentration of a gas is 41 mmol, so that when the nitrogen has a concentration of 2.2 mmol/L in the gas phase, its partial pressure is 2.2/41 = 0.05 atm. Comparing the calculation results of the partial pressure of carbon dioxide, water and nitrogen (Sections 6.4.2 and 6.4.3), surprisingly it is observed that for anaerobic digestion of normal sewage in the produced biogas, the partial pressures for carbon dioxide, nitrogen and water vapour are all of the same order of magnitude.

### 6.4.5 Oxygen

Different from the case of nitrogen, the oxygen concentration in the influent is generally low, because microorganisms can consume the oxygen transferred from the atmospheric gas to the liquid phase of the bioreactor. Oxygen consumption can occur or not during the permanence of the wastewater in the sewerage network, but, in any case, its utilization by facultative hydrolytic and acidogenic bacteria in the anaerobic reactor will cause its rapid depletion. In case of feeding devices causing air dragging into the digester, it is possible to observe a reduction in methane production because the organic material will be selectively oxidized by oxygen and subtracted to methanogenesis. Therefore, it is important to avoid air dragging into the feeding system.

### 6.4.6 Ammonium

Ammonium is present in the influent and also generated in the anaerobic digester as a result of mineralization of organic nitrogen. In sewage normally the concentration is in the range of 40–60 mg N/L or 3–4 mmol/L. As in the case of nitrogen, ammonium is distributed between the gas and liquid phases. The ratio between the number of moles per day, leaving the system in the gas and liquid phases can be expressed as

$$N_g / N_e = (Q_g C_g) / (Q_l C_s) \tag{6.9a}$$

Combining the previous equation with Equation (6.1) it is obtained:

$$N_g / N_e = (Q_g / Q_l) / k_d \tag{6.9b}$$

where $N_g$ and $N_e$ = the number of moles of $NH_3$ in the biogas and in the effluent, respectively.
For a ratio of gas and liquid flows $(Q_g/Q_e) = 0.22$ from the previous example, and knowing the value of the distribution coefficient ($k_d \approx 0.7$ in Table 6.1) it is estimated that

$$N_g/N_e = 0.19/0.7 = 0.3$$

On the contrary, in the neutral pH range the fraction that gaseous ammonium ($NH_3$) in the liquid phase is small (1%, since $pK_{NH_3} = 9.1$ at 25°C); almost all ammonium (~99% at a neutral pH) is in the ionic form ($NH_4^+$). Therefore, the ammonium fraction in the biogas will be small of the order of 0.3% of the total ammonium mass in the system. Thus, for practical purposes it can be considered that ammonium practically remains in the liquid phase in anaerobic reactors. Even in concentrated wastes where ammonium and the ratio $(Q_g/Q_l)$ are high, the ammonium fraction in the biogas remains low. For example, in the case of anaerobic digestion of manure or sludges, ammonium concentrations of $C_l = 3$ g N/L and $(Q_g/Q_l) = 3$ may be found, so that $N_g/N_l = 3/0.7 = 4.3$ and $N_g = 4.3N_l$. The $NH_3$ concentration in the effluent will be 1% of the total ammonium: $N_l = 0.03$ g/L and the concentration in the biogas $4.3 \times 0.03/3 = 0.043$ g/L = 3 mmol/L. This means that still almost all ammonium is in the liquid phase, but the partial pressure of ammonium is no longer negligible: $p_{NH_3} = 3/41 = 0.07$.

### 6.4.7 Hydrogen sulphide

Sulphate reduction by sulphate-reducing bacteria (SRBs) is a microbiological process, which occurs in anaerobic reactors and usually predominates over methanogenesis. Sulphate reduction in anaerobic systems may have advantages and disadvantages (Lens & Hulshoff Pol, 2000). Drawbacks are as follows:

(1) Hydrogen sulphide is a strong toxic gas, dangerous to human life at concentrations as low as 100 ppm in air.

(2) It has unpleasant odour of rotten eggs, detectable by nose at concentrations as low as 0.5 ppb, thus increasing odour problems in sewage treatment plants (STPs).
(3) At high concentrations, it is a toxic compound for methanogens and other microorganisms.
(4) It could cause the corrosion of concrete and steel of the infrastructure.
(5) Upon biogas combustion hydrogen sulphide may be converted into sulphuric acid, which corrodes metals in internal combustion engines.
(6) After combustion, it enters the atmosphere as sulphur dioxide, a harmful gas.
(7) Increased BOD and COD of the effluent because of the oxygen demand of sulphide.
(8) In the event of considerable presence of metals, inorganic sludge accumulation occurs in the form of precipitated sulphides.
(9) Deterioration of aerobic post-treatment (filamentous sludge, nitrification inhibition).

Among potential advantages that sulphate reduction can bring about are as follows:

(1) Removal of sulphate and biological transformation into sulphur and water recovery.
(2) Degradation of xenobiotic compounds that otherwise could not be removed biologically.
(3) Possibility of removal and recovery of heavy metals from the liquid phase.
(4) The precipitated metal sulphide may serve as granulation nucleus.

The sulphide generated in anaerobic reactors is physically distributed into three phases: (1) in the biogas, (2) in the anaerobic effluent and (3) in the emissions from the surface of anaerobic reactors or other points of escape. To prevent odour from spreading due to release of gas from the surface, anaerobic reactors can be covered while a partial vacuum (1–2 cm water column) applied by forced ventilation, which removes the gas phase including desorbing biogas. This mixture must be treated to remove hydrogen sulphide. In quantitative terms the mass of hydrogen sulphide in the contaminated air is small.

The formed sulphide in a digester is distributed among the three forms of reduced sulphur: dissolved hydrogen sulphide ($H_2S$), bisulphide ($HS^-$) and sulphide ions ($S^{2-}$) which can be estimated from the below chemical equilibrium:

$$H_2S \leftrightarrow HS^- + H^+ \leftrightarrow S^{2-} + 2H^+ \tag{6.10}$$

Hydrogen sulphide is a weak acid, as its dissociation constants are: $pK_1 = 7.0$ and $pK_2 = 14.0$. As the dissociation constants are known, the relative concentrations of the three species can be calculated. In Figure 6.4a, the profile distribution of sulphide, bisulphide and sulphide in water is plotted as a function of pH. Figure 6.4b shows the corresponding pH–pX diagram. It can be observed that in water

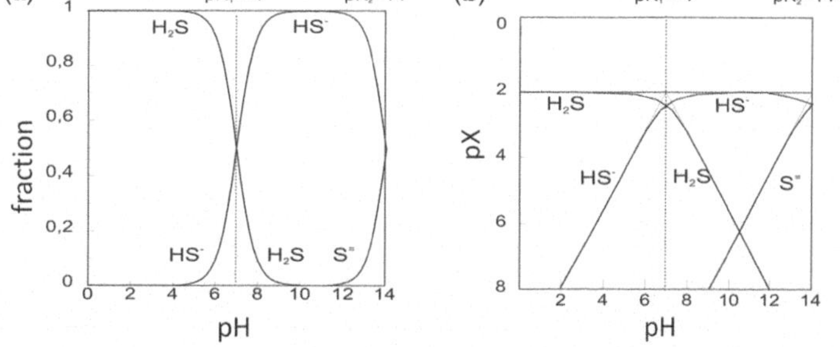


**Figure 6.4** (a) Fractions of hydrogen sulphide, bisulphide and sulphide as a function of pH and (b) pX–pH plot for a sulphidric system (10 mmol/L of reduced sulphur).

the form of non-dissociated hydrogen sulphide prevails when the pH is below the neutral value. The form of sulphide ion ($S^{2-}$) only appears in a significant way when the pH is extremely high (pH >12).

The ratio of the concentration of dissolved hydrogen sulphide and dissociated bisulphide can be calculated by the below dissociation equation:

$$C_{be}/C_{he} = 10^{(pH-pK_1)} \tag{6.11}$$

where $C_{be}$, $C_{he}$ = concentration of bisulphide and dissolved hydrogen sulphide in the effluent (and in the liquid phase of a digester).

Hydrogen sulphide gas, being a volatile compound, tends to establish an equilibrium between the fraction dissolved in the liquid phase (the content of a reactor) and the gas phase (biogas). When it is assumed that the biogas produced and the digester contents are in equilibrium according to Henry's law (Equation (6.1)), the concentration of hydrogen sulphide ($H_2S$ without the dissociated forms $HS^-$ and $S^=$) can be calculated as:

$$C_s = (N_{he}/Q_e) = k_d C_g = k_d (N_g/Q_g) \tag{6.12a}$$

or

$$N_g/N_{he} = (Q_g/Q_e)/k_d \tag{6.12b}$$

where $N_{he}$, $N_g$ = flux (daily number of moles) of $H_2S$ in the effluent and the biogas.

Hence the ratio between the flows of $H_2S$ in the biogas and in the effluent depends on two factors:

(1) the $Q_g/Q_l$ ratio and
(2) the value of the distribution coefficient.

In Figure 6.5 the division between species of sulphide has been plotted as a function of the $Q_g/Q_l$ ratio for different pH values in the liquid phase. It shows that the fraction of the sulphide that is in the biogas depends heavily on the $Q_g/Q_l$ ratio. In the case of wastewaters with a low concentration of organic material like sewage ($Q_g/Q_l \approx 0.2$) almost all sulphide remains dissolved, but in concentrated wastes ($Q_g/Q_l \approx 10$) a large fraction of sulphide migrates to the gas phase. The curves were built according to the following equations:

$$N_{be}/N_{he} = 10^{(pH-pK_1)}$$

$$N_g / N_{he} = (Q_g / Q_e) / k_d$$

$$N_t = N_g + N_{he} + N_{be}$$
$$= N_g + N_{he}\left(1 + 10^{(pH-pK_1)}\right) N_{he}\left[(Q_g/Q_e)/k_d + 1 + 10^{(pH-pK_1)}\right]$$

$$N_{he} / N_t = 1/\left[(Q_g / Q_e) / k_d + 1 + 10^{(pH-pK_1)}\right]$$

$$N_g / N_t = (Q_g / Q_e) / k_d / \left[(Q_g / Q_e) / k_d + 1 + 10^{(pH-pK_1)}\right]$$

$$N_{be}/N_t = 1 - (N_g/N_t) - (N_{he}/N_t)$$

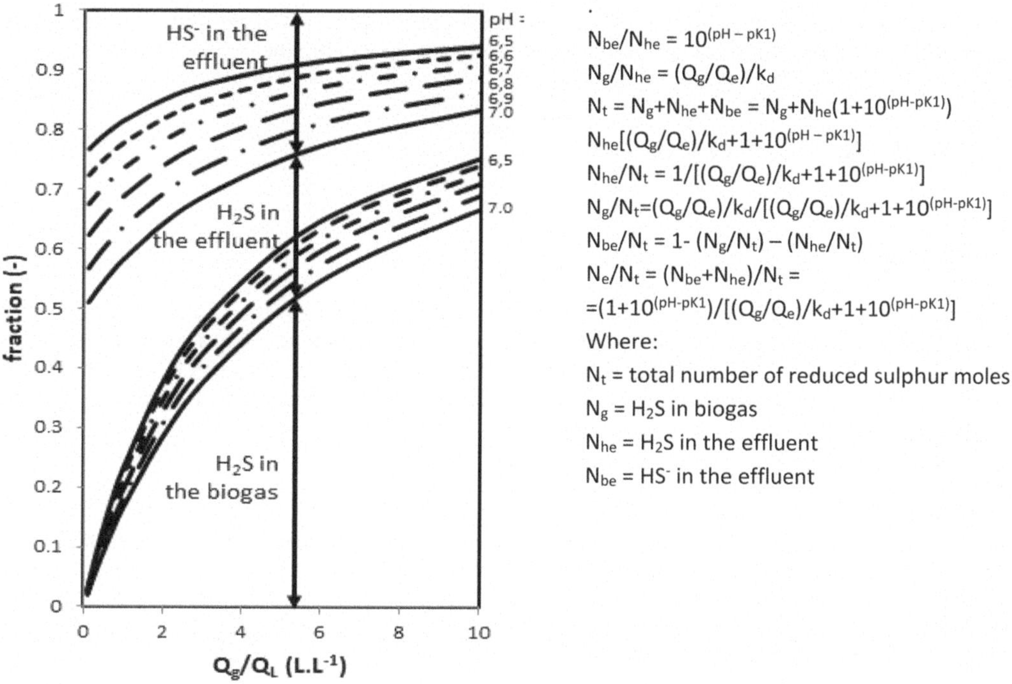


**Figure 6.5** Fraction of the sulphide species as a function of the $Q_g/Q_l$ ratio for different pH values.

$$N_e/N_t = (N_{be} + N_{he})/N_t$$
$$= \left(1+10^{(pH-pK_1)}\right)/\left[(Q_g/Q_e)/k_d + 1 + 10^{(pH-pK_1)}\right]$$

where $N_t$ = total number of reduced sulphur moles; $N_g = H_2S$ in biogas; $N_{he} = H_2S$ in the effluent; $N_{be} = HS^-$ in the effluent.

***Example 6.5***

For the conditions of Examples 6.2, estimate the fraction of the generated hydrogen sulphide gas remaining in the effluent when the sulphate concentration in the influent is 20 mg S/L and all sulphate in a digester is reduced.

***Solution***

It was estimated that the ratio $(Q_g/Q_l) = 23/120 = 0.19$ and for a distribution coefficient $k_d = 2.5$ (Table 6.1) it is calculated that:

$$N_g / N_e = (Q_g / Q_e) / k_d = 0.19 / 2.5 = 0.08$$

Therefore, the number of moles of hydrogen sulphide in the biogas is 8% of the number of $H_2S$ moles in the effluent. On the contrary, in the liquid phase the pH is approximately neutral, so that there is a partial dissociation of the hydrogen sulphide gas to bisulphite:

$$H_2S \leftrightarrow HS^- + H^+ \ (pK_1 = 7 \text{ at } 25°C)$$

For pH = 7, the number of moles of dissolved hydrogen sulphide will be equal to the number of moles of disulphide. Thus the fraction of hydrogen sulphide in the biogas is estimated as:

$$f_g = N_g/N_t = N_g/(N_g + 2N_e)$$
$$= N_g/N_e/(N_g/N_e + 2) = 0.08/2.08 = 0.04$$

and

$$f_l = 1 - f_g = 0.96$$

Therefore, for a generated sulphide concentration in the digester of 20 mg S/L, the concentration of sulphide in the effluent will be 20 × 0.96 = 19.2 mg/L. The concentration of hydrogen sulphide in the biogas $C_g = C_s/k_d = 0.5 \times 19.2/2.5 = 3.8$ mg/L or 3.8/32 = 0.12 mmol/L. As in the biogas total concentration of the components is 41 mmol/L, the partial pressure of hydrogen sulphide is $p_{H_2S} = 0.12/41 = 0.003$.

***Example 6.6***
For the conditions of Example 6.4 but with a ratio $Q_g/Q_l = 10$, estimate the fraction of the hydrogen sulphide gas which remains in the effluent generated when the sulphate concentration in the influent is 20 mg S/L and all the sulphate in the digester is reduced (pH = 6.8).

***Solution***
For $(Q_g/Q_a) = 10$ and a distribution coefficient $k_d = 2.5$ (Table 6.1) it is calculated that

$$N_g / N_l = (Q_g / Q_l) / k_d = 10 / 2.5 = 4$$

Hence, the number of moles per day in the biogas is four times larger than the number in the effluent. On the contrary, for pH = 6.8 with Equation (6.10): $N_{be} = 0.63N_{he}$ and one can estimate the fraction of sulphide ions in the biogas as

$$f_g = N_g / N_t = N_g / (N_g + 1.63N_l)$$
$$= (N_g / N_l) / (N_g / N_l + 1.63) = 4 / (4 + 1.63) = 0.70$$

and

$$f_l = 1 - f_g = 0.30$$

Hence, for a generated sulphide concentration of 20 mg S/L in the digester, the concentration of sulphide in the effluent will be 20 × 0.3 = 6 mg/L. The concentration of hydrogen sulphide in the biogas is: $C_g = C_s/k_d = 0.5 \times 6/2.5 = 1.2$ mg/L or 1.2/32 = 0.04 mmol/L. As the total concentration is 41 mmol/L in the biogas, the partial pressure of hydrogen sulphide is $p_{H_2S} = 0.04/41 = 0.001$.

Note that much more $H_2S$ is stripped from the liquid phase than in Example 6.5 (70% against 4%), but the partial pressure is actually smaller (0.001 vs 0.003), that is, the biogas is 'cleaner' because there is so much more methane and carbon dioxide produced.

Although some production of sulphide may be occurred due to the mineralization of proteins, generally the major part comes from sulphate reduction by organic material:

$$8e + 8H^+ + SO_4^{2-} \rightarrow S^{2-} + 4H_2O$$

In this half reaction, sulphate acts as an oxidant in the oxidative catabolic reaction in a similar way to oxygen in an aerobic environment:

$$4e + 4H^+ + O_2 \rightarrow 2H_2O$$

It is concluded that 1 mol of sulphate (32 g) has the same oxidation capacity as 2 mol of oxygen (64 g), that is, 1 g of sulphur in the sulphate form is equivalent to 2 g of oxygen and can therefore oxidize 2 g of COD, when it is reduced to hydrogen sulphide. As 2 g of COD have the potential of producing 0.5 g $CH_4$ it follows that, per mg of formed sulphide, there is a reduction of methane production of 0.5 mg.

The sulphate concentration in natural waters and therefore in municipal wastewater, varies widely from <10 mg S/L to more than 100 mg S/L. This has a very important effect on anaerobic treatment: not only is the methane production potential is half reduced, but the concentration of hydrogen sulphide in the biogas will increase, making its productive use more complicated. The presence of hydrogen sulphide in the liquid phase is not only potentially toxic to methanogens, but the sulphide in the effluent is also toxic to many aquatic organisms and consumes oxygen in the oxidation of sulphur to sulphate, which may reduce the dissolved oxygen in the receiving water body. Thus, the post-treatment of the effluent to remove or reduce sulphide concentration is usually necessary when the sulphate concentration in sewage is high. The recovery of sulphur from the effluent may be interesting if water must be reused in industrial processes such as cellulose pulp production.

## 6.5 ITERATIVE CALCULATION PROCEDURE FOR BIOGAS PRODUCTION AND COMPOSITION

In the previous sections, estimates were made of the partial pressure of different components of biogas. In these preliminary calculations there are two flaws: (1) it was tacitly admitted that the methane fraction in the biogas was very high and (2) it was supposed sulphide formation did not have a significant effect on the methane production potential. In this section, an iterative calculation procedure is developed that corrects these inaccuracies, considers the fraction of methane in the biogas is variable and accounts that a considerable oxidation of the organic material by sulphate reduction instead of methane production can occur. This requires iterative calculations according to the below procedures. One can distinguish essentially two cases: (1) the sulphate concentration is low and its reduction does not significantly affect the methane production or (2) the sulphate concentration is high and has a significant effect on the entire process of anaerobic digestion.

Some waters have naturally a much higher sulphate concentration (e.g. >100 or even >300 mg S/L and this may seriously limit the applicability of anaerobic digestion of sewage. Another source of high sulphate may be infiltration of sea water that sometimes may occur.

### 6.5.1 Low-sulphate concentration

In this case, it is not necessary to consider the reduction in methane production due to oxidation of organic material by sulphate, which simplifies the calculations. The iterative calculation procedure for estimating the biogas composition is as follows (data of Examples 6.2, 6.4 and 6.5):

(a) New estimate of the biogas composition
Having estimated the partial pressures of $CO_2$ (0.04), $H_2O$ (0.03) and $N_2$ (0.06), a new estimate can be made of the methane pressure. In the previous examples:

$$p_{CH_4} = 1 - p_{CO_2} - p_{H_2O} - p_{N_2} = 1 - 0.04 - 0.03 - 0.06 = 0.87 \quad (6.13)$$

(b) New estimate of the methane production in the biogas
The new estimate of the methane partial pressure leads to a new estimate of the dissolved methane in the liquid phase. In the example above the partial pressure was 0.87 and the

methane gas concentration in the biogas is now 0.87 × 41 = 36 mmol/L, whereas the saturation concentration in the liquid phase will be $C_s = 0.03 \times 36 = 1.08$ mmol/L or 17 mg/L. Therefore, the COD concentration of digested organic material, which is found as dissolved methane in the liquid phase, is 4 × 17 = 68 mg COD/L. For a per capita contribution of 120 L/day, a mass of 68 × 120 = 8160 mg COD/day, equivalent to $M_{CH4} = 8.16/4 = 2$ g $CH_4$/day, remains in the effluent. Moreover, for the assumed sulphate concentration of 20 mg S/L, the oxidation of 2 × 20 = 40 mg COD/L or 0.04 × 120 = 4.8 g COD/day is used for sulphate reduction, which reduces the production of methane by 4.8/4 mg $CH_4$/L or 1.2 g $CH_4$/day. Therefore, the mass of methane in the biogas can now be expressed as:

$$\begin{aligned} M_{CH4} &= (MS_{ti} - MS_{te} - MS_{xv} - MS_{CH4d} - MS_{oxi}) / 4 \\ &= (100 - 20 - 10 - 8.2 - 4.8) / 4 = 57 / 4 = 14.25\,g\,CH_4/inh/day \end{aligned} \tag{6.14a}$$

The new volume can now be calculated, duly not accounting the dissolved methane and the methane, which is not produced due to oxidation of organic material by sulphate:

$$\begin{aligned} V'_{CH4} &= 22.4 \times (273 + t) / 273 \times M_{CH4} / 16 \\ &= 22.4(273 + t) / 273 \times 14.25 / 16 = 23.2\,L/inh/day \end{aligned} \tag{6.14b}$$

(c) New estimate of the biogas/effluent flow ratio
Once the new estimates of the methane flow and partial pressure are obtained, the biogas/effluent flow ratio, $Q_g/Q_a$, can be calculated as

$$Q_g = V'_{CH4} / p_{CH4} = 23.2 / 0.87 = 26.7\,L/inh/day \tag{6.14c}$$

and

$$Q_g / Q_l = 26.7 / 120 = 23.8 / 0.87 / 120 = 0.22\,L/L \tag{6.14d}$$

(d) New estimate of the partial pressure of nitrogen
Having a new estimate for the $Q_g/Q_l$ ratio, the new values for the partial pressure of nitrogen is obtained, using the same procedure as in Example 6.3:

$$p_{N_2} = C_a / (k_d + (Q_g/Q_l)) / 28 / 41 = 0.05$$

(e) Compare the recalculated value of the composition and flow of the biogas with the values obtained in the previous iteration.
Knowing that the values of the partial pressures of $CO_2$ (=0.04) and $H_2O$ (=0.03) remain the same because they depend only on the pH + alkalinity and the temperature respectively, the new values of the partial pressure of $CH_4$ (=1 − 0.04 − 0.03 − 0.05 = 0.87) and the biogas volume (=26.7 L/inh/day) are now compared with the values of the previous iteration (1 and 26.7 L/inh/day respectively). It can be observed that these values are different and, for this reason, a new calculation cycle is repeated from steps (a) to (d) and this procedure iterated until the calculated values of two successive iterations are very close, that is, different only for an established acceptable error.

(f) Continue iteration calculations until a constant value is obtained for the variables
Normally, two or three iterations are sufficient. Table 6.3 shows a series of iterative calculations relative to the data of Examples 6.2 and 6.4. It can be noted that the values become constant after the third iteration (bold), when four decimal numbers are used. The values of the partial pressures of $CH_4$ and $N_2$ show the slowest convergence.

**Table 6.3** Interactive calculations to establish the values of the volume and composition of biogas if there is little sulphate reduction.

| Iteration | $p_{CH_4}$ | $M_{CH_4}$ | $Q_g$ | $Q_g/Q_a$ | $p_{N_2}$ |
|---|---|---|---|---|---|
| 0 | 1 | 17.5 | 26.6875 | 0.222396 | 0.052616 |
| 1 | 0.8732 | 14.2500 | 21.7313 | 0.2074 | 0.0562 |
| 2 | 0.8697 | 14.2500 | 21.7313 | 0.2082 | 0.0559 |
| **3** | **0.8699** | **14.2500** | **21.7313** | **0.2082** | **0.0560** |
| 4 | 0.8699 | 14.2500 | 21.7313 | 0.2082 | 0.0560 |

$A_q$ = 120 L/inh/day, temp = 25°C, $MS_{inh}$ = 100 g COD/day (70% digestion), methane production = 17.5 g $CH_4$/inh/day, $p_{CO_2}$ = 0.04, $p_{H_2O}$ = 0.03.

### 6.5.2 High-sulphate concentration

When the sulphate concentration is high, the methane production is affected because the organic material will be preferentially oxidized. In that case, the above iterative calculation procedure is extended in two aspects:

(a) Methane production is now estimated by decreasing the availability of organic material for anaerobic digestion of 2 mg COD/mg S. For example, at a concentration of 100 mg/L of $SO_4$–S or 120/1000 × 100 = 12 g S/inh/day:

$$V''_{CH_4} = 22.4(273+T)/273(MS_{ti} - MS_{te} - MS_{xv} - 2MS_{SO_4})/64$$
$$= 24.6 \times (100 - 20 - 10 - 2 \times 12) = 17.7 \text{ L/inh/day.}$$

(b) Part of the formed hydrogen sulphide will pass to the gas phase, but the magnitude of this part depends on the ratio between the established biogas flow and the effluent flow. Figure 6.4 clearly shows that the fraction of hydrogen sulphide in the biogas depends on $Q_g/Q_l$ ratio and pH. Once the sulphide fraction which migrates to the gaseous phase is established, the volume and the partial pressure of hydrogen sulphide can be calculated as well as the concentration of this component in the effluent.

***Example 6.7***

Estimate for the data in previous examples the production of methane and biogas and the residual sulphide concentration in the liquid phase as a function of the sulphate concentration over the range from 0 to 200 mg S/L.

***Solution***

For each sulphate concentration, the problem requires an iterative calculation procedure that consists of the following steps:

(1) Initially, it is assumed that the sulphate is converted to sulphide, the organic material demand to effect this redox process is estimated and the corresponding reduction in methane production is calculated (column 2 of Table 6.4):

$$MS_{oxid} = 2MSO_4$$
$$MS_{CH_4tot} = (MS_{ti} - MS_{te} - MS_{xv} - MS_{oxid})/4$$

In reality, this estimate is imperfect because SRBs will grow and in this anabolic process SRBs, being heterotrophic, use organic material. At high-$SO_4$ concentrations, this growth will not be insignificant because oxidative catabolism occurs, which generates much energy and thus anabolism is considerable.

(2) Estimate the partial pressure of methane in the biogas using the values of the partial pressures of $CO_2$, $H_2O$, $N_2$ and $H_2S$. For $N_2$ and $H_2S$ this will require an iterative calculation procedure.
(3) Estimate the methane mass in the effluent as the difference between the total methane production and the mass that is transferred to the gaseous phase (column 4):

$$MS_{CH4,efl} = MS_{CH4tot} - p_{CH4} \times 0.020 \times 120$$
$$MCH_{4,efl} = MS_{CH4,efl} / 4$$

(4) Estimate the methane volume corresponding to the daily mass of the produced methane in the gas phase (column 5):

$$V_{CH_4} = 22.4(273+t)/273\left(MS_{CH_4tot} - MS_{CH_4,efl}\right)/64$$

(5) With the aid of the estimates of the pressure and volume of methane, calculate the ratio between the gas and liquid flows, $Q_g/Q_l$ (column 6):

$$Q_g / Q_l = \left(V_{CH_4} / p_{CH_4}\right)/Q_l$$

(6) Obtaining the $Q_g/Q_l$ ratio, calculate the partial pressure of nitrogen and hydrogen sulphide (columns 7 and 8).
(7) Knowing that the partial pressures of $CO_2$ and water vapour do not depend on the flow of biogas, calculate the partial pressure of these two components.
(8) Now, having the estimates for the partial pressures of $CO_2$, $H_2O$, $N_2$ and $H_2S$ calculate a new estimate for the partial pressure of $CH_4$.
(9) With the aid of the new $p_{CH_4}$, for each sulphate concentration repeat steps 2–8 until the same result is obtained in two consecutive iterations. At the end of each iterative calculation the flow and composition of the biogas are obtained.

Table 6.4 shows the results of the iterative calculations for different sulphate to sulphide conversions. Figures 6.6 and 6.7 are graphical presentations of Table 6.4. When analysing the values in Table 6.4 and Figures 6.6 and 6.7, the dramatic influence of sulphate reduction on the production of methane and biogas composition can be noted: the mass of the produced methane is reduced by a factor of 4 when the concentration of sulphate is increased from 0 to 200 mg S/L. For the admitted values of 100 g COD/inh/day and a per capita volume of 120 L/day the calculated COD of the influent is 100/120 = 0.833 g/L and a biodegradable concentration of 0.7 × 833 = 583 (70% efficiency). The data indicate that for a ratio S-$SO_4$/COD in the influent greater than 200/583 ≈ 1/3 mg S/mg COD methane production is not feasible, which has also confirmed by the practical experience. Under these conditions, all the organic material is used for sulphate reduction.

As for the estimation of concentration of hydrogen sulphide in the liquid phase, firstly the mass of sulphur in the biogas is calculated: for fixed influent concentration of 200 mg S/L, the partial pressure of $H_2S$ is 0.0275 and the concentration in the gas is 41 × 0.0275 = 1.15 mmol/L or 37 mg/L biogas. The ratio biogas/influent flow rate is $Q_g/Q_l = 0.048$ (column 6 of Table 6.4), thus the daily production per capita is 0.048 × 120 = 5.8 L/inh/day with a sulphur mass of 37 × 5.8 = 213 mg S/day. This is <1% of the influent sulphur mass (calculated by 213/(120 × 200)). For this reason, the mass of sulphur in the effluent, in good approximation, will be equal to the mass of sulphide generated in the anaerobic reactor or present in influent: 200 mg S/L. The concentration of 200 mg/L in the effluent will produce bisulphide and hydrogen sulphide gas. For a pH = 7.0 this ratio is equal, that is, $C_{H_2Se} = C_{HS-} = 100$ mg/L. The concentration of hydrogen sulphide in the effluent and the reactor is very high: many researchers (Buisman & Lettinga, 1990; Rinzema, 1988) have shown that a

**Table 6.4** Estimates of the production of biogas and the partial pressures of methane, nitrogen and hydrogen sulphide as functions of the reduced sulphate concentration for the same suppositions as in Table 6.3.

| 1 | 2 | 3 | 4 | 5 | 6 | 7 | 8 |
|---|---|---|---|---|---|---|---|
| $SO_4^=$ mg S/L | $MS_{CH_4tot}$ g/inh/day | $p_{CH_4}$ (dimensionless) | $MS_{CH_4g}$ g/inh/day | $V_{CH_4}$ L/inh/day | $Q_g/Q_a$ (dimensionless) | $p_{N_2}$ (dimensionless) | $p_{H_2S}$ (dimensionless) |
| 0 | 17.5 | 0.873 | 15.40 | 23.54 | 0.1962 | 0.0569 | 0.0000 |
| 20 | 16.3 | 0.866 | 14.22 | 21.73 | 0.1811 | 0.0613 | 0.0026 |
| 40 | 15.1 | 0.858 | 13.04 | 19.93 | 0.1661 | 0.0663 | 0.0053 |
| 60 | 13.9 | 0.849 | 11.86 | 18.12 | 0.1510 | 0.0723 | 0.0079 |
| 80 | 12.7 | 0.840 | 10.68 | 16.33 | 0.1361 | 0.0794 | 0.0106 |
| 100 | 11.5 | 0.828 | 9.51 | 14.53 | 0.1211 | 0.0881 | 0.0134 |
| 120 | 10.3 | 0.815 | 8.34 | 12.75 | 0.1063 | 0.0988 | 0.0161 |
| 140 | 9.1 | 0.798 | 7.18 | 10.98 | 0.0915 | 0.1123 | 0.0189 |
| 160 | 7.9 | 0.778 | 6.03 | 9.22 | 0.0768 | 0.1301 | 0.0217 |
| 180 | 6.7 | 0.751 | 4.90 | 7.48 | 0.0624 | 0.1541 | 0.0246 |
| 200 | 5.5 | 0.714 | 3.79 | 5.79 | 0.0482 | 0.1881 | 0.0275 |

concentration of 100 mg/L $H_2S$ severely inhibits methane production in anaerobic digesters. Thus, the production of methane could be even less than the calculations indicate since it was evaluated without considering the toxic effect of hydrogen sulphide gas.

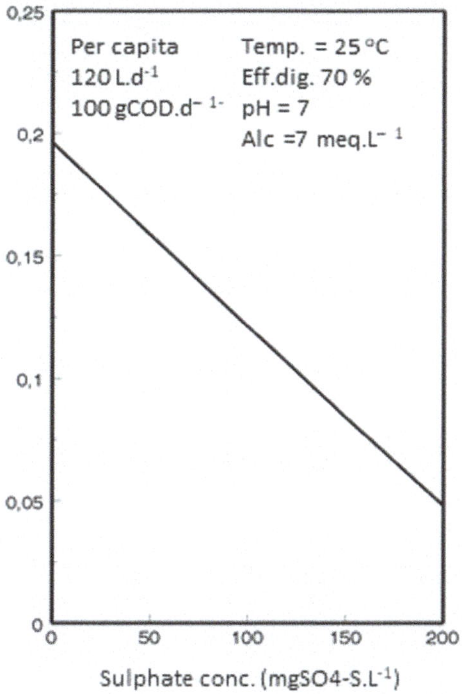


**Figure 6.6** Influence of the sulphate concentration on the biogas production (see also Table 6.4).

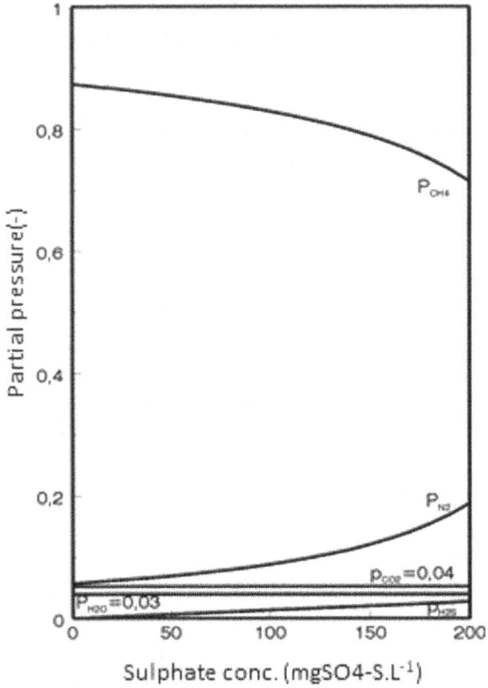


**Figure 6.7** Influence of the sulphate concentration (reduced to sulphide) on production and composition of biogas (see also Table 6.4).

***Example 6.8***

A system composed of two UASB reactors, with a total volume of 7500 m$^3$ treats municipal sewage. The water supply system is precarious and for that reason, the population tends to use also ground water, which is very rich in sulphate (1 g/L). As a result the average concentration of sulphate in the sewage is 200 mg/L (as $SO_4^=$) or 67 mg S/L. Table 6.5 shows operational data. Evaluate the influence of sulphate on the methane production.

***Solution***

To estimate the effective production of biogas, it must be considered that the organic material of the influent is distributed into four fractions in the UASB reactor:

(1) not removed and released into the effluent
(2) transformed into volatile solids and released as such in excess sludge
(3) oxidized in the redox process with sulphate reduction
(4) digested forming methane.

Therefore, one can calculate the methane production potential as:

$$CH_4 = S_d / 4 = (S_{ti} - S_{te} - S_{xv} - S_{ox}) / 4 \quad (6.15)$$

where $CH_4$ = mass of the produced methane per L of influent (mg $CH_4$/L); $S_d$ = concentration of digested COD (mg COD/L); $S_{ti}$ = influent COD concentration (mg COD/L); $S_{te}$ = effluent COD

**Table 6.5** Operational data of an anaerobic treatment system for sewage with 200 mg/L of sulphate in the influent.

| Variable | Unit | Value |
|---|---|---|
| Population | Inh | 90,000 |
| Flow | $m^3$/day | 9000 |
| Organic load | kg COD/day | 6300 |
| Influent COD concentration | mg/L | 650 |
| Effluent COD concentration | mg/L | 307 |
| Sludge production | kg TSS/day | 1200 |
| Volatile fraction in the sludge | mg $SO_4$/L | 0.67 |
| Sulphate in the effluent | | ≈0 |

concentration (mg COD/L); $S_{xv}$ = COD transformed into volatile sludge in the UASB reactor (mg COD/L); $S_{ox}$ = oxidized COD by sulphate (mg COD/L).

If there is 200 mg $SO_4^{2-}$/L or 200/3 = 67 mg S-$SO_4^{2-}$/L in the influent, with an oxidation capacity of 2 × 67 = 133 mg COD/L, it can prevent the production of 133/4 = 33 mg $CH_4$/L.

The $S_{xv}$ concentration is estimated knowing that 1 g of volatile sludge has a COD of ~1.5 g. According to Table 6.5, there is a production of 1200 kg Total Suspended Solids (TSS) sludge per day for an organic load of 6300 kg COD/day. Table 6.5 also shows that the anaerobic sludge has a volatile fraction of 2/3 so that the sludge production, in terms of Volatile Suspended Solids (VSS), is 2/3 × 1200 = 800 kg VSS/day, representing 1.5 × 800 = 1200 kg COD/day. As the applied load is 6300 kg COD/day the COD incorporated into the sludge is a fraction of 1200/6300 = 0.19 of the COD input, so it accounts for 0.19 × 650 = 124 mg/L. In the absence of sulphate in the influent, the expected anaerobic sludge production would be lower, since the sludge yield coefficient generated from the sulphate reduction is much greater (due to the oxidative catabolism).

Hence, the methane production can now be calculated as:

$$CH_4 = S_d/4 = (S_{ta} - S_{te} - S_{xv} - S_{ox})/4$$
$$= (650 - 307 - 124 - 133)/4 = 86/4 = 22\,\text{mg}\,CH_4/L$$

However, not all of the methane produced will desorb: methane solubility in water is ~20 mg/L when the partial pressure of methane in the gas phase is 1 atm. Therefore, in that case the concentration of methane that could desorb is only:

$$CH_{4g} = CH_4 - 20 = 22 - 20 = 2\,\text{mg/L}$$

Knowing that the density of methane is ~2/3 kg $CH_4/m^3$, the produced methane is calculated:

$$V_{CH_4} = 1.5 \times CH_{4g} = 3\,\text{mL}_{CH_4}\,/\,\text{L influent.}$$

It can be concluded that there is practically no biogas production, not only due to the direct cause, the oxidation of organic material by sulphate, but also due to indirect reasons. The first one is that organic material is also lost to the higher sludge production. Another indirect effect of the presence of sulphate and the consequent reduction of biogas production is that the mixing of the bioreactor is reduced because the contribution of rising biogas bubbles is absent. The generation of biogas is actually the main driver of the reactor mixing and facilitates uniform distribution of influent organic material and contact between the biomass and the substrate. Since there is no contribution of biogas to mix

**Table 6.6** Comparison between the performance of UASB reactors treating sewage with and without the presence of 67 mg S-$SO_4^{2-}$/L in the influent, assuming 100% $SO_4$ conversion.

| Mass of COD | Sulphate: 67 mg S/L | | No Sulphate | | Assumptions |
|---|---|---|---|---|---|
| | kg COD/day | Fraction | kg COD/day | Fraction | |
| In the influent | 6300 | 1 | 6300 | 1 | |
| In the effluent | 2980 | 0.47 | 1575 | 0.25 | Efficiency = 75% |
| In the sludge | 1200 | 0.19 | 630 | 0.10 | Sludge production = 10% |
| Dissolved $CH_4$ | 720 | 0.11 | 720 | 0.11 | 20 mg/L of $CH_4$ |
| Oxidized | 1200 | 0.19 | 0 | 0.00 | 2 kg COD/kg S |
| In the biogas | 200 | 0.03 | 3375 | 0.54 | |
| Methane mass (kg/day) | 50 | | 844 | | 0.25 kg $CH_4$/kg COD |
| Vol $CH_{4g}$ ($m^3$/day) | 75 | | 1265 | | 1.5 $m^3$/kg $CH_4$ |
| Electric energy (kW) | 10 | | 169 | | 0.2 kW/(kg $CH_4$/day) |

the reactor and thus to improve the contact between sludge mass and incoming organic material, it is conceivable that preferential channels will be formed, resulting in hydraulic short circuits. Under these conditions the fraction of the sludge present in stagnant zones receives little substrate and operates below its full capacity, whereas in other zones, where the flow of the liquid phase is accelerated, overload can occur. Both features tend to give a reduction of the organic material removal efficiency and particularly of methanogenesis. On the contrary, the concentration of 67 mg/L of sulphide itself can be toxic for methanogens, especially if the pH is slightly reduced, which again would lead to a decrease of methanogenic activity.

It is concluded that the presence of 200 mg/L of sulphate has a disastrous effect on the efficiency of anaerobic sewage digesters. The problems extend beyond the performance of UASB reactors: post-treatment in either polishing ponds or aerobic reactors is very complicated. The high-sulphide concentration makes algae growth in polishing ponds almost impossible and exerts a large oxygen demand in aerobic reactors, thus increasing energy consumption. In fact, if aerobic post-treatment is applied, it would be preferable not to use the first anaerobic step and apply the aerobic treatment directly to the influent.

Table 6.6 and Figure 6.8 show the dramatic influence of a high-sulphate concentration on anaerobic sewage treatment. Although without sulphate at least 65% of the influent COD is converted into methane with 11% remaining as dissolved methane in the liquid phase, with the high-sulphate concentration there is virtually no biogas production.

## 6.6 TREATMENT OF SULPHIDE PRODUCED IN ANAEROBIC DIGESTERS

It is apparent from Figure 6.8 that the division of reduced sulphur mass depends primarily on the ratio $Q_g/Q_l$ and also on the pH in the digester. As the ratio $Q_g/Q_l$ depends on the concentration of the influent organic matter, this will be the main factor that determines whether the generated sulphide is found predominantly in the liquid phase or in the gas phase. Depending on the organic load, it can be necessary to remove the sulphide from the biogas or the effluent. These two alternatives will be discussed below. The removal of sulphide from contaminated gas above the liquid surface in the digester is also discussed. Figure 6.9 shows schematically the auxiliary units of the anaerobic digester to treat reduced sulphur that is generated in the anaerobic reactor. For the treatment of biogas three methods are distinguished: biological (Lin *et al.*, 2018), physical and chemical (Zhang & Zhang, 2002). For removal of sulphide in the liquid phase the same methods can be applied.

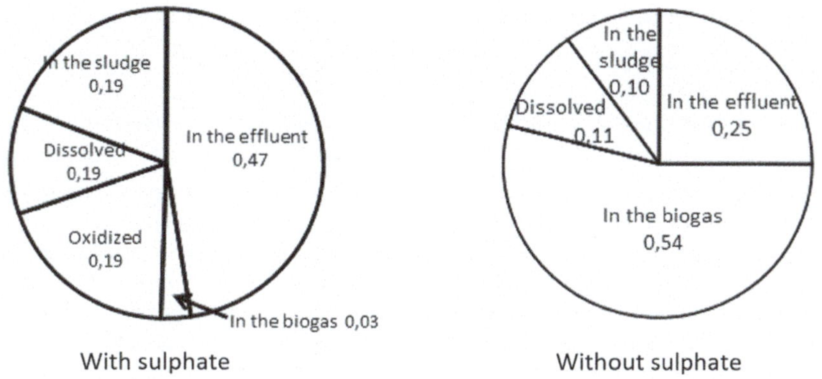


**Figure 6.8** Distribution of the influent COD in different fractions with (left) and without (right) sulphate.

### 6.6.1 Sulphide in biogas

The most widely applied unit for hydrogen sulphide removal from biogas is probably the Thiopaques process in which the hydrogen sulphide is transformed into elemental sulphur. The process consists of three steps (see Figure 6.9):

(1) In a counter current scrubber the biogas is placed in contact with an alkaline solution and the $H_2S$ in the biogas is absorbed and dissociates:

$$H_2S + OH^- \rightarrow HS^- + H_2O \tag{6.16a}$$

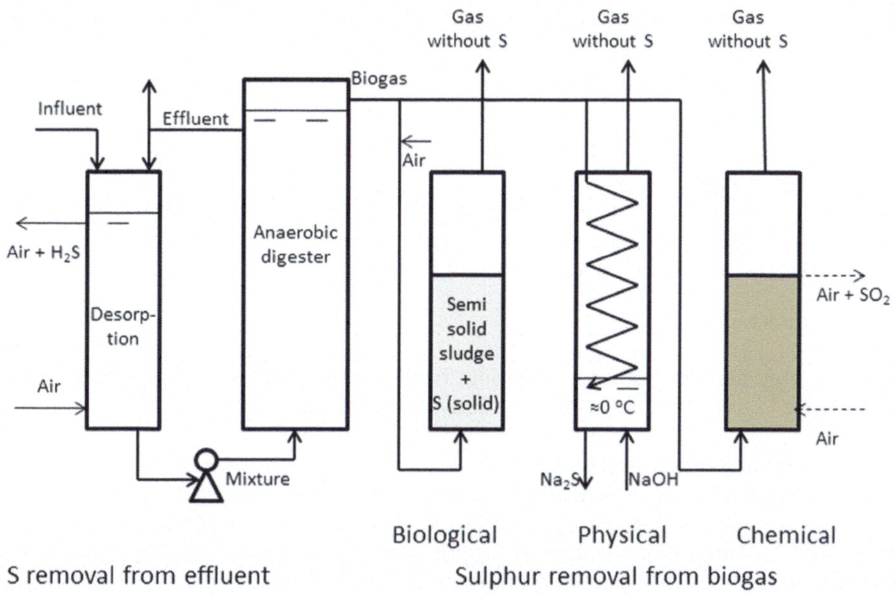


**Figure 6.9** Schematic representation of an anaerobic digestion system with possible auxiliary units for sulphide removal from biogas.

(2) In a biological unit and a controlled aerobic environment $HS^-$ is oxidized to sulphur:

$$HS^- + 0.5O_2 \rightarrow S\downarrow + OH^- \tag{6.16b}$$

(3) The solid sulphur is separated from the liquid phase and becomes available as a nutrient or raw material.

An alternative to remove smaller quantities of hydrogen sulphide in biogas is to add a small amount of oxygen to the mixture of biogas and then pass it through a unit operated with attached sulphide-oxidizing bacteria able to convert aerobically the $H_2S$ in the biogas into sulphur. Physically, this biological unit can be located in the biogas accumulation zone above the liquid phase in the digester. This option is sometimes used for cleaning biogas in manure digesters, where biogas may be used directly to power generators for electricity production. Support media with a rough surface and high superficial area for better adhesion of biomass are employed for biogas treatment. The sulphide absorbed onto the solid surface accommodating the biomass will be oxidized by oxygen.

The formed sulphur accumulates onto the solid surface and occasionally chunks of sulphur fall off and can be withdrawn for use of recovered sulphur or mixed with the effluent for agricultural reuse. Care must be taken not to add too much oxygen since it can make a potentially explosive mixture with the biogas when the volumetric ratio $O_2/CH_4$ is above 1:15–1:20. It is worth noting that the oxygen demand for the sulphide removal from the biogas is much smaller than this critical ratio.

To inoculate biomass for the process, the solid surface can be covered with liquid manure, which always contains bacteria that oxidize hydrogen sulphide. The size of the hydrogen sulphide oxidation unit is small relative to the size of the digester. For a 1000 $m^3$ digester, just a few square metres are enough to remove the hydrogen sulphide in the biogas.

Physical removal of sulphides is possible because the hydrogen sulphide solubility in water increases with decreasing temperature. When the water vapour in the biogas has to be removed, the biogas may be cooled to $\sim$0°C: condensation of water will occur and hydrogen sulphide from the biogas will tend to dissolve in the formed liquid phase. To facilitate this dissolution, alkalinity is added to the condensed water. This alternative is attractive in regions with a cold climate, where ambient temperature is low for most of the year.

In the chemical removal process (iron sponges process) the principle of operation is based on the reaction between iron oxide and hydrogen sulphide producing ferrous sulphide. In a reactor with partially oxidized iron filings the following reaction occurs:

$$FeO + H_2S \rightarrow FeS + H_2O \tag{6.17a}$$

When the removal capacity is exhausted, due to the conversion of ferrous oxide to ferrous sulphide, it can be regenerated by oxidation with air as indicated in Figure 6.10:

$$FeS + 3/2O_2 \rightarrow FeO + SO_2\uparrow \tag{6.17b}$$

Other oxidants may also be used for chemical treatment of the biogas, for example hypochlorite solutions, but in these cases, regeneration is usually not possible.

### 6.6.2 Sulphide removal from the effluent

The methods employed to remove hydrogen sulphide from the effluent, as already mentioned, are biological, physical or chemical. A biological method consists of the oxidation of sulphide to sulphur with oxygen. A physical method is based on the application of a desorption unit to transfer the hydrogen sulphide dissolved in the effluent to air in an open tank or in a closed unit with air in counter current.

For sulphide oxidation in an anaerobic reactor, it is possible to add oxygen directly to the reactor bulk phase through micro-aeration. In this case, a population of sulphide-oxidizing bacteria (*Thiobacillus*

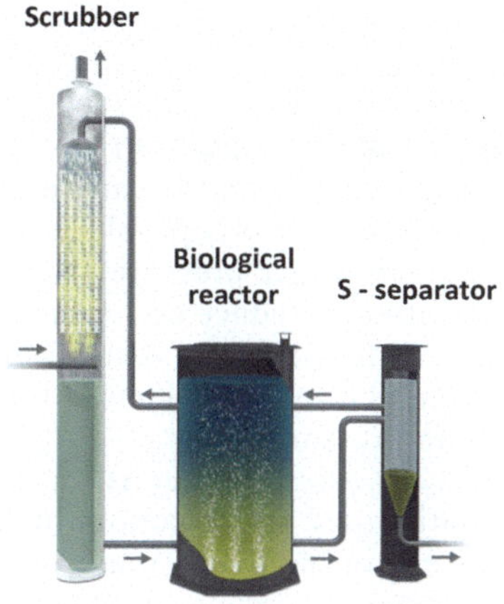


**Figure 6.10** Removal of $H_2S$ from biogas by absorption and partial oxidation.

sp.) is established, and the sulphide in the liquid phase is oxidized to sulphur. The sulphur suspension is mixed with the sludge mass in the reactor and then discharged with the excess sludge. In this way, accumulation of a considerable concentration of sulphide in the liquid phase is avoided and thus indirectly the desorption of hydrogen sulphide to biogas will be limited. This method is particularly useful if much of the sulphur remains in the liquid phase, that is, when ratio $Q_g/Q_l$ is small and when the sulphate concentration is high.

An alternative is to operate a separate reactor where a low concentration of dissolved oxygen is maintained. Under these conditions a specific bacterial population (*Thiobacillus* sp.), which generates a suspend sulphur from sulphide will develop. Control of the dissolved oxygen concentration must be accurate to avoid oxidation to sulphate, which requires four times more oxygen (and therefore energy) and does not remove the sulphur from the liquid phase. Janssen *et al.* (1999) demonstrated the advantage of using the redox potential of the reactor to control the concentration of dissolved oxygen. If the oxygen transfer is carried out in such a way that there is not high turbulence, sulphur aggregates in large flocs that are removable by simple settling. Marked turbulence tends to generate a colloidal suspension and separation of the sulphur from the liquid phase may become problematic.

The removal of the liquid phase can also be effected by desorption, a physical process taking place in a separate unit, where the digester effluent is transferred and placed in a counter current flux in the presence of air. This method is particularly effective when the influent has a low pH. In that case, mixing effluent (neutral pH) and influent (acid conditions) produces a mixture with low pH, in which hydrogen sulphide ion is converted into $H_2S$. Like carbon dioxide, hydrogen sulphide gas is supersaturated in the mixture and it is transferred to the air phase. An industrial unit (tray aerator or scrubber) or simply an accumulation tank can be used to mix influent and effluent. Mechanical stirring or bubble aeration may be applied to accelerate desorption of volatile supersaturated compounds ($CO_2$ and $H_2S$). In case of a high-sulphide concentration in the effluent, the desorption unit may be closed to avoid spreading of bad odours. This implies that a solution must be given for the hydrogen sulphide transferred to the air. In the desorption unit there is also air oxygen transfer to the liquid phase of the mixture. This micro-aeration can facilitate sulphide oxidation in the digester itself when the influent and recirculation mixture is introduced. Sulphur in colloidal form, which may be formed, will be extracted with the effluent or with excess sludge.

Chemical removal of sulphide from the liquid phase has been applied mostly for wastewaters with high-sulphate concentration (Song *et al.*, 2004). Ferric chloride may be added to the influent and used to precipitate sulphur as ferrous sulphide and, in addition, it will act as a coagulant improving the settleability of the anaerobic sludge.

### 6.6.3 Hydrogen sulphide in air used for odour collection

An important source of odour due to hydrogen sulphide can be the liquid surface of UASB reactors, where gas may escape from the liquid phase. For this problem, sometimes UASB reactor is covered. In this case it is possible to apply a small under pressure (1–2 mbar) and treat the air containing desorbed biogas. The amount of air required to reduce odour depends on the under pressure that it is applied but it should not be >1% of the wastewater flow. A small motor can pump the air to the odour removal unit. Another important source of odour problems is the discharge of effluent, especially if this discharge leads to turbulence due to pressure drops.

The treatment of the contaminated air essentially consists of $H_2S$ removal. There are different mechanisms to achieve odour elimination: (1) use of contaminated air as part of the oxidation air for the combustion of the biogas, (2) use of biological odour control systems to promote sulphide abatement, (3) use of iron oxide for sulphide oxidation and (4) use of limestone (seashells).

In the first option, air is mixed with the biogas for its combustion. This option is particularly attractive when using a flare for burning biogas rather than its use as fuel. In this case, the hydrogen sulphide in the biogas and the contaminated air will be oxidized to sulphur dioxide ($SO_2$):

$$2H_2S + 3O_2 \rightarrow 2SO_2 + 2H_2O \tag{6.18}$$

This option is less suitable in case of productive use of biogas because it introduces a certain amount of sulphide (though not too high), which tends to accelerate corrosion of the generator engine.

The oxidation of sulphide gas using biological odour control systems can be carried out using three different kinds of bioreactor configurations (Cooper & Alley, 2011): biofilters; biotrickling filters and bioscrubbers. Among them, biofilters are probably the most popular, in which a fixed bed with a porous biofiltering material of 1–1.5 m depth is used to support a wet biofilm. Natural porous media such as wooden chips, bark, peat or bagasse are used as media to develop a specific bacterial mass under adequate conditions of humidity. The solid medium itself should contain trace inorganic compounds and nutrients as nitrogen and phosphorus, to support biological sulphide oxidation. Biological oxidation generates sulphuric acid; thus, a pH buffer should be provided to avoid pH drop and loss of biological activity of the biofilter. Therefore, the production of some leachate is desired to avoid sulphate salt accumulation in the solid media. Biofilters are simple and compact and can handle more than 10 $m^3/h$ of air per $m^3$ reactor. To inoculate the bed material with a population of oxidizing bacteria it can be mixed with manure or wetted with treated sewage or secondary sludge.

Chemical oxidation of hydrogen sulphide with iron oxide is a spontaneous process that occurs at ambient temperature when biogas passes through a solid which contains iron oxide (FeO) for example the oxidized form of iron filings or soil containing iron oxide (Equation (6.16a)). The oxidation capacity of the unit can be recovered by flowing air (oxygen) through the solid, thus oxidizing sulphide to $SO_2$. The experience of a full-scale system with a soil rich in iron at Pirassununga (Brazil) is that it requires a very small area (~100 $m^2$ for 100,000 inhabitants), obtaining a removal of more than 90% of $H_2S$ and practically eliminating the problem of odours which was serious at the plant site. The soil depth is 0.5 m and air is introduced by perforated polyvinyl chloride (PVC) pipes placed with a spacing of 0.5 m. Another original method was applied successfully in Bucaramanga (Colombia), where seashells in a suspension were used to fix $H_2S$. The oxygen in air oxidizes the hydrogen sulphide to sulphuric acid and the acid reacts with the calcium carbonate that is the main component of seashells, forming gypsum:

$$H_2S + 2O_2 + CaCO_3 \rightarrow CaSO_4 + H_2O + CO_2 \tag{6.19}$$

The only operation in this case is the occasional withdrawal of $CaSO_4$ to replace it with seashells.

Regardless of the manner of removing odours from air in the case of sewage treatment, a large part of sulphide will not desorb from the liquid phase and eventually it will be discharged from UASB reactors with the reactor effluent for the post-treatment, which can be aerobic. In this case, the formed sulphide will be oxidized back to sulphate. This oxidation is rapid and complete. If ponds are used for post-treatment, the presence of sulphide may reduce the growth of algae and hence the photosynthetic production of oxygen, which is an essential aspect of the pond performance as has been shown in Chapter 5. On the contrary, under normal conditions the produced oxygen in the pond will oxidize the sulphide forming sulphur or sulphate.

## 6.7 BIOGAS COLLECTION

The biogas generated in an UASB reactor has a pressure of ~0.3–0.5 m water column (30–50 mbar), which is normally sufficient for direct use in power generators. A gas collection system in each phase separator element is composed of a network of rigid PVC pipes with diameters much larger than necessary for the biogas transport, but necessary to avoid blockages due to introduction of solid particles from the reactor into the collection system. For this reason, it is important that the biogas outlet from the gas chamber is situated as far as possible from the gas–liquid interface in the phase separator. It is convenient that washing of the pipes is foreseen to remove solids periodically from the interior of the pipes.

The biogas pressure is usually controlled by a hydraulic seal, used to set the level of the liquid–gas interface in the separator elements as shown in Figure 6.11a: the height *H* of the water column above the point of biogas release in hydraulic seal sets the biogas pressure in the gas chamber and sets the level difference between the level of the gas–liquid interface in the separator elements and the free surface of the UASB reactor. The biogas pressure can be read on the gauge as shown in Figure 6.11a. The hydraulic seal can also be used to retain condensed water that is in the biogas when the temperature of the UASB is higher than environmental temperature (the water in the seal). The condensed water is automatically drained from a siphon placed at the desired pressure level. In Figure 6.11b, a hydraulic seal for 10,000 $m^3$/day of biogas is shown.

In the case of domestic sewage treatment, biogas is normally burned in a flare, which can be a single unit as shown in Figure 6.12a. The flare in Figure 6.12a is designed to burn biogas generated in a system for 70,000 inhabitants. It is concluded that it is a small unit of low cost.

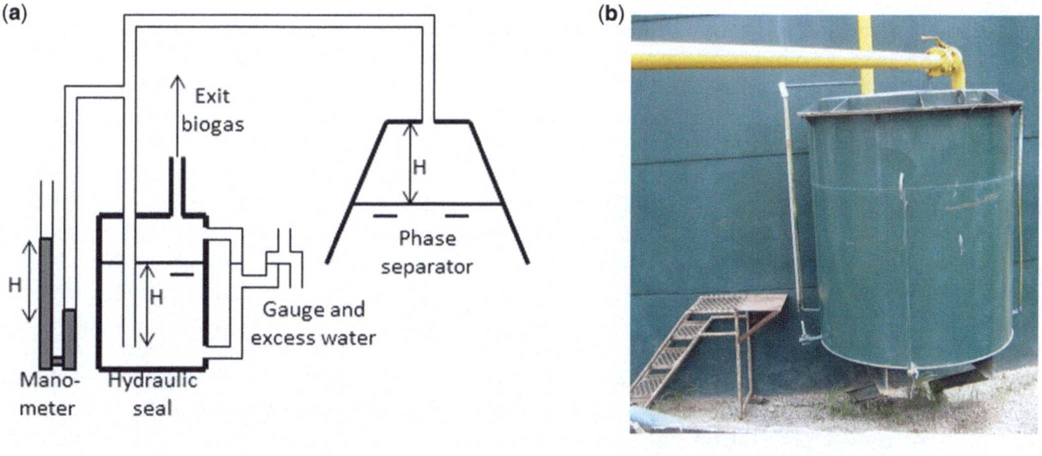


**Figure 6.11** (a) Schematic representation of gas withdrawal from the anaerobic reactor and (b) constructed hydraulic seal with manometer and gauge to read the biogas pressure.

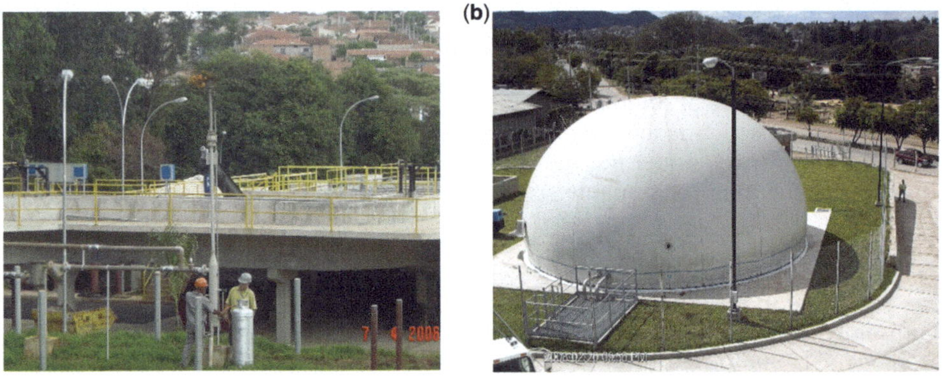

**Figure 6.12** (a) Flare of the UASB reactor in Matão (SP, Brazil) and (b) plastic semi-sphere for biogas accumulation in Tegucigalpa (Honduras).

If the goal is to use biogas for the production of electric energy, it is advisable to install a biogas storage unit to accommodate the daily variations in production rate. The volume of this unit should be such that it can store at least 1 h of gas production. Traditionally, biogas has been stored in a gasholder which was constructed by polyester reinforced with fibreglass or steel covered with material to resist corrosion. These units are rigid and the gas volume varies with changing production and usage rates. On top of the gasholder a weight was placed so that the desired pressure is automatically set. Nowadays, semi-spheres of inflatable plastic material ('biogas balloons') are used more frequently as they have a lower construction cost. These units have an outer layer that is inflated by a small fan to the desired pressure (typically 30 mbar) and an inner gas membrane impermeable for $CH_4$. This inner membrane fluctuates with the biogas production variations. Figure 6.12b shows the biogas storage unit.

## 6.8 GENERATION OF ELECTRIC ENERGY FROM BIOGAS

Ideally, all produced biogas should be used in generators, but it is difficult to continuously operate generators and use exactly the amount of biogas that is produced. Therefore, it is realistic to expect that the average biogas production rate will be higher than the average consumption rate by generators. In this case, it is necessary to have a flare to burn the excess production. This unit must have a device to automatically initiate combustion every time biogas starts to flow through it. The flare may be made using masonry or stainless steel. Its burning capacity should be enough to burn all biogas, when no electric energy is generated.

Power generation from biogas produced in the treatment plant is particularly attractive because the cost of generation will be low. Normally, the major cost generation is the operational, that is, the cost of acquisition of a generator. In the case of biogas generation in a wastewater treatment system, in principle, the fuel is free. Energy production becomes particularly interesting when there is a direct use in the treatment system, for example to drive pumps or aerators in post-treatment units. Also there is a strong incentive for electricity production if the legislation that regulates sale of electricity to distributors and is favourable for the producers or if carbon credits can be obtained.

The potential of electric power production is directly proportional to the methane production, which in turn depends on the organic load and the efficiency of anaerobic digestion. To calculate the potential of power generation the lower combustion heat is used, which is the combustion energy required when it is considered that the water formed in the oxidation is not condensed and therefore the condensation energy is lost. The amount of methane combustion heat is 50 MJ/kg $CH_4$ or

50,000/3600 = 13.9 kWh/kg $CH_4$. However, only part of the methane combustion energy is effectively converted into electrical energy in a power generator. The most widely used type of generator for power generation is a gas motor in which biogas and air are directly injected. Modern units have a power conversion efficiency of 35 (small units) to 40% (units for more than 1 MW). For this range the efficiency of electricity production is 4.9–5.6 kWh/kg $CH_4$.

***Example 6.9***

Estimate the per capita production potential of electric energy with the aid of the following data:

(1) Contribution per capita: 100 g COD/day, 120 L/day
(2) Anaerobic digestion efficiency: 80% of the COD (70% digested and 10% transformed into sludge)
(3) Losses due to dissolved methane: 20 mg/L effluent
(4) Biogas losses due to collection, transport and storage: 30%
(5) Conversion efficiency of the generator: 40%

***Solution***

Methane production is one-fourth of the digested COD mass, so the per capita production of methane: $0.7 \times 100/4 = 17.5$ g $CH_4$/day. For a per capita contribution of 120 L/day the dissolved methane is $20 \times 0.120 = 2.4$ g $CH_4$/day. Hence, the biogas production will be $17.5 - 2.4 = 15.1$ g $CH_4$/inh/day. Now, taking into account the loss of 30%, the available production for generating energy is $0.7 \times 15.1 = 10.6$ g $CH_4$/inh/day.

For a generator efficiency of 40%, the energy production is estimated to be 5 kWh/kg $CH_4$ or 5 Wh/g $CH_4$. As there is a per capita production of 10.6 g $CH_4$/day, per capita energy production will be $10.6 \times 5 = 53$ Wh/day, that is, there is a potential to produce a power of $53/24 = 2.2$ W/inh. This power is usually sufficient to meet the entire demand of a treatment system, including aerobic post-treatment.

On the contrary, the energy produced may not be enough for a water company economic convenience, unless for very large systems. One of the largest anaerobic sewage treatment systems in the world is the Onça plant at Belo Horizonte with a volume of 48,000 $m^3$. This system operates with a residence time of 8 h so the treated volume is close to 150,000 $m^3$/day with an organic load estimated at 120 t COD/day (assumed COD = 800 mg/L). Considering that 60% of this load will be effectively used for power generation one would have a mass of $0.6 \times 80/4 = 12$ t $CH_4$/day or ~60,000 kWh/day, that is, power of 60,000/24 = 2500 kW. An analysis will be needed to evaluate if the generation of this power is economically feasible. In many cases, smaller systems probably do not have economic feasibility in the Brazilian context, where the price for energy is low, thanks to ample possibility hydropower generation.

## 6.9 SLUDGE MANAGEMENT STRATEGIES

### 6.9.1 Sludge composition

The sludge generated in anaerobic reactors is characterized by the accumulation of inorganic solids, with a 55–70% of volatile solids. This value is slightly lower than the 60–85% observed for aerobic activated sludge systems. The wasted anaerobic sludge contains a moderate amount of nitrogen, ~3–6%, and phosphorus, 1–2%. The concentration of certain chemical elements in sludge, such as metals and metalloids, is often regulated for its application as agricultural fertilizer in different countries (Table 6.7).

When comparing the results of metal analyses, it is observed that anaerobic sludge is characterized by a high iron and zinc content. This is probably due to the abundance of these elements in raw wastewater and the precipitation of insoluble Fe and Zn sulphides in anaerobic bioreactors. In

**Table 6.7** Presence of metals and metalloids in the sludge generated in different anaerobic STPs (columns 1–3) as well as the sludge generated in the 40 largest cities of China, using aerobic-based processes (column 4). dw: Dry Weight.

| | Metals and Metalloids in STP Sludge (mg/kg dw) | | | | Legal Requirements (mg/kg dw) | |
|---|---|---|---|---|---|---|
| | (1) | (2) | (3) | (4) | Brazil (5) | Europe (6) |
| Cd | 0.74 ± 0.02 | | | 1.66 ± 2.07 | 39 | 20–40 |
| Cu | 707 ± 37 | 714 | 161 ± 10 | 163 ± 140 | 1500 | 1000–1750 |
| Ni | 38 ± 3 | 157 ± 60 | 53.9 ± 3.4 | 51.1 ± 76.9 | 420 | 300–400 |
| Pb | 110 ± 1 | 28 ± 3.1 | 6.8 ± 1 | 44.5 ± 28.1 | 300 | 750–1200 |
| Zn | 3048 ± 853 | 1300 ± 1210 | 1400 ± 98 | 609 ± 372 | 2800 | 2500–4000 |
| Hg | 2.1 ± 0.1 | – | – | 2.84 ± 2.88 | 17 | 16–25 |
| Cr | 720 ± 19 | 132 ± 42 | – | 180 ± 555 | 1000 | – |
| As | 11 ± 1 | – | – | 20.8 ± 14.4 | 41 | – |
| Se | – | 21 ± 1.1 | – | – | 100 | – |
| Fe | 11,766 ± 684 | 10,560 ± 2900 | 31,500 ± 1900 | – | – | – |
| Mn | 162 ± 10 | 98 ± 12 | 200 ± 6 | – | – | – |

(1) De la Varga *et al.* (2013); one anaerobic UASB pilot plant treating sewage, Spain.
(2) Braga *et al.* (2017); six full-scale anaerobic STP, Brazil.
(3) Lombardi and Garcia (2002) and Souza *et al.* (2014); Campinas full-scale anaerobic STP, Brazil.
(4) Geng *et al.* (2021); sludge from 40 different STP representing 22% of the China population.
(5) Brasil (2006) Conselho Nacional do Meio Ambiente. Resoluçao n 375.
(6) Council Directive 86/278/EEC of 12 June 1986; European Union limits for metals in STP sludge.
Furthermore, this information is complemented with the legal limits in Brazil (5) and Europe (6) for the safe use of sewage sludge in agriculture.

general, and except for zinc that is higher, the metal content of anaerobic STP sludge is not very different from the sludge generated in aerobic processes. The presence of heavy metals in waste sludge is influenced by the discharge of industrial wastewaters into sewerage systems, especially from the metallurgy and metalworking industry, as well as urban runoff, carrying pollutants from diffuse sources such as the ones originated from the washout of roads. In fact, particles emitted from wear and tear of brakes contain Fe, Cu, Zn, Sn, Zr and Sb, among others (Grigoratos & Martini, 2014) which may end up in sewage.

### 6.9.2 Quantities of generated sludge

Sludge management in STPs is usually one of the three main operational cost items, along with personnel and energy costs. One of the main advantages of anaerobic sewage treatment, in comparison to the aerobic alternative, is the lower amount of sludge generated, which reduces the costs associated with sludge disposal. Typically, the observed yields in anaerobic UASBs are between 0.1 and 0.18 g TSS/g COD, which are much lower than the values observed in aerobic sewage treatment, that is, 0.3–0.5 g TSS/g COD. Additionally, the purged solids' concentration from anaerobic UASB systems is between 3 and 6% TSS, a value much higher than that of 0.4–1.2% reported for aerobic systems. This reduces the amount of the sludge stream to be treated in the sludge line.

### 6.9.3 Sludge dewatering and drying

One of the main characteristics of waste sludge from anaerobic UASB reactors is the high solids' concentration. Additionally, due to the high sludge residence times applied, the waste sludge becomes stabilized, and can be easily dewatered, diminishing its fermentation potential and simplifying sludge treatment. For these reasons, thickening and stabilization operations, common for primary or secondary aerobic sludge treatment, are not required. The percentage of solids in waste anaerobic sludge can be increased to 20–45% in terms of dry solid percentage using mechanical dewatering processes or sludge-drying beds.

Dewatering is a physical operation employed to remove the excess water from the thickened sludge, producing a residue with high solid content known as 'cake'. If mechanical operational units are used, a liquid contaminated stream will be generated, which has to be returned to the water line to be treated adequately. There are many reasons to obtain the cake with high solids' concentration in STPs:

(1) Sludge transportation costs are substantially decreased.
(2) The dewatered cake is easier to handle and transport.
(3) Dewatering is required if the sludge cake is to be landfilled, thus diminishing the potential of leachate production.
(4) The dewatered cake has a higher calorific value when incineration is the destination for the sludge, as water is eliminated. In any case, it is very important to reduce the moisture as much as possible to increase the calorific value of the sludge. Sludge cakes with over 50% volatile solids can often be incinerated without auxiliary fuel.
(5) Dewatering is required if the sludge cake is to be composted.

Dewatering the waste sludge generated by anaerobic bioreactors can be accomplished utilizing different technologies such as filter presses, screw presses, belt-filters and centrifuges (Metcalf & Eddy, 2014). The efficiency of sludge dewatering in these technologies could be improved by adding inorganic coagulants as aluminium sulphate, ferric chloride or poly-aluminium chloride, and especially synthetic polymeric flocculants, such as cationic polyacrylamide, or natural flocculants derived from starch, chitosan, cellulose or tannin, in which the chemical structure of the natural product was partially modified to improve the product performance. Doses between 2 and 15 g of polymer per kg of dehydrated TSS are normally used for this purpose.

#### 6.9.3.1 Filter presses

Filter presses are equipment that operates intermittently, with sequential stages of sludge loading, pressure filtration, discharge of the dehydrated cake and a final idle stage until the cycle is restarted again. During the operation, the sludge to be dewatered is introduced into the plates, on which filter cloths have been arranged and fitted. In the first stage, the increase in pressure, resulting from pumping, forces the water to be filtered through the filter cloths. In the second stage, the compression caused by a hydraulic cylinder helps to achieve higher degrees of dehydration. With filter presses, very high solids' concentrations can be achieved, which can exceed 35% TSS. Unlike the other sludge dewatering technologies, polymers are barely used for sludge conditioning whereas inorganic coagulants, ferric chloride or aluminium sulphate are usually preferred. In some cases, lime is added after the coagulant to improve the filtration capacity. The energy requirement of this technology is between 20 and 40 kWh/t TSS.

#### 6.9.3.2 Screw presses

A screw press is a unit that operates at low speed and consists of an Archimedean screw rotating at a low speed inside a metal screen confined in a cylindrical casing. The aperture of the screen is below 0.5 mm, using normally wedge-wire metal filter. The whole device is usually inclined to facilitate the separation of the sludge cake from the drained water. Once chemically conditioned, the sludge is fed at low pressure into one end of the unit. The rotating action of the screw causes the sludge to advance forward upwards. The friction forces originated during the slow transport of the sludge, between the Archimedean screw and the screen, create a pressure that facilitates water filtering, collecting the cake at the opposite end to the feeding side.

Screw press systems can directly dewater sludge without requiring any previous thickening stage, which could result in substantial reductions in capital and operating expenses. However, this is not an advantage for dewatering anaerobic sludge, as it does not require any thickening step, unlike conventional waste activated sludge. In contrast to other mechanical dewatering technologies, this technology exhibits lower dry solid content in the cake (17–25%). Nevertheless, it provides several benefits, including reduced energy consumption, lower operational and maintenance costs, compactness and decreased noise levels.

#### 6.9.3.3 Belt filters

Belt filters operate continuously and consist of three sections: a gravity drainage section, in which water is removed by gravity, a second section where moderate pressure is exerted, as the sludge is confined between the band cloths and finally, a high-pressure area where the sludge is subjected to higher pressures as the force exerted by the cloths and the rollers increases, thus generating a cake of dehydrated sludge. The solid content in these systems can achieve between 20 and 28% TSS. The energy requirement of this technology is between 10 and 25 kWh/t TSS, being lower than that required for centrifuges or filter presses.

#### 6.9.3.4 Centrifuges

Centrifugation process is widely used in STPs to separate excess water and generate a thickened cake with a higher solid content using bowl centrifuges. A polymer-conditioned sludge stream is constantly fed into a rotating bowl, where a solid cake and a diluted liquid stream are produced. The separated liquid, called centrate, must be returned to the STP water line, while the thickened sludge cake has a solids' concentration often ranging between 20 and 30%. The energy requirement of this technology is between 30 and 60 kWh/t TSS.

#### 6.9.3.5 Sludge-drying beds

A sludge-drying bed is one of the oldest and simplest units for sludge dewatering and it has been applied to dewater the sludge purged from anaerobic sewage bioreactors (Figure 6.13). It gradually dewaters the sludge over the periods of several weeks by evaporation from the sludge surface in contact with the air and by drainage, using a system composed by a sand and gravel layer with drainage tubing located at the bottom of the bed. The collected water is usually recycled to the water line of the STP. The sludge to be dried is placed on the system in a 200–300 mm depth layer. Once dried, the solids are transported out of the STP to be used in agriculture or are landfilled. Conventional drying beds typically have rectangular shapes, with outer walls normally constructed with concrete blocks. Although the use of

**Figure 6.13** Sludge-drying bed in a municipal STP containing fresh sludge (left) and with dehydrated sludge (right).

drying beds is associated with small plants in areas where cheap land is available and with hot or dry climates, it is possible to use them in temperate climates, always considering the rainfall in the area.

Odour generation is usually very low when anaerobic sludge from UASB reactors is processed in drying beds since the sludge is stabilized as a result of the high sludge age in anaerobic bioreactors. Drying periods of 30–45 days can be applied, and the solids' concentration generally lies between 30 and 45% and may achieve more than 50–60% TSS under dry climate conditions, attaining a partial sludge sanitization in terms of microbial indicators (Lopes *et al.*, 2020). This technology requires more time and is less compact than mechanical devices. However, operational and maintenance costs are very low, its operation is simple and the percentage of solids in the dehydrated sludge is very high.

#### 6.9.3.6 Sludge-drying lagoons

Similar to drying beds, lagoons can be viewed as a cost-effective sludge dewatering technology in regions characterized by hot and dry climates, low population density and with cheap land availability. While it shares similarities to drying beds in terms of requiring vast land areas and labour-intensive mechanical removal of dewatered solids, its construction is simpler due to the absence of the need for filtrate drainage. However, drying lagoons are significantly less utilized than drying beds to treat digested sludge. The depth of the lagoon commonly ranges between 0.75 and 1.25 m. The main difference between this process and drying beds lies in the fact that evaporation is the primary mechanism driving the dewatering process, since no drainage can occur. This results in the need for longer drying times, ranging from several months to more than a year. Typically, the cake is removed with a concentration in the range of 25–30% of TSS (Table 6.8).

**Table 6.8** Comparison among the most common technologies used for dewatering waste sludge.

| | Filter Press | Screw Press | Belt Filter | Centrifuge | Drying Beds | Lagoons |
|---|---|---|---|---|---|---|
| Consumption (kWh/t TSS) | 20–40 | 5–15 | 10–25 | 30–60 | – | – |
| Land occupied | Low | Low | Low | Low | High | High |
| Cake solids (%) | Up to 35 | 17–25 | 20–28 | 20–30 | 50–60 | 25–30 |
| Operation mode | Discontinuous | Continuous | Continuous | Continuous | Discontinuous | Discontinuous |
| Operational complexity | High | Intermediate | Intermediate | Intermediate | Low | Low |
| Capital costs | Intermediate | Low | Intermediate | High | Low | Low |
| Dewatering time | Intermediate | Short | Short | Short | Long | Very Long |
| Maintenance | High | Low | High | Intermediate | Low | Low |
| Chemical requirements | High | High | High | High | Low or none | – |
| Odour problems | – | – | – | – | Potential | Potential |
| Weather impact | – | – | – | – | High | High |
| Groundwater pollution | Low | Low | Low | Low | Intermediate | High |
| Noise | Intermediate | Low | Intermediate | High | – | – |

*Source:* Adapted from Von Sperling and Chernicharo (2005), and Metcalf and Eddy (2014).

## 6.10 ENERGY GENERATION FROM SLUDGE COMBUSTION

Sludge generation in anaerobic treatment systems has a production of ~0.15 kg TSS/kg COD applied, with an organic fraction of 55–70% and a concentration of 50–80 g TSS/L. The produced sludge may be separated from the water by mechanical processes (filtration, centrifugation), whereupon it becomes a semi-solid cake with a solid fraction of ~20% and thus a residual humidity of ~80%. If further drying is desired, thermal processes such as sludge-drying beds or external heat application must be applied.

Depending on the use of the dried sludge and the cost of providing a final destination of the dried sludge it may be interesting to increase the solid fraction beyond the value that can be achieved by mechanical methods (15–25%). If the area for drying beds is not available, thermal drying may be applied, which is a process that produces a final sludge with no or very little water. It requires an external energy source, which can supply the heat required to evaporate the water from a sludge cake. Thermal drying can be completed with combustion of the dried sludge with air in an incinerator; in that case the energy source of the process is the combustion heat of the volatile sludge itself (see Figure 6.1). In that case, an important issue to be considered for sewage sludge incineration is the emission of toxic pollutants, due to the potential emission, among others, of dioxins, furans and nitrogen oxides, making the in-situ sludge incineration unaffordable. For this reason, centralized sludge incineration or co-incineration plants are used in many European countries, USA, and Japan.

The possibility of producing electric energy from the dried sludge is limited based on the following analysis: in anaerobic treatment systems roughly 2/3 of the influent organic material are transformed into biogas and the remaining fraction (1/3) is equally divided between organic material residual in the effluent and organic material transformed into sludge, both accounting for a fraction of 1/6 of the influent organic load. Thus, the fraction of the organic material transformed into biogas (methane) is of the order of four times greater than the fraction transformed into volatile sludge ((2/3)/1/6) = 4). When it is desired to use energy to generate a usable product e.g. electric energy the difference becomes even greater for two reasons:

(1) The biogas transformation efficiency is greater than that of sludge, because biogas is a gaseous fuel that can be used in an Otto-type explosion generator, characterized by an efficiency of 35–40%. In the case of sludge combustion, it is necessary to use a combination of a boiler to generate steam and a turbine to generate electricity. The efficiency of the boiler–turbine combination is at most ~15%. Knowing that the chemical energy per unit mass of COD is ~13.7 MJ, it is estimated that per unit mass of COD of the influent the potential of electricity generation is:

$$2/3 \times (0.35 \text{ to } 0.40) \times 13.7 \approx 3.4 \text{ MJ/kg COD}$$

whereas the energy generated from the sludge is no more than

$$1/6 \times 0.15 \times 13.7 = 0.34 \text{ MJ/kg COD.}$$

Therefore, it is estimated that the potential of electric energy generation from the biogas is ~10 times higher than the one generated from the sludge combustion.

(2) The sludge combustion energy is even lower because the sludge after drying still has residual humidity, which must be removed and it requires ~2.5 MJ/kg $H_2O$. Thus, if the sludge after drying has a humidity $U_e$ and a volatile fraction $f_v$, and knowing that $f_{cv}$ = 1.5 g VSS/g COD, it is calculated that the generation of the combustion heat is given by

$$Q_c = (1 - U_e)\, f_v f_{cv} 13.7 / 6 \text{ MJ / kg COD} \tag{6.20}$$

On the contrary, the value to evaporate water from the dried sludge with a residual humidity $U_e$ is given by

$$Q_e = 2.5U_e / 6 / (1 - f_v)\,\mathrm{MJ / kg\,COD} \tag{6.21}$$

In Figure 6.14, the values of heat production by combustion ($Q_c$) and the heat demand by water evaporation ($Q_e$) are plotted as functions of the residual humidity $U_e$ for organic sludge fractions of 60 and 70%. In the figure, an 80% heat utilization efficiency as well as the difference between the effective heat generated and the heat demand for evaporation are also reported. It can be seen that the difference is positive for a residual humidity of ~60%, that is, the heat generated is sufficient to sustain sludge combustion when the residual moisture is <60%. This value can be easily obtained in drying beds, but not with the aid of mechanical methods of sludge dehydration.

***Example 6.10***

Evaluate if the biogas production in a UASB reactor is sufficient to dry the produced sludge after mechanical dehydration to 80% humidity. Assume a sludge production of 0.1 g VSS/g COD, a VSS/TSS ratio of 0.67 for the sludge and a digestion efficiency of 70%. Evaporation heat of water is 2.5 MJ/L and combustion heat is 13.7 MJ/kg COD

***Solution***

Since 1 g VSS has 1.5 g TSS and the humidity is 80%, the sludge production is 0.1 g VSS or 0.15 g TSS per g COD with 0.15/0.8 = 0.19 kg of wet sludge per kg COD. The required heat for evaporation of (0.19 – 0.15) = 0.04 kg $H_2O$/kg COD is 0.04 × 2.5 = 0.1 MJ/kg COD. On the contrary, the digested COD is 0.7 g/g COD with a combustion heat for the formed methane of 0.7 × 13.7 ≈ 10 MJ/kg COD. However, only ~50% of this biogas is actually captured and used, so that a more realistic energy production of 5 MJ/kg COD is estimated. It is concluded that the combustion heat of the biogas is ~50 times greater than the energy demand for evaporation of the water in the sludge cake, so that even with low heat transmission efficiency, it should be possible to dry the sludge using only the combustion heat of the biogas to dry sludge, even though this does not seem to be applied in practice.

***Example 6.11***

Evaluate if the heat production of sludge combustion from a UASB reactor is enough to have auto-sufficiency in an incinerator. Adopt the same assumptions as in the previous example.

***Solution***

The required heat for evaporation is the same as in the previous example. The combustion heat depends on the volatile fraction of the sludge. In Figure 6.13, the evaporation heat and the combustion heat are

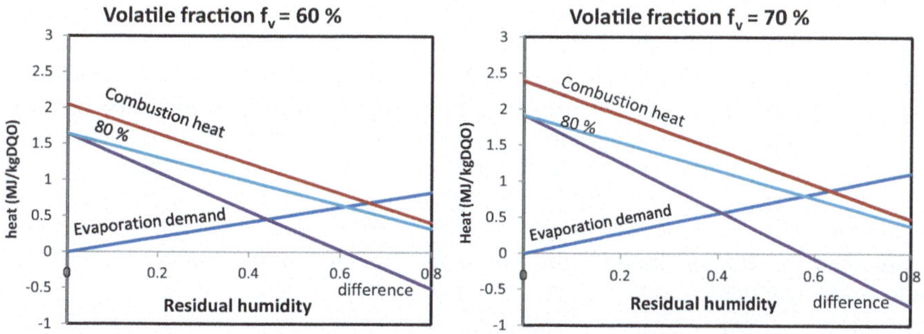


**Figure 6.14** Evaporation and combustion energy upon combustion of sludge cake as a function of humidity. The energy balance for 80% efficiency is also indicated, showing the need for external energy for humidity of ≲ 60%.

shown as functions of sludge humidity for volatile sludge fractions of 0.6 and 0.7. In Figure 6.13 also shown is the energy balance for an assumed energy transmission efficiency of 80%. The figure shows that under these conditions the maximum humidity of the sludge is ~60%, depending on the VSS/TSS ratio, which is much less than the humidity obtained from mechanical dehydration, so that some external energy must be supplied, if it is desired that completely dry sludge is produced for maximum heat production by combustion. This external energy can be biogas in which case the energy balance is positive even for a very high initial humidity of the sludge.

***Example 6.12***

Dry sludge is produced, and combustion is applied to generate steam to feed a turbine to produce electric energy. Estimate the production potential per capita using a top-quality boiler–turbine combination, which has an efficiency of 16%. The combustion heat of volatile sludge is 13.7 MJ/kg COD = 20 MJ/kg VSS. Volatile sludge production is 0.1 g VSS/g COD. Per capita contribution: 100 g COD/day.

***Solution***

The volatile sludge production per capita is 100 × 0.1/1000 = 0.010 kg VSS/inh/day with a per capita combustion heat of 20 × 0.01 = 0.2 MJ/day = 2.3 W. If the conversion efficiency is 16%, the per capita production potential is 0.16 × 2.3 = 0.37 W. This production potential is much smaller than the potential from gas, not only because the organic material is much smaller (16% of the influent COD for volatile sludge against 68% for methane), but also because the conversion efficiency of the solid fuel (16% for dry sludge) is much smaller than that for the gaseous fuel (40% for methane).

## 6.11 NUTRIENT RECOVERY FROM SLUDGE AND WASTEWATER STREAMS

Phosphorus and nitrogen compounds are essential nutrients required by modern agriculture to improve yields. The world's population increased from 2.5 billion inhabitants in 1950 to 8 billion at the end of 2022, and it is expected that it will reach 9.8 billion by around 2050. A large fraction of the phosphorus used in agriculture worldwide is obtained from a non-renewable resource, phosphate rock, mostly mined in Morocco (including Western Sahara), China, the USA, and Russia. In contrast, the production of nitrogen fertilizers is not restricted to a small number of countries and depends on the availability of energy to produce them through the Haber–Bosch and Ostwald processes. A large part of the developing countries is currently facing a great challenge: the increase in the prices of phosphorus fertilizers is limiting their use in agriculture and impacting food production in these regions (Brownlie *et al.*, 2023). Thus, the recovery of nutrients from STPs and other waste resources would be important to guarantee the necessary food for the entire world population.

***Example 6.13***

Estimate the phosphorus and nitrogen daily loads in a 90,000 inhabitants anaerobic STP and determine the fractions associated with the treated effluent and the anaerobic sludge generated. Consider a generation of 100 g COD/day, 10 g total nitrogen (TN)/day and 1.2 g total phosphorus (TP)/day and 100 L/day of sewage produced. Sludge yields 0.1 g TSS/g COD removed, COD removal percentage is 80. Nutrient content in the sludge is 0.05 kg N/kg TSS and 0.015 kg P/kg TSS.

***Solution***

Considering the above indicated figures, the amount of the incoming mass of TN and TP present in the raw sewage is (daily basis)

$$TN_{in} = 90{,}000\,\text{inh} \times 10\,\text{g TN / inh/day} = 900{,}000\,\text{g TN/day or } 900\,\text{kg TN/day}$$

$$TP_{in} = 90{,}000\,\text{inh} \times 1.2\,\text{g TP/inh/day} = 108{,}000\,\text{g TP/day or } 108\,\text{kg TP/day}$$

For anaerobic sewage treatment, the only mechanism contributing to N and P removal from wastewater is associated with the sludge formation. Thus, N and P associated with sludge generation should be:

$$MS_X = 0.8\,g\,COD(removed) \times g\ COD(fed) \times 0.1\,g\,TSS/g\,COD(removed) \times 90{,}000\,inh$$
$$= 720{,}000\,g\,TSS/day \approx 720\,kg\,TSS/day$$

$$TN_X = 0.05\,kg\,N/kg\,TSS \times 720\,kg\,TSS/day = 36\,kg\,TN/day$$

$$TP_X = 0.012\,kg\,P/kg\,VSS \times 720\,kg\,TSS/day = 10.8\,kg\,TP/day$$

where $MS_X$ = anaerobic sludge generation (kg TSS/day); $TN_{SV}$ = total nitrogen in the sludge (kg N/day); $TP_{XV}$ = total phosphorus in the sludge (kg P/day).

The percentage of TN and TP associated with the sludge formation represents 4 and 10% of the daily nutrient load, respectively. In those STPs using anaerobic technologies, the recovery of phosphorus or nitrogen associated with the sludge is low, due to the low biomass yield of the process. In any case, the sludge generated in anaerobic STPs contains all the nutrients removed from the treated sewage. In fact, the use of sludge from STPs as agricultural fertilizer is a common practice in many countries around the world.

Anaerobic sewage treatment is not the adequate solution to remove nutrients from the water phase. In recent years, many studies have tried to improve the removal of nutrients in these systems. One approach is the promotion of struvite precipitation, a phosphate salt containing ammonia and magnesium ions ($MgNH_4PO_4{\cdot}6H_2O$). This strategy is only effective when the treated wastewater contains high concentrations of orthophosphate ions, as is often the case with black water. The precipitation of struvite only compensates if the phosphorus concentration greatly exceeds 100 mg P/L, a level far from that of 5–15 mg P/L observed in sewage.

In Table 6.9, the calculated concentrations of pollutants of raw and treated sewage for the analysed scenario are shown. As indicated, the calculated nutrient concentrations in the effluent and raw sewage are similar. In fact, there are many studies in the literature indicating that the observed removal of nitrogen and phosphorus in anaerobic bioreactors is almost negligible.

Most of the nutrients contained in anaerobically treated sewage, without further post-treatment, remain in the effluent. The eutrophication of the receiving water bodies is one of the main problems of the discharge of anaerobically treated sewage. Nutrients contained stimulate algae blooms, harming water life. An alternative to take advantage of the nutrients and the water itself could be to store the treated waters in accumulation ponds, and use them for agricultural irrigation, as long as the distance between the fields and the STP does not make this application unfeasible. Water reuse is treated in Chapter 8, and could be a way to recover nutrients, and diminishing the use of chemical fertilizers. This practice, as well as the use of sewage sludge as fertilizer, is regulated by law in several countries.

The World Health Organization publishes guidelines for the safe use of wastewater in agriculture (WHO 2006), to reduce the risks associated with the presence of pathogens and certain chemical compounds associated with wastewater and sewage sludge. For these reasons, in most cases, anaerobically treated sewage should undergo post-treatments to ensure adequate water quality, including a partial reduction of nutrients, especially nitrogen compounds, to guarantee a good food crop.

**Table 6.9** COD, TN and TP concentrations and nutrients loads of the analysed scenario.

| | mg COD/L | mg TN/L | mg TP/L | kg TN/day | kg TP/day |
|---|---|---|---|---|---|
| Raw | 833.3 | 83.3 | 10.0 | 900 | 108 |
| Effluent | 166.7 | 80.0 | 9.0 | 864 | 97 |

One of the questions that could be raised is the comparison of the amounts of nutrients present in sewage, with the amount of fertilizer consumed. According to the International Fertilizer Association (IFA, 2023) during 2020, 21.8 and 112.3 Mt of phosphorous- and nitrogen-based fertilizers were consumed worldwide, which implies an average annual consumption per world inhabitant of 2.8 kg P and 14.4 kg N. It can be quickly calculated for our case study that an inhabitant generates 0.438 kg P and 3.65 kg N annually. Assuming that all nutrients from wastewater could be effectively reused, ideally 15% of P and 25% of N could be recovered, thereby reducing the consumption of chemical fertilizers. This is obviously a very simple calculation; surely the real figures that can be achieved are more modest, highlighting the great potential associated with the nutrients contained in urban wastewater.

## 6.12 CONCLUDING REMARKS

It has been shown that anaerobic wastewater systems can be operated without the need of external energy, even when wastewater is diluted such as the case of sewage. The required energy for anaerobic treatment is obtained by using its organic material as an energy source. In the case of sewage ~2/3 of the influent organic material is converted into methane and may be used for energy generation. About 1/6 of the influent organic material is converted into organic sludge and 1/6 is discarded in the effluent. Organic sludge may be incinerated to produce electric energy and heat, but the alternative of using it as an organic fertilizer (after drying) may be more attractive from the economic point of view.

Nitrogen and phosphorus recovery from anaerobically treated wastewater can be carried out by using the reclaimed water and sludge generated in agriculture, according to local regulations. In most cases anaerobically treated sewage should undergo further treatment to ensure adequate water quality to guarantee a good crop.

## REFERENCES

Braga A. F. M., Zaiat M., Silva G. H. R. and Fermoso F. G. (2017). Metal fractionation in sludge from sewage UASB treatment. *Journal of Environmental Management*, **193**, 98–107, https://doi.org/10.1016/j.jenvman.2017.01.070

Brasil (2006). Conselho Nacional do Meio Ambiente. Resoluçao n 375.

Brownlie W. J., Sutton M. A., Cordell D., Reay D. S., Heal K. V., Withers P. J. A., Vanderbeck I. and Spears B. M. (2023). Phosphorus price spikes: a wake-up call for phosphorus resilience. *Frontiers in Sustainable Food Systems*, **7**, 1–8, https://doi.org/10.3389/fsufs.2023.1088776

Buisman C. J. N. and Lettinga G. (1990). Sulphide removal from anaerobic waste treatment effluent of a papermill. *Water Research*, **24**(3), 313–319, https://doi.org/10.1016/0043-1354(90)90006-R

Cooper C. D. and Alley F. C. (2011). Air Pollution Control: A Design Approach, 4th edn. Waveland Press, Long Grove, IL.

Council Directive 86/278/EEC of 12 June 1986. European Union limits for metals in STP sludge.

Cozma P., Wukovits W. and Mamaliga J. (2013). Analysis and modelling of the solubility of biogas components in water for physical absorption processes. *Environmental Engineering and Management Journal*, **12**(1), 147–162, https://doi.org/10.30638/eemj.2013.017

De la Varga D., Díaz M. A., Ruiz I. and Soto M. (2013). Heavy metal removal in an UASB-CW system treating municipal wastewater. *Chemosphere*, **93**, 1317–1323, https://doi.org/10.1016/j.chemosphere.2013.07.043

dos Santos S. L. (2013). Pós-tratamento de esgoto digerido visando reuso agrícola ou proteção de corpos receptores. PhD thesis, Federal University of Campina Grande (in Portuguese).

Geng Y., Zhang C., Zhang Y., Huang D., Yan S., Sun T., Wang J. and Mao Y. (2021). Heavy metal(loid)s in sewage sludge in China: concentrations and spatial-temporal variations. *Environmental Science and Pollution Research*, **28**, 29146–29156, https://doi.org/10.1007/s11356-021-12762-8

Grigoratos T. and Martini G. (2014). Non-exhaust Traffic Related Emissions – Brake and Tyre Wear PM. EUR 26648. Publications Office of the European Union, Luxembourg (Luxembourg). JRC89231.

Henze M. and Harremoes P. (2001). Wastewater Treatment: Biological and Chemical Processes. Springer, Berlin.

IFA, International Fertilizer Association (2023). https://www.ifastat.org/ (accessed on 23 May 2023)

Janssen A. J. H., Lettinga G. and de Keizer A. (1999). Removal of hydrogen sulphide from wastewater and waste gases by biological conversion to elemental sulphur: colloidal and interfacial aspects of biologically produced sulphur particles. *Colloids and Surfaces A: Physicochemical and Engineering Aspects*, **151**(1–2), 389–397, https://doi.org/10.1016/S0927-7757(98)00507-X

Lens P. N. L. and Hulshoff Pol L. W. (2000). Environmental Technologies to Treat Sulphur Pollution. IWA Publishing, London (UK).

Lin S., Hamish H. R., Tianwei H., Guo G., van Loosdrecht M. and Guanghao Chen C. (2018). Biological sulfur oxidation in wastewater treatment: a review of emerging opportunities. *Water Research*, **143**(15), 399–415, https://doi.org/10.1016/j.watres.2018.06.051

Loewenthal R. E. and Marais G. V. R. (1976). Carbonate Chemistry of Aquatic Systems: Theory and Application. Ann Arbor Science Publishers, Michigan (USA).

Lombardi A. T. and Garcia O. (2002). Biological leaching of Mn, Al, Zn, Cu and Ti in an anaerobic sewage sludge effectuated by *Thiobacillus* ferrooxidans and its effect on metal partitioning. *Water Research*, **36**, 3193–3202, https://doi.org/10.1016/S0043-1354(02)00008-8

Lopes B. C., Rodrigues H. F., Costa T. E. R., Batista A. M. M., Filho C. R. M., Araújo J. C. and de Matos A. T. (2020). Air-drying bed as an alternative treatment for UASB sludge under tropical conditions. *Journal of Water, Sanitation and Hygiene for Development*, **10**, 458–470, https://doi.org/10.2166/washdev.2020.041

Metcalf & Eddy (2014). Wastewater Engineering: Treatment and Resource Recovery, 5th edn. McGraw Hill, New York (USA).

Rinzema A. (1988). Anaerobic Treatment of Wastewater with High Concentrations of Lipids or Sulphate. Doctoral thesis, Wageningen Agricultural University, Wageningen, The Netherlands.

Song Z., Williams C. J. and Edyvean R. G. J. (2004). Treatment of tannery wastewater by chemical coagulation. *Desalination*, **164**(3), 249–259, https://doi.org/10.1016/S0011-9164(04)00193-6

Souza L. C. F., Canteras F. B. and Moreira S. (2014). Analyses of heavy metals in sewage and sludge from treatment plants in the cities of Campinas and Jaguariúna, using synchrotron radiation total reflection X-ray fluorescence. *Radiation Physics and Chemistry*, **95**, 342–345, https://doi.org/10.1016/j.radphyschem.2013.01.025

Van Haandel A. C. and van der Lubbe J. (2019). Anaerobic Sewage Digestion: Theory and Applications. International Water Association, London, UK.

Von Sperling M. and Chernicharo C. A. D. L. (2005). Biological Wastewater Treatment in Warm Climate Regions. IWA Publishing, London (UK).

WHO Guidelines for the safe use of wastewater, excreta and greywater (2006). World Health Organization. ISBN: 92 4 154686 7.

Zhang S. and Zhang C. Z. (2002). Novel properties of ionic liquids in selective sulphur removal from fuels at room temperature. *Green Chemistry*, **4**, 376–379, https://doi.org/10.1039/b205170m

doi: 10.2166/9781789063479_0207

# Chapter 7

# Dissolved methane

E. Centeno Mora[1]*, C. L. Souza[2], T. Bressani-Ribeiro[2] and C. A. L. Chernicharo[2]

[1]Civil Engineering School and Centre of Research of Sustainable Development, University of Costa Rica, San José, Costa Rica
[2]Sanitary and Environmental Department, Federal University of Minas Gerais, Belo Horizonte, Brazil
*Correspondence: erick.centenomora@ucr.ac.cr

**ABSTRACT**

This chapter addresses the fate of dissolved methane (D-$CH_4$) in mainstream anaerobic-based sewage treatment plants (STPs). As much as 30–40% of the total $CH_4$ generated during anaerobic digestion can be lost in the effluent of anaerobic reactors treating sewage. This implies a loss of energetic potential, possible emission of $CH_4$ downstream the anaerobic reactor with an increase in the carbon footprint, greenhouse gas emissions and safety risks for operators. The first section presents the fundamentals that explain the origin of D-$CH_4$ and the physical principle behind its emission and desorption. Next, a review of the measuring methods for D-$CH_4$, reported values for different anaerobic reactor configurations applied to sewage treatment and the energetic and environmental effects of D-$CH_4$ are presented. Later sections explore different techniques and bioprocesses that have been studied to mitigate D-$CH_4$ in anaerobic effluents, describing the general principles behind its functioning, experimental performance and current state of development. Four management strategies are depicted to mitigate D-$CH_4$ in anaerobic effluents: (a) desorption-based techniques followed by the oxidation of waste gas; (b) desorption-based techniques aiming at recovering an energetically valuable gas; (c) bioprocesses to oxidize D-$CH_4$ in post-treatment unit and (d) bioprocesses that use D-$CH_4$ in the post-treatment unit for other treatment objectives, such as denitrification or electricity generation. Some of these techniques and bioprocesses are available and already applied in full-scale anaerobic-based STPs, whereas others seem very promissory but require further research to validate their application with real anaerobic effluents.

**Keywords:** carbon footprint, desorption techniques, energy loss, energy recovery, fugitive emissions, methane bio-oxidation.

## 7.1 INTRODUCTION

Mainstream anaerobic sewage treatment offers many advantages, as highlighted in previous chapters. However, certain challenges need to be addressed to enhance the suitability and sustainability of anaerobic-based sewage treatment plants (STPs) (Chernicharo *et al.*, 2015). One of the major issues is the presence of high concentrations of dissolved methane (D-$CH_4$) in liquid effluent, which has been extensively reported over the last 15 years.

In the case of upflow anaerobic sludge blanket (UASB) reactors treating sewage, up to 36–41% of the total $CH_4$ produced during an anaerobic digestion process can be lost as D-$CH_4$ in the liquid phase (Souza *et al.*, 2011). This condition poses some problems:

- It represents a significant loss of the energetic potential of anaerobic-based STPs, equivalent to the percentage of $CH_4$ lost in the anaerobic effluent (Lobato *et al.*, 2012).
- Depending on the flowsheet and hydraulic profile of the STP, D-$CH_4$ can be emitted to the atmosphere downstream the reactor. Considering that $CH_4$ is a greenhouse gas (GHG) with a global warming potential of ~28 times that of $CO_2$ (IPCC, 2021), its uncontrolled emission in anaerobic-based STPs tend to represent a major carbon footprint contributor (Wu *et al.*, 2022).
- The emission and blending of $CH_4$ with atmospheric air in confined spaces (e.g. closed tanks, downstream pipes such as sewers) can create a flammable atmosphere with a $CH_4$ content between the lower flammability level (LFL) of 5.0%, and the upper flammability level of 15.0% in air (Crowl, 2003), which poses a safety risk for operators in anaerobic-based STPs and even in the sewerage network.

In the last decade, various techniques and strategies have been proposed and experimentally studied to mitigate or recover/use D-$CH_4$, and this topic has been frequently addressed in the literature during this period (Centeno-Mora *et al.*, 2020; Crone *et al.*, 2016; Heile *et al.*, 2017; Li *et al.*, 2021; Stazi & Tomei, 2021). This chapter discusses the nature of D-$CH_4$ in anaerobic-based STPs and the available strategies and techniques to mitigate it. Section 7.2 reviews the fundamentals that explain the presence of dissolved gases in anaerobic effluent and the desorption of dissolved gases. Section 7.3 examines the methods for measuring D-$CH_4$, the reported values in the literature for different anaerobic reactors applied to sewage treatment and the significance of its presence for anaerobic-based STPs. In Section 7.4, the available techniques and strategies that have been tested are discussed, taking into account different approaches for $CH_4$ dissolved in anaerobic effluents: (1) its transfer to a waste gas in low concentration, which should be abated downstream; (2) its transfer to a $CH_4$-rich gas that could be energetically valorized; (3) its direct biological oxidation in the liquid phase and (4) its direct beneficial use in the liquid phase with different bioprocesses. The management of waste/recovered gas is also addressed. Finally, some perspectives for future research on this topic are presented.

## 7.2 BACKGROUND

### 7.2.1 Equilibrium between liquid and gas phases

$CH_4$ and other gases are produced during the anaerobic digestion of sewage. Under normal conditions of temperature and pressure (e.g. 20°C and 1 atm), a theoretical maximum yield of 350 mL of $CH_4$ per g of removed chemical oxidation demand (COD) can be obtained based on a simplified mass balance. This $CH_4$ is generated in the sludge bed or in the bulk liquid of the reactor, where the substrate (sewage) and the biomass come into contact. Due to their lower density with respect to water, $CH_4$ and other gases produced during the anaerobic digestion (e.g. $CO_2$ and $H_2S$) are transported through the bulk liquid to the upper part of the reactor, and then to the gas phase compartment or reactor headspace. The resulting mixture of gases is known as biogas, a $CH_4$-rich gas, and a source of renewable energy. However, a fraction of these gases remains dissolved in the liquid effluent of anaerobic reactors, following a thermodynamic equilibrium.

The thermodynamic equilibrium between a gas dissolved in water and its content in the gas phase (headspace) is defined by the Henry's law, as shown in Equation (7.1):

$$C_{dis,g} = K_H P_{p,g} \tag{7.1}$$

where $C_{dis,g}$ is the concentration of the gas dissolved in the liquid phase (mg/L); $K_H$ is the Henry's law constant (mg/L/atm) and $P_{p,g}$ is the partial pressure of the gas in the gas phase (atm). Different forms of Henry's law are available in the literature, and a cautious selection of $K_H$, which considers the appropriate units for the equilibrium equation, should be considered in every case. It should be noticed that $K_H$ is temperature-dependent, and the reported values must be adapted to the ambient conditions of anaerobic reactors. This is normally done with the Van't Hoff equation. In addition, in practice, values of $K_H$ for clean water are used for anaerobic effluents, though the presence of different compounds (e.g. solids, surfactants, salts, among others) can affect this equilibrium. More information on this topic can be found elsewhere (Sander, 2015; Staudinger & Roberts, 2009).

The value of $K_H$ for $CH_4$ to be used in Equation (7.1) is 23.6 mg/L/atm at 20°C (Sander, 2015). $CH_4$ can be considered a gas with a low solubility in water. As a reference, at 20°C, $H_2S$ and $CO_2$ are highly soluble in water, presenting $K_H$ values of 3911 mg/L/atm and 1628 mg/L/atm, respectively. According to the Henry's law, when a gas is present in a headspace or in the atmosphere, a fraction of this gas is expected to be dissolved in the liquid phase under equilibrium conditions. Figure 7.1 shows the interaction between the liquid phase and the gas phase in anaerobic reactors (UASB reactors and anaerobic membrane bioreactors, AnMBRs, with completely stirred tank reactors (CSTRs)).

As shown in Figure 7.1, a gas (e.g. $CH_4$, $CO_2$, $H_2S$, $N_2$) exchange will occur in the interface of the liquid phase and biogas compartment inside anaerobic reactors. This exchange will take place inside a gas–liquid–solid separator and in the settling compartment surface in UASB reactors (Figure 7.1a), and in the headspace of AnMBRs (Figure 7.1b). The thermodynamic equilibrium established by the

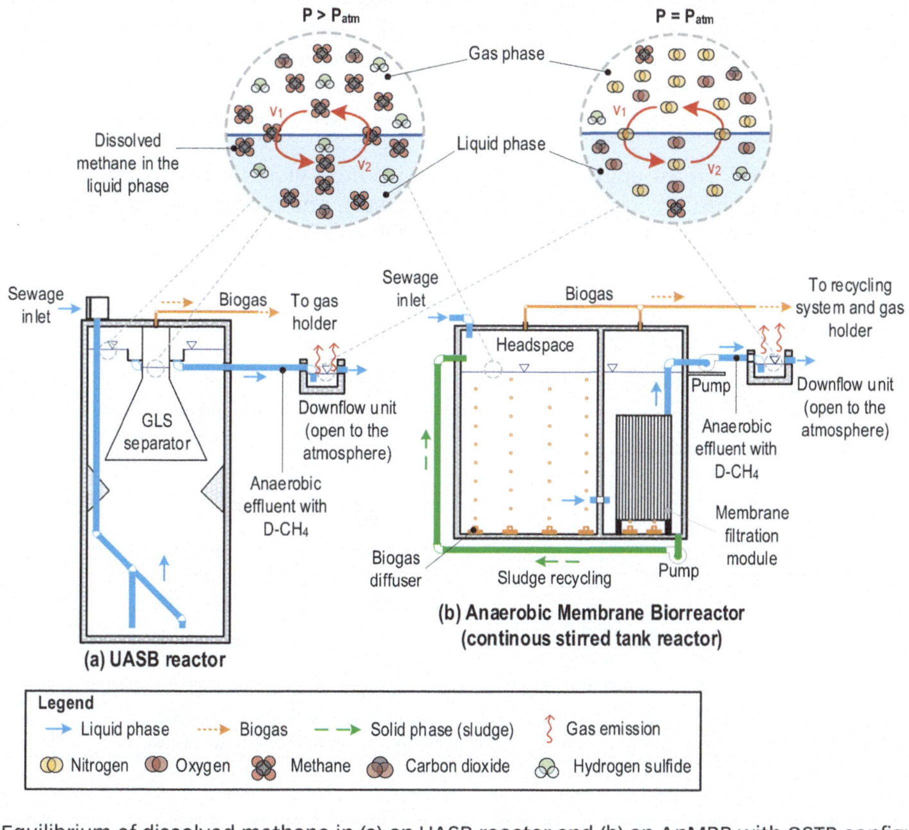


**Figure 7.1** Equilibrium of dissolved methane in (a) an UASB reactor and (b) an AnMBR with CSTR configuration.

Henry's law (Equation 7.1) occurs when the flux of gas transfer to the liquid phase (V1) equals the flux of gas transfer to the gas phase (V2).

As an example, if a content of $CH_4$ in biogas of 75% v/v (Noyola *et al.*, 2006) is considered, for a pressure of 1 atm, at a temperature of 15°C and 30°C ($K_H$ values of 26.4 and 19.8 mg/L/atm, respectively), the concentration of D-$CH_4$ at the equilibrium (Equation 7.1) would be 19.8 and 14.8 mg/L, respectively. However, this equilibrium does not occur frequently in anaerobic reactors, and concentrations of D-$CH_4$ above the value obtained with Equation (7.1) have been repeatedly reported in the literature (see Section 7.3.2). This is known as the supersaturation of $CH_4$ in anaerobic effluents, and it is due to mass-transfer limitations in the system (Hartley & Lant, 2006; Pauss *et al.*, 1990).

As shown in Figure 7.1, when an anaerobic effluent, saturated or oversaturated with $CH_4$, is conducted to an open unit or a turbulent flow unit downstream (e.g. a non-confined settler compartment of UASB reactors, an open channel with weirs, a distribution arm of a trickling filter, among others), this $CH_4$ will be emitted to the atmosphere. Considering that the concentration of this gas in the atmosphere is negligible, the equilibrium between the phases will indicate a concentration of D-$CH_4$ is also negligible in the liquid phase for this new condition.

### 7.2.2 Principle of desorption of dissolved gases

As explained in Section 7.2.1, in a system open to the atmosphere with a $CH_4$-saturated anaerobic effluent, a mass-transfer of $CH_4$ (and other dissolved gases) from the liquid to the gas phase (atmosphere) will occur to attain the thermodynamic equilibrium between both the phases (Equation 7.1). This is a diffusion-driven process (Cussler, 2007), in which the driving force for mass transfer is the difference of concentration between both the phases, and it is also known as gas desorption.

The same principle can be applied to engineered processes to force the transference of the $CH_4$ dissolved in the liquid phase of anaerobic reactors to a gas phase in a controlled system, as presented in Figure 7.2.

As shown in Figure 7.2, in an engineered gas desorption device, there are two fluxes: the liquid phase (anaerobic reactor effluent containing dissolved gases), and a $CH_4$-free sweeping gas ($N_2$ or air is commonly used). Due to the different $CH_4$ concentrations between both the phases, the system

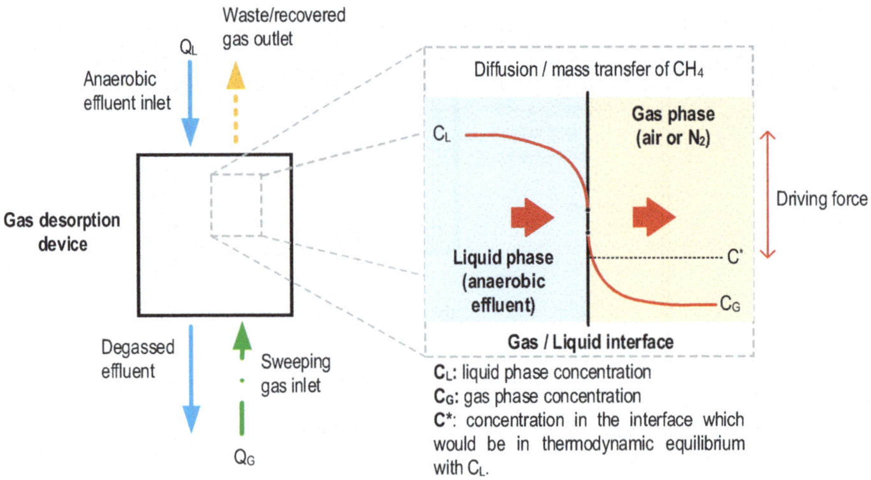


**Figure 7.2** Principle of dissolved methane desorption from the effluent of an anaerobic reactor. *Source*: Adapted from Brandt *et al.* (2021a, 2021b, 2021c).

tends to its thermodynamic equilibrium, which forces the mass transfer of $CH_4$ (and other gases, such as $CO_2$ and $H_2S$) to the sweeping gas. The continuous renewal of the liquid and gas phases allows the operation under a steady-state condition, producing a continuous transference (and recovery) of $CH_4$ in the gas phase, which is known as waste gas or recovered gas.

Depending on many parameters of the system (liquid flow – $Q_L$, gas flow – $Q_G$, $Q_G/Q_L$ ratio, interface area, temperature, concentration difference between both the phases), the removal efficiency (and the transference of $CH_4$ to the gas phase) can be affected. In addition, for the renewal of the gas phase, it is possible to use vacuum instead of sweeping gas. This allows the recovery of a more concentrated $CH_4$ in the recovered gas, avoiding its excessive dilution with air or $N_2$.

Different techniques have been studied for the D-$CH_4$ desorption from anaerobic reactor effluents (Centeno-Mora *et al.*, 2020; Heile *et al.*, 2017): desorption chambers (with and without packing media, with sweeping gas or vacuum), membrane contactors, enclosed biological reactors of downflow hanging sponge (DHS) and stripping tanks. These techniques are described in detail in Section 7.4.2.

## 7.3 DISSOLVED METHANE IN THE EFFLUENT OF MAINSTREAM ANAEROBIC REACTORS TREATING SEWAGE

### 7.3.1 Methods for measuring dissolved methane in wastewater

Dissolved methane in wastewater can be measured with different methods. Figure 7.3 shows the main measuring principles reported in the literature (Boulart *et al.*, 2008, 2010; Mampaey *et al.*, 2015).

As shown in Figure 7.3, there are laboratory and in-situ methods for measuring D-$CH_4$. In the case of laboratory methods, the usual procedure consists in taking samples of anaerobic effluents using septum vials (40–120 mL), which should be stored at low temperature (e.g. 4°C) until they can be analysed. Then, the samples are conditioned to attain a standard temperature (usually 25°C) to establish the equilibrium between the liquid and gas phases (usually 1–2 h). When this condition is fulfilled, a sample of the gas headspace is obtained with a gas-tight syringe (e.g. 10–500 μL) and injected into a gas chromatograph (GC). Flame-ionizing detector is normally employed due to its

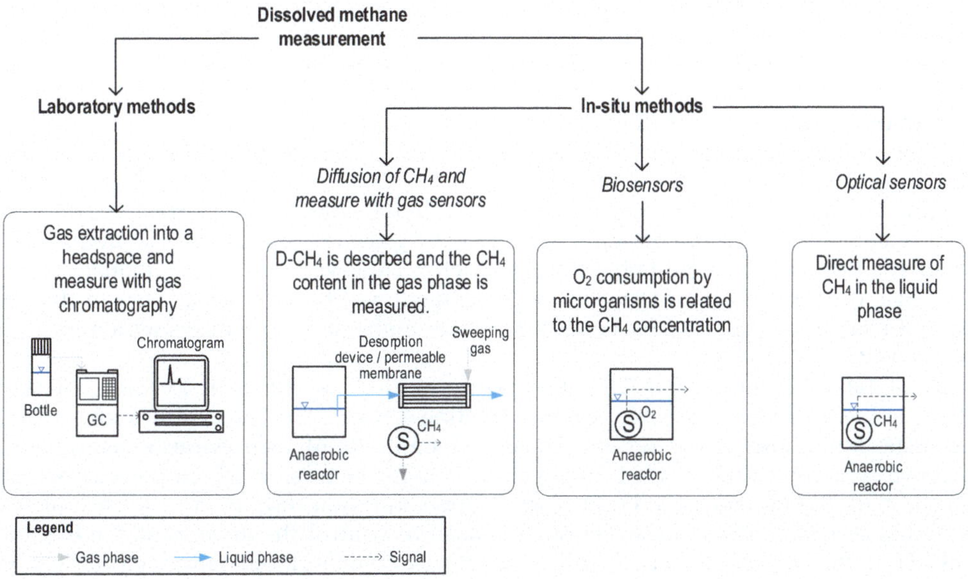


**Figure 7.3** Methods for measuring dissolved methane in anaerobic effluents.

high sensitivity, which allows the measurement of low contents of $CH_4$ in the gas sample. When the $CH_4$ content of the headspace is known, the D-$CH_4$ concentration can be calculated by assuming the thermodynamic equilibrium between the gas and liquid phases and using an adaption of Equation (7.1). This method has been extensively used for the measurement of D-$CH_4$ in anaerobic effluents. It should be noted that, to date, there is not a standardized procedure for this method, and different adaptions are reported in the literature: type and volume of the bottle; bottle prefilling with nitrogen or argon (Bandara *et al.*, 2011); use of vacuum tubes (Daelman *et al.*, 2012); use of a biological inhibitor (e.g. $HgCl_2$) (Bandara *et al.*, 2011); use of high salt concentration (e.g. NaCl) to reduce gas solubility, also inhibiting microbial activity (Daelman *et al.*, 2012), known as 'salting-out' method; time to attain the standard temperature and equilibrium; presence of agitation to attain this equilibrium and GC specifications such as the type of detector or equipment configuration. Differences in the values of D-$CH_4$ obtained from different procedures are expected. As an example, Nelting *et al.* (2015) reported values of D-$CH_4$ to be 11% higher when the salting-out method (NaCl) was used instead of a simpler method without salts. Therefore, the definition of a standardized method should make easier and more precise the comparison of the values reported in the literature.

In the case of in-situ methods, they can be advantageous to evaluate the spatial and temporal variability of D-$CH_4$ in anaerobic reactors. Three principles are reported in the literature (Boulart *et al.*, 2010), as shown in Figure 7.3: measure of $CH_4$ in the gas phase, biosensors and optical sensors. For the desorption devices, the same principle explained in Section 7.2.2 can be applied to transfer D-$CH_4$ to a gas (air or $N_2$). The content of $CH_4$ in the recovered gas is measured with a gas probe, and D-$CH_4$ is calculated with a mass balance. This principle can be used in situ, as described in Liu *et al.* (2015) and Mampaey *et al.* (2015). An alternative is the use of a passive system, in which a probe with a permeable membrane is in contact with the liquid phase. $CH_4$ diffuses through this membrane to a gas phase, where it is measured. Some of the commercially available sensors use this principle and it has been applied for the measurement of D-$CH_4$ in anaerobic effluents (Rongwong *et al.*, 2019; Sethunga *et al.*, 2019; Velasco *et al.*, 2022; Wongchitphimon *et al.*, 2017).

The measuring principle of biosensors quantifies the $O_2$ consumption of methane-oxidizing bacteria (MOBs) in a gel matrix to estimate the D-$CH_4$ concentration, whereas optical sensors directly measure the D-$CH_4$ concentration in the liquid phase with optical principles such as infrared-spectroscopy, Raman spectroscopy or surface-plasmon resonance (Boulart *et al.*, 2010). To date, none of these methods has been used to report D-$CH_4$ concentrations in anaerobic effluents.

### 7.3.2 Experimental values reported in the literature

Table 7.1 shows reported data for D-$CH_4$ concentrations in different types of mainstream anaerobic reactors treating sewage.

As explained in Section 7.2.1, the presence of D-$CH_4$ in the effluent of anaerobic reactors is related to the thermodynamic equilibrium between the liquid and gas phases. A part of the produced $CH_4$ will be dissolved and lost with the liquid phase of the reactor. However, in the case of anaerobic reactors treating sewage, the percentage of $CH_4$ lost in the effluent will be around 30–50% of the total $CH_4$ generated, with values as high as 88% being reported at ambient temperatures with a mean value of 18°C (Cookney *et al.*, 2016).

In addition, Table 7.1 shows that $CH_4$ supersaturation indexes between 1.0 and 1.66 have been reported in the literature for anaerobic reactors treating sewage. This supersaturation in $CH_4$ is due to mass-transfer limitations in the system (Hartley & Lant, 2006), as previously mentioned. In this regard, passive systems (Crone *et al.*, 2016) such as UASB reactors, in which there is not an active mixing of the bulk liquid, tend to present higher supersaturation indexes. In the case of AnMBRs, where a mixing system such as biogas sparging is used to control the membrane fouling, the mass-transfer of $CH_4$ to the biogas is favoured, and values close to the thermodynamic equilibrium are more common in the liquid effluent of these systems.

**Table 7.1** Reported values of D-$CH_4$ concentration in the effluent of mainstream anaerobic reactors during the sewage treatment.

| Type of Reactor | Scale of the Reactor | Measurement Method | D-$CH_4$ (mg/L) | $CH_4$ Supersaturation index[a] (dimensionless) | Percentage of the Produced $CH_4$ Lost in the Effluent | Temp. (°C) | Reference |
|---|---|---|---|---|---|---|---|
| UASB reactor | Pilot | Laboratory | 19.6–22.0 | 1.64–1.66 | 39–41 | 25 | Souza *et al.* (2011) |
| UASB reactor | Demo (14 m³) | Laboratory | 19.2 | 1.37 | 36 | 24 | |
| UASB reactor | Full | Laboratory | 20.0 | NR | NR | 25 | |
| EGSB reactor[b] | Pilot | Laboratory | 25.4 | 1.57 | 45 | 16 | Cookney *et al.* (2012) |
| SAF-MBR[b] | Bench | Laboratory | 17.9 | 1.44 | NR | 25 | Yoo *et al.* (2012) |
| AnMBR | Pilot | Laboratory | 10.9 | 1.01 | 46 | 33 | Giménez *et al.* (2012) |
| AnMBR | Pilot | Laboratory | 8.5 | 1.01 | 43 | 21 | |
| AnMBR | Bench | Laboratory | NR | 1,5 | 30–40 | 15 | Smith *et al.* (2013) |
| UASB reactor | Pilot | Laboratory | 18.4 | 1.13 | 46–54 | 21-28 | Matsuura *et al.* (2015) |
| ABR | Pilot | Laboratory | 21.0 | 1.33 | 39 | 12–23 | Hahn and Figueroa (2015) |
| UASB reactor[b] | Bench | Laboratory | 25.4 | 1.3 | 88 | 18 | Cookney *et al.* (2016) |
| AnMBR[b] | Bench | Laboratory | 8.8 | 1.0 | 45 | 14.2 | |
| UASB reactor | Demo (14 m³) | Laboratory | 15.0 | NR | NR | 22 | Glória *et al.* (2016) |
| UASB reactor | Full | Laboratory | 12.2 | 1.14 | NR | 25 | Huete *et al.* (2018) |
| AnMBR | Demo (40 m³) | Laboratory | 12.0 | ≤1.0 | NR | 18 | Sanchis-Perucho *et al.* (2020) |
| UASB reactor | Pilot | Commercial probe | 16.5 | 1.25 | NR | 23.2 | Velasco *et al.* (2022) |
| UASB reactor | Demo (14 m³) | Laboratory | 17.5 | 1.2 | NR | 25 | Centeno Mora and Chernicharo (2022) |

*Temp.:* reported temperature; UASB: upflow anaerobic sludge blanket; EGSB: expanded granular sludge bed; SAF-MBR: staged anaerobic fluidized membrane bioreactor; AnMBR: anaerobic membrane bioreactor; ABR: anaerobic baffled reactor; NR: not reported.

[a]*The supersaturation index can be defined as the ratio of the actual D-$CH_4$ to the saturated value from the Henry's law constant for clean water.*

[b]*Pre-settled sewage.*

*Source:* Adapted from Crone *et al.* (2016), Centeno-Mora *et al.* (2020) and Stazi and Tomei (2021).

Table 7.1 indicates that similar D-$CH_4$ conditions were obtained for anaerobic reactors regardless the system scale, and the experimental procedure mostly used was the laboratory analysis with the headspace method. Only in one case a commercial probe was used (Velasco *et al.*, 2022), and it was verified that this device had the capacity to measure supersaturated concentrations.

Another parameter that affects the D-$CH_4$ and the supersaturation index is temperature (Crone *et al.*, 2016; Li *et al.*, 2021). At lower temperatures the Henry's constant is higher, which implies that a higher concentration of D-$CH_4$ should be expected in anaerobic effluents. In addition, the biological activity of anaerobic reactors and $CH_4$ production will be slowed at lower temperatures. Both conditions contribute to increase the loss of $CH_4$ in anaerobic effluents in winter at subtropical and temperate climates.

It should be noted that in all these studies the focus was on D-$CH_4$ concentration in anaerobic reactor effluents, and that the behaviour of this parameter inside and along the reactor depth has been scarcely reported. Nelting *et al.* (2015) measured higher D-$CH_4$ concentrations at higher depths in UASB reactors, and also in enclosed units (covered settlers) compared to open units (uncovered settlers). This suggests that possible interventions in anaerobic reactor configurations could lead to lower concentrations of D-$CH_4$ in effluents, at least for passive systems such as UASB reactors.

### 7.3.3 Effects of dissolved methane in anaerobic effluents

As introduced in Section 7.1, D-$CH_4$ in effluents of anaerobic reactors has at least three consequences: (1) loss of energetic potential of anaerobic-based STPs; (2) emission of GHGs and (3) risk of formation of a flammable atmosphere.

Regarding the energetic potential loss, this value will be proportional to the fraction of $CH_4$ dissolved in the liquid phase. Therefore, for anaerobic reactors treating sewage, it would represent 30–40% of the total energetic potential of the produced $CH_4$. According to Lobato *et al.* (2012), for UASB reactors treating sewage, the current energetic potential of these systems varies depending on many variables, such as per capita COD contribution, sulphate concentration in the raw sewage and quality of the three-phase separator and the biogas conduction pipeline (e.g. leaks). For a 'typical' scenario of operation, the authors estimated an energetic potential (due to the biogas use) of 1.2–3.7 $MJ/m^3$ of treated sewage, with a mean value of 1.5 $MJ/m^3$ of treated sewage. The loss of energy would be of 0.6–1.0 $MJ/m^3$ of treated sewage. Different desorption techniques have been explored in the literature to recover this $CH_4$ for its further energetic use (Centeno-Mora *et al.*, 2020), as detailed in Section 7.4.2.

Concerning the emission of GHGs, if D-$CH_4$ is not abated or recovered, the possibility of its release or emission due to the turbulence in downstream structures is significant (Centeno-Mora *et al.*, 2020). This emission will occur even in quiescent surfaces exposed to the atmosphere. Souza *et al.* (2012) measured an emission rate of 11.0–17.8 g $CH_4/m^2$/day in the settler compartment surface of a demo-scale (14 $m^3$) UASB reactor treating sewage, which is equivalent to a carbon footprint of 308–500 g $CO_{2eq}/m^2$/day, considering a global warming potential of $CH_4$ of 28 g $CO_2$/g $CH_4$ (IPCC, 2021). For UASB reactors treating sewage, the carbon footprint associated with D-$CH_4$ lost in effluents and released to the atmosphere, without any control measure, would vary between 386 and 520 g $CO_{2eq}/m^3$ of treated sewage (Centeno Mora *et al.*, 2023a, 2023b). In this regard, $CH_4$ emissions can be considered as the largest direct GHG emission in anaerobic-based STPs (scope 1 emissions). Some studies (Centeno Mora *et al.*, 2023a, 2023b; Chernicharo *et al.*, 2017; Wu *et al.*, 2022) showed that, when adequate measures to mitigate D-$CH_4$ are applied, this direct emission can be reduced, and the biogas can replace fossil fuels to offset the carbon footprint of the system (reduction of scope 2 emissions). If D-$CH_4$ is recovered for its energetic use, the carbon footprint of anaerobic-based STPs could be decreased up to 94% (acting as a carbon sink) if this additional energy is used for electricity generation, and up to 100% if it is used in substitution to liquified petroleum gas (Centeno Mora *et al.*, 2023a, 2023b). It should be noted that these results will depend upon site-specific conditions (e.g. GHG emission of electrical grids, among other conditions), although this general trend should be maintained. Chernicharo *et al.* (2017) recommended that the carbon footprint should be considered the main decision factor to evaluate alternatives for the management of gaseous emissions (e.g. biogas and D-$CH_4$). Regarding other environmental impacts of D-$CH_4$, after performing a life-cycle assessment, Pretel *et al.* (2013) determined that, for AnMBR systems treating sewage discharging effluents to surface waters, the environmental effect on human toxicity and freshwater aquatic ecotoxicity would be significant unless measures to mitigate or recover this D-$CH_4$ are applied.

In the case of the safety risk in confined spaces (e.g. sewers or enclosed tanks), due to the mixture of $CH_4$ with atmospheric $O_2$, Cookney *et al.* (2016) indicated the necessity of a D-$CH_4$ concentration in anaerobic effluents of only 1.4 mg/L (0.14 mg/L if a safety factor of 10 is considered) to obtain a $CH_4$ content in the confined space (assuming the thermodynamic equilibrium between the liquid and gas phases, 1 atm and 15°C) of 5.0% v/v, corresponding to the LFL of $CH_4$ in air (Crowl, 2003). In this regard, if a D-$CH_4$ of 15–25 mg/L in anaerobic effluents is considered, a removal efficiency of D-$CH_4$ up to 91–94% in a treatment unit should be required before its discharge to avoid the flammable atmosphere formation risk.

## 7.4 STRATEGIES TO MITIGATE DISSOLVED METHANE IN ANAEROBIC EFFLUENTS

### 7.4.1 Preliminary

Figure 7.4 shows a scheme with the main strategies reported in the literature for the mitigation, recovery or beneficial use of D-$CH_4$ present in effluents of anaerobic reactors treating sewage, with their main associated techniques or processes. As shown in Figure 7.4, these techniques/processes can be organized into four D-$CH_4$ management strategies:

(A) *Desorption/removal of $CH_4$ for its further oxidation*: $CH_4$ is transferred to a diluted waste gas, to reduce the D-$CH_4$ concentration in anaerobic effluents. The main techniques are simplified desorption chamber, packed desorption chamber, vacuum-assisted chamber, membrane contactors and diffuse aeration. The waste gas should be adequately treated (e.g. thermally or biologically oxidized) to avoid its direct discharge to the atmosphere.

(B) *Desorption/recovery of $CH_4$ for its further energetic use*: $CH_4$ is recovered into a $CH_4$-rich gas stream, which could be used for energetic purposes, alone or blended with biogas. The techniques that have been reported in the literature for this purpose are vacuum-assisted chambers, membrane contactors and enclosed DHS reactors.

(C) *Direct biological oxidation of $CH_4$ in the post-treatment unit*: D-$CH_4$ is biologically oxidized/destroyed in an aerobic reactor. In this case, the D-$CH_4$ oxidation consumes $O_2$ in the aerobic reactor. The main available processes are activated sludge (AS) and its variants (providing adequate conditions to avoid its desorption with the reactor aeration), and enclosed DHS reactors.

(D) *Direct beneficial use of $CH_4$ in the post-treatment unit*: D-$CH_4$ oxidation is purposely used as a source of carbon or electrons for other biological processes such as heterotrophic denitrification, microbial fuel cells (MFCs) or methanol production.

It should be noted that the strategy to be implemented in a full-scale anaerobic-based STP will depend on the specific context and the state of development of each process/technique. Many of these processes are still on a development stage and scaling-up experiments are required. However, at least strategies A or C should be considered in full-scale anaerobic-based STPs to reduce their carbon footprint.

Each one of these techniques is addressed in the following sections, with a discussion on the reported efficiencies and experiences for their application with D-$CH_4$ from mainstream anaerobic reactors treating sewage.

## 7.4.2 Desorption-based techniques

### *7.4.2.1 Overview of techniques*

Different desorption-based techniques are available to transfer D-$CH_4$ to a gas. The general principle of these techniques is explained in Section 7.2.2. For each of them, a liquid and a gas flow will be established. The D-$CH_4$ desorption efficiency is defined below:

$$E = \frac{C_{L,I} - C_{L,O}}{C_{L,I}} 100\% \tag{7.2}$$

where $E$ is the desorption efficiency (%) and $C_{L,I}, C_{L,O}$ are the D-$CH_4$ concentrations at the desorption unit inlet and outlet, respectively.

The desorption efficiency of each technique will depend on many factors (Cussler, 2007): contact time between the liquid and gas phases, driving force intensity (difference between the actual and equilibrium concentrations in the liquid phase), fluid behaviour in each phase which influences the overall mass-transfer resistance, presence of a third phase between the liquid and gas phases (e.g. membrane, in membrane contactors), operating parameters such as the liquid flow ($Q_L$), gas flow ($G_L$) and vacuum pressure if vacuum is applied. In the case of $Q_L$ and $Q_G$, the recommended gas-to-liquid ($Q_G/Q_L$) ratio will depend on the specific compound to be desorbed, as shown below (Heile *et al.*, 2017):

$$SF = \frac{Q_G}{Q_L} H \tag{7.3}$$

where SF is the stripping factor; $Q_G/Q_L$ is the gas-to-liquid ratio and $H$ is the dimensionless Henry's law constant (gas/water). The recommended minimum SF to maximize the desorption efficiency and avoid an excessive energy consumption is 3.5 (Crittenden *et al.*, 2012). In the case of $CH_4$ ($H = 30.38$ at 1 atm and 25°C), the resulting $Q_G/Q_L$ ratio would be 0.115. In addition, the theoretical minimum $Q_G/Q_L$ ratio to desorb 100% of D-$CH_4$ from the liquid effluent (at 25°C) is 0.033 (Heile *et al.*, 2017), determined as the ratio that would produce the equilibrium between the liquid and gas phases.

In practice, for the desorption of D-$CH_4$ from anaerobic effluents, two options (Section 7.4.1) have been tested, depending on the operating conditions for the desorption units:

- *Transference of $CH_4$ to a diluted waste gas which must be treated (oxidized) downflow (strategy A in Figure 7.4):* in this case the desorption units have been tested at elevated $Q_G/Q_L$ ratios (above 0.15, up to 2–4) to increase the driving force for desorption (due to the negligible content of $CH_4$ in waste gas).
- *Transference of $CH_4$ to a rich gas which can be energetically used, alone or mixed with biogas (strategy B in Figure 7.4)*: in this case the desorption unit can be operated at very low $Q_G/Q_L$ ratios (between 0.003 and 0.1) which avoids the excessive dilution of the recovered gas with sweeping gas ($N_2$ or atmospheric air), or vacuum can be used instead or as a complement of sweeping gas.

In both cases the waste/recovered gas should be treated or upgraded downstream. Section 7.4.3 presents the waste/recovered gas characteristics.

The main techniques reported in the literature for D-$CH_4$ desorption are shown in Figure 7.5. Other desorption techniques, such as diffuse aeration or tray aerators are not presented in this discussion due to the lack of studies with anaerobic effluents (Heile *et al.*, 2017).

Table 7.2 presents some experimental results of each of these techniques reported in the literature.

The state of development varies with each desorption technique, from a laboratory-scale system to commercial prototypes and full-scale applications. In addition, the use of vacuum instead of (or as a complement to) sweeping gas avoids the excessive dilution of the recovered $CH_4$ and allows an elevated desorption efficiency at the same time. Each of these techniques is discussed in the following sections.

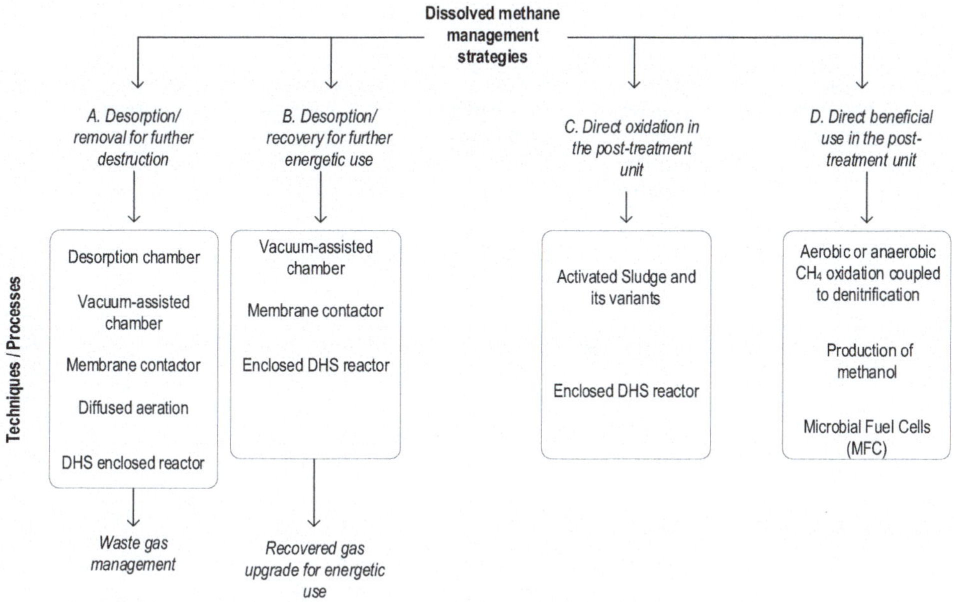


**Figure 7.4** Available strategies to deal with dissolved methane in anaerobic reactors treating sewage.

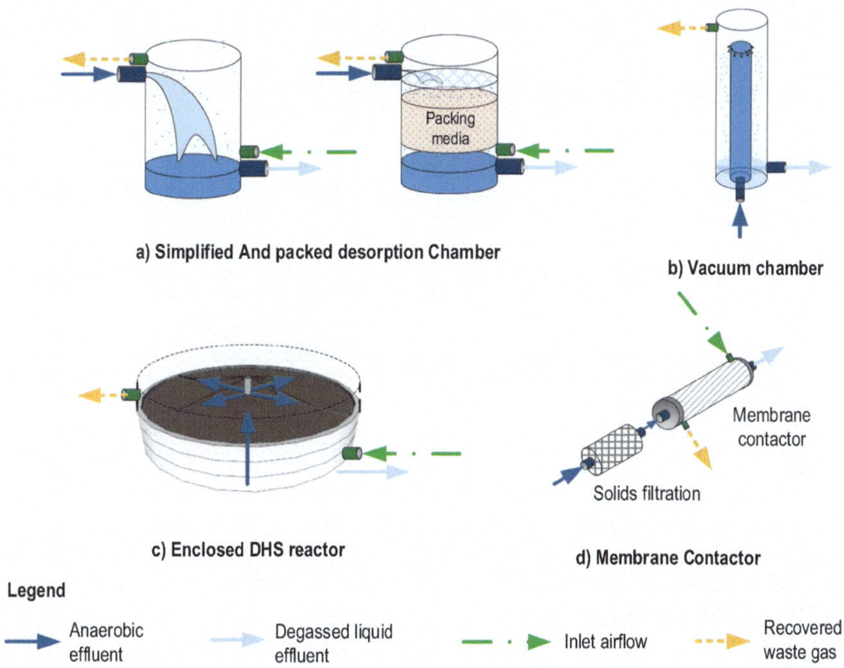


**Figure 7.5** Desorption techniques to remove/recover the dissolved methane and transfer it into a waste gas. *Source*: Adapted from Centeno-Mora *et al.* (2020).

**Table 7.2** Experimental studies with desorption-based techniques to remove/recover the dissolved methane from anaerobic effluents.

| Technique | D-$CH_4$ Management Strategy | Operating Mode ($Q_G/Q_L$ Ratio or Vacuum Pressure, mbar) | D-$CH_4$ Desorption Efficiency (%) | Observations | Technology Application | Reference |
|---|---|---|---|---|---|---|
| Simplified desorption chambers | A | $Q_G/Q_L$ between 2 and 4 | 60–65 | Very simple system | Pilot | Glória *et al.* (2016)<br>Santo (2017) |
| Packed desorption chambers | A and B | 2–4 (strategy A)<br>0.03–0.05 (strategy B) | 75–85<br>50–70 | Different types of packing media: pall rings, electrical tubes | Pilot | Huete *et al.* (2018)<br>Machado (2018)<br>Marinho (2019) |
| Enclosed DHS reactors in two stages | A and C | First unit (strategy B): 0.048–0.071<br>Second unit (strategy C): 0.48 | 75<br>100 | Two units in series, the first to recover D-$CH_4$ and the second to oxidize it | Pilot | Matsuura *et al.* (2015) |
| Membrane contactors | A and B | >0.15 and vacuum up to 900 mbar (strategy A)<br>0.03–0.10 (strategy B) | Up to 95<br>100 | Very compact unit, requires an effluent conditioning (solid removal) | Pilot | Centeno Mora and Chernicharo (2020) |
| Vacuum-assisted chambers | A and B | Vacuum | Up to 80 in one stage | Different configurations operated with vacuum are possible. | Commercial prototype DiMeR | Nelting *et al.* (2021) |
| | | | Up to 94% in three stages | In-series units, all vacuum-assisted at 800 mbar | Laboratory | Lee *et al.* (2020) |

Management strategies: A – desorption to a diluted gas, which must be oxidized; B – recovery of a $CH_4$-rich gas, which can be energetically used; C: oxidation in the biological post-treatment unit.
*Source:* Adapted from Heile *et al.* (2017); Centeno Mora and Chernicharo (2020) and Centeno-Mora *et al.* (2020).

It should be noted that many of the techniques described in Table 7.2 have been also studied for the removal of $H_2S$ from anaerobic reactor effluents. $H_2S$ is another gas of concern in anaerobic effluents due to its toxicity, corrosivity (concrete and steel) and odorous characteristics. More information on this issue and on the performance of these techniques for the removal of $H_2S$ can be found elsewhere (Brandt *et al.*, 2019; Centeno-Mora *et al.*, 2020).

### 7.4.2.2 Simplified and packed desorption chambers

Simplified and packed desorption chambers are easy-to-install units for the desorption of dissolved methane and removal of other gases (e.g. $H_2S$). Figure 7.6 shows a diagram and photographs of these

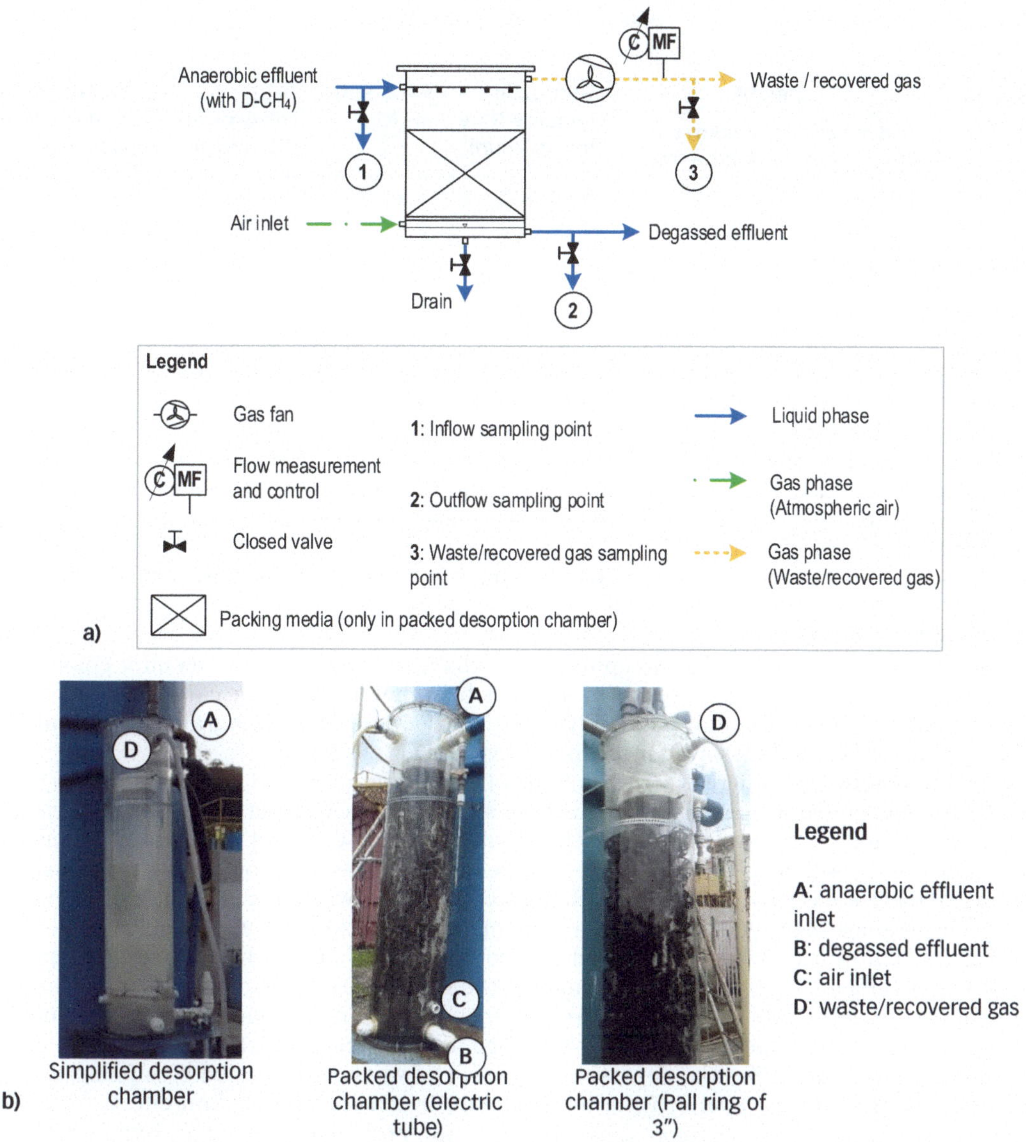


**Figure 7.6** Simplified and packed desorption chamber: (a) process flow diagram and (b) photograph of pilot-scale systems. *Source*: Adapted from Souza *et al.* (2021).

units. This technology has been successfully applied in Brazil in both pilot- and full-scale treatment systems. They have the advantage of being able to receive the anaerobic effluent directly from UASB reactors (with high solid concentration) without the need of a pre-treatment.

In the desorption chambers, there are two counter-flowing streams: the anaerobic effluent with dissolved methane in the downward direction and the sweeping air in the upward direction. Air can be introduced under pressure (using an air blower) or drawn-in using a gas extractor connected to the residual gas outlet (point D in Figure 7.6b). The effluent outlet will be at the bottom of the system to

**Table 7.3** Design parameters for simplified and packed desorption chambers for the desorption of D-$CH_4$ from anaerobic effluents.

| Desorption Chamber Type | Height | | Hydraulic Loading Rate ($m^3/m^2/min$) | Flow Ratio ($Q_G/Q_L$)[b] | D-$CH_4$ Desorption Efficiency | Concentration of $CH_4$ in the Waste Gas (% v/v) |
|---|---|---|---|---|---|---|
| | Water Fall[a] (m) | Packing Media (m) | | | | |
| Simplified | 1.2–1.5 | – | 0.15–0.30 | 2–4 | 60–65 | 0.4–0.8 |
| Packed | 1.2–1.5 | 0.8–1.0 | 0.15–0.30 | 2–4 | 75–85 | 0.4–0.8 |

[a]Useful height of hydraulic fall.
[b]Air/liquid flow ratio ($Q_G/Q_L$).
*Source:* Adapted from Souza *et al.* (2021).

continue its path to the post-treatment unit. The waste gas will be transported to an additional unit for its biological or thermal oxidation (see Section 7.4.3).

The difference between the simplified and packed desorption chambers is the existence of an inert packing material in the latter case. The use of this material increases the contact area between the anaerobic effluent and the sweeping gas, increasing the desorption efficiency. Various packing materials have been tested in pilot tests (Figure 7.6b), such as electrical conduit and 3″ pall rings (Marinho, 2019).

Table 7.3 shows the main design parameters for simplified and packed desorption chambers.

#### 7.4.2.3 Vacuum chambers

When vacuum is used instead of a sweeping gas, a chamber operates as a vacuum chamber. The interest of this technique is the recovery of a gas rich in methane, which can be used for energy purposes. Figure 7.7 shows different possibilities for this process and a photograph of a commercially available system known as a dissolved methane recovery (DiMeR) system.

As shown in Figure 7.7a, different configurations of desorption units can be operated with vacuum. Lee *et al.* (2020) operated a free-fall unit with three stages, with a vacuum of 800 mbar in each unit. With this configuration, they could remove up to 94 and 88% of D-$CH_4$ and $S_2^-$ from the effluent of an AnMBR, respectively. Although they did not report the recovered gas $CH_4$ content, with a mass balance they estimated that the electricity generated with the recovered $CH_4$ would compensate and overcome the electricity consumption of the equipment (vacuum pumping), producing an energy surplus in the system.

Another free-fall unit that has been tested for the recovery of D-$CH_4$ from anaerobic reactors is a DiMeR system (Figure 7.7b). In this technology, the effluent from anaerobic reactors flows to a DiMeR reactor through a thin layer with a large contact surface and produces turbulence, enabling a degassing process. At the same time, vacuum is applied to the reactor, which reduces the gas pressure in the headspace and intensifies the degassing process. The vacuum generated by a pump drives the sewage to be treated into the reactor from the upstream treatment unit (siphon effect). The effluent from the degassing reactor flows freely downstream to the post-treatment stage. Experimental data of a DiMeR system (Nelting *et al.*, 2021) reported a D-$CH_4$ desorption efficiency up to 80%, with a $CH_4$ content up to 50% v/v. In addition, results from pilot-scale operation of this system have shown a consumption of ~20–40% of the electric power potential of the recovered biogas (due to the vacuum pump). Therefore, the system is energy self-sustainable and enables an effective energy gain in the overall balance of anaerobic-based STPs.

#### 7.4.2.4 Enclosed DHS reactors

DHS reactors can be used as a sustainable post-treatment for the effluent of UASB reactors to remove additional organic matter and solids, as sponge-filled trickling filters (Bressani-Ribeiro *et al.*, 2018). However, if the units are operated with a forced ventilation and are enclosed, they will also work as desorption units. Figure 7.8 shows a diagram of this system.

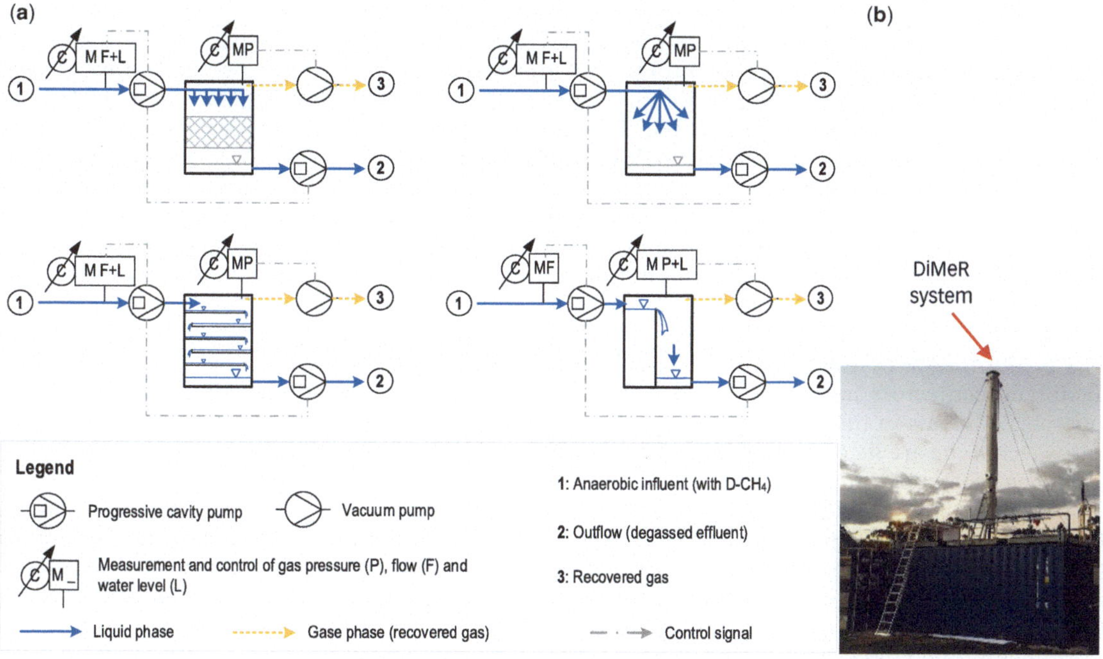


**Figure 7.7** Vacuum-assisted chamber: (a) process flow diagram and (b) photograph of a commercially available system (DiMeR). *Source*: Adapted from Nelting *et al.* (2021).

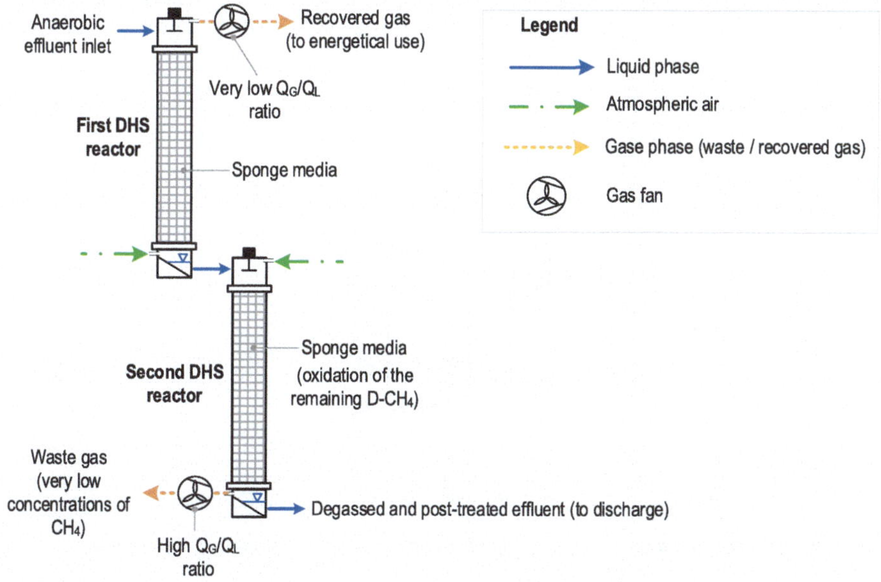


**Figure 7.8** Diagram of enclosed DHS reactors for the desorption of D-$CH_4$ from anaerobic effluents. *Source*: Adapted from Matsuura *et al.* (2015).

Matsuura *et al.* (2015) operated two in-series enclosed DHS reactors (80 L, with a percolation time of 2 h, each unit) as a post-treatment unit for the effluent of an UASB reactor treating sewage, and to mitigate the D-$CH_4$ in anaerobic effluent. The first unit was operated at a very low $Q_G/Q_L$ ratio (0.048–0.071) to recover a $CH_4$-rich gas at the top, whereas the second unit was operated at a $Q_G/Q_L$ ratio above 0.48 to oxidize the remaining D-$CH_4$. The authors reported a D-$CH_4$ removal efficiency between 57–88 and 99% in the first and second enclosed DHS reactors, respectively. The recovered gas in the first unit had a $CH_4$ content up to 39% v/v, which is energetically usable. In addition, they reported the presence of $CH_4$-oxidizing bacteria in the second DHS unit (Matsuura *et al.*, 2017), which proves that oxidation and desorption are the removal mechanisms of D-$CH_4$ in these reactors (see Section 7.4.4).

These studies show that the desorption unit and aerobic post-treatment can be integrated in the same system.

#### 7.4.2.5 Membrane contactors

Membrane contactors have been extensively studied for the desorption and recovery of $CH_4$ from anaerobic effluents in the last few years (Centeno Mora & Chernicharo, 2020). The main advantages of this technique are its compacity and elevated efficiency, whereas its disadvantages include its high investment cost, operating complexity and long-term stability of the system. In addition, in the case of anaerobic effluents with a high concentration of solids (e.g. UASB reactors, expanded granular sludge bed (EGSB) reactors), a conditioning process (solid removal) is necessary to prevent the membrane contactor clogging. This pre-treatment can be omitted for the effluent of AnMBRs, in which a microfiltration or ultrafiltration membrane is used to separate the liquid effluent from the biomass and other solids in biological processes.

Although different configurations of membrane contactors have been tested in the literature, the commercially available units are hollow-fibre membrane contactors (HFMCs). The fibre material can be microporous (pore size close to 0.5 μm) or dense (without pores). The general operating principle of HFMCs is shown in Figure 7.9.

As presented in Figure 7.9, the same principle for the $CH_4$ desorption presented in Section 7.2.2 is applied to HFMCs. The $CH_4$ dissolved in the liquid phase will flow through one side of the HFMC (the shell side is more common for the liquid phase in large-scale systems) and the gas phase will flow

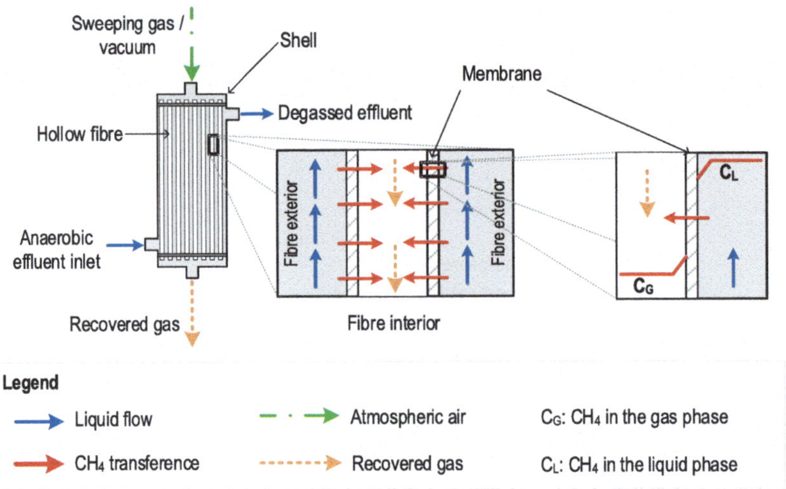


**Figure 7.9** Operating principle of an HFMC for the D-$CH_4$ desorption from anaerobic effluents. *Source*: Adapted from Centeno Mora *et al.* (2021).

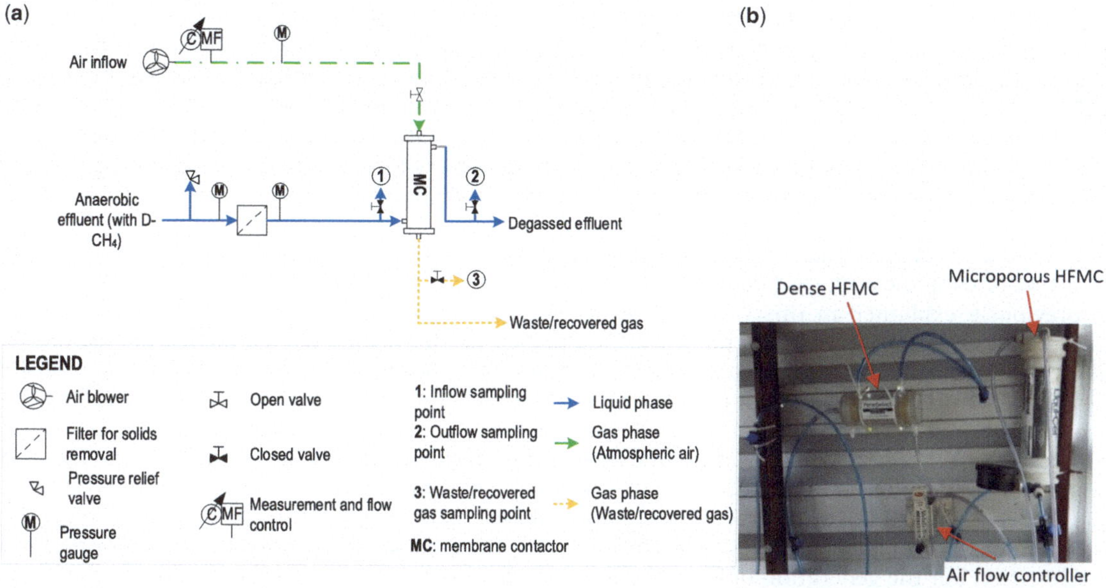


**Figure 7.10** Membrane contactor (MC): (a) process flow diagram and (b) photograph of a pilot-scale system. *Source*: Adapted from Centeno Mora *et al.* (2021).

through the other one (lumen side). In this system, the liquid and gas phases are not directly in contact, and an additional layer (the membrane) separates both the phases. Thus, an additional resistance to mass transfer (the membrane) will limit the $CH_4$ desorption. Still, for the D-$CH_4$ desorption application, the desorption system tends to be liquid-phase controlled, and the mass-transfer resistance from the membrane phase tends to represent <5–10% of the overall resistance in the case of dense membrane, and it is negligible for microporous membranes (Centeno Mora & Chernicharo, 2020).

Figure 7.10 shows a diagram and a photograph of a pilot-scale system for an HFMC used to recover D-$CH_4$ from the effluent of an UASB reactor in an experimental system in Brazil. A filtration with dynamic membrane was used in this case, and a microporous and a dense HFMC were evaluated with sweeping gas (atmospheric air) and with vacuum (at 900 mbar).

Efficiencies of D-$CH_4$ desorption up to 90–99% have been reported in a single HFMC unit (Cookney *et al.*, 2016; Henares *et al.*, 2016) from the effluent of UASB reactors and from AnMBRs with microporous and dense membranes. In addition, HFMCs have been operated with a sweeping gas (nitrogen and atmospheric air), vacuum and a combination of both (Centeno Mora & Chernicharo, 2020). In an experiment with the effluent of an UASB reactor and a dense membrane HFMC, Centeno Mora *et al.* (2023a, 2023b) showed that, to obtain a $CH_4$-rich recovered gas, $Q_G/Q_L$ ratios below 0.10 were required to avoid an excessive dilution of $CH_4$; and to obtain a $CH_4$ content in the recovered gas above 40%, vacuum (900 mbar) or $Q_G/Q_L$ ratios below 0.03 were required. The energetic self-sufficiency of this technique during the recovery of D-$CH_4$ has been reported in many studies (Centeno Mora *et al.*, 2023a, 2023b; Cookney *et al.*, 2012; Rongwong *et al.*, 2018; Sanchis-Perucho *et al.*, 2020).

One of the main concerns with membrane contactors is their long-term stability. The interaction of different substances in the anaerobic effluent (e.g. organic or inorganic solids) with the membrane can affect the membrane surface and the pores hydrophobicity (in the case of microporous membrane contactors) in the long term (Centeno Mora & Chernicharo, 2020). Various studies have reported the loss of efficiency of microporous membrane contactors due to membrane wetting (Centeno Mora *et al.*, 2023a, 2023b; Henares *et al.*, 2018; Sethunga *et al.*, 2019), which is the penetration of the liquid phase into the pores (which should be kept filled with gas). This problem does not occur with dense membranes, although many studies do report a loss of efficiency in the system due to biological, organic and inorganic fouling of the membrane (Centeno Mora *et al.*, 2023a, 2023b; Henares *et al.*, 2017; Velasco *et al.*, 2022). In that sense, appropriate measures for fouling control, such as proper conditioning (filtration of the effluent in the case of UASB or EGSB reactors), as well as the application of physical and chemical fouling control measures, are necessary to maintain the membrane desorption efficiency in the long term (Centeno Mora & Chernicharo, 2020).

### 7.4.3 Waste/recovered gas treatment and valorization

#### *7.4.3.1 Waste/recovered gas characteristics*

As previously explained in this chapter, the waste gas and recovered gas are the outflows of the desorption unit, after the D-$CH_4$ (and other dissolved gases) desorption from anaerobic effluents. The difference between them is its content of $CH_4$. In the case of waste gas, due to its low content of $CH_4$ (e.g. the system is operated at high $Q_G/Q_L$ ratios), this gas cannot be energetically used, and it should be oxidized (or used in a way other than energy generation) before its discharge into the atmosphere to reduce the carbon footprint of the anaerobic-based STPs. Sections 7.4.3.2 and 7.4.3.3 describe the biological and thermal oxidation of waste gas, respectively. On the contrary, the recovered gas would have a content of $CH_4$ high enough to allow its energetic use, alone or mixed with biogas. The recovered gas or its blend with biogas (recovered gas + biogas) can be used for the generation of heat or electricity. For the generation of heat, the lower content of $CH_4$ in the gas is of ~20% v/v, whereas a content of $CH_4$ of 30–40% v/v can be considered for the electricity generation in micro-turbines or combined heat and power units (Centeno-Mora *et al.*, 2020). Table 7.4 shows the main characteristics of the waste/recovered gas obtained during the desorption of D-$CH_4$ from anaerobic effluents reported in the literature.

As shown in Table 7.4, reported data show that the recovered gas has a $CH_4$ content as high as 50% v/v, which can be used for heat or/and electricity generation. In the case of the sweeping gas operation mode, very low $Q_G/Q_L$ ratios below 0.10 are necessary, which will limit the D-$CH_4$ desorption efficiency. In this sense, a choice should be made in a single unit operation when sweeping gas is used: the operation of the system at a high $Q_G/Q_L$ ratio to allow a maximum D-$CH_4$ desorption efficiency with the obtention of a diluted waste gas; or the recovery of a $CH_4$-rich gas which would limit the D-$CH_4$ desorption efficiency. When vacuum is used instead of sweeping gas, D-$CH_4$ desorption efficiencies up to 95% with a $CH_4$-rich recovered gas can be achieved, particularly with HFMCs.

Regarding the recovered gas flow, Centeno Mora *et al.* (2023a, 2023b) reported, for HFMCs, an increase in volume of 1–1.8 times the inlet flow for $Q_G/Q_L$ ratios between 0.02 and 0.15; and a flow of recovered gas of ~4% of $Q_L$ when HFMCs were operated at 900 mbar. Sanchis-Perucho *et al.* (2020) and Bandara *et al.* (2011) obtained a flow of recovered gas between 1 and 12% of $Q_L$ for HFMCs operated at vacuums between 200 and 800 mbar.

When the composition of the recovered gas is considered, Table 7.4 shows that this gas will be composed by sweeping gases ($N_2$ and $O_2$ if atmospheric air is used), and gases desorbed from the anaerobic effluent ($CH_4$, $H_2S$, $CO_2$ and $N_2$). Thus, the recovered gas will have a composition similar to that of biogas (with a lower content of $CH_4$). In this sense, the recovered gas and biogas mixing should be expected to obtain a gas mixture which would require an upgrade ($H_2S$ removal, at least) before its energetic use.

#### *7.4.3.2 Energetic use and thermal oxidation of the recovered/waste gas*

The combination of recovered gas and biogas should extend the $Q_G/Q_L$ operating ratio in the desorption unit if the gas mixture is used for energetic purposes (heat or electricity).

Figure 7.11 shows the effect of the D-$CH_4$ desorption efficiency and the $Q_G/Q_L$ ratio on the $CH_4$ content in the recovered gas and gas mixture (recovered gas + biogas). For this case, a D-$CH_4$ concentration in the anaerobic effluent of 20 mg/L, and the 'typical' conditions specified in Lobato

**Table 7.4** Characteristics of the waste/recovered gas obtained from the desorption of D-$CH_4$ from anaerobic effluents.

| Desorption Technique | Gas phase Renewal Method | $Q_G/Q_L$ Ratio or Vacuum Pressure | $CH_4$ Content in the Recovered Waste Gas (% v/v) | Other Gas Characteristics | Reference |
|---|---|---|---|---|---|
| HFMC | Vacuum | 500 mbar | 63 | $CO_2$: 10% v/v | Bandara *et al.* (2011) |
| HFMC | Sweeping gas (pure $N_2$) | 213 | 0.028 | – | Cookney *et al.* (2012) |
| HFMC | Vacuum | 500 mbar | 23 | – | Henares *et al.* (2018) |
| HFMC | Vacuum | 200–500 mbar | 5–42 | – | Sanchis-Perucho *et al.* (2020) |
| Enclosed DHS reactor (first unit) | Sweeping gas (atmospheric air) | 0.048–0.071 | 38.4 | $O_2$: 0.1% v/v<br>$CO_2$: 6.3% v/v<br>$N_2$: 55.2% v/v | Matsuura *et al.* (2015) |
| Simplified and packed desorption chambers | Sweeping gas (atmospheric air) | 2–4 | 0.4–0.8 | $H_2S$: 200–550 ppmv | Souza *et al.* (2021) |
| Vacuum chamber | Vacuum | NE | 20–50 | – | Nelting *et al.* (2021) |
| HFMC | Vacuum | 900 mbar | 28–52 | $O_2$: <10% v/v<br>$CO_2$: 14–25% v/v<br>$N_2$: 20–45% v/v<br>$H_2S$ >5000 ppmv | Centeno Mora *et al.* (2023a, 2023b) |
| HFMC | Sweeping gas (atmospheric air) | 0.018–0.10 | 21–51 | $O_2$: 5–13% v/v<br>$CO_2$: <5% v/v<br>$N_2$: 40–60% v/v<br>$H_2S$ >1000–2000 ppmv | Centeno Mora *et al.* (2023a, 2023b) |

HFMC: hollow-fibre membrane contactor; DHS: downflow hanging sponge; NE: not specified.
*Source:* Adapted from Centeno Mora and Chernicharo (2020).

*et al.* (2012) were considered, with a $CH_4$ content in biogas of 75% v/v, and a biogas production of 85.6 NL/m$^3$ of treated sewage.

As shown in Figure 7.11, $Q_G/Q_L$ ratios below 0.1 would be required to obtain a $CH_4$ content of at least 20% v/v to burn the recovered gas alone (to produce heat or for his thermal oxidation in a flare) for desorption efficiencies above 70%. If the recovered gas alone is to be used to produce electricity, for which a $CH_4$ content above 40% v/v is required, $Q_G/Q_L$ ratios below 0.04 and desorption efficiencies above 80% should be applied to the desorption unit. When the gas mixture (recovered gas + biogas) is considered, the operating span is extended, and $Q_G/Q_L$ ratios up to 0.2 and 0.3 allow the generation of a gas mixture usable for electricity and heat generation, respectively, for desorption efficiencies above 40%.

Therefore, to burn and destroy (thermal oxidation) the recovered gas, $Q_G/Q_L$ ratios in the desorption units below 0.1 and 0.3 should be considered for the gas alone and mixed with biogas, respectively.

In addition, the operation at higher $Q_G/Q_L$ ratios will increase the gas mixture volume. Operating at $Q_G/Q_L$ ratios of 0.03, 0.10 and 0.3 would increase the biogas flow by ~30, 130 and 580% (Brandt *et al.*,

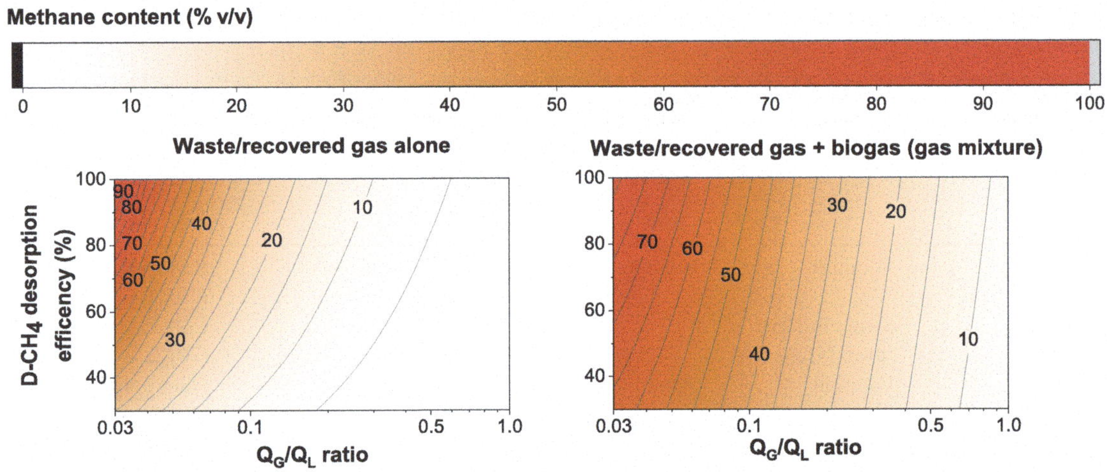


**Figure 7.11** Content of $CH_4$ in the recovered gas and gas mixture (biogas + recovered gas) derived from the recovery of dissolved methane from anaerobic reactors for a 'typical' scenario of biogas characteristics. *Source*: Adapted from Brandt *et al.* (2021a, 2021b, 2021c).

2021a, 2021b, 2021c), respectively, increasing the gas storage costs; and the recovered gas mixture with biogas would increase the energetic potential of anaerobic-based STPs between 30 and 40%. It should be noted that vacuum-assisted units (e.g. vacuum chamber and HFMCs operated with vacuum) would produce a lower gas volume increase, of ~23–120%, with a similar energetic potential increase.

Finally, different studies (Centeno Mora *et al.*, 2023a, 2023b; Cookney *et al.*, 2012; Sethunga *et al.*, 2020) demonstrated that an energetically positive balance (e.g. the electricity generation with the recovered gas compensates and overpass the electricity consumption of the desorption unit) is attainable with the recovered gas energetic use.

### 7.4.3.3 Biological oxidation of waste gas

When a waste gas is too diluted and its $CH_4$ content is too low (as in simplified and packed desorption chambers with high $Q_G/Q_L$ ratios, or in the case of gas extraction systems such as in enclosed pre-treatment units or UASB reactors with enclosed settlers), the thermal oxidation (burn) of this gas is not feasible (Section 7.4.3.2) and another process for the $CH_4$ oxidation is required. One option that has been explored in the literature is the biological oxidation in biomass growth reactors, such as biofilters, biotrickling filters and bioscrubbers (Brandt *et al.*, 2016, 2019, 2021a, 2021b, 2021c). These units can be used to oxidize the $H_2S$ and $CH_4$ in a waste gas. Among them, biofilters have been tested at pilot-scale systems for the oxidation of $H_2S$ and $CH_4$ in two stages.

The necessary conditions for the biofiltration of $H_2S$ and $CH_4$ are quite different, as the mass transfer of gas to the biofilm is usually not a limiting factor for biofiltration processes of more soluble and ionizable gases in water (e.g. $H_2S$), although it can be a key factor in the biofiltration of less soluble gases in water (e.g. $CH_4$). Therefore, much longer gas residence times (and empty bed residence times) are required for the treatment of $CH_4$, resulting in much larger biofilters. If the interest is in the combined treatment of odours ($H_2S$) and $CH_4$ present in a residual gas, the installation of two biofilters in series is recommended, with the first one intended for the treatment of odour compounds and the second for the treatment of $CH_4$. In the biofiltration of gas mixtures containing $H_2S$ and $CH_4$, attention should be paid to possible inhibitory effects on the activity of methanotrophic microorganisms ($CH_4$ oxidizers) due to the acidification of the filter bed caused by the biological oxidation of $H_2S$, which can be minimized by a sequential bioreactor system, where the oxidation of $H_2S$ is carried out in the first

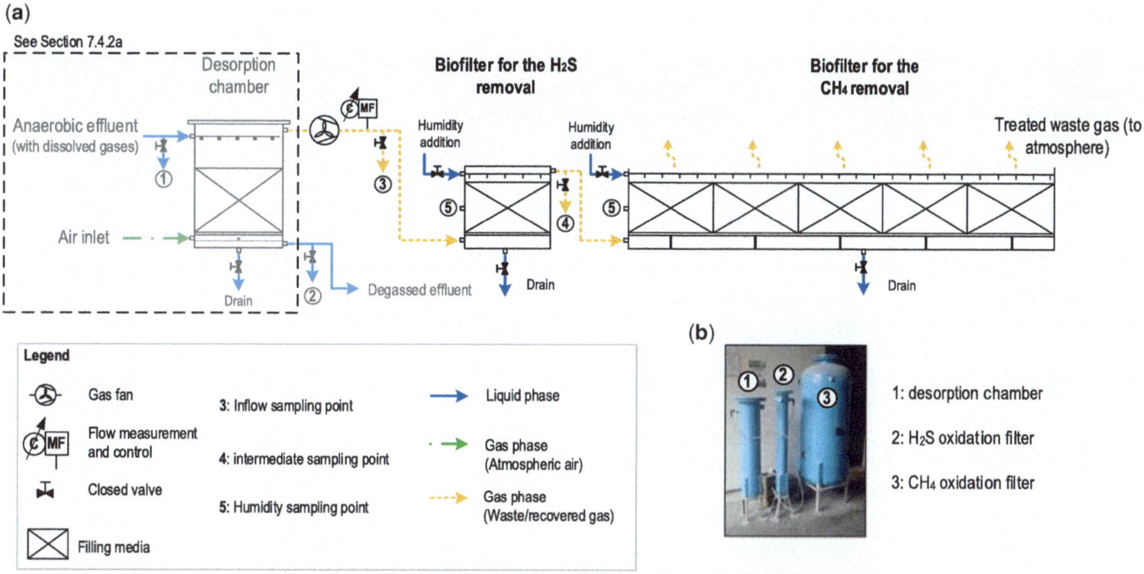


**Figure 7.12** Biofilter for the biological oxidation of $H_2S$ and $CH_4$ in two stages: (a) process flow diagram and (b) photograph of a pilot system. *Source*: Adapted from Brandt *et al.* (2021a, 2021b, 2021c).

reactor (with consequent acidification of the filter bed) and the biological oxidation of $CH_4$ is carried out in the second reactor at a pH close to neutrality (Brandt *et al.*, 2016). Figure 7.12 shows a flow process diagram and a photograph of a pilot-scale biofilter system for the oxidation of $H_2S$ and $CH_4$ in a waste gas in two stages.

Regarding the filling material for biofilters, fibrous/inert/synthetic materials can be used to improve the porosity and structure of the filter bed. Design and operating specifications for biofilters applied to the oxidation of $H_2S$ and $CH_4$ can be found elsewhere (Brandt *et al.*, 2019).

### 7.4.4 Dissolved methane treatment or valorization in biological post-treatment units

#### *7.4.4.1 Preliminary*

Instead of being desorbed and transferred to a gas stream, D-$CH_4$ can be directly biologically oxidized in a biological unit downstream the anaerobic reactor. For this, a biological reactor is required with adequate conditions for the growth of MOBs.

This oxidation can occur under aerobic and anaerobic conditions, following different biological paths. In this section, two approaches are revised, according to the D-$CH_4$ mitigation strategies discussed in Section 7.4.1:

- Strategy C: biological oxidation of D-$CH_4$ under aerobic conditions, with the only purpose of its destruction to avoid its emission to the atmosphere.
- Strategy D: biological oxidation D-$CH_4$ under aerobic or anaerobic conditions, using it in a regulated environment as a substrate for other biological processes (e.g. denitrification, energy generation).

Table 7.5 shows some literature studies of experiments carried out to biologically oxidize to eliminate or beneficially use the D-$CH_4$ present in anaerobic effluents.

As shown in Table 7.5, different biological reactors have been tested for the oxidation and beneficial use of D-$CH_4$. Adequate conditions (e.g. dissolved oxygen concentration, presence of electron acceptors

**Table 7.5** Experimental studies with the biological oxidation of D-$CH_4$ of anaerobic effluents.

| Biological Reactor Configuration | D-$CH_4$ Management Strategy | Biological Condition for the D-$CH_4$ Oxidation | D-$CH_4$ Removal Efficiency (%) | Observations | Technology Application | Reference |
|---|---|---|---|---|---|---|
| AS reactor | C | Aerobic oxidation | Up to 80% | D-$CH_4$ generated in sewers and primary settlers | Full-scale | Daelman *et al.* (2012) |
| Enclosed DHS reactor | C | Aerobic oxidation | Up to 98% | Percolation time of 2.0 h | Pilot | Matsuura *et al.* (2010) |
| IFAS reactor | D | Aerobic oxidation of $CH_4$ coupled to denitrification (AME-D) | Up to 85% | Removal of TN of 33–44%. HRT of 9 h | Pilot | Arias *et al.* (2018) |
| MBR | D | Aerobic oxidation of $CH_4$ coupled to denitrification (AME-D) | Up to 95% | Synthetic sewage. Removal of TN up to 60% by heterotrophic denitrification. $CH_4$ used as an electron donor. HRT of 4.6–5.6 h | Pilot | Silva-Teira *et al.* (2017) |
| Biofilm reactor (polypropylene membrane as support) | D | Anaerobic oxidation of $CH_4$ coupled to denitrification. (AnME-D) | Up to 85% | Synthetic effluent and enriched media. anaerobic ammonia oxidation (anammox) also occurred. The TN removal was up to 99% | Laboratory | Liu *et al.* (2020) |
| Biofilm reactor (sponge-based) | D | Partial oxidation of $CH_4$ by AOB | Up to 85% | Synthetic enriched effluent. Valuable methanol ($CH_3OH$) is produced | Laboratory | Gupta and Goel (2019) |
| MFCs | D | Exoelectrogens microorganisms generate electricity with the D-$CH_4$ aerobic oxidation by-products | Up to 85% | Synthetic effluent. Conversion to electricity of 17%. HRT of 4.0–16 h. | Laboratory | Chen and Smith (2018) |

Management strategy: C – simple oxidation in the biological post-treatment unit; D – oxidation coupled to another biological process in the post-treatment unit; HRT – hydraulic retention time.
*Source:* Adapted from Centeno-Mora *et al.* (2020) and Stazi and Tomei (2021).

or donors) for the maintenance of the specific microorganisms which carries out the desired biological process are necessary in these cases. For the D-$CH_4$ oxidation process without beneficial use (strategy C) and for its use during the heterotrophic denitrification in AME-D (as an electron donor, strategy D), full-scale and pilot experiments have been tested. For the AnME-D process, methanol production and MFCs, only laboratory-scale experiments, with synthetic effluents (not anaerobic effluent) and in many cases enriched reactors have been tested to date, showing that there is a need for more research. These processes are further detailed in the following sections.

#### 7.4.4.2 Direct oxidation of D-$CH_4$ without valorization

Under aerobic conditions, $CH_4$ can be biologically oxidized to $CO_2$ in a sequence of bioreactions (Daelman *et al.*, 2014), as shown below:

$$CH_4 + \frac{1}{2}O_2 \rightarrow CH_3OH \tag{7.4}$$

$$CH_3OH + \frac{1}{2}O_2 \rightarrow HCHO + H_2O \tag{7.5}$$

$$HCHO + O_2 \rightarrow CO_2 + H_2O \tag{7.6}$$

Overall reaction:

$$CH_4 + 2O_2 \rightarrow CO_2 + 2H_2O \tag{7.7}$$

In the first stage (Equation 7.4), D-$CH_4$ is oxidized to methanol ($CH_3OH$) by methane monooxygenase enzymes produced by methanotrophic bacteria (*Methylocystis* and *Methylosinus*). Then, methanol is oxidized to formaldehyde (HCHO) by methanol dehydrogenase enzymes produced by the same methanotrophic bacteria (Equation 7.5). In the third stage (Equation 7.6), formaldehyde is oxidized to carbon dioxide by formaldehyde dehydrogenase enzymes produced by various types of bacteria, including methanotrophs. In each stage, $O_2$ is consumed and $CH_4$ or other by-products are not valorized during this process. These reactions also occur during the oxidation of $CH_4$ in biofilters (Section 7.4.3.3).

Different aerobic biological reactor configurations can be used for the oxidation of D-$CH_4$ from anaerobic effluents. Figure 7.13 shows the main systems reported in the literature. An important risk in these processes is $CH_4$ stripping and transference to a waste gas (Section 7.4.2) instead of being oxidized. Therefore, appropriate operating conditions should be provided to stimulate the $CH_4$ oxidation rather than its desorption.

In the case of the AS systems (Figure 7.13a), when this process is considered as a post-treatment unit for anaerobic effluents, it should be possible to biologically oxidize the D-$CH_4$ in this unit, as long as proper conditions to avoid methane stripping are provided, such as fine bubble aeration and submersed inlet. Daelman *et al.* (2012) reported oxidation efficiencies up to 80% in a full-scale STP. In a later article, Daelman *et al.* (2014) determined the conditions which could optimize the D-$CH_4$ oxidation process in AS systems with a mathematical model. They concluded that an accurate aeration control, compatible with optimal conditions for the effluent quality, should be provided. An aeration too low or too high would stimulate the $CH_4$ stripping to the gas phase. In addition, deeper tanks and CSTRs would produce the best results in terms of $CH_4$ oxidation. These recommendations must be experimentally validated in full-scale systems. Although it has not been reported in the literature to date, similar conditions should be observed in other AS variants or submerged aerobic reactors (e.g. membrane bioreactors (MBRs), moving bed biofilm reactors (MBBRs), submerged aerobic biofilters, among others).

In the case of an enclosed DHS reactor (Figure 7.13b), when it is operated at high $Q_G/Q_L$ ratios (~0.48), D-$CH_4$ can be oxidized in the unit (percolation time of 2 h), as already analysed in Section 7.4.2.3. D-$CH_4$ oxidation efficiency up to 90% (only in this unit) was reported in Matsuura *et al.* (2015).

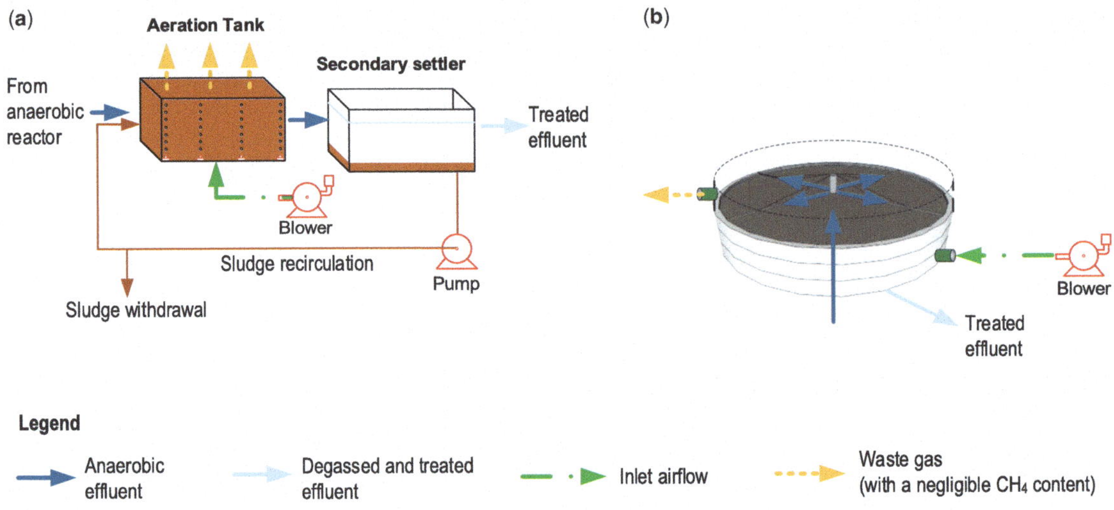


**Figure 7.13** Available biological reactor configurations for the aerobic oxidation of D-$CH_4$ in the liquid phase: (a) AS reactor and (b) enclosed DHS reactor.

#### *7.4.4.3 Valorization of D-$CH_4$ in the biological post-treatment*

D-$CH_4$ can be used by different groups of microorganisms as a source of carbon or energy during its oxidation. Figure 7.14 shows some biological processes that have been experimentally studied and reported in the literature (Gupta & Goel, 2019; Stazi & Tomei, 2021).

Figure 7.14a shows the process of aerobic oxidation of methane coupled with denitrification (AME-D). In this case, a wide group of methanotrophs aerobically oxidizes $CH_4$ with $O_2$, and the intermediate by-products of this oxidation are used as electron donors for the heterotrophic denitrification of $NO_3^-$. The anoxic conditions for this denitrification are possible due to the $O_2$ consumption. The overall reaction is presented below:

$$5CH_4 + 5O_2 + 4NO_3^- + 4H^+ \rightarrow 2N_2 + 12H_2O + 5CO_2 \quad (7.8)$$

Pre-denitrification MBRs (Silva-Teira *et al.*, 2017) and an integrated fixed film activated sludge (IFAS) reactors (Arias *et al.*, 2018) have been tested in pilot-scale experiments for the application of AME-D. D-$CH_4$ removal efficiencies of 95 and 85%, respectively, with total nitrogen (TN) removal efficiencies of 60 and 44% were reported.

It should be noted that this process has an important consumption of $O_2$ (for the complete nitrification of ammonia, and for $CH_4$ oxidation), and that the aeration process could produce the $CH_4$ stripping from the liquid phase (Stazi & Tomei, 2021). For this reason, a good strategy to reduce methane desorption to the environment is to carry out the process under anoxic conditions, either in a continuous bioreactor with a pre-anoxic compartment or a sequencing batch reactor (SBR) in which anaerobically treated sewage is introduced in a non-aerated stage. Moreover, there is the possibility that under hypoxic conditions, anaerobic $CH_4$ oxidation processes could develop, even without the presence of any known denitrifying anaerobic methane oxidizing (DAMO) microorganism, as has been referred to previously (Silva-Teira *et al.*, 2017; Xie *et al.*, 2023).

Figure 7.14b presents the process of anaerobic oxidation of methane coupled with denitrification (AnME-D). According to Sánchez *et al.* (2016), two pathways are possible for AnME-D. In the first one, anaerobic methanotrophic archaea an anammox bacteria oxidize $CH_4$ using nitrate as an electron

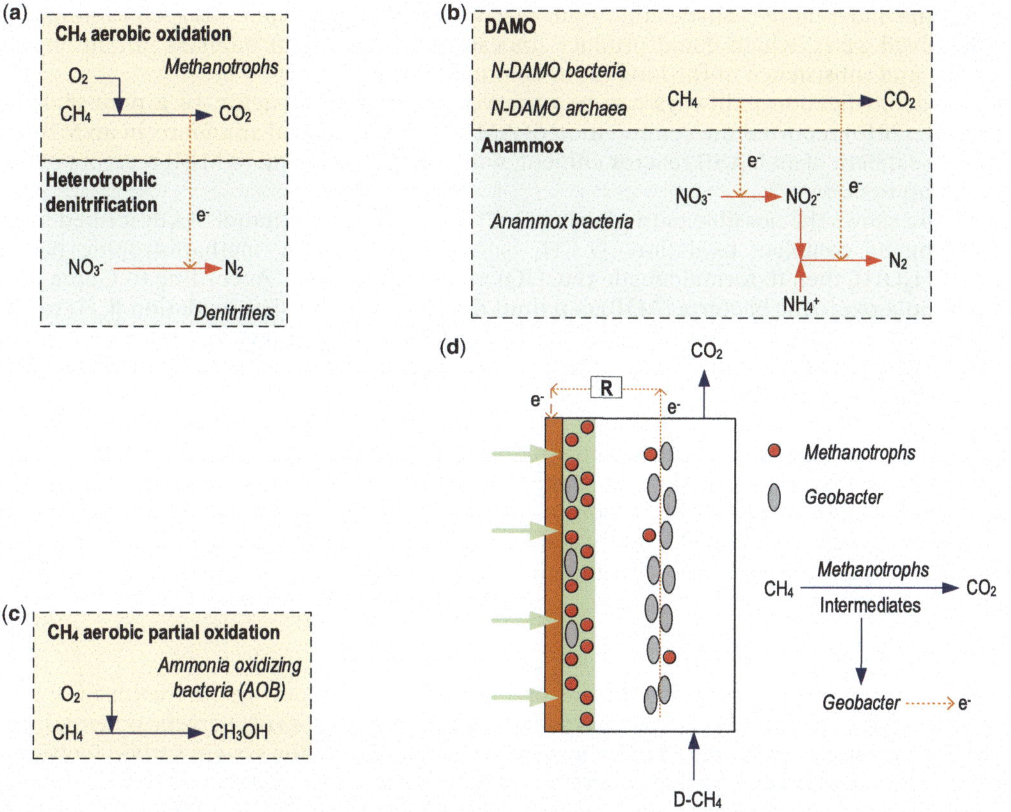


**Figure 7.14** Available biological processes for the valorization of D-$CH_4$ in the biological post-treatment: (a) aerobic oxidation of methane aerobic coupled with denitrification (AME-D); (b) anaerobic methane oxidation coupled with denitrification (AnME-D); (c) partial oxidation of $CH_4$ to produce methanol and (d) MFCs. *Source*: Adapted from Stazi and Tomei (2021).

donor and nitrite, which is produced by archaea is reduced to $N_2$ by anammox bacteria. The overall reaction is shown below (Sánchez *et al.*, 2016; Stazi and Tomei, 2021):

$$5CH_4 + 16.1NO_3^- + 15.15NH_4^+ \rightarrow 15.45N_2 + 5CO_2 \tag{7.9}$$

In the second pathway, nitrate and nitrite can both act as electron acceptors. The nitrate reduction is done by DAMO archaea, meanwhile the nitrite reduction is done by DAMO bacteria, following the overall reactions shown below, respectively:

$$5CH_4 + 8NO_3^- + 8H^+ \rightarrow 5CO_2 + 4N_2 + 14H_2O \tag{7.10}$$

$$3CH_4 + 8NO_2^- + 8H^+ \rightarrow 3CO_2 + 4N_2 + 10H_2O \tag{7.11}$$

AnME-D has been studied in the last few years as a promissory path for the simultaneous D-$CH_4$ mitigation and the denitrification in anaerobic effluents. D-$CH_4$ removal efficiencies up to 85% with a TN removal of 99% have been reported in an enriched biofilm reactor with synthetic effluents (Liu *et al.*, 2020). However, more experiments with AnME-D process in pilot-scale systems, with real anaerobic effluent and at long-term operation are required to determine the process viability and stability in time under more realistic conditions (e.g. with different compounds such as solids, organic matter, dissolved gases, which could produce undesirable reactions or biomass inhibition; biomass maintenance and subsistence in the long term, among others).

Different denitrification pathways can be observed in the same reactor. In a pilot-scale system, Sánchez *et al.* (2016) reported the combination of AME-D, AnME-D and anammox in an MBR reactor for the post-treatment of an UASB reactor effluent, with a D-$CH_4$ removal in MBR reactors of 95% and TN removal up to 60%.

Figure 7.14c shows the possible path of using D-$CH_4$ to produce methanol. As described in Section 7.4.4.2, during its complete oxidation, D-$CH_4$ is first converted by methanotrophic bacteria in methanol ($CH_3OH$), then to formaldehyde (HCHO) and finally to $CO_2$. According to Gupta and Goel (2019), ammonia-oxidizing bacteria (AOB) can limit the process to the first oxidation ($CH_4$ to $CH_3OH$, shown in Equation (7.4)), allowing the accumulation and recovery of methanol, which is a valuable by-product. This process should also be studied in pilot-scale experiments to evaluate its operation under more realistic conditions.

Finally, Figure 7.14d shows MFCs which can use the D-$CH_4$ oxidation to generate electricity. In this case, the by-products of the $CH_4$ aerobic oxidation (performed by methanotrophs) are used by exoelectrogenic microorganisms (such as geobacter), which can produce electricity during the $CH_4$ oxidation. Chen and Smith (2018), in a laboratory-scale experiment with a synthetic effluent, reported a D-$CH_4$ removal of 85% with an electricity conversion of 17%. This process should also be studied in the future at a larger scale and with real effluent.

## 7.5 FINAL REMARKS

The presence of methane dissolved in effluents of anaerobic reactors during the treatment of sewage is inherent to anaerobic-based STPs. This dissolved methane represents: loss of energetic potential, potential safety risk for operators and source of $CH_4$ emissions which increases the system carbon footprint.

However, this chapter showed that different strategies for the management of D-$CH_4$ are available to mitigate these effects, and even to recover or beneficially use D-$CH_4$. Physical and biological processes were discussed and the current state of development of each of the available technologies was described. Particularly, the desorption-based techniques and some biological processes based on the aerobic oxidation of D-$CH_4$ (coupled to denitrification or not) seems to be at a more advanced stage of development, and more evidence of its application for anaerobic effluents at larger scales is available. Other processes, particularly for the AnME-D and MFCs seems very promising technologies to promote a sustainable use of D-$CH_4$, although more experimentation at larger scales is necessary to validate its viability and stability with real anaerobic effluents and to define more suitable reactor configurations.

In addition, for every management strategy and technique presented in this chapter, more information related to their application at full-scale plants (e.g. performance and costs) is necessary to support decision-making processes during the planning and the design of anaerobic-based STPs. In this regard, to date, a management tool that guides decision makers during the selection of the more appropriate D-$CH_4$ mitigation strategy should be developed, considering that many different techniques and technologies described in this chapter could be more or less suitable for an specific anaerobic-based STP, depending on many variables such as its scale (population served), the type of anaerobic reactor, flow sheet of the system, temperature variation and many other site-specific conditions.

## REFERENCES

Arias A., Alvarino T., Allegue T., Suárez S., Garrido J. M. and Omil F. (2018). An innovative wastewater treatment technology based on UASB and IFAS for cost-efficient macro and micropollutant removal. *Journal of Hazardous Materials*, **359**(March), 113–120, https://doi.org/10.1016/j.jhazmat.2018.07.042

Bandara W. M. K. R. T. W., Satoh H., Sasakawa M., Nakahara Y., Takahashi M. and Okabe S. (2011). Removal of residual dissolved methane gas in an upflow anaerobic sludge blanket reactor treating low-strength wastewater at low temperature with degassing membrane. *Water Research*, **45**(11), 3533–3540, https://doi.org/10.1016/j.watres.2011.04.030

Boulart C., Mowlem M. C., Connelly D. P., Dutasta J.-P. and German C. R. (2008). A novel, low-cost, high performance dissolved methane sensor for aqueous environments. *Optics Express*, **16**(17), 1260712617, https://doi.org/10.1364/OE.16.012607

Boulart C., Connelly D. P. and Mowlem M. C. (2010). Sensors and technologies for in situ dissolved methane measurements and their evaluation using technology readiness levels. *TrAC – Trends in Analytical Chemistry*, **29**(2), 186–195, https://doi.org/10.1016/j.trac.2009.12.001

Brandt E. M. F., Duarte F. V., Vieira J. P. R., Melo V. M., Souza C. L., Araújo J. C. and Chernicharo C. A. L. (2016). The use of novel packing material for improving methane oxidation in biofilters. *Journal of Environmental Management*, **182**, 412–420, https://doi.org/10.1016/j.jenvman.2016.07.075

Brandt E. M. F., Noyola A. and McAdam E. J. (2019). Control of diffuse emissions in UASB reactors treating sewage. In: Anaerobic Reactors for Sewage Treatment: Design, Construction and Operation, C. A. L. Chernicharo and T. Bressani-Ribeiro (eds.), IWA Publishing, London, England, pp. 245–283.

Brandt E., Centeno Mora E., Mattos J. and Chernicharo C. (2021a). Nota Técnica 6 – Uso energético e queima de gases residuais. *Cadernos Técnicos eng Sanit Ambient*, **1**(2), 77–90, https://doi.org/10.5327/276455760102006

Brandt E., Souza C., Centeno E., Mattos J. and Chernicharo C. (2021b). Nota Técnica 5 – Sistemas de biofiltração para tratamento de gases residuais. *Cadernos Técnico eng Sanit Ambient*, **1**(2), 63–76, https://doi.org/10.5327/276455760102005

Brandt E. F., Souza C. L., Centeno Mora E. J., Bohrer Santos J. and Chernicharo C. A. de L. (2021c). Part B. Advances in gas emission control techniques for anaerobic-based STPs technical note 1. *Cadernos Técnicos eng Sanit Ambient*, **1**(2), 5–22, https://doi.org/10.5327/276455760102001

Bressani-Ribeiro T., Almeida P. G. S., Volcke E. I. P. and Chernicharo C. A. L. (2018). Trickling filters following anaerobic sewage treatment: state of the art and perspectives. *Environmental Science: Water Research and Technology*, **4**(11), 1721–1738, https://doi.org/10.1039/c8ew00330k

Centeno Mora E. and Chernicharo C. de L. (2020). Use of membrane contactors for removing and recovering dissolved methane from anaerobic reactors effluents: state-of-the-art, challenges, and perspectives. *Reviews in Environmental Science and Biotechnology*, **19**(3), 673–697, https://doi.org/10.1007/s11157-020-09546-w

Centeno Mora E. and Chernicharo C. A. L. (2022). Simultaneous removal of dissolved sulphide and dissolved methane from anaerobic effluents with hollow fibre membrane contactors. *Environmental Science and Pollution Research*, **29**, pages 90549–90566, https://doi.org/10.1007/s11356-022-22074-0

Centeno-Mora E., Fonseca P., Andreão W. L., Brandt E., Souza C. L. D. and Chernicharo C. A. L. (2020). Mitigation of diffuse $CH_4$ and $H_2S$ emissions from the liquid phase of UASB-based sewage treatment plants: challenges, techniques, and perspectives. *Environmental Science and Pollution Research*, **27**(29), 35979–35992, https://doi.org/10.1007/s11356-020-08644-0

Centeno Mora E. J., Fonseca P. R., Brandt E. M. F. and Chernicharo C. A. de L. (2021). Parte B: avanços nas técnicas de controle de emissões gasosas em ETEs com reatores anaeróbios Nota Técnica 4 – Recuperação de metano dissolvido em efluentes de reatores anaeróbios por meio de contactores de membrana desgaseificadora. *Cadernos Técnicos Engenharia Sanitária e Ambiental*, **1**(2), 49–62, https://doi.org/10.5327/276455760102004

Centeno Mora E., Paranhos A. G. de O., Aquino S. F. de and Chernicharo C. A. de L. (2023a). Use of hollow fibre membrane contactors to remove dissolved gases from effluents of UASB reactors treating sewage after its conditioning with dynamic membrane filtration. *Journal of Water Process Engineering*, **53**, 103593, https://doi.org/10.1016/j.jwpe.2023.103593

Centeno Mora E., Souza C. L. de, Neves T. de A. and Chernicharo C. de L. (2023b). Characterisation and perspectives of energetic use of dissolved gas recovered from anaerobic effluent with membrane contactor. *Bioresource Technology*, **367**, 128223, https://doi.org/10.1016/j.biortech.2022.128223

Chen S. and Smith A. L. (2018). Methane-driven microbial fuel cells recover energy and mitigate dissolved methane emissions from anaerobic effluents. *Environmental Science: Water Research and Technology*, **4**(1), 67–79, https://doi.org/10.1039/C7EW00293A

Chernicharo C. A. L., van Lier J. B., Noyola A. and Bressani Ribeiro T. (2015). Anaerobic sewage treatment: state of the art, constraints and challenges. *Reviews in Environmental Science and Biotechnology*, **14**(4), 649–679, https://doi.org/10.1007/s11157-015-9377-3

Chernicharo C. A. L., Brandt E. M. F., Bressani-Ribeiro T., Melo V. R., Bianchetti F. J., Motafilho C. R. and McAdam E. (2017). Development of a tool for improving the management of gaseous emissions in UASB-based sewage treatment plants. *Water Practice and Technology*, **12**(4), 917–926, https://doi.org/10.2166/wpt.2017.097

Cookney J., Cartmell E., Jefferson B. and McAdam E. J. (2012). Recovery of methane from anaerobic process effluent using poly-di-methyl-siloxane membrane contactors. *Water Science and Technology*, **65**(4), 604–610, https://doi.org/10.2166/wst.2012.897

Cookney J., Mcleod A., Mathioudakis V., Ncube P., Soares A., Jefferson B. and McAdam E. J. J. (2016). Dissolved methane recovery from anaerobic effluents using hollow fibre membrane contactors. *Journal of Membrane Science*, **502**, 141–150, https://doi.org/10.1016/j.memsci.2015.12.037

Crittenden J. C., Trussel R. R., Hand D. W., Howe K. J. and Tchobanoglous G. (2012). MWH' S Water Treatment Principles and Design (Third). John Wiley & Sons, Inc., New Jersey, USA, https://doi.org/10.1002/9781118131473.ch14

Crone B. C., Garland J. L., Sorial G. A. and Vane L. M. (2016). Significance of dissolved methane in effluents of anaerobically treated low strength wastewater and potential for recovery as an energy product: a review. *Water Research*, **104**, 520–531, https://doi.org/10.1016/j.watres.2016.08.019

Crowl D. (2003). Understanding Explosions. American Institute of Chemical Engineers, New York, USA, https://doi.org/10.1016/S0950-4230(03)00051-2

Cussler E. L. (2007). Diffusion: Mass Transfer in Fluid Systems (Third). Cambridge University Press, Cambridge, UK.

Daelman M. R. J., van Voorthuizen E. M., van Dongen U. G. J. M., Volcke E. I. P. and van Loosdrecht M. C. M. (2012). Methane emission during municipal wastewater treatment. *Water Research*, **46**(11), 3657–3670, https://doi.org/10.1016/j.watres.2012.04.024

Daelman M. R. J., van Eynde T., van Loosdrecht M. C. M. and Volcke E. I. P. (2014). Effect of process design and operating parameters on aerobic methane oxidation in municipal WWTPs. *Water Research*, **66**, 308–319, https://doi.org/10.1016/j.watres.2014.07.034

Giménez J. B., Martí N., Ferrer J. and Seco A. (2012). Methane recovery efficiency in a submerged anaerobic membrane bioreactor (SAnMBR) treating sulphate-rich urban wastewater: evaluation of methane losses with the effluent. *Bioresource Technology*, **118**, 67–72, https://doi.org/10.1016/j.biortech.2012.05.019

Glória R. M., Motta T. M., Silva P. V. O., Da Costa P., Brandt E. M. F., Souza C. L. and Chernicharo C. A. L. (2016). Stripping and dissipation techniques for the removal of dissolved gases from anaerobic effluents. *Brazilian Journal of Chemical Engineering*, **33**(4), 713–721, https://doi.org/10.1590/0104-6632.20160334s20150291

Gupta V. and Goel R. (2019). Managing dissolved methane gas in anaerobic effluents using microbial resource management-based strategies. *Bioresource Technology*, **289**, 121601, https://doi.org/10.1016/j.biortech.2019.121601

Hahn M. J. and Figueroa L. A. (2015). Pilot scale application of anaerobic baffled reactor for biologically enhanced primary treatment of raw municipal wastewater. *Water Research*, **87**, 494–502, https://doi.org/10.1016/j.watres.2015.09.027

Hartley K. and Lant P. (2006). Eliminating non-renewable $CO_2$ emissions from sewage treatment: an anaerobic migrating bed reactor pilot plant study. *Biotechnology and Bioengineering*, **95**(3), 384–398, https://doi.org/10.1002/bit.20929

Heile S., Chernicharo C. A. L., Brandt E. M. F. and McAdam E. J. (2017). Dissolved gas separation for engineered anaerobic wastewater systems. *Separation and Purification Technology*, **189**(May), 405–418, https://doi.org/10.1016/j.seppur.2017.08.021

Henares M., Izquierdo M., Penya-Roja J. M. M. and Martínez-Soria V. (2016). Comparative study of degassing membrane modules for the removal of methane from expanded granular sludge bed anaerobic reactor effluent. *Separation and Purification Technology*, **170**, 22–29, https://doi.org/10.1016/j.seppur.2016.06.024

Henares M., Izquierdo M., Marzal P. and Martínez-Soria V. (2017). Demethanization of aqueous anaerobic effluents using a polydimethylsiloxane membrane module: mass transfer, fouling and energy analysis. *Separation and Purification Technology*, **186**, 10–19, https://doi.org/10.1016/j.seppur.2017.05.035

Henares M., Ferrero P., San-Valero P., Martínez-Soria V. and Izquierdo M. (2018). Performance of a polypropylene membrane contactor for the recovery of dissolved methane from anaerobic effluents: mass transfer evaluation, long-term operation and cleaning strategies. *Journal of Membrane Science*, **563**(June), 926–937, https://doi.org/10.1016/j.memsci.2018.06.045

Huete A., de los Cobos-Vasconcelos D., Gómez-Borraz T., Morgan-Sagastume J. M. and Noyola A. (2018). Control of dissolved $CH_4$ in a municipal UASB reactor effluent by means of a desorption–biofiltration arrangement. *Journal of Environmental Management*, **216**, 383–391, https://doi.org/10.1016/j.jenvman.2017.06.061

IPCC (2021). Climate Change 2021: The Physical Science Basis. Contribution of Working Group I to the Sixth Assessment Report of the Intergovernmental Panel on Climate Change (6th edn). Cambridge University Press.

Lee E., Rout P. R., Kyun Y. and Bae J. (2020). Process optimization and energy analysis of vacuum degasifier systems for the simultaneous removal of dissolved methane and hydrogen sulfide from anaerobically treated wastewater. *Water Research*, **182**, 115965, https://doi.org/10.1016/j.watres.2020.115965

Li X., Lee H. S., Wang Z. and Lee J. (2021). State-of-the-art management technologies of dissolved methane in anaerobically-treated low-strength wastewaters: a review. *Water Research*, **200**, 117269, https://doi.org/10.1016/j.watres.2021.117269

Liu Y., Sharma K. R., Fluggen M., O'Halloran K., Murthy S. and Yuan Z. (2015). Online dissolved methane and total dissolved sulfide measurement in sewers. *Water Research*, **68**, 109–118, https://doi.org/10.1016/j.watres.2014.09.047

Liu T., Li J., Khai Lim Z., Chen H., Hu S., Yuan Z. and Guo J. (2020). Simultaneous removal of dissolved methane and nitrogen from synthetic mainstream anaerobic effluent. *Environmental Science and Technology*, **54**(12), 7629–7638, https://doi.org/10.1021/acs.est.0c00912

Lobato L. C. S., Chernicharo C. A. L. and Souza C. L. (2012). Estimates of methane loss and energy recovery potential in anaerobic reactors treating domestic wastewater. *Water Science and Technology*, **66**(12), 2745–2753, https://doi.org/10.2166/wst.2012.514

Machado L. M. (2018). Avanços no emprego de Câmaras de Dessorção para remoção e recuperação dos gases metano e sulfeto de hidrogênio dissolvidos em efluente de reator UASB no tratamento de esgoto doméstico. Master Thesis, Federal University of Minas Gerais.

Mampaey K. E., van Dongen U. G. J. M., van Loosdrecht M. C. M. and Volcke E. I. P. (2015). Novel method for online monitoring of dissolved $N_2O$ concentrations through a gas stripping device. *Environmental Technology (United Kingdom)*, **36**(13), 1680–1690, https://doi.org/10.1080/09593330.2015.1005029

Marinho T. D. (2019). Estudo da microaeração e sua associação com câmara de dessorção na remoção e recuperação de sulfeto de hidrogênio e metano em reatores UASB no tratamento de esgoto doméstico. Master Thesis, Federal University of Minas Gerais.

Matsuura N., Hatamoto M., Sumino H., Syutsubo K., Yamaguchi T. and Ohashi A. (2010). Closed DHS system to prevent dissolved methane emissions as greenhouse gas in anaerobic wastewater treatment by its recovery and biological oxidation. *Water Science and Technology*, **61**(9), 2407–2415, https://doi.org/10.2166/wst.2010.219

Matsuura N., Hatamoto M., Sumino H., Syutsubo K., Yamaguchi T. and Ohashi A. (2015). Recovery and biological oxidation of dissolved methane in effluent from UASB treatment of municipal sewage using a two-stage closed downflow hanging sponge system. *Journal of Environmental Management*, **151**, 200–209, https://doi.org/10.1016/j.jenvman.2014.12.026

Matsuura N., Hatamoto M., Yamaguchi T. and Ohashi A. (2017). Methanotrophic community composition based on pmoA genes in dissolved methane recovery and biological oxidation closed downflow hanging sponge reactors. *Biochemical Engineering Journal*, **124**, 138–144, https://doi.org/10.1016/j.bej.2017.05.003

Nelting K., Weichgrebe D., Caicedo C. and Rosenwinkel K.-H. (2015). Influence of UASB reactor configuration and operation conditions on the dissolved methane quantities in the effluent. IWA, 14th World Congress on Anaerobic Digestion, 1–5 November.

Nelting K., Possetti G. R. C., Rietow J. C., Paula A. C., Costa F. J., Gallo E. M. and Trautmann N. (2021). Parte B: avanços nas técnicas de controle de emissões gasosas em ETEs com reatores anaeróbios Nota Técnica 3 – Recuperação de gases dissolvidos em efluentes de reatores anaeróbios por meio de câmaras a vácuo. *Cadernos Técnicos eng Sanit Ambient*, **1**(2), 35–47.

Noyola A., Morgan-Sagastume J. M. and López-Hernández J. E. (2006). Treatment of biogas produced in anaerobic reactors for domestic wastewater: odor control and energy/resource recovery. *Reviews in Environmental Science and Biotechnology*, **5**(1), 93–114, https://doi.org/10.1007/s11157-005-2754-6

Pauss A., Andre G., Perrier M. and Guiot S. R. (1990). Liquid-to-gas mass transfer in anaerobic processes: inevitable transfer limitations of methane and hydrogen in the biomethanation process. *Applied and Environmental Microbiology*, **56**(6), 1636–1644, https://doi.org/10.1128/aem.56.6.1636-1644.1990

Pretel R., Robles A., Ruano M. V., Seco A. and Ferrer J. (2013). Environmental impact of submerged anaerobic MBR (SAnMBR) technology used to treat urban wastewater at different temperatures. *Bioresource Technology*, **149**, 532–540, https://doi.org/10.1016/j.biortech.2013.09.060

Rongwong W., Goh K. and Bae T. H. (2018). Energy analysis and optimization of hollow fiber membrane contactors for recovery of dissolve methane from anaerobic membrane bioreactor effluent. *Journal of Membrane Science*, **554**(February), 184–194, https://doi.org/10.1016/j.memsci.2018.03.002

Rongwong W., Goh K., Sethunga G. S. M. D. P. and Bae T. H. (2019). Fouling formation in membrane contactors for methane recovery from anaerobic effluents. *Journal of Membrane Science*, **573**(December 2018), 534–543, https://doi.org/10.1016/j.memsci.2018.12.038

Sánchez A., Rodríguez-Hernández L., Buntner D., Esteban-García A. L., Tejero I. and Garrido J. M. (2016). Denitrification coupled with methane oxidation in a membrane bioreactor after methanogenic pre-treatment of wastewater. *Journal of Chemical Technology and Biotechnology*, **91**(12), 2950–2958, https://doi.org/10.1002/jctb.4913

Sanchis-Perucho P., Robles Á, Durán F., Ferrer J. and Seco A. (2020). PDMS membranes for feasible recovery of dissolved methane from AnMBR effluents. *Journal of Membrane Science*, **604**, 118070, https://doi.org/10.1016/j.memsci.2020.118070

Sander R. (2015). Compilation of Henry's law constants (version 4.0) for water as solvent. *Atmospheric Chemistry and Physics*, **15**(8), 4399–4981, https://doi.org/10.5194/acp-15-4399-2015

Santo B. C. E. (2017). Avaliação de câmara de dessorção de metano e sulfeto de hidrogênio dissolvidos em efluentes de reatores uasb tratando esgoto doméstico. Master Thesis, Federal University of Minas Gerais.

Sethunga G. S. M. D. P., Karahan H. E., Wang R. and Bae T. H. (2019). PDMS-coated porous PVDF hollow fiber membranes for efficient recovery of dissolved biomethane from anaerobic effluents. *Journal of Membrane Science*, **584**(May), 333–342, https://doi.org/10.1016/j.memsci.2019.05.016

Sethunga G. S. M. D. P., Lee J., Wang R. and Bae T.-H. (2020). Influences of operating parameters and membrane characteristics on the net energy production in dense, porous, and composite hollow fiber membrane contactors for dissolved biomethane recovery. *Journal of Membrane Science*, **610**, 118301, https://doi.org/10.1016/j.memsci.2020.118301

Silva-Teira A., Sánchez A., Buntner D., Rodríguez-Hernández L. and Garrido J. M. (2017). Removal of dissolved methane and nitrogen from anaerobically treated effluents at low temperature by MBR post-treatment. *Chemical Engineering Journal*, **326**, 970–979, https://doi.org/10.1016/j.cej.2017.06.047

Smith A. L., Skerlos S. J. and Raskin L. (2013). Psychrophilic anaerobic membrane bioreactor treatment of domestic wastewater. *Water Research*, **47**(4), 1655–1665, https://doi.org/10.1016/j.watres.2012.12.028

Souza C. L., Chernicharo C. A. L. and Aquino S. F. (2011). Quantification of dissolved methane in UASB reactors treating domestic wastewater under different operating conditions. *Water Science and Technology*, **64**(11), 2259–2264, https://doi.org/10.2166/wst.2011.695

Souza C. L., Chernicharo C. A. L. and Melo G. C. B. (2012). Methane and hydrogen sulfide emissions in UASB reactors treating domestic wastewater. *Water Science and Technology*, **65**(7), 1229–1237, https://doi.org/10.2166/wst.2012.854

Souza C. L., Santos B., Brandt E., Mattos J. and Chernicharo C. (2021). Nota técnica 2 – câmaras de dessorção para remoção de sulfeto de hidrogênio e metano dissolvidos em efluentes de reatores anaeróbios. *Cadernos Técnicos Engenharia Sanitária e Ambiental*, **1**(2), 23–33, https://doi.org/10.5327/276455760102002

Staudinger J. and Roberts P. V. (2009). A critical review of Henry's law constants for environmental applications. *Critical Reviews in Environmental Science and Technology*, **26**(3), 205–297, https://doi.org/10.1080/10643389609388492

Stazi V. and Tomei M. C. (2021). Dissolved methane in anaerobic effluents: a review on sustainable strategies for optimization of energy recovery or internal process reuse. *Journal of Cleaner Production*, **317**, 128359, https://doi.org/10.1016/j.jclepro.2021.128359

Velasco P., Jegatheesan V. and Othman M. (2022). Effect of long-term operations on the performance of hollow fiber membrane contactor (HFMC) in recovering dissolved methane from anaerobic effluent. *Science of the Total Environment*, **841**, 156601, https://doi.org/10.1016/j.scitotenv.2022.156601

Wongchitphimon S., Rongwong W. W., Chuah C. Y., Wang R. and Bae T.-H. H. (2017). Polymer-fluorinated silica composite hollow fiber membranes for the recovery of biogas dissolved in anaerobic effluent. *Journal of Membrane Science*, **540**(June), 146–154, https://doi.org/10.1016/j.memsci.2017.06.050

Wu Z., Duan H., Li K. and Ye L. (2022). A comprehensive carbon footprint analysis of different wastewater treatment plant configurations. *Environmental Research*, **214**, 113818, https://doi.org/10.1016/j.envres.2022.113818

Xie T., Liu X., Xu Y., Bryson S., Zhao L., Huang K., Huang S., Li X., Yang Q., Dong H. and Winkler M. K. H. (2023). Coupling methanotrophic denitrification to anammox in a moving bed biofilm reactor for nitrogen removal under hypoxic conditions. *Science of the Total Environment*, **856**, 158795, https://doi.org/10.1016/j.scitotenv.2022.158795

Yoo R., Kim J., McCarty P. L. and Bae J. (2012). Anaerobic treatment of municipal wastewater with a staged anaerobic fluidized membrane bioreactor (SAF-MBR) system. *Bioresource Technology*, **120**, 133–139, https://doi.org/10.1016/j.biortech.2012.06.028

doi: 10.2166/9781789063479_0237

# Chapter 8

# Water reuse

V. Lazarova*

*Water Globe Consultants, 361 rue du Chateau des Bain, 01630 Sergy, France*
**Correspondence: valentina.lazarova@orange.fr*

**ABSTRACT**

With the increasing concerns of water scarcity, water reuse is becoming the cornerstone of circular water economy and sustainable development. One of the potential solutions to resolve another important environmental challenge, the water–energy nexus, is the introduction of anaerobic processes in wastewater treatment and reclamation trains. Despite the numerous advantages of anaerobic treatment, effluent water quality is lower compared to conventional aerobic secondary effluents and requires adequate post-treatment to meet current wastewater discharge and reuse standards. On the basis of the most popular water reuse standards for non-potable uses and the overview of the characteristics and efficiency of proven treatment processes, this chapter provides guidance on the selection of appropriate wastewater treatment and reclamation alternatives for treatment of urban and industrial wastewater from food and agro-industries which reflect the current practical experience worldwide. The proposed decision-making process for selection of the most suitable treatment scheme for a given project takes into account one additional important parameter – the treatment capacity (plant size) which is not taken into account in current water reuse regulations. Nevertheless, treatment capacity and ease of operation define the feasibility and economic viability of the most suitable post-treatment of anaerobic effluents and tertiary treatment with disinfection for water reuse applications. The presented design and performance parameters for selecting proven and mature treatment technologies are given only for information – professional judgement is necessary for accurate design taking into account all local conditions and requirements. Innovative technologies under investigation or with limited full-scale application which are still not validated for water reuse applications are not discussed. More research efforts are needed for their development and full-scale implementation.

**Keywords:** post-treatment of anaerobic effluents, reuse regulations, selection of treatment trains, tertiary treatment, water quality parameters, water reuse.

## 8.1 INTRODUCTION

The long-term experience in water reuse – with recent great successes of new water reuse schemes and applications – demonstrates that water reclamation provides a unique and viable opportunity to augment traditional water supplies (Asano *et al.*, 2007; Lazarova & Asano, 2013). Indeed, water reuse is becoming a cornerstone of sustainable water management and urban planning, and is considered to be a key chain link of the emerging circular economy (IWA, 2016; Lazarova, 2022).

Circular economy and rational, responsible, renewable and sustainable use of water resources are closely intertwined. Looking beyond the current take–make–dispose extractive linear approach, a circular economy aims to redefine growth, focusing on positive society-wide benefits by recovering and reusing resources thereby creating a closed-loop and valorization of water resources and side streams from various treatment processes. This is also considered one of the most efficient ways to solve another important challenge of the 21st century – the water–energy nexus – preserving better our environment and maximizing energy and resource recovery from wastewater. With the development of innovative technologies, wastewater treatment plants are evolving into environmental energy neutral facilities with the potential for recovering a wide range of other valuable resources such as organic matters, nutrients (mainly nitrogen and phosphorus), metals, proteins, polysaccharides and so on (Pikaar *et al.*, 2022). In this respect, anaerobic wastewater treatment is a proven and beneficial route to energy positive treatment facilities.

Producing reclaimed water of a specified quality to fulfil multiple water-use objectives is now a reality due to the progressive evolution of water reclamation technologies, regulations and environmental and health risk protection measures (Lazarova *et al.*, 2013). This chapter presents basic principles and technologies enabling water reuse mainly for non-potable purposes. It provides guidance for the selection of appropriate tertiary treatment trains depending on water quality requirements, plant size, costs and easiness of operation, based on state-of-the-art and full-scale water reuse experience worldwide. The key challenges related to the introduction of anaerobic technologies for wastewater mainstream treatment are also discussed.

### 8.1.1 Water reuse practices and applications

Water reuse practices can be classified into two main categories: non-potable and potable water reuse. The dominant non-potable applications of recycled water include agricultural irrigation, landscape irrigation, industrial reuse, environmental enhancement and various urban uses. Among them, agricultural and landscape irrigation are widely practiced throughout the world with well-established health protection guidelines and good agronomic practices (Asano *et al.*, 2007; Lazarova & Bahri, 2004). Industrial wastewater recycling is also a well-established current practice.

It is important to underline that the terms 'water reuse', 'wastewater reclamation' and 'water recycling' are used in this chapter as synonyms. The importance to specify 'reclamation', which in the past indicated that wastewater should be treated before its reuse is losing its relevance due to the more stringent water reuse standards requiring appropriate treatment to fit-to-purpose water quality.

In addition, depending on the source of recycled water, two main categories of water reuse should be dissociated:

(1) Reuse of treated municipal wastewater named also urban wastewater reuse, which also include industrial uses for different purposes, and
(2) Reuse of treated industrial wastewater, which include internal water recycling for various purposes and external water reuse, mainly for agricultural irrigation and/or on-site landscape irrigation.

Agricultural irrigation was, is and will remain the largest recycled water consumer with recognized benefits and contribution to food security. Both municipal and industrial reclaimed water (mainly from food industry and cooling systems) are used for irrigation. As a rule, urban water recycling is practiced using effluents from municipal wastewater reclamation plants, in particular for landscape irrigation, car washing, firefighting and street cleaning. Other relevant and cost-effective applications are also emerging such as environmental enhancement and recreation, in-building recycling and industrial uses of reclaimed urban wastewater.

Indirect potable reuse, in particular groundwater (aquifer) recharge, after complementary polishing and storage of recycled water in an environmental buffer, has been implemented in many countries as an efficient response to the need to increase water supply. Finally, direct potable reuse, practiced since 1968 in Namibia and recently introduced in Texas and Arizona, is emerging as a solution to the challenges which some countries will face in the next 20–50 years.

The dominant application of industrial reclaimed water is plant-internal water recycling for cooling, boiler make-up, process water, transport, cleaning and on-site irrigation. For some industries, internal reuse of wastewater can reduce water consumption by up to 50–95% (Rosenwinkel *et al.*, 2013). In some industries, additional benefits of water recycling include the recovery of heat, raw materials and other valuable products.

### 8.1.2 Water quality requirements for wastewater treatment and reuse

The majority of water reuse standards and guidelines worldwide are focused principally on health protection. Contact, inhalation and in particular ingestion of reclaimed water containing pathogenic microorganisms or toxic chemicals create the potential for adverse health effects for humans and animals.

The principal contaminants of urban wastewater are total suspended solids (TSS) with typical concentration range of ~100–300 mg/L and organic matter expressed by its chemical oxygen demand (COD) of ~200–1000 mg/L or biological oxygen demand (BOD) in the range of 100–500 mg/L. Accordingly, the treatment of municipal wastewater is typically designed to meet water quality objectives based on suspended solids (SS) (<5–35 mg SS/L), organic content (<10–45 mg BOD/L or <50–125 mg COD/L), nutrient levels (<10–20 mg N/L and <0.1–2 mg P/L) and biological indicators (total or faecal coliforms or *Escherichia coli* in the range of <10–$10^4$ cfu/100 mL, helminth eggs <0.1–5/L, enteroviruses) and, in some cases, chlorine residual (>0.5–5 mg Cl/L).

Water quality requirements vary with the type of application, regional context and overall risk perception. In practice, these factors are expressed through different water quality requirements, as well as criteria for design, operation and reliability of treatment facilities. The most stringent guidelines and regulations operate on the basis of the precautionary principle, which demands high water quality and intensive treatment, leading to lower health risks without any additional specific measures. However, similar health protection can be achieved by means of additional health protection barriers, as demonstrated by the World Health Organization (WHO) wastewater reuse guidelines (WHO, 2006). The application of additional health protection barriers and codes of good practices (the multibarrier approach) could form an essential part of a risk mitigation strategy, as also underlined by USEPA guidelines (2012).

Table 8.1 illustrates the microbial water quality and treatment requirements of the most important cornerstone water reuse guidelines and regulations for irrigation followed worldwide. In many countries, unrestricted irrigation of crops eaten raw is considered to present the greatest potential for disease transmission associated with water reuse. However, this is not always the case – for example, some regulations recommend more stringent standards for public lawns where crops are eaten raw.

The third edition of the WHO guidelines for agricultural irrigation with reclaimed water (WHO, 2006) and the European Water Reuse Regulation (EU Regulation 2020/741, 2020) were inspired by the Australian Water Recycling Guidelines (NRMMC, 2006) which aims to provide adequate health protection and attain the health-based target of $\leq 10^{-6}$ disability-adjusted life years per person per year. The microbial concentration levels for verification monitoring recommended for unrestricted irrigation of food crops (Table 8.1) vary from 1 to 10 *E. coli*/100 mL, which means full disinfection as this is the range of variation of detection limits of the most common analytical methods. The WHO guidelines (WHO, 2006) indicate the value of 10 *E. coli*/100 mL for the treatment alone, which can rise to 10,000 *E. coli*/100 mL with additional health protection barriers, for example, drip irrigation of high-growing crops. It should be noted that these requirements apply to all treatment works, including small facilities.

**Table 8.1** Comparison of water quality required for agricultural irrigation of selected guidelines and regulations.

| | | **WHO (2006)** | **USEPA guidelines (2012)** | **State of California (2000)** | **NRMMC (2006) Australian guidelines** | **EU Regulation (2020)** |
|---|---|---|---|---|---|---|
| | | ***Guidelines*** | ***Guidelines*** | ***Regulation*** | ***Guidelines*** | ***Regulation*** |
| | **Microbial Indicator** | ***E. coli*** | **Faecal Coliforms** | **Total Coliforms** | ***E. coli*** | ***E. coli*** |
| Unrestricted irrigation of food crops/consumed raw | Coliforms, number (cfu or MPN)/100 mL | 10–$10^5$ depending on treatment, additional health barriers and type of crops | Not detected (daily, 7-day median, 14 max) | ≤2.2 (daily, 7-day median, 23 max in 30 days, 240 max) | ≤1 (weekly) | ≤10 (weekly, 90%) |
| | Helminths (eggs/L) | ≤1 | NS | NS | NS[a] | NS |
| | $BOD_5$ (mg/L) | NS | ≤10 (weekly) | NS | NS[b] | ≤10 (weekly) |
| | TSS (mg/L) | | NS | | | |
| | Turbidity (NTU) | | ≤2 (on-line, avg. 24 h, 5 max) Membranes ≤0.2 any time | ≤2 (on-line, avg. 24 h, 5 max) Membranes ≤0.2 (max 0.5) | | ≤5 |
| | Chlorine residual | | >1 mg/L (on-line) | >1 mg/L | | NS |
| | Log removal requirements (pathogens) | 6–7 logs in total via various combinations of pathogen barriers, water treatment and natural die-off | NS | 5 logs of MS2 bacteriophages of disinfection process | 6 log viruses 5 logs for bacteria and protozoa | >5 log *E. coli* >6 log coliphages[c] >4 log *Clostridium perfringens*[d] |
| | Minimum treatment requirements | NS | Secondary, filtration, disinfection | Tertiary + disinfection; chlorination $C_t = 450$ mg min/L | Secondary, filtration and disinfection[e] | Secondary, filtration, disinfection |
| Others | Coliforms/100 mL | Restricted irrigation: 3–4 log removal requirement | Processed/non-food crops ≤200 FC (median, 800 max) | NS | Commercial food crops, pastures, fodder ≤100 to ≤1000 *E. coli* | |
| | Helminths (eggs/L) | ≤1 | NS | | NS | ≤1 for pastures or fodder[f] |

cfu: colony forming units; MPN: most probable number; NS: not specified.

[a]For highest exposures, the verification monitoring includes also *C. perfringens*, somatic and F-specific bacteriophages (weekly), as well as adenovirus and *Cryptosporidium* oocysts (monthly).

[b]The state regulations in Australia requires 10 mg/L of $BOD_5$ and TSS for category A, highest exposure and highest water quality.

[c]Total coliphages or alternatively F-specific or somatic coliphages.

[d]Spore-forming sulphate-reducing bacteria can be as an alternative.

[e]A number of specific operational parameters are recommended depending on the given treatment process.

[f]The monitoring frequency depends on the initial count of helminth eggs.

The revised U.S. Environmental Protection Agency (USEPA) guidelines (USEPA guidelines, 2012) and the Californian Water Recycling Criteria (State of California, 2000) for irrigation and other urban uses also require a high level of disinfection for almost total coliform inactivation and high level of treatment, which includes secondary treatment, filtration and disinfection for food crops irrigation. In addition, conservative requirements are included for water quality monitoring, treatment train design and process operation. For example, turbidity requirements for conventional tertiary treatment with disinfection state that turbidity should be <2 NTU (max: 5 NTU), and if membranes are used the turbidity cannot exceed 0.2 NTU more than 5% of the time within a 24-h period or exceed 0.5 NTU at any time.

In California, the same stringent disinfection requirements are applied for industrial uses of recycled municipal wastewater for industrial cooling and air conditioning that involves use of cooling towers, evaporative condensers, spraying or any other devices that creates a mist, as well as industrial process water that may come into contact with workers.

Concerning trace elements, agronomic and physico-chemical parameters and compounds, the Food and Agriculture Organization guidelines (Ayers & Westcot, 1985; Pescod, 1992; FAO & IWMI, 2023) are the key document of reference for the water reuse standards, guidelines and regulations of other organizations and countries. These parameters are of critical importance for the implementation of safe agricultural and landscape irrigation with recycled water due to their influence on crops quality and yield, as well as soil properties and productivity. The maximum threshold values of agronomic parameters depend on type of crops and soil, for example, salinity (<450–4000 mg total dissolved solids (TDS)/L), sodium adsorption ratio (<3–9), boron concentration (<1 mg/L) and content of some heavy metals (Cd, Ni, Hg, Zn, etc.) and phytotoxic compounds (Lazarova & Bahri, 2004; FAO & IWMI, 2023).

As a rule, environmental and agronomic risks of water reuse for irrigation are related to uncontrolled or large industrial wastewater discharge in municipal sewers (e.g. high concentrations of heavy metals or organic micropollutants). In coastal areas, the major risk is the high salinity and sodicity of wastewater due to seawater intrusion into municipal sewers, for example, high rates of infiltration, in particular under high-tide conditions. The feedback from acquired experience shows that in biological wastewater treatment (e.g. activated sludge), the major part of heavy metals and refractory organic micropollutants are concentrated in sludge (biosolids). Consequently, the land application of polluted wastewater sludge and reclamation of industrial wastewater represent the highest risk for agriculture and landscape irrigation.

Water recycling criteria for industrial water reuse are specific for each industry and depend on the final application. The most important water quality issues for industrial uses with few examples of threshold limits are summarized in Table 8.2. Corrosion, scaling and biofouling are the major issues related to salinity (TDS or conductivity), hardness, ammonia, phosphorus, SS and residual organics. Water quality requirements are the most stringent for boiler make-up water (low and high pressure) and electronic industry. Cooling water also must be treated to low threshold values for ammonia, silica, hardness and conductivity.

The development and enforcement of water reuse standards is an essential step in the social acceptance of water recycling. However, in some cases, regulations could represent a challenge and a burden for water reuse, as for example in the case of very restrictive requirements based on the precautionary principle. For example, health risk-based regulations for irrigation, such as those developed in Australia and used as the basis for the new European regulations require an additional health risk assessment (qualitative or quantitative microbial risk assessment) and validation of log removal of treatment technologies, in addition to water quality monitoring for class A recycled water. These new requirements lead to significantly higher permit and operation costs without any guarantee of lower health risks or better process reliability. A comprehensive review performed by a large group of leading experts (Olivieri *et al.*, 2016) demonstrated that a treatment-based approach used for years in the USA, in particular in California, does not increase public health risks compared to Australian regulations and that modifying the standards to make them more restrictive will not improve public health.

**Table 8.2** Water quality requirements and related issues for industrial water reuse.

| No. | Parameter | Adverse Impacts and Concerns | Examples of Water quality Limits[a] (mg/L) | Treatment Processes |
|---|---|---|---|---|
| 1 | TDS | Corrosion | <0.1; 5; 35; 60 up to 1000 | RO, electrodialysis, blending |
| 2 | Conductivity | | 0.05; 300–800 μS/cm | |
| 3 | TSS | Clogging, biofouling | 5; 10; 30 | Sand filtration, UF, MF |
| 4 | Ammonia | Corrosion, microbial growth, biofouling | <0.1; 1.9; 4; 10 as N | Nitrification, ion exchange, air stripping |
| 5 | Ca, Mg, Fe, silica | Scaling | <0.01; 1.0; 1.5 (up to 65) silica<br><1 Ca and Mg<br>0.3; 1 Fe | Softening, ion exchange, RO |
| 6 | Hardness | Scaling | <0.05; 0.3; 70, 120 up to 200 | Softening, ion exchange |
| 7 | Phosphorus | Scaling, algal growth, biofouling | 2 as $P_{tot}$ | Biological or chemical P removal, ion exchange |
| 8 | Organic matters | Microbial growth, biofouling | 0.7 TOC<br>10; 20; 30 $BOD_5$<br>40; 60 COD | Biofiltration, activated carbon, RO, biocides |

UF, ultrafiltration; MF, microfiltration; RO, reverse osmosis; TOC, total organic carbon
[a]*Source:* Adapted from Lazarova *et al.* (2013) on the basis of feedback from full-scale facilities.

## 8.2 TYPICAL TREATMENT SCHEMES FOR MUNICIPAL AND INDUSTRIAL WATER REUSE

The starting point for consideration of water reuse for any specific application is ensuring the public health, biological and chemical safety of using recycled water by applying appropriate treatment technologies. Consequently, the choice and design of treatment schemes to meet reclaimed water quality objectives is a critical element of any water reuse system.

### 8.2.1 Main water quality parameters of concern

The key parameter of concern in the first step of wastewater treatment, named also primary treatment, is TSS which when present in effluents can plug distribution systems and protect microorganisms, thus decreasing the disinfection efficiency of most subsequent treatment processes. The main objective in the next step of wastewater treatment is carbon and nutrient removal (BOD, COD, nitrogen and phosphorus), known also as secondary treatment. Even if organics removal is not strictly needed for the reuse of effluents for agricultural irrigation, organic carbon removal is required for effluent discharge, as well as it is of great importance for the reduction of the regrowth potential of residual microorganisms in reclaimed water distribution systems.

Nutrients removal is generally recommended for environmental purposes and protection of sensitive water bodies. In addition, their presence is not desirable if the reclaimed water is used for groundwater recharge and most industrial applications. Further treatment includes tertiary polishing, such as removal of residual SS, nutrients, some specific compounds including organic micropollutants, salts, heavy metals and finally disinfection.

Disinfection is the most important treatment step for almost all water reuse applications. The removal of organic micropollutants, such as pesticides, polychlorinated biphenyls (PCBs), polycyclic aromatic hydrocarbons (PAHs), pharmaceuticals and health-care products, which are also known to be potentially harmful for human health, is generally required only for potable reuse applications. In some cases, heavy metals and dissolved salts (measured as TDS and/or conductivity) can also be of health and environmental concern, as well as for industrial uses. In addition to microbiological pollution, agronomic parameters such as salinity, sodium and boron are very important parameters for the use of recycled water for irrigation.

### 8.2.2 Main treatment processes used for wastewater treatment and reuse

The choice of wastewater treatment trains includes the evaluation of treatment performance and reliability of the available technologies, their ability to consistently achieve target water quality limits, as well as various local conditions such as plant size, energy efficiency, existing infrastructure and equipment, distribution systems and so on. The final step is the techno-economic evaluation, including the financial resources for the capital and operating costs.

The main removal mechanism and fields of application for the most common wastewater treatment technologies are given in Table 8.3.

As a rule, proven technologies are selected for water reuse systems in order to consistently meet water reuse quality requirements. It should be noted that the performances and economic viability of tertiary treatment technologies are highly influenced by secondary effluent quality, load variations and other specific factors. For this reason, for innovative and advanced technologies, pilot-plant testing is highly recommended for the establishment of process efficiency and adequate design criteria, in particular for the production of high-quality recycled water.

Figure 8.1 summarizes the main treatment steps and the most common individual processes used in wastewater treatment and reuse. Such typical treatment trains for reuse purposes are common for municipal treatment plants, as well as for many industrial wastewater treatment facilities, as for example food and paper industries. As mentioned earlier, additional advanced treatment of secondary effluents for the production of high-quality recycled water with the removal of organic micropollutants, salinity and some specific compounds is required for potable reuse and several industrial uses such as process, cooling and boiler make-up water.

It is important to underline that usually in municipal wastewater treatment plants additional polishing is carried out only on a portion of secondary effluents depending on water reuse demand. This concept is also common for many industries where industrial and domestic wastewater is treated in the same facility. Nowadays, closing water cycle and zero liquid discharge are becoming an environmentally friendly objective for many industries and an increasing number of municipalities, in particular in coastal and environmentally sensitive areas. A new more sustainable design of water reuse systems is decentralized treatment and building of satellite treatment units near the point of use, especially for the supply of high-quality recycled water (Lazarova *et al.*, 2013).

Currently, anaerobic wastewater treatment is a common technology for food processing and agro-industrial wastewater with increasing interest for pulp and paper industries. In municipal wastewater treatment plants, anaerobic digestion is used mainly for sludge treatment (Van Lier *et al.*, 2020) and domestic sewage in some warm-climate countries, especially in Brazil, Colombia and India (Chernicharo, 2006). Only very recently anaerobic systems have been developed for treating cold and low-strength wastewater, including mainstream municipal sewage. Their main advantages are energy savings, reduced sludge production and lower investment costs (see Chapter 1), as well as additional benefits for agricultural irrigation due to elevated nutrient conservation (C, N and P). A number of advanced biotechnologies were developed for anaerobic treatment, but only upflow anaerobic sludge blanket (UASB) technology achieved the most wide and successful application for industrial and urban wastewater (Van Lier *et al.*, 2020).

Adequate pre- and post-treatment is essential for the success of wastewater treatment trains including anaerobic processes (van Lier, 2006). Anaerobic technologies are achieving only a partial treatment, requiring a proper post-treatment for the removal of organic matter and other constituents little affected by the anaerobic treatment, such as nutrients (N and P) and pathogenic organisms (viruses, bacteria, protozoa and helminths) in order to comply with discharge and/or reuse standard.

Proven disinfection technologies are chlorination, which is not recommended in some European countries (e.g. in UK) for wastewater treatment because the generation of toxic by-products, ultraviolet (UV) irradiation and ozonation. Reverse osmosis (RO), applied for salt removal, is also very efficient for the removal of organic micropollutants and microorganisms. As a rule, for the majority of reuse applications, final chlorination is required also for the RO permeate.

**Table 8.3** Main mechanisms and applications of treatment processes used for wastewater treatment and reuse.

| Process | Description | Application | Comments |
|---|---|---|---|
| *Conventional wastewater treatment processes* | | | |
| Sedimentation | Elimination of SS, including particles, sand and flocs | Pre-treatment, primary and secondary settling | Colloidal solids should be transformed into SS |
| Aerobic biological treatment | Transformation of biodegradable organic matters to $CO_2$ and $H_2O$ by microorganisms in the presence of oxygen | Wastewater containing significant quantities of organic matter | Several systems are available: lagoons, activated sludge, oxidation ditches, biofilm reactors |
| Anaerobic biological treatment | Transformation of biodegradable organic matters and biomass to $CH_4$, $CO_2$, $H_2O$ by microorganisms in the absence of oxygen | Wastewater containing high quantities of organic matter, high temperature or excess biomass from aerobic systems (sludge treatment) | Several systems are available: low-tech anaerobic lagoons and high-rate bioreactors with high biomass concentration in granular anaerobic-activated sludge and biofilm reactors – predominance at full-scale UASB and EGSB reactors with only a few full-scale AnBRM and other biotechnologies |
| Combination of aerobic/anaerobic/ anoxic systems | Combination of different types of microorganisms to reduce the amount of N and P | Usually applied as advanced treatment to achieve discharge standards, not for irrigation | Intensive interest and research on combined annamox systems and investigation of other advanced technologies such as FO, membrane distillation, microbial fuel cells, etc. |
| *Advanced treatment processes used for additional polishing before reuse* | | | |
| Coagulation–flocculation | Increase of solid size through addition of chemicals and particle aggregation | Usually before specific disinfection systems in combination with granular media filtration | Chemicals destabilize particles and improve SS removal |
| Filtration | Removal of particle and colloidal solids by retention in granular media or surface filtration (discs or micro-screens) | Removal of particles above a certain level defined by media characteristics | Mainly sand and activated carbon are used as granular filtration media with increasing interest in clothy, inox and plastic media for surface filtration, e.g. self-cleaning microfilters |
| Disinfection processes | Removal or inactivation of pathogens using heat, caustic chemicals, physical separation, chemicals, UV light, etc. | Crucial process for health protection before reuse | Chlorination and UV are the most common method with increasing interest in ozonation |
| Membrane processes | Pressure- or vacuum-driven membrane processes based on size exclusion or molecular diffusion | Removal of impurities: bacteria, viruses, dissolved salts, colloids, etc. and production of high-quality recycled water | Main systems: MF, UF, nanofiltration and RO – characterized by high-energy requirements |
| *Extensive and low-tech processes used in water reuse schemes* | | | |
| Extensive processes in liquid media | Biological degradation of organic matters in natural systems and disinfection by sunlight; macrophytes and/or algae could be a part of the system | Appropriate low cost treatment, in particular for small-sized treatment plants – high residence time and high land footprint | Main systems include stabilization and maturation ponds (lagooning), wetlands and algae ponds – lagooning has the additional advantage to ensure the storage of recycled water |
| Extensive treatments using low-rate infiltration | Biological degradation of organic matters and retention of solids and pathogens in solid media | Low cost treatment that can be used for small (IP) or large soil aquifer treatment (SAT) units | Main systems include SAT and infiltration–percolation (IF) |

EGSB, Expanded Granular Sludge Bed

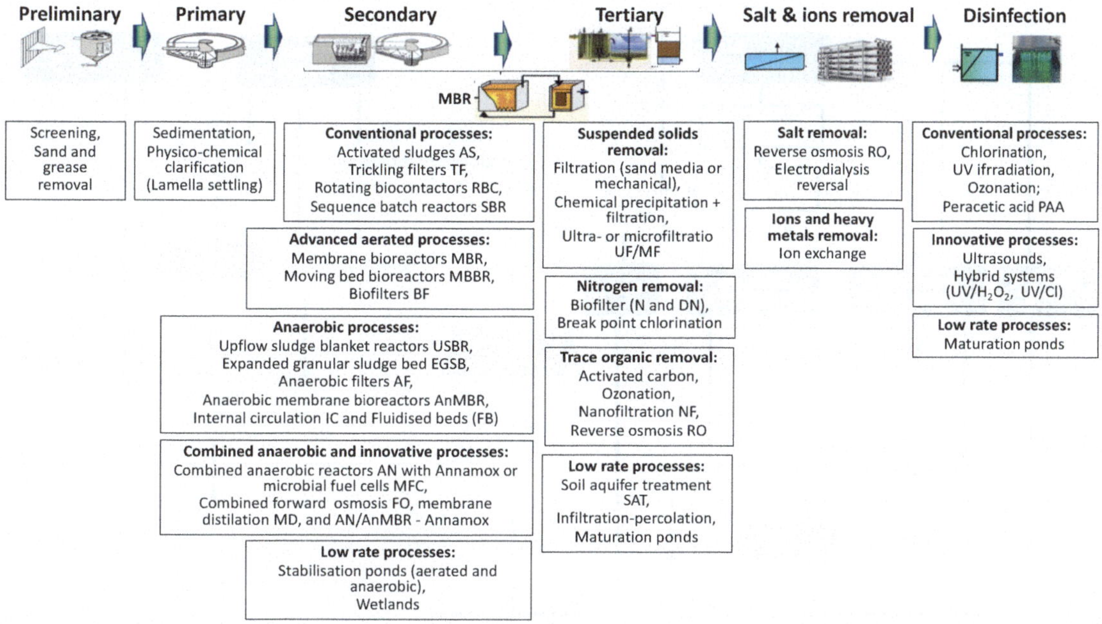


**Figure 8.1** Typical combinations of wastewater treatment processes for beneficial reuse of urban and industrial wastewater (*source*: adapted from Lazarova, 2005).

Although almost any treatment level can be attained with currently available technologies, it is important to remember that costs involved and sludge production increase almost exponentially as the treatment level rises. It is common for construction costs to double from primary to secondary treatment and to rise again by 50–100% from secondary to more advanced treatment. Similarly, energy demand and operation costs increase for high-rate and high-tech tertiary treatment technologies. Energy-intensive processes such as RO or other advanced hybrid technologies such as $UV/H_2O_2$ polishing should be limited to use for production of very high-quality water whereas alternative solutions might be more energy and cost effective.

## 8.3 SELECTION OF APPROPRIATE POLISHING TREATMENT FOR WATER REUSE

As mentioned earlier, tertiary treatment includes additional removal of SS, organic carbon and nitrogen, phosphorus precipitation, disinfection and in some cases complementary salt removal.

Nitrogen and phosphorus removal are not required for irrigation, but could be mandatory for discharge purposes, which is the case in Europe and many other countries for discharge in sensitive water bodies. More stringent standards for environmental and recreational uses such as <2–5 mg TN (total nitrogen)/L and <0.1–0.2 mg TP (total phosphorus)/L are unlikely to be feasible using traditional treatment processes. New advanced treatment processes, including for example oxidation and carbon adsorption, would be necessary to remove residual dissolved organic and nitrogen constituents to consistently achieve very low-nutrient levels.

Figure 8.2 shows a simplified decision-making flowchart of some typical tertiary treatment processes depending on the type of secondary effluents (aerobic or anaerobic), treatment capacity (plant size expressed in p.e., people equivalent) and water quality requirement for the removal of residual SS, carbon, nutrients, salts and for disinfection.

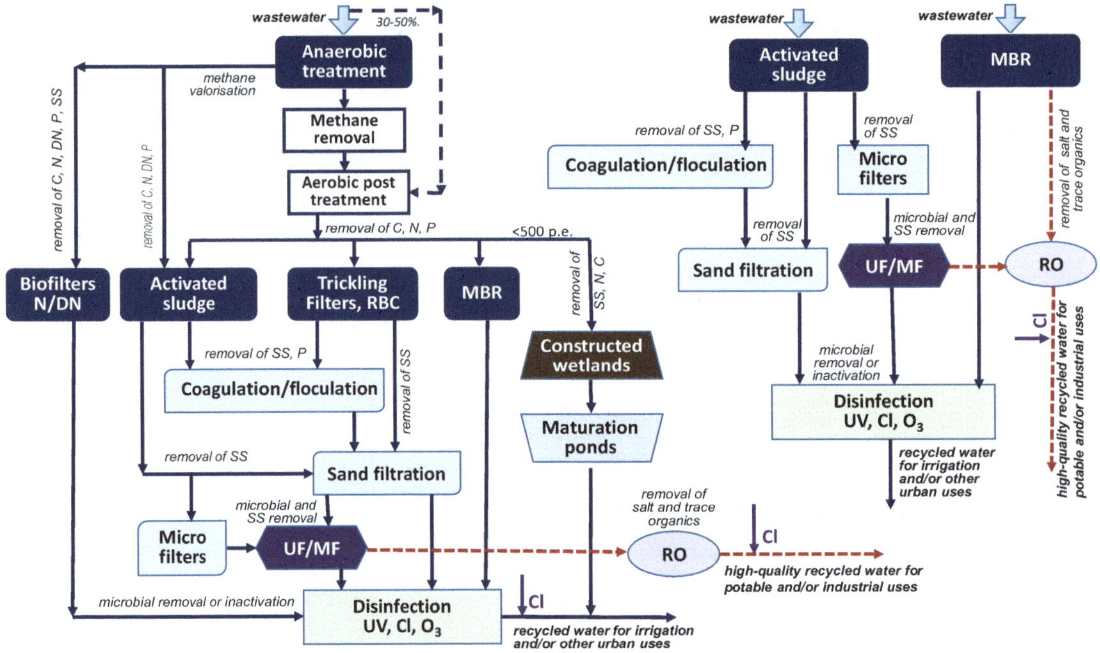


**Figure 8.2** Decision-making flowchart for the selection of appropriate wastewater tertiary treatment for non-potable reuse – illustration of typical treatment trains as a function of water quality requirements for carbon and nutrient removal.

In addition to target consent, other factors such as the quality of secondary effluents, plant size and other local specificities need to be taken into account to select an adequate and cost-competitive option.

Table 8.4 provides a summary of capacity, main design parameters, typical achievable water quality limits, costs, advantages and constraints of the most common treatment processes used for tertiary polishing of secondary effluents.

## 8.3.1 Specific features of post-treatment of anaerobic effluents

Compared to conventional secondary effluents, anaerobic effluents still contain a variety of residual compounds which require adequate post-treatment. The main water quality parameters of concern are as follows:

- Higher residual organic concentrations with COD well above the common threshold limits for wastewater discharge and reuse.
- Higher concentrations of organic micropollutants, including industrial chemicals, pesticides, pharmaceuticals, endocrine disruptors and so on.
- Higher content of nitrogen and phosphorus up to 50 mg N/L and 20 mg P/L, respectively.
- Presence of dissolved methane up to 25 mg/L that should be removed or recovered.

Combined anaerobic–aerobic systems provide good removal of organic carbon due to enhanced biodegradation of selected compounds under anaerobic or aerobic conditions. As a rule, residual COD and nitrogen from anaerobic effluents are removed by conventional and advanced aerobic biotechnologies such as activated sludge, biofiltration, trickling filters (TFs), membrane bioreactors

**Table 8.4** Main characteristics of selected mature aerobic post-treatment and tertiary treatment technologies.

| Process | Working Size | Design Parameters | Treated Water Quality (mg/L) | | | Costs | | Advantages | Disadvantages |
|---|---|---|---|---|---|---|---|---|---|
| | | | SS/BOD | COD/N-NH$_4$ | N/P | Capex | Opex | | |
| Activated sludge (SBRs, oxidation ditches) | >2000 p.e. (>500 p.e.) | Organic loading rates <0.25–0.5 kg BOD/m$^3$ day<br>F/M 0.05–0.15 kg BOD/kg MLSS day<br>HRT 24–48 h<br>Sludge age >8–20 days<br>Oxygenation rate >1.8–2 kg O$_2$/kg BOD (C + N removal)<br>Power for mixing 3–10 W/m$^3$ | <20–30 mg SS/L<br><10–20 mg BOD/L | <40–90 mg COD/L<br><2 mg N-NH$_4$/L | <20–60 mg N/L<br><6–15 mg P/L | High | High | Well approved<br>Very good effluent quality, including N removal and simultaneous P removal<br>Low footprint 0.1–0.25 m$^2$/p.e. | Need qualified operators and high surveillance<br>Need clarification, sludge disposal and telemetry to control aeration<br>Low tolerance of low temperatures and load variations<br>SBR systems require good automation and control of process cycles |
| MBRs | >500 p.e. | MLSS 8–12 g/L<br>Flux 25–60 L/m$^2$ h<br>Organic loading rates 0.05–0.66 kg BOD/m$^3$ day | <1 mg SS/L<br><0.1–0.5 NTU<br><1–5 mg BOD/L | <5–50 mg COD/L | <3–20 mg N/L<br><10 mg P/L<br><0.1 mg P/L *with chemicals* | Very high | Very high | Small footprint<br>High efficiency<br>Excellent effluent quality | High maintenance (require regular maintenance and membrane cleaning)<br>Qualified operators are needed |
| BAFs<br>*Can be combined with DN biofilters* | >1000 p.e. | C + N: 0.4–0.6 kg N-NH$_4$/m$^3$ day<br>Flow velocity 3–12 m$^3$/m$^2$ h<br>N: 1.2–1.6 kg NH$_4$/m$^3$ day<br>Flow velocity 3–12 (3–20) m$^3$/m$^2$ h | <5–10 mg SS/L | <0.2–1 mg N-NH$_4$/L | <0.5 mg P/L *with chemicals* | High | High | High efficiency<br>Simultaneous removal of SSs and biological oxidation<br>Easy automation<br>Good tolerance of shock loading | Need good pre-treatment to avoid any accidental sludge losses from clarifiers<br>Need qualified operation<br>Relatively high-energy demand |
| Tertiary TFs with humus tank<br>*Need to be combined with sand filtration* | >100–50,000 p.e. | Hydraulic loading rates <1 m$^3$/m$^2$ day<br>Surface loading rates 10–40 m$^3$/m$^2$ day<br>Max organic loading rates 0.07–0.4 kg BOD/m$^3$ day<br>Ammonia loading rates <1.2 g N-NH$_4$/m$^2$ day<br>Depth (min) of medium 4 m (plastic medium) | <15 mg BOD/L *only for two-stage*<br><15 mg SS/L *with additional filtration* | <5 mg N-NH$_4$/L *only for tertiary TFs* | Not removed<br>*15/2 only with additional tertiary treatment by filtration* | Relatively low | Relatively low | Well known<br>Free oxygen<br>Low maintenance | High residual SSs<br>Need clarification<br>Not resistant to load fluctuation (can be improved by recirculation rates: 2–2.5) |

(*Continued*)

**Table 8.4** Main characteristics of selected mature aerobic post-treatment and tertiary treatment technologies. (*Continued*)

| Process | Working Size | Design Parameters | Treated Water Quality (mg/L) | | | Costs | | Advantages | Disadvantages |
|---|---|---|---|---|---|---|---|---|---|
| | | | SS/BOD | COD/N-$NH_4$ | N/P | Capex | Opex | | |
| SAFs with integrated humus tank *Need to be combined with sand filtration* | <200–30,000 p.e. | Hydraulic loading rates 2–8 $m^3/m^2$ day<br>Organic loading rates <3 kg BOD/$m^3$ day<br>Oxygen supply rates 1–1.5 kg $O_2$/kg BOD<br>Loading rates of humus tank 2–8 $m^3/m^2$ day | <15 mg BOD/L if good design | <5 mg N-$NH_4$/L | Not removed<br>*15/2 only with additional tertiary treatment by filtration* | Relatively low | Moderate | Relatively low maintenance<br>Resistance to load fluctuations<br>Rapid start-up | Need clarification<br>Need backwashing<br>Aeration cannot be controlled<br>Restricted treatment capacity<br>High residual SSs |
| High-rate clarification with lamella settling | >10,000 p.e. | Range of velocity 20–140 m/h (20–25 m/h for P removal)<br>Need of microsand or sludge for floc stabilization | <5–10 mg SS/L | Not removed | N not removed<br><1 mg P/L | Moderate | High | Proven technology<br>Low power and maintenance requirements<br>Very short start-up<br>High reliability of operation | Extra sludge production<br>Lost of alkalinity<br>Operational problems related to microsand |
| Pressurized sand filters | >200–10,000–20,000 p.e. | Filtration rates 15–20 up to 25 $m^3/m^2$ h<br>Head loss up to 9 m | <15 mg BOD/L<br><10 mg SS/L | Not removed | N not removed<br><1 mg P/L *only with chemicals* | High | High | Very compact<br>Good automation<br>Proven technology | Complex and difficult inspection<br>Need good operation and maintenance |
| Rapid sand filters | >200 to >100 000 p.e. | Surface loading rates 5–15 $m^3/m^2$ h<br>(1) conventional: depth 0.75–1.0 m; sand size 0.4–0.8 mm<br>(2) deep bed: depth up to 1.8–2.0 m; sand size 0.9–1.5 mm | <15 mg BOD/L<br><10 mg SS/L<br><5 mg SS/L *with chemicals* | Not removed | N not removed<br><1 mg P/L *only with chemicals* | Moderate-to-high | Moderate | Proven technology<br>Good automation<br>No size limitation<br>Low backwash volumes (1–3%) | Need backwashing<br>Need good operation |
| Continuously cleaned sand filters | >200–50,000 p.e. | Surface loading rates 5–10 $m^3/m^2$ h<br>Media depth 1–4 m<br>Sand size 0.9–1.3 mm | <5–10 mg SS/L | Not removed | N not removed<br><10–15 mg N/L *with methanol*<br><0.5–1 mg P/L *only with chemicals* | High | Moderate | Small footprint<br>No backwash holding and pumps required<br>Proven technology | Complex operation<br>Sand abrasion<br>High backwash volumes 5–15%<br>Sensitive to high-solid loading |
| Microscreens | >200–50,000 p.e. | Typical filtration velocity 15–20 m/h (10–35 m/h)<br>Opening size 10–35 μm<br>Head loss <50 cm (8–20 cm)<br>Rotation speed 4–5 m/min | <15–25 mg SS/L | Not removed | Not removed<br><2 mg P/L *with chemical precipitation and adapted devices* | Moderate (lower than sand filters) | High (higher than sand filtration) energy 18 Wh/$m^3$ | Low footprint<br>Low head loss<br>Higher loading rate than sand filters<br>Easy integration into the piezometric line | Incomplete solid removal<br>High sensibility to solid fluctuation<br>Chemical cleaning is periodically necessary<br>High backwash volumes of up to 15% |

(*Continued*)

**Table 8.4** Main characteristics of selected mature aerobic post-treatment and tertiary treatment technologies. (*Continued*)

| Process | Working Size | Design Parameters | Treated Water Quality (mg/L) | | | Costs | | Advantages | Disadvantages |
|---|---|---|---|---|---|---|---|---|---|
| | | | SS/BOD | COD/N-NH$_4$ | N/P | Capex | Opex | | |
| Stabilization ponds | >200 to max 5000 p.e. | Anaerobic: $H = 2.5$–5 m, 20–50 days, 220–560 kg BOD/ha day<br>Facultative: 1.5–2.5 m, 25–180 days, 50–200 kg BOD/ha day<br>Aerated: 3–5 m, 5–20 days<br>Maturation: 0.5–1.5 m, 2.5 m$^2$/p.e., 15–30 days | <20–120 mg SS/L<br><5–20 mg BOD/L | <2 mg N-NH$_4$/L | <5–25 mg N/L<br><2–6 mg P/L | Relatively high depending on land cost | Low | Easy to operate<br>Low operating costs<br>Good efficiency for carbon removal<br>Maturation ponds: provide disinfection and storage capacity<br>Cost efficient for small units | Large footprint<br>Effluent quality is a function of hydraulic loads, climate and season<br>High evaporation rates<br>Sludge disposal requires infrequent but intensive work<br>Nuisance (odour, mosquitos) |
| Constructed wetlands (vertical flux) | 100–500 p.e. In specific cases up to 1000 p.e. | 20–25 g BOD/m$^2$ day<br>4 plants/m$^2$<br>2.3 m$^2$/hab<br>nitrification: nitrogen load 7 g NTK/m$^2$ day<br>hydraulic load 180 mm/day | <10–20 mg SS/L<br><10 mg BOD/L | <10 mg N-NH$_4$/L | – | Relatively low | Low (no power needed if topography allows it) | Simultaneous sludge treatment<br>Simple operation<br>Good tolerance to seasonal variations<br>Good nitrification | Large footprint and land requirement<br>Annual removal of vegetation·<br>Complete cleaning of the bed every 10–15 years·<br>Risk of insects and odours<br>Thermal protection is necessary for long periods of freezing |
| RBC – rotating biological contactors (multi-stage RBCs) | Between 100 and 2000 p.e. | Hydraulic loading rates 0.02–0.12 m$^3$/m$^2$<br>Organic loading rates 0.04–0.05 kg BOD/m$^3$ day<br>Full nitrification: <2.5–5 g BOD/m$^2$ day<br>HRT min 1 h<br>Rotor speed <0.35 m/s | <15–30 mg SS/L<br><15–30 mg BOD/L<br>(<7–15 mg BOD/L) | (<2 mg N-NH$_4$/L) | – | Low | Low | Compact technology<br>Low power consumption·<br>Good nitrification<br>Good landscape integration·<br>Good adaptation to cold climates | Good maintenance and rotor lubrication<br>Adequate sludge removal (once every 4 weeks)<br>Weak denitrification<br>Moderate capital costs<br>Need for sludge storage and evacuation |

MLSS, mixed liquor suspended solids; HRT, hydraulic residence time

(MBRs) and moving bed bioreactors (MBBRs). For small systems, low-tech natural treatment systems are more appropriate, including constructed wetlands and polishing ponds. Biofilm processes such as biofilters, TFs and constructed wetlands, as well as MBBRs are characterized by higher removal efficiency and better tolerance of variations of hydraulic and organic loads.

### 8.3.2 Removal of SS and nutrients

The most common tertiary treatment technologies for additional removal of SS and nutrients are as follows:

- High-rate clarification using chemical precipitation followed by lamella clarification (TSS and TP removal).
- Sand filtration with or without chemical dosing (TSS and TP removal, removal of organic N).
- Biofiltration: aerated biofilters (nitrification) or denitrifying biofilters with methanol addition.
- TFs or submerged-aerated filters (SAFs) with nitrification.

In the case of anaerobic mainstream wastewater treatment with requirements for nitrogen removal, potential deficit of biodegradable organic carbon for denitrification can be avoided by methane recovery or by-pass of 30–50% of the influent raw sewage directly to biological post-treatment (Chernicharo, 2006). Phosphorus removal in treatment plants using anaerobic reactors is only possible by physico-chemical treatment using chemicals for P precipitation.

TFs and SAFs are less complex and less expensive treatment processes; however, they only provide additional nitrification. To comply with the requirements for SS and phosphorus removal, both TFs and SAFs should be combined with sand filtration. Because of their ease of operation, SAFs are well adapted for small and intermediate-to-medium sized wastewater reclamation plants.

Granular biological-aerated filters (BAFs) could achieve simultaneously good nitrification and solid removal, including phosphorus (with addition of chemicals). Nevertheless, BAFs are more sensitive to solid overloading compared to TFs and SAFs (sludge loss in secondary clarifiers).

The most compact filtration technologies are microscreens and pressure sand filters. Both technologies require good operation, repair and maintenance, which are more complex for pressure filters. The main constraint of microscreens is that P removal is not possible and SS content in effluents is higher, thus limiting its application to less stringent standards.

Good P removal can be achieved by chemical precipitation and/or filtration. In the case of high-solid content in secondary effluents (>30 mg TSS/L), high-rate clarification with lamella settling is the best option. Rapid sand filters could be the best option for well-clarified secondary effluents (<25 mg TSS/L). Continuously backwashed filters have the additional advantage to avoid filter bed clogging, but these devices have more complex operation, higher susceptibility to failure, production of higher volume of backwash water and additional operational problems associated with sand abrasion of pipes and pumps.

If total nitrogen removal is required with additional denitrification, BAFs are a good option as a combination of denitrifying and nitrifying filters.

### 8.3.3 Membrane technologies

Tertiary membrane filtration, microfiltration (MF) and ultrafiltration (UF), are the most costly treatment option which provides, however, the best effluent quality with almost complete solid and pathogen removal. In the case of stringent requirement for disinfection, this solution could be cost-competitive compared to complex conventional tertiary treatment by coagulation/flocculation, sand filtration and UV or ozone disinfection.

SS, bacteria, parasites and viruses are readily removed by UF and MF, in exception of viruses that had passed through the more porous MF membranes (0.2 μm cut-off), in particular after chemical cleaning. The use of MF or UF produces recycled water quality which is suitable for unrestricted irrigation and urban uses. In many cases, a pre-treatment of secondary effluent by sand or cloth filtration has been implemented before tertiary UF and MF systems in order to improve their operating efficiency.

Nowadays, the combination of MF or UF with RO is recognized as a 'multiple barriers' polishing of municipal and industrial effluents and is considered as the best available technology for high-quality water reuse applications, including potable and industrial reuse. The main advantages of combined membrane systems are the high efficiency of removal of organic micropollutants, pathogens and dissolved solids (desalination). Despite the development of new membrane technologies, pre-treatment step plays a critical role in membrane performance, for example, cleaning, energy consumption and operating costs.

Among the membrane processes used for wastewater treatment, MBR technology is advancing most rapidly and is becoming a mature proven technology. Currently, the majority of MBR systems are implemented for municipal wastewater treatment using submerged membranes for solid/liquid separation of the mixed liquor, whereas side-stream MBR configuration (external recirculation of the mixed liquor) is the preferred technology for industrial wastewater treatment. It is important also to underline that MBRs are considered as the state-of-the-art technology for decentralized systems, in particular in-building wastewater treatment and recycling plants. Compared to tertiary MF or UF, MBR systems have demonstrated a better reliability of operation, in particular lower fouling.

Recent studies demonstrated that the combination of MBRs and RO could be a competitive option for advanced water reuse, ensuring improved removal of micropollutants. As MBR systems are increasingly implemented for municipal wastewater treatment instead of conventional activated sludge, it is easier and cheaper to add RO polishing for the treatment of only to the part of the effluent that is needed for water reuse.

New membrane technologies and hybrid combinations are intensively studied in laboratory and pilot scale in order to increase treatment and cost efficiency. Forward osmosis (FO) and pressure-retarded osmosis are investigated as an energy-efficient solution. These membrane processes use natural osmosis phenomenon for water transport through a highly selective FO membrane. Various combinations of FO/MBR or FO/MBR/NF and new configurations of MBRs are also under investigation.

### 8.3.4 Disinfection

Despite significant advances in science and technology, the three mature disinfection technologies used for wastewater disinfection worldwide remain to be chlorination, UV disinfection and ozonation. Although the use of chlorination has been declining over the last decade, this technology remains the most commonly used, in particular in the USA. In the same time, UV disinfection has grown at an accelerated rate in Europe and all over the world. New technologies such as electro-chlorination and hybrid disinfection processes have been developed and implemented.

The success of UV technology is largely attributable to low costs, the absence of toxic by-products and the ability to consistently achieve almost complete wastewater disinfection (Lazarova *et al.*, 2011). Only a few wastewater treatment plants are currently using ozonation, but the interest in this technology is growing because this is the only disinfection alternative capable of partially or completely remove trace organics, including pharmaceuticals and endocrine disruptors at typical disinfection doses.

UV disinfection is the most sensitive process compared to chlorination and ozonation. As a rule, for efficient UV disinfection, the average values of influent transmittance should be more than 50% and residual concentration of SS <15 mg/L (Savoye *et al.*, 2001). Stringent standards for almost total disinfection (coliforms below the detection limits) could be consistently achieved only under the condition of complete removal of SS (TSS <5 mg/L) and with high-UV doses up to 60–120 mJ/cm$^2$.

Even if ozonation is less impacted by SS, designed ozone doses are also strongly influenced by upstream treatment. Depending on the target disinfection level, the required transferred ozone doses for municipal wastewater disinfection could be from 5 to 15 mg $O_3$/L with contact times of <10 min and good mixing (Lazarova *et al.*, 2011).

Additional advantage of ozonation for treatment of industrial wastewater and anaerobic effluents is the removal of refractory organics measured by the COD (up to 50% of total COD depending on chemical compounds), enhanced biodegradability of residual organics and very efficient colour removal. Advanced oxidation (ozonation, $H_2O_2$/UV, Fenton, etc.) should be considered for stringent standards with requirements for low-residual COD concentrations.

## 8.4 CONCLUSIONS

During the last few years, water reuse has become a significant alternative to conventional water supply sources to meet water needs and respond to climate changes. This applies either to large-scale urban areas, rural or small-scale on-site recycling schemes and industrial facilities worldwide. With the increasing concerns of energy efficiency and rising costs, the role of anaerobic treatment is growing.

Despite the numerous proven advantages, anaerobic biotechnologies require adequate post-treatment to comply with wastewater discharge and reuse standards. Compared to conventional wastewater treatment trains based on aerobic biological treatment, tertiary treatment technologies that can be used are similar but are implemented after the aerobic post-treatment of anaerobic effluents.

This chapter provides guidance on the selection of suitable wastewater treatment and reclamation schemes for urban wastewater and industrial wastewater from food and agro-industries which reflects the current practical experience worldwide. Both urban and industrial water reuse standards are presented with indication of achievable treatment level of the most common and proven technologies enabling to consistently meet water quality limits for SSs, carbon, nutrients, pathogens and some other parameters such as salinity and organic micropollutants.

As underlined in this chapter, the selection of adequate tertiary treatment options should include a number of other considerations, including treatment capacity (plant size), local conditions, existing infrastructure, land availability, costs, ease of operation and so on. The presented design and performance parameters for selected tertiary treatment technologies are given only for information. Professional judgement and experience are necessary to apply them for the specific conditions of a given project.

Rapid advances in science and technology are greatly contributing to the development of new breakthrough technologies such as innovative biotechnologies, FO, microbial fuel cells, membrane distillation, new advanced oxidation methods and so on, but their scale-up and full-scale validation need further research. There is no doubt that the combination of anaerobic and innovative technologies will contribute to the implementation of energy neutral and energy positive wastewater treatment plants with recovery of many other beneficial resources and water reuse for various uses currently supplied with drinking water from conventional water resources.

## REFERENCES

Asano T., Burton F. L., Leverenz H., Tsuchihashi R. and Tchobanoglous G. (2007). Water Reuse: Issues, Technologies, and Applications. McGraw Hill, New York.

Ayers R. S. and Westcot D. W. (1985). Water Quality for Agriculture; FAO Irrigation and Drainage Paper R1. Food and Agriculture Organization of the United Nations, Rome, Italy.

Chernicharo C. A. L. (2006). Post-treatment options for the anaerobic treatment of domestic wastewater. *Reviews in Environmental Science and Bio/Technology*, **5**, 73–92, https://doi.org/10.1007/s11157-005-5683-5

EU Regulation (2020). Regulation (EU) 2020/741 of the European Parliament and of the Council of 25 May 2020 on minimum requirements for water reuse. Official Journal of the European Union, https://eur-lex.europa.eu/eli/reg/2020/741/oj (accessed August 2023).

FAO and IWMI (2023). Water Quality in Agriculture: Risks and Risk Mitigation. In: The Food and Agriculture Organization of the United Nations and International Water Management Institute, Drechsel P., Marjani Zadeh S. and Pedrero Salcedo F. (eds.), Rome, 186pp.

IWA (2016). Water Utility Pathways in a Circular Economy, http://www.iwa-network.org/wp-content/uploads/2016/07/IWA_Circular_Economy_screen.pdf (accessed 15 June 2018).

Lazarova V. (2005). Wastewater treatment for water recycling. In: Irrigation with recycled water: agriculture, turfgrass and landscape, V. Lazarova and A. Bahri (eds.), CRC Press, Boca Raton, FL, USA.

Lazarova V. (2022). Water reuse – a pillar of the circular water economy. In: Resource Recovery from Water: Principles and Application, Pikaar *et al.* (ed.), IWA Publishing, London, UK.

Lazarova V. and Asano T. (2013). Milestones in water reuse: main challenges, keys to success and trends of development. An overview. In: Milestones on Water Reuse: The Best Success Stories, V. Lazarova, T. Asano, A. Bahri and J. Anderson (eds.), IWA Publishing, London, UK, pp. 1–22.

Lazarova V. and Bahri A. (2004). Water Reuse for Irrigation: Agriculture, Landscapes, and Turf Grass. CRC Press, Boca Raton, USA.

Lazarova V., Feran B. and Savoye Ph. (2011). Influence of wastewater quality on disinfection efficiency and reliability of operation of UV disinfection and ozonation. Proceedings of IOA IUVA World Congress & Exhibition, 23–27 May 2011, Paris, France, 9pp.

Lazarova V., Asano T., Bahri A. and Anderson J. (eds) (2013). Milestones on Water Reuse: The Best Success Stories. IWA Publishing, London, UK.

NRMMC (2006) Australian guidelines for water recycling: managing health and environmental risks (phase 1). Natural Resource Management Ministerial Council (NRMMC), Environment Protection and Heritage Council (EPHC), Australian Health Ministers' Conference, Canberra, Australia.

Olivieri A. W., Crook J., Anderson M. A., Bull R. J., Drewes J. E., Haas C. N., Jakubowski W., McCarty P. L., Nelson K. L., Rose J. B., Sedlak D. L. and Wade T. J. (2016). Expert Panel Final Report: Evaluation of the Feasibility of Developing Uniform Water Recycling Criteria for Direct Potable Reuse. Prepared August 2016 by the National Water Research Institute for the State Water Resources Control Board, Sacramento, CA.

Pescod M. (1992). Irrigation and Drainage Paper 47, Wastewater Treatment and Use in Agriculture. Food and Agriculture Organization of the United Nations (FAO), Rome, Italy.

Pikaar I., Guest J., Ganigué R., Jensen P., Rabaey K., Seviour T., Trimmer J., Van der Kolk O., Vaneeckhaute C. and Verstraete W. (2022). Resource Recovery from Water: Principles and Application. IWA Publishing, London, UK, 445pp.

Rosenwinkel K. H., Borchmann A., Engelhart M., Eppers R., Jung H., Marzinkowski J. and Kipp S. (2013). Closing loops – industrial water management in Germany. In: Milestones on Water Reuse: The Best Success Stories, V. Lazarova, T. Asano, A. Bahri and J. Anderson (eds.), IWA Publishing, London, UK, pp. 243–258.

Savoye P., Janex M. L. and Lazarova V. (2001). Wastewater disinfection by low-pressure UV and ozone: a design approach based on water quality. *Water Science and Technology*, **43**(10), 163–171, https://doi.org/10.2166/wst.2001.0609

State of California (2000). Water Recycling Criteria. Title 22, Division 4, Chapter 3, California Code of Regulations. California Department of Health Services, Drinking Water Program, Sacramento, CA.

USEPA guidelines (2012). Guidelines for Water Reuse. U.S. Environmental Protection Agency, EPA/600/R-12/618, Washington, DC, USA.

Van Lier J. B. (2006). Anaerobic industrial wastewater treatment; perspectives for closing water and resource cycles. Paper presented at Invited Key-Note Speaker. https://edepot.wur.nl/39480

Van Lier J. B., Mahmoud N. and Zeeman G. (2020). Anaerobic wastewater treatment. In: Biological Wastewater Treatment: Principles, Modelling and Design, G. Chen, M. C. M. van Loosdrecht, G. A. Ekama and D. Brdjanovic (eds.), IWA Publishing, London, UK, pp. 701–756.

WHO (2006). Guidelines for the Safe Use of Wastewater, Excreta and Greywater; WHO-UNEP-FAO. World Health Organization, Paris, France, Volume II.

doi: 10.2166/9781789063479_0255

# Chapter 9

# Fate of micropollutants

G. H. D. de Oliveira[1], G. Lovato[2], J. A. D. Rodrigues[2] and M. Zaiat*[1]

[1]São Carlos School of Engineering, São Paulo University (EESC/USP), Av. Trabalhador São Carlense 400, 13.566-590, São Carlos, SP, Brazil

[2]Mauá School of Engineering, Mauá Institute of Technology (EEM/IMT), Praça Mauá 1, 09.580-900, São Caetano do Sul, SP, Brazil

*Corresponding author: zaiat@sc.usp.br

**ABSTRACT**

Micropollutants (MPs) are biological or chemical compounds resulting from human activities that make their way into water bodies in trace quantities. They cause adverse effects on aquatic environments and their complexity and costly quantification makes them difficult to monitor and, consequently, the implementation of legislation for controlling their disposal. Thus, the aim of this chapter is to describe occurrence, environmental and health impacts, and current regulatory frameworks of MPs. The fate and removal of these contaminants in anaerobic reactors treating domestic wastewater is discussed and strategies for enhancing MP removal are presented. Studies on MP removal in anaerobic systems are still emerging and a great deal of work should be carried out to evaluate whether conventional anaerobic reactors applied to domestic wastewater treatment under usual operating conditions are able to effectively remove contaminants of emerging concern. Transferring the mechanistic understanding of the anaerobic biotransformation of MPs to feasible changes to be implemented in mainstream anaerobic domestic wastewater treatment remains a major challenge. Moreover, the study of new operating strategies and reactor configurations seems to be mandatory to comply with the requirements of removing organic matter, nutrients, and MPs, as well as generating energy (biogas), thus resulting in robust, safe, and sustainable units.

**Keywords:** anaerobic digestion, domestic sewage, emerging contaminants, legislation, micropollutant, pesticides, pharmaceutical residues, sweeteners.

## 9.1 MICROPOLLUTANTS IN DOMESTIC WASTEWATER

Micropollutants (MPs) or emerging contaminants include new pollutants that, until recently, were not detected. They are biological or chemical compounds resulting from human activities that make their way into water bodies in trace quantities (at or below $\mu$g/L). The complexity of MPs and their costly quantification makes them difficult to monitor and, consequently, the implementation of legislation for controlling their disposal (Aquino *et al.*, 2013). A few examples of common MPs are pharmaceutical residues, endocrine disruptors, plasticizers, pesticides, sweeteners, and personal hygiene products.

Sources of MPs in the environment are diverse. Pharmaceuticals mainly come from domestic wastewater (from excretion), hospital effluents, and surface run-off from concentrated animal feeding

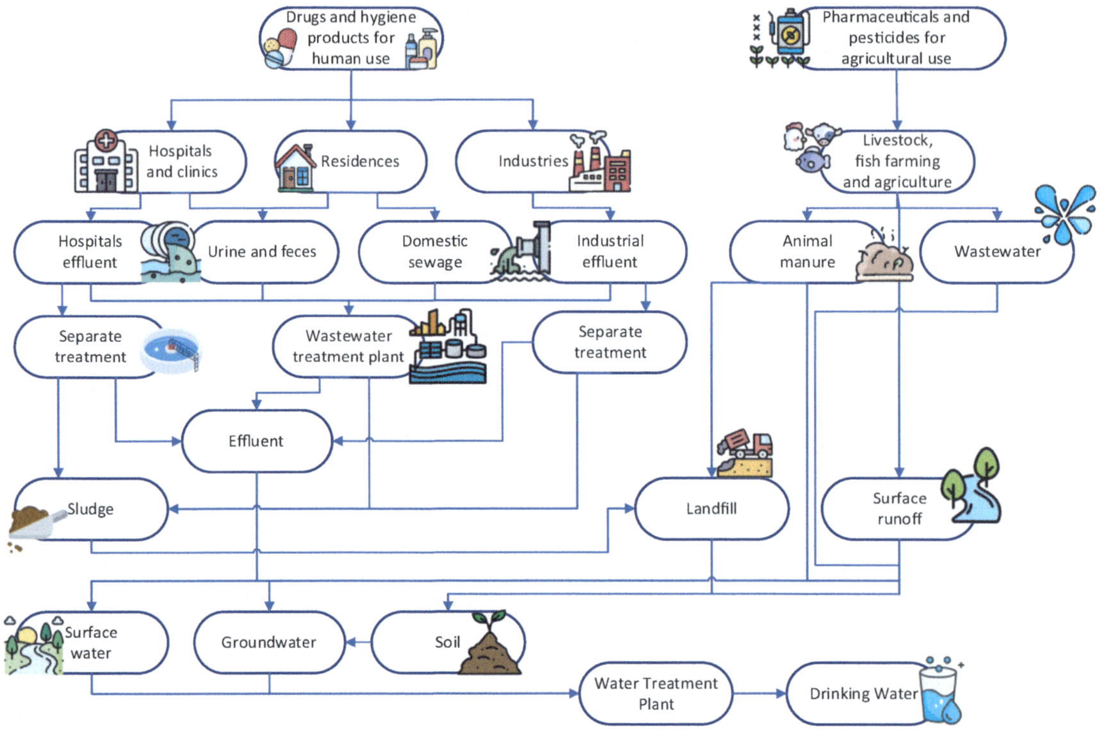


**Figure 9.1** Contamination pathway for MPs.

operations and aquaculture. MPs from personal care products, such as fragrances, disinfectants, and insect repellents, mostly come from household sources, such as bathing, shaving, spraying, and swimming. Steroid hormones come from human excretion and livestock farming. Non-ionic surfactants, plasticizers, and fire retardants come from industrial and domestic wastewater (from production and by leaching out of the material, respectively). MPs from pesticides, such as insecticides, herbicides, and fungicides can come from improper cleaning, run-off from gardens, lawns, roadways, and agricultural run-off (Luo *et al.*, 2014). Figure 9.1 shows the contamination pathways for MPs.

Contamination of water and sewage by MPs has adverse effects, such as resistance in pathogenic bacteria and reduced diversity of bacteria and algae in running water (Carey *et al.*, 2016). These substances cause changes in the behavior and morphology of aquatic biota (Chen *et al.*, 2014; Corcoll *et al.*, 2015; Petersen *et al.*, 2014) and interfere with the hormonal system of wildlife and humans. Several negative effects are often associated with short- and long-term toxicity, regardless of concentrations in the order of ng/L. Even with the partial elucidation of the consequences and toxicity of exposure to these MPs, most countries do not have legislation to impose a maximum limit for the concentration of MPs in water bodies.

Therefore, the aim of this chapter is to describe the occurrence, environmental, and health impacts, and current regulatory frameworks for MPs. The fate and removal of MPs in anaerobic reactors treating domestic wastewater is then discussed and strategies for enhancing MP removal are presented.

### 9.1.1 Occurrence

Data on the presence of MPs in water bodies can be categorized into: (1) occurrence in wastewater treatment plant (WWTP) influents and effluents; (2) occurrence in surface water; (3) occurrence in groundwater; and (4) occurrence in drinking water. Due to the large number of existing MPs and the cost of their analysis, a limited number of compounds is usually monitored in environmental samples. Frequently, the targeted compounds are chosen based on their frequency of occurrence, or as representatives of particular MP classes. The reported concentrations of MPs in WWTPs vary significantly in space and time due to the rate of production of MPs, sales and practices of a given community, water consumption and use, and size and efficiency of WWTPs (Jelic *et al.*, 2011; Luo *et al.*, 2014). Particularly, the excretion rate plays a crucial role in determining the introduction of pharmaceuticals into raw wastewater: drugs such as aspirin and carbamazepine have an excretion rate lower than 5%; medicines such as amoxicillin, ciprofloxacin, and tetracycline excretion rates are over 70% of the ingested mass. However, the massive use of the former drugs usually leads to high concentrations in domestic effluents. Pharmaceuticals found with the highest concentration in WWTP influents, according to a comprehensive review performed by Luo *et al.* (2014), are ibuprofen (0.004–603 $\mu$g/L), followed by caffeine (0.22–209 $\mu$g/L) and diclofenac (0.001–94.2 $\mu$g/L). Insect repellant *N,N*-diethyl-meta-toluamide has been tracked in concentrations between 2.56 and 3.19 $\mu$g/L and triclosan, an MP found in personal care products, was found in concentrations of 0.03–23.9 $\mu$g/L. In contrast, steroid hormones were found in wastewater at much lower levels (below 100 ng/L). However, their occurrence even at low concentrations is a concern because of their high estrogenic effect. The use of pesticides can be seasonal due to the prevalence of pests under different climatic conditions, and rainfall and sunlight affect the flow pattern of these MPs. Herbicides were found in the range of 0.02–28 $\mu$g/L, insecticides between 0.0007 and 4.16 $\mu$g/L and fungicides below 1.89 $\mu$g/L (Kasprzyk-Hordern *et al.*, 2009; Luo *et al.*, 2014).

The release of WWTP effluents into water bodies has been considered a main cause of the presence of MPs in surface water. Following treatment processes in WWTPs, MPs are subjected to dilution in surface water, sorption onto suspended solids and sediments, direct and indirect photolysis, and biodegradation. Due to river water dilution, pharmaceutical compounds may occur at levels at least one order of magnitude lower than effluent levels (Gros *et al.*, 2007; Luo *et al.*, 2014). Non-steroidal anti-inflammatory drugs, carbamazepine, sulfamethoxazole, and triclosan are the most frequently reported compounds in surface water. However, caffeine is an MP reported at the highest concentration in countries such as Costa Rica, Taiwan, and the USA (up to $10^6$ ng/L) (Luo *et al.*, 2014).

Groundwater has been found to be less contaminated with MPs than surface water. Better characterization of MPs in groundwater has been only done in some parts of Europe and North America. MP contamination of groundwater mainly results from landfill leachate, groundwater–surface water interaction, infiltration of contaminated water from agricultural land or seepage of septic tanks, and sewer systems. In France, Germany, Spain, and the USA, most MPs were detected at below 100 ng/L (Luo *et al.*, 2014).

Publications reporting the occurrence of MPs in drinking water are scarce. Some recent studies showed that most MPs in tap water were below the limit of quantitation or limit of detection and that drinking water treatment plays a significant role in eliminating MPs from drinking water. The maximum occurrence concentrations of most MPs were reported to be below 100 ng/L in France, Spain, the USA, and Canada except for carbamazepine (1000 ng/L) and caffeine (200 ng/L) (Luo *et al.*, 2014). However, two recent studies conducted in China report the presence of pesticides and flame retardants in the region of Taihu Basin, in which even advanced treatment procedures were inefficient in removing these contaminants (Ren *et al.*, 2020). In the region of northeast China (Heilongjiang, Jilin, and Liaoning provinces), 19 kinds of pesticides, 6 kinds of organophosphates, 2 kinds of phthalates, and 22 kinds of pharmaceutical and personal care product pollutants were found in drinking water samples (Wang *et al.*, 2022).

### 9.1.2 Environmental and health impacts

So far, little is known about the possible chronic health effects associated with the long-term ingestion of MP mixtures in drinking water. Furthermore, after entering an aquatic environment, MPs are subject to natural attenuation processes, such as microbial degradation, photodegradation, sorption in suspended particulate matter, and deposition in sediments. All these processes influence the fate and toxicity of contaminants in water bodies (Gibs *et al.*, 2013; Yang *et al.*, 2011). One must also consider the predicted no-effect concentrations of these MPs, which represent the concentration of a substance below which an unacceptable effect most likely will not occur. The known environmental and health impacts of a few major MPs are described below.

(1) Ibuprofen: a drug from the group of non-steroidal anti-inflammatory drugs, used to treat pain, fever, and inflammation. It is a popular over-the-counter drug with a high daily therapeutic dose (600–1200 mg/day), of which 70–80% is excreted. Along with diclofenac, it was considered a priority for studies by the *Global Water Research Coalition*. While ibuprofen readily degrades under aerobic conditions, it exhibits resistance to anaerobic degradation. Among the adverse effects caused by its contamination, the induction of vitellogenin (a protein present only in females) in mussels exposed to the drug was observed (Gonzalez-Rey & Bebianno, 2014).

(2) Diclofenac: a popular drug, also belonging to the class of non-steroidal anti-inflammatory drugs, with analgesic, antipyretic, and anti-inflammatory properties. Diclofenac has the highest acute toxicity of this type of drug and has been detected in water and sewage treatment plants due to its high level of consumption and resistance to biodegradation in the aquatic environment. Considered a priority for study, its presence in water bodies causes adverse effects, such as hormonal changes in aquatic beings. It has a significant deleterious effect on the survival and reproduction of fish and zooplanktonic and benthic organisms (an effect also caused by propranolol) (Fent *et al.*, 2006).

(3) Propranolol: an antihypertensive drug used for the treatment and prevention of myocardial infarction, angina pectoris, and cardiac arrhythmias. Although designed for human use, it is equally effective when used in physiological studies in fish and causes growth disorders in invertebrates at concentrations above 0.5 mg/L (Fent *et al.*, 2006).

(4) Triclosan: bactericide used in medicines, cosmetics, and personal hygiene products. In Brazil, the National Sanitary Agency allows a concentration of up to 0.3% of triclosan in personal care products, although it is prohibited in the USA. There are indications that prolonged exposure to products with bactericides can generate bacterial resistance and hormonal changes. It can cause oxidative stress in fish and mollusks, and at 500 ng/L of triclosan, changes were observed in the levels and activities of some enzymes in such organisms (Ku *et al.*, 2014).

(5) Carbamazepine: a medicine that belongs to the group of antiepileptics. Carbamazepine is one of the most persistent drugs of concern in water bodies receiving effluents and is resistant to both conventional and advanced wastewater treatment. In addition to being stable in the environment, studies show that it is captured by plants irrigated with wastewater, as well as its bioaccumulation in aquatic organisms. Furthermore, as a psychoactive drug, carbamazepine has the potential to make fish more vulnerable to predators (Keen *et al.*, 2012).

(6) Atenolol: indicated for cardiac patients, atenolol is mainly used to control arterial hypertension and angina pectoris. Atenolol is part of the group of beta-blocker drugs and is widely prescribed worldwide. Atenolol is a drug regularly found in wastewater, as 90% of it is eliminated in feces and urine and is captured by plants and vegetables (Beltrán *et al.*, 2020).

(7) Ciprofloxacin: fluoroquinolone antibiotic is usually applied to treat bacterial infections. Given the high usage of ciprofloxacin, it can be found in hospital and industrial wastewater at over 31 mg/L. Ciprofloxacin is of particular concern since genotoxic and mutagenic effects have been reported in bacteria (Kümmerer *et al.*, 2000; Richard *et al.*, 2014).

(8) Bisphenol A: a chemical produced in large quantities used primarily in the production of polycarbonate plastics. It is found in various products including shatterproof windows, eyewear, water bottles, and epoxy resins that are used to coat some metal food cans, bottle tops, and water supply pipes. Bisphenol A is a suspected anthropogenic endocrine disruptor. Although it

is less estrogenic toward aquatic organisms than natural hormones, it has been reported to be able to induce feminization phenomena in various species of animals at high concentrations (Metcalfe *et al.*, 2001). Golub *et al.* (2010) concluded that bisphenol A mainly affects offspring viability, sex differentiation, immune hypersensitivity, and gender-differentiated morphology, thus affecting the endocrine system when prenatally exposed (Richard *et al.*, 2014).

(9) Caffeine: in the surface water quality report, caffeine appears as a chemical quality indicator, used as a tracer of the presence of human excreta and some substances from the contaminant group (Bernegossi, 2019). The mutagenic effects of caffeine can cause increased cell death (inactivation of DNA regeneration), and interference with enzymatic reactions and replication mechanisms in cells and microorganisms (Fernandes *et al.*, 2017; Kihlman, 1974).

(10) Cyclamate, aspartame, and sucralose (sweeteners): artificial sweeteners are synthetic chemicals used as a substitute for sugar. These molecules have been found both in surface water and in sewage treatment plants. They are potential indicators of anthropogenic contamination of surface water by sanitary sewage (Zirlewagen *et al.*, 2016). Studies on the consequences arising from the presence of sweeteners in water bodies are scarce, but it has been shown that the presence of sucralose causes changes in behavior and feeding pattern in crustaceans (Li *et al.*, 2018).

(11) Chlorpyrifos: the most widely used pesticide on crops, including corn, soybeans, broccoli, and apples, and it is also widely used in non-agricultural settings such as golf courses. Chlorpyrifos was invented as an alternative to the pesticide dichloro-diphenyl-trichloroethane and has become part of a pattern known as 'regrettable substitution.' Chlorpyrifos works by attacking insects' nervous systems. High doses, for instance, what farmers are exposed to when they spray pesticides, can cause people to experience nausea, dizziness, and confusion. The most disconcerting effect of chronic exposure to chlorpyrifos, however, is its potential to impair children's developing brains (Burke *et al.*, 2017; Hu, 2018).

### 9.1.3 Current regulatory frameworks

Political awareness of water quality has been growing over the last few decades, especially as WWTPs have been identified as a major source of MP pollution (Rogowska *et al.*, 2020). However, very few countries have implemented legal norms to regulate the maximum concentration of MPs that can be detected in water samples. A lack of bi-directional communication between scientists and policymakers has contributed to fragmentation and inconsistencies in chemical inventories and environmental regulations (Sanganyado, 2022). The identified current regulatory frameworks of major countries are described and summarized in Figure 9.2.

(1) The European Union: Regulations have been adopted for limiting the presence of micro-contaminants in water bodies throughout the territory of the European Union. Environmental quality standards for a few MPs have been regulated by the European Parliament through Directive 2008/105/EC. Annex I of this directive lists the annual average and the maximum allowable concentration of 33 MPs, including chlorpyrifos and a few pesticides. In October 2022, the European Commission reviewed and updated the lists of pollutants in surface water and groundwater, including additional MPs such as carbamazepine, sulfamethoxazole, diclofenac, ibuprofen, erythromycin, triclosan, and others. If the proposal is agreed by the European Union Council and the European Parliament, Member States will be required to take measures to meet the quality standards for additional pollutants, and to make their monitoring data available on a more frequent basis (European Commission, 2023). Currently, the legislation of the European Union is the best worldwide for emerging MPs.

(2) The United States of America: The United States Environmental Protection Agency regulates the pollution of water bodies. Currently, the agency has a strong program for screening and regulating endocrine disruptors in water samples. The National Primary Drinking Water

**Figure 9.2** Current regulatory frameworks for MPs in water samples around the world.

Regulation enforces primary standards for a few microorganisms, disinfection by-products, disinfectants, inorganic chemicals, radionuclides, and organic chemicals (MPs listed include majorly pesticides, herbicides, and chemicals from the discharge of industrial chemical factories) (The United States Environmental Protection Agency, 2023).

(3) Canada: Environment and Climate Change Canada and the Environmental Protection Agency have reported that MPs are increasingly and consistently being found in groundwater, surface water, municipal wastewater, drinking water, and food sources. The MPs being tracked are mostly polybrominated diphenyl ethers (fire retardant), acetaminophen, ibuprofen, chlortetracycline, caffeine, bisphenol A, triclosan, and a few others. However, although water quality criteria have been established for a few emerging contaminants, they are not legally enforced. The Guidelines for Canadian Drinking Water Quality regulate the maximum amount of a few pesticides/herbicides, in which chlorpyrifos is among them (Government of Canada, 2023).

(4) China: In 2014, China declared a 'war on pollution.' Since then, air quality has improved significantly, but China's standards for environmental quality and pollutant emissions mainly focus on conventional pollutants, as do its lists of managed chemicals. Most feature on one of the two lists: for air pollutants and for water pollutants. Once listed, chemicals can be controlled under laws on air and water pollution. A first list of 'new pollutants for priority control' is expected to be published in 2023. Substances on those lists will either be banned or subjected to usage and emission restrictions. Meanwhile, regulators are currently revising guidelines on industrial restructuring to gradually phase out some pesticides, veterinary drugs, cosmetics, and industrial chemicals (Zi, 2022).

(5) Southeast Asia: Although research of emerging contaminants in southeast Asia is insignificant in comparison to other regions globally, it has gained substantial momentum in recent years.

Except for Timor-Leste, all southeast Asian countries are part of the Association of Southeast Asian Nations (ASEANs), a collaborative intergovernmental organization that started in 1967. To enhance water quality and better water management practices, the ASEAN working group on Water Resource Management was created to increase regional cooperation and provide frameworks to assist member countries. However, there are no ASEAN-specific water quality guidelines similar to those of the European Union, and only some member countries have developed or adopted a country-specific water quality index to evaluate the quality of their water resources. Even then, the parameters measured were focused on basic indicators and microbial risk (ammoniacal nitrogen, biochemical oxygen demand, total coliform bacteria, pH, etc.), and there are no water quality guidelines created for MPs (Lee *et al.*, 2022).

(6) Latin America: The class of MPs that is most measured in Latin America is pharmaceuticals, followed by personal care products and endocrine disruptors. The lack of information about MPs in countries such as Cuba, El Salvador, Haiti, Honduras, Nicaragua, Panama, Paraguay, Peru, and the Dominican Republic is largely economical, due to the high cost of quantification of these compounds and environmental political interests. Currently, representatives from Argentina, Brazil, Chile, Cuba, Costa Rica, Guatemala, Ecuador, Nicaragua, Panama, Paraguay, and Uruguay have participated in the project RLA/7/019 'Elaboration of indicators to determine the effects of pesticides, heavy metals, and emerging pollutants in continental aquatic ecosystems important for agriculture and agro-industry.' The main objective of this project is to improve quality of life, food security, and agricultural resources in Latin America and the Caribbean through the proper management of water resources. However, legislation in this context in Latin America is almost non-existent (Peña-Guzmán *et al.*, 2019).

## 9.2 FATE AND REMOVAL OF MPS IN ANAEROBIC REACTORS TREATING DOMESTIC WASTEWATER

### 9.2.1 Fate of MPs during wastewater treatment

Traditional domestic wastewater treatment systems are typically designed and operated to remove influent solids, organic, and macronutrient loads. Organic MPs present in wastewater, which are not specifically targeted, exhibit varying degrees of removal, due to the combined effect of different biotic and abiotic mechanisms. Broadly speaking, the concentration of MPs in wastewater can be attenuated by abiotic degradation (due to photolysis, chemical oxidation, and hydrolysis reactions), volatilization, a partition to the solid phase (adsorption), and biotransformation reactions. The extent of the contribution of each of these pathways to the overall MP removal is highly dependent on the physicochemical properties of MPs, wastewater treatment technology, environmental conditions at treatment plants, and their operational parameters.

The widespread environmental presence of MPs originating from domestic sewage is indicative of their chemical stability and resistance to abiotic degradation under ambient conditions. As volatilization is only relevant for compounds with high vapor pressures, sorption to the particulate phase and biotransformation are considered the main removal mechanisms of these recalcitrant MPs during secondary (biological) treatment of domestic wastewater.

Adsorption of MPs to the particulate phase may occur due to hydrophobic interactions between aliphatic and aromatic groups of MPs and the lipidic fractions of biomass and sludge; electrostatic interactions between charged functional groups and the superficial charge of solids; as well as the formation of complexes between the MPs and metal ions present in the solid phase (Tran *et al.*, 2018). The properties of MPs that influence their adsorption include molecular size; charge; hydrophobicity; and the presence of specific functional groups in their chemical structure. Naturally, the properties of sorbents (which depend on, among other things, the sludge type and sludge age) and environmental factors such as pH, temperature, and ionic strength of the medium are equally relevant in determining the extent of adsorption, particularly for MPs harboring ionizable moieties. A large number of parameters involved in the characterization of the phenomena makes the theoretical prediction of MP sorption unfeasible. Thus, the combined effect of these parameters is condensed in liquid–solid partitioning coefficients ($K_d$), which can be experimentally determined for the sorbate–sorbent pair and are often used as predictors of the sorption potential of an MP onto the particulate phase in WWTPs (Berthod *et al.*, 2016).

The strength of the sorbate–sorbent interaction determines the reversibility of the adsorption, which in turn influences the MP mobility in the environment once the particulate phase is removed from a WWTP. Reversible adsorption of MPs onto sludge raises concerns of environmental spread in cases where sewage sludge is disposed of on agricultural land, in the form of biosolids. In these situations, MP removal by adsorption would not be a desired mechanism, as the environmental risk would remain unchanged, simply being transferred out of the treatment plant.

The reversibility of the adsorption also affects MPs availability to undergo further biological transformations. As most of the anabolic and catabolic microbiological reactions are mediated by intracellular enzymes, MPs should be in solution and available for uptake by microorganisms (and therefore, not physically adhered to a surface – adsorbed) in order to be biotransformed. This, however, is not a consensus in the literature, as some authors consider the adsorbed MPs to be bioavailable, whether due to biotransformation mediated by extracellular enzymes or the consideration of reversible adsorption as an initial transport step to biodegradation (Gonzalez-Gil *et al.*, 2018a).

Finally, the significance of sorption as an MP removal mechanism depends on sludge-wasting practice in WWTPs, as measured by solid retention times (SRTs) in bioreactors. For aerobic reactors based on the activated sludge technology, which are usually operated under short SRTs (3–10 days), there is a considerable removal of MPs adsorbed to wasted sludge. Anaerobic processes, on the contrary, have lower biomass yield in comparison with the aerobic ones, and bioreactors are typically operated under much longer SRTs (hundreds of days) and therefore with minimal sludge wasting. The time necessary for saturation of MPs sorbed onto the anaerobic sludge is insignificant relative to the SRT, and therefore adsorption is negligible as a removal pathway, when compared to biotransformation reactions (Harb *et al.*, 2019). This is not to say that adsorption does not play a role in MP removal in anaerobic reactors, as the retention time of some MPs is increased due to adsorption, which could favor the occurrence of biological transformation.

Biotransformation is a broad term that encompasses different types of reactions mediated by microorganisms. Biotransformed MPs can be partially degraded to other organic compounds of lower molecular mass (which is often termed biodegradation); conjugated to form products of higher molecular mass; or completely mineralized to $CO_2$, $H_2O$, $CH_4$, and other inorganic compounds. Conjugation is a biotransformation process in which functional groups are added to an MP, usually rendering it less toxic, more water-soluble, or more amenable to biodegradation. The resulting conjugated products can be reverted back to the parent compound, which is often observed in WWTPs (Verlicchi *et al.*, 2012).

The organic products of biotransformation, whether from biodegradation or conjugation, are commonly referred to as transformation products (TPs). TPs of some MPs might exhibit higher toxicity or recalcitrance than the parent compound, posing different environmental risks (Escher & Fenner, 2011). During secondary treatment, TPs might accumulate in the medium as 'dead-end' metabolites, or they might be further transformed by different microbial groups as part of a wider metabolic network (Fischer & Majewsky, 2014). Thus, simply following the disappearance of an MP is not enough to assess the effectivity of treatment systems in reducing potential environmental risks (Stadler *et al.*, 2012). Complete mineralization or the formation of innocuous TPs should be the desirable endpoints of MP biotransformation in WWTPs.

Similar to adsorptive removal, the susceptibility of an MP to biotransformation is contingent on a wide range of factors, such as temperature, pH, redox potential, bioavailability, microbial diversity, presence of other substrates, hydraulic retention time (HRTs), and the physicochemical properties of MPs (Falås *et al.*, 2016). Particularly for MPs, the concentration level is a key aspect in determining the type of biological transformation the MPs will undergo during secondary treatment, as it will define the metabolic pathways acting on them. MPs might be part of the central metabolic routes of the microorganisms or they might be fortuitously transformed due to low substrate specificity of some enzymes (co-metabolism). Considering that MPs are found in domestic wastewater in the ng/L to $\mu$g/L range, it is not expected that their presence alone would be enough to induce the production of enzymes and cofactors necessary for their metabolization. Carbon and energy derived from the degradation of MPs would not be sufficient to support biomass growth. Thus, the biotransformation of MPs is understood to be necessarily dependent on other growth-supporting substrates (Kennes-Veiga *et al.*, 2022).

The complexity of removal mechanisms and their dependence on a multitude of factors explain the wide variation in overall removal efficiency found for similar compounds in similar wastewater treatment technologies (Alvarino *et al.*, 2018). Recognition and characterization of the mechanisms by which MPs are removed from domestic wastewater during mainstream biological treatment are essential to substantiate the proposition of novel treatment technologies constituting more effective barriers in containing the release of MPs to the environment.

### 9.2.2 MP removal in anaerobic reactors

Most of the investigations on MP removal during secondary treatment of domestic wastewater have focused on aerobic reactors, mainly those based on activated sludge technology. As mainstream anaerobic wastewater treatment is increasingly used for domestic wastewater, the literature on the fate of MPs in these systems is much scarcer. The transformation of MPs in biological reactors revolves around the action of enzymes on functional groups present in the contaminant's molecules. While some of these enzymatic reactions are known to proceed in both aerobic and anaerobic environments, as the enzymes that catalyze these reactions are redox-independent (e.g., hydrolases and transferases; Gonzalez-Gil *et al.*, 2019), some occur exclusively or at least preferentially under anaerobic conditions. This, in turn, results in distinct MP removal capabilities between anaerobic and aerobic reactors.

Several studies have shown that most MPs are biotransformed under anaerobic conditions, albeit to variable extents (Ghattas *et al.*, 2017), as presented in Table 9.1. There is, however, significantly

**Table 9.1** Apparent removal of selected MPs during mainstream anaerobic treatment of real domestic wastewater.

| Selected MPs | Removal Range (%) | References |
|---|---|---|
| Bisphenol A | <0–87 | [1–3, 14] |
| Caffeine | 0–98 | [4–6] |
| Carbamazepine | 5–96 | [5–8] |
| Ciprofloxacin | 84–100 | [5, 9] |
| Diclofenac | <0–78 | [1, 2, 4, 5, 7, 9–14] |
| Ibuprofen | <0–91 | [4–7, 9–12, 14] |
| Propranolol | 46–100 | [8, 12] |
| Sulfamethoxazole | <0–100 | [1, 2, 5, 8] |
| Triclosan | 18–89 | [3, 6, 12, 13] |
| Trimethoprim | 33–100 | [1, 2, 5, 9] |

[1] Queiroz *et al.* (2012); [2] Brandt *et al.* (2013); [3] Mladenov *et al.* (2022); [4] Arrubla *et al.* (2016); [5] Dutta *et al.* (2014); [6] Reyes-Contreras *et al.* (2011); [7] Chen *et al.* (2019); [8] McCurry *et al.* (2014); [9] Butkovskyi *et al.* (2015); [10] de Graaff *et al.* (2011); [11] Pirete *et al.* (2022); [12] Granatto *et al.* (2021); [13] Butkovskyi *et al.* (2018); [14] Vassalle *et al.* (2020).

little understanding of the fundamental mechanisms that lead to that removal. Obtaining mechanistic insight is difficult owing to the complexity of biological treatment systems. Thus, much of the work has been focused on unraveling the various factors (reactor-related or MP-related) that directly or indirectly influence the underlying microbiological process.

Among the MP-related properties (chemical structure, molecular weight, hydrophobicity, dissociation constants, etc.), the types of functional groups present in the molecule are of particular interest in assessing their removal in anaerobic reactors. It has been proposed, mostly agreeing with experimental data, that MPs containing electron-donating groups (such as amines, methyl, and hydroxyl) are more susceptible to anaerobic biotransformation, whereas MPs with electron-withdrawing groups (such as chloro, sulfonyl, amide, carboxyl, nitrile, and aldehyde) are more recalcitrant to anaerobic treatment (Wijekoon *et al.*, 2015). This, however, is not a universal rule, as structurally similar MPs often exhibit distinct transformation rates in the same bioreactor (Luo *et al.*, 2014). The interaction with other factors must be considered.

Diverse environmental (temperature, pH, ionic strength) and reactor-related properties (HRT, organic loading rate – OLR, SRT, redox potential, presence of growth-supporting matrices, wastewater composition) shape the composition of the microbial consortium present in anaerobic reactors, and thus the metabolic potential for MP biotransformation. It is believed that increased microbial diversity is beneficial to MP removal due to an increased number of functional traits in the community (Falås *et al.*, 2016). In this context, using longer SRTs would be beneficial: not only does it allow for the establishment of slow-growing organisms, but also extends the exposure of adsorbed MPs to the microbial population (Harb *et al.*, 2019). However, the exposure of the non-sorbed fraction remains limited by the HRT.

The putative existence of key microbial populations involved in MP biotransformation, which could be selectively enriched in anaerobic reactors to enhance overall MP removal, has motivated recent research efforts to isolate the contribution of the different stages of an anaerobic process (hydrolysis, acidogenesis, acetogenesis, methanogenesis). Experimental evidence so far has discussed acidogenesis, acetogenesis, and methanogenesis as the main steps involved in MP removal (Carneiro *et al.*, 2020; Gonzalez-Gil *et al.*, 2018b). However, some MPs are removed to similar extents under both acidogenic and methanogenic conditions, which might indicate that either the reactions are carried out by enzymes common to the different microbial groups, or that the biotransformation of MPs is not necessarily linked to the main metabolic pathways in a reactor, but to specific microbial communities common to all anaerobic environments (Kennes-Veiga *et al.*, 2022).

Transferring the mechanistic understanding of anaerobic biotransformation of MPs, which is still in its infancy, to feasible changes to be implemented in mainstream anaerobic domestic wastewater treatment remains a major challenge.

### 9.2.3 Development of anaerobic reactors for MP removal

Most of the research focusing on the application of anaerobic reactors for MP removal is performed with lab-scale units. The option for such reduced prototypes is a scientific strategy to best control the process and to search for fundamentals for the removal process. Moreover, lab-made (simulated or synthetic) domestic wastewater is frequently applied to avoid interferences of a cocktail of compounds found in real domestic sewage, thus facilitating the mass balance and, consequently, the evaluation of anaerobic removal pathways. However, some studies also use real domestic sewage to prevent extreme simplifications from masking the obtained results.

The studies on MP removal in anaerobic lab-scale reactors can be divided into two categories: reproduction of conventional anaerobic reactors applied to domestic wastewater, that is, upflow anaerobic sludge blanket (UASB) reactors, and proposition of new configurations of reactors to enhance the removal of compounds. The first type aims at searching for the best operating conditions, as mainstream anaerobic WWTPs are designed to remove influent solids, as well as organic and macronutrient loads. Moreover, a combination of conventional anaerobic reactors with biological or physical–chemical processes is addressed in some studies, often using anaerobic technology as the core of the process. The second type of study is focused on design procedures, mainly related to the increment of the SRT over the HRT, and consequently, the achievement of a high concentration of microorganisms, while at the same time trying to maximize the diversity of the cell population. Invariably, these studies are based on immobilized-cell technologies or with advanced techniques of cell retention, such as membrane processes, whose configurations are meant to comply with such requirements. Additionally, some studies in this line have sought to evaluate the influence of the phase separation of anaerobic processes in specific reactors, such as hydrolytic, acidogenic, and predominantly methanogenic ones as further discussed.

The challenge of developing more efficient and reliable bioreactors is subject to exhaustive research that begins on a laboratory scale. At this stage, the main variables that interfere with reaction rates and the stability of the system must be evaluated. These data can lead to the proposal of a model that allows a rational increase of the scale for a pilot unit. This new unit must also be studied to validate the model or modify it with the purpose of a new increase in scale. The progressive increase in scale, based on rational data, ensures security to the configuration of development process; however, it is more time-consuming. For this reason, this approach should be more unpretentious and freer from market pressures for quick results. It is an experimental–mechanistic approach, in which the experiment provides subsidies for models based on fundamental phenomena that will serve for scale-up, design, simulation, and optimization of the units.

Contrary to the experimental–mechanistic approach, there is a more traditional and more common approach in the development of anaerobic reactor configurations. This approach, which can be called purely experimental or empirical, is based on smaller-scale, pilot, or even on real-scale studies, focusing on directly observable and manipulable parameters, such as HRT and OLR. These parameters encompass physical, chemical, and biochemical phenomena and can be the main factors for scale-up or reactor design. This type of approach allows for faster evolution of the development of a given configuration but is generally less secure because it approaches the system as small 'black boxes,' which involve many physical, chemical, microbiological, and biochemical phenomena that are not properly elucidated.

The opposition of these two approaches is presented by Wentzel and Ekama (1997), who call the empirical approach the 'black-box approach,' in which applications are restricted to the experimental conditions that generated the model. According to the authors, the empirical approach only allows interpolation of results, whereas the experimental–mechanistic approach (called only mechanistic by the authors) would allow interpolations and extrapolations.

There would still be a third approach fundamentally based on mathematical modeling. However, the lack of mechanistic models that adequately cover all steps of the anaerobic process, parameters, and historical series of operational data means that this approach still encounters little or no applicability in this field.

It is clear that these three approaches are not completely independent and, even within the same research group, there is a need to balance the approaches well, to allow research to result in real advances in the area of anaerobic reactors. However, the experimental–mechanistic approach, although more complex and more time-consuming to generate practical results, should be the one that can make the greatest contribution to the consolidation of anaerobic treatment processes as a viable and reliable alternative. Furthermore, this approach allows studies on the microbiology and biochemistry of anaerobic digestion to be included in the generated models, allowing such studies to gain practical meaning.

In general, studies on MP removal in anaerobic reactors are based on the experimental–mechanistic approach, mainly because the mechanisms involved in the degradation of different compounds have to be elucidated to better understand the system. A black-box approach in this area may lead not only to sub-optimized units as occurs in the case of organic matter or nutrient removal, but also to unsafe systems incapable of removing MPs in the required levels. In addition, using a scientific method of trial and error, in this case, can be disastrous not only because it releases compounds that are harmful to the environment, but also because it releases degradation products into the environment that can be even more toxic than the original compound.

Conventional anaerobic reactors commonly applied for the treatment of domestic wastewater, such as expanded granular sludge bed (EGSB) and UASB, have been studied on lab scale, mainly focusing on the HRT required to degrade some MPs, as performed by Granatto *et al.* (2021), who evaluated degradation of diclofenac (35–37% removal), ibuprofen (43–44% removal), propranolol (46–51% removal), triclosan (51–72% removal), and linear alkylbenzene sulfonate (63–65% removal). In the same configuration, Gutiérrez *et al.* (2022) evaluated the removal of nonylphenol ethoxylate (48–82% removal) in the co-digestion of domestic sewage and commercial laundry wastewater. In both studies, such a conventional reactor was evaluated under specific operating conditions and not related to the design parameters commonly used for this technology when organic matter removal is the main objective. For UASB reactors, the literature also explores HRT beyond the conventionally applied or conjugated technologies, aiming at MP removal. For example, Mora-Cabrera *et al.* (2021) studied a UASB reactor followed by a membrane electrochemical bioreactor to treat domestic wastewater containing ibuprofen (24% removal), carbamazepine (23% removal), diclofenac (29% removal), and 17$\alpha$-ethinylestradiol (26% removal), whereas Vassalle *et al.* (2020) combined UASB with high-rate algal ponds to remove ibuprofen (65% removal), diclofenac (65% removal), naproxen (71% removal), paracetamol (65% removal), gemfibrozil (39% removal), estrone (95% removal), 17$\beta$-estradiol (91% removal), 17$\alpha$-ethinylestradiol (92% removal), estriol (89% removal), nonylphenol (70% removal), and bisphenol A (43% removal). In both studies, the HRT applied to UASB reactors was commonly applied in domestic WWTPs, in which the focus of the studies was on combining the systems, searching for reliable post-treatment units.

The proposition of new configurations of anaerobic reactors focusing on MP removal is based on the enhancement of the SRT/HRT ratio and the achievement of a high-diversity microbial population with increased concentration. One way to achieve this goal is to use cells immobilized on an inert support, as obtained in anaerobic filters, or even self-immobilized in the form of granules, as observed in UASB reactors. In addition, the use of adhered or immobilized cell technology allows for obtaining cell concentrations greater than those obtained in systems with non-adhered cells, with obvious benefits for the treatment of wastewater and for MP removal. In these systems, however, the kinetic analysis becomes more complex, as the phenomena of mass transfer from the liquid phase to the biological solid phase effectively influence the global rates of conversion of organic matter.

In the case of anaerobic reactors, in addition to the objective of increasing the cell retention time, the immobilization of the biomass can be used to improve the relationships between the different microorganisms, facilitating the transfer of primary and intermediate substrates between the various groups that participate in the complex process of anaerobic digestion. In addition, resistance to mass transfer in systems containing immobilized cells may represent protection for organisms when potentially toxic compounds or inhibitors are present.

More than an alternative, Speece (1996) considers the retention of biomass essential for the maintenance of methanogenic organisms, which have a low growth rate. The classic way of treating reactors containing immobilized biomass derives from the way of treating heterogeneous reactors containing immobilized enzymes. Two resistances to mass transfer are considered in this approach: transfer in the liquid phase, in the stagnant liquid film around a bioparticle, and transfer in the solid or intraparticle phase as classically presented by Bailey and Ollis (1986).

Uncertainty regarding granulation (self-immobilization) in UASB or EGSB reactors and the empiricism involved in the design of conventional anaerobic filters were some of the main motivators for proposing new reactor configurations containing cells immobilized on different and varied support materials, which allow adhesion and growth of diverse or specific biomass, depending on the application. Support material can be used as a selection or enrichment factor for some desirable organisms and the elimination of undesirable ones.

Immobilized-cell anaerobic bioreactors have been proposed in different configurations for MP removal from domestic wastewater. Carneiro *et al.* (2019) compared two ways of packing the support material in a fixed-bed reactor. An anaerobic packed-bed reactor was compared to a structured-bed one for sulfamethoxazole and ciprofloxacin removal from domestic sewage. In this study, a structured bed showed to be potentially more viable as it presented the same performance for a quantity of support material 50% lower than that in a packed bed. Moreover, the structure of the bed, thus providing high bed porosity, may prevent hydrodynamic misbehavior, common in fixed-bed reactors, such as channeling and dead zones. Besides the fixed-bed configuration, immobilized cell reactors can be configured as expanded or even fluidized-bed, leading to the improvement of mass transfer fluxes, leading to benefits in the overall reaction rate although with higher energy demands.

Another way of achieving high SRT/HRT ratios is to separate suspended biomass and recirculate it to the reactor, thus decoupling the SRT from the HRT. Using efficient separation systems, such as processes using membranes, allows not only high HRT/SRT ratios but also the possibility of obtaining higher concentrations of microorganisms in a reactor. This alternative has been studied in some research groups as presented by Sawaya *et al.* (2022) and Arcanjo *et al.* (2022).

As mentioned earlier, adopting combined acidogenic–methanogenic reactors can be a feasible strategy to enhance MP removal. Carneiro *et al.* (2020) observed that the acidogenic phase is crucial in the biotransformation of some MPs, whereas the hydrolysis of carbohydrates does not significantly contribute to the metabolic transformation of such compounds. The removal of some compounds, such as galaxolide, celestolide, tonalide, erythromycin, and roxithromycin was favored under acidogenic conditions compared to acetogenesis/methanogenesis whereas the removal of other compounds, such as triclosan, fluoxetine, bisphenol A, and carbamazepine depends on acetogenesis/methanogenesis. Macêdo *et al.* (2021) evaluated tetrabromobisphenol A removal in two anaerobic structured-bed reactors under acidogenic and methanogenic conditions and observed that the biodegradation occurred during acidogenesis via co-metabolism. Pirete *et al.* (2022) found that acidogenic bacteria were responsible for diclofenac and ibuprofen biodegradation in an anaerobic fluidized-bed reactor.

A combination of anaerobic and aerobic processes may be relevant for MP removal from domestic wastewater. This approach is known as a hybrid anaerobic–aerobic process. In a hybrid process, the anaerobic step is used to convert complex organic compounds into simpler compounds, such as volatile fatty acids, which are then fed to an aerobic bioreactor for further degradation. An aerobic step can be used to remove residual organic matter and to provide additional treatment of MPs that are not effectively removed by the anaerobic process. This combination of anaerobic and aerobic processes can result in a more efficient removal of MPs compared to using either process alone. Anaerobic processes can remove MPs that are biodegradable and have low solubility, whereas aerobic processes can remove MPs that are more recalcitrant and have higher solubility.

### 9.2.4 Strategies for enhancing MPs removal in anaerobic reactors

The development of anaerobic reactors for MP removal from wastewater faces several challenges, including: (1) the selection of suitable microorganisms: it is important to select microorganisms that can efficiently degrade the MPs present in wastewater without generating toxic by-products; (2) organic load control: the quantity and quality of organic matter in wastewater can affect the performance of anaerobic reactors. The organic load needs to be controlled to avoid system overload; (3) MP monitoring: MPs in wastewater can be difficult to detect and quantify. It is important to monitor them to assess the reactor's efficiency and identify possible issues; (4) reactor scaling-up: reactor sizing should consider various factors, such as the type of MPs present in wastewater, the organic load, and the treatment system's characteristics; (5) cost–benefit: the development of anaerobic reactors for MP removal must be economically viable, considering the cost of materials, energy required for system operation, and the reactor's lifespan; and (6) legislation: environmental legislation may impose limits on the amount of MPs allowed in treated wastewater. It is essential to develop treatment systems that meet these legal requirements.

Different approaches have been considered to enhance MP removal in anaerobic reactors. In addition to manipulating traditional operational variables known to be crucial to the performance of anaerobic reactors (such as pH, OLRs, solids, and HRTs), other more structurally demanding alternatives have also shown promise in enhancing the removal of MPs or enabling the removal of certain MPs. Recent studies have highlighted the potential of two-phase anaerobic reactors; a combination of anaerobic–aerobic or anaerobic–physicochemical processes; the use of membrane separation; and the addition of external substrates to domestic wastewater, which would serve as additional electron donors.

The utilization of two-phase anaerobic reactors, which consist of separate hydrolytic–acidogenic and acetogenic–methanogenic stages, has been demonstrated to be advantageous for the treatment of domestic wastewater over traditional single-phase reactors. Initially developed to treat high-strength waste, this approach has been shown to lead to a more stable process, increased methane production rates and yields, and greater microbial diversity than that of single-phase reactors (Carneiro *et al.*, 2022; Rajagopal *et al.*, 2019; Smith *et al.*, 2017). The improved performance is a result of providing optimal environmental conditions for the spatially separated microbial groups active in an anaerobic process. Current research suggests that these modifications can have positive effects on MP removal (Carneiro *et al.*, 2020), although further investigation is needed to fully assess this potential benefit.

Treatment systems incorporating biological processes that occur under different redox conditions have been proposed to enhance MP removal. As aerobic and anaerobic bioreactors mainly target different moieties of MPs, a combination of the two environments would broaden the range of compounds removed during biological treatment. Theoretically, the aerobic environment can remove MPs that are recalcitrant to anaerobic degradation and vice versa. In the context of domestic wastewater treatment, this combination of redox conditions can be achieved in biological nutrient removal setups consisting of alternating anaerobic, aerobic and anoxic tanks, or through direct microaeration of anaerobic reactors. Falås *et al.* (2016) investigated the removal of 31 organic MPs in 15 biological reactors incorporating different redox conditions and found that although different redox conditions led to an increased removal of a few compounds (venlafaxine, diatrizoate, tramadol, codeine, and trimethoprim), a large number of MPs persisted in the treated wastewater, regardless of the operational conditions. Similar results were found by Wolff *et al.* (2018), who evaluated reactor cascades combining aerobic and strictly anaerobic treatments and observed a significant improvement in the removal of selected MPs (diatrizoate (20–70 percentage points increase), venlafaxine (90–98 percentage points increase), and diclofenac (10–70 percentage points increase)), but only a slight improvement in the overall MP removal. do Nascimento *et al.* (2021) investigated the effects of the injection of small amounts of oxygen directly into a UASB reactor treating domestic wastewater on the removal of seven MPs (estrone, 17$\beta$-estradiol, 17$\alpha$-ethinylestradiol, bisphenol A, diclofenac, sulfamethoxazole, and trimethoprim). The authors found significant enhancements of MP removals for all evaluated MPs (67–81 percentage points increase) and no significant deterioration in reactor performance. The microaeration prompted a gradual alteration of the microbial community inside the reactor, without compromising the archaeal community and therefore, the methanogenesis. Thus, microaeration presents a practical and promising alternative for retrofitting existing treatment plants to reduce MP emissions.

Due to the typical concentration range of MPs in domestic wastewater, their removal is often reliant on the presence of other growth-supporting organic substrates. Therefore, for low-strength wastewater such as domestic sewage, the presence of sufficient degradable organic matter might be a limiting factor for MP removal. Studies have shown that the addition of external electron donors can significantly enhance MP degradation (Oliveira *et al.*, 2016). This is also why dosing external electron donors in post-denitrifying reactors for residual nitrogen removal can increase MP degradation (Torresi *et al.*, 2017). Thus, it can be hypothesized that co-treating domestic sewage with other wastewaters could favor the overall removal of MPs. However, this approach is still in its early stages, and it is necessary to evaluate the impacts of the subsequent increase in organic load on the treatment system's efficiency. Additionally, practical applicability must be considered, including aspects such as the availability and transportation costs of appropriate co-substrates.

The previously discussed emerging strategies for enhancing MP removal rely on the biological degradation route, which might still be insufficient for acutely recalcitrant MPs. In these cases, oxidative physicochemical post-treatments constitute compelling alternatives. There are numerous established and innovative oxidative physicochemical processes capable of removing MPs, such as ozonation, ultraviolet-peroxide, chlorination, hydrodynamic cavitation, sonolysis, Fenton-based processes, electrochemical oxidation, and ferrate oxidation. These processes are non-selective and generally capable of high-MP transformation rates. Some processes, such as chlorination and ozonation, are already found in domestic WWTPs containing tertiary stages for effluent polishing. While it is not within our current scope to provide an in-depth review of the performance of each of these technologies in MP removal, it is important to discuss the applicability of upgrading existing treatment plants aiming at MP oxidation. Oxidative chemical post-treatments typically entail elevated operating costs associated with high-energy requirements and costly inputs. While complete mineralization of some MPs is achievable for some of the processes, in some cases this might demand unfeasible reaction times, resulting in incomplete transformation of MPs. In some cases, TPs (deriving from MPs or from the wastewater matrix) can be more resistant or even more toxic than the initial target compounds (Ribeiro *et al.*, 2015). Furthermore, the efficacy of oxidative chemical treatments is dependent upon the overall quality of the preceding wastewater treatment, as residual organic matter and common inorganic ions (sulfide, carbonate, bicarbonate, and nitrate) act as scavengers for the oxidant species, increasing the process cost. Nevertheless, cost-effective MP removal through ozonation has been implemented on full scale in Germany and Switzerland (Rizzo *et al.*, 2019).

The strategies discussed to enhance MP removal from domestic wastewater involve adding a subsequent unit process to an anaerobic reactor. Naturally, these alternatives should achieve superior MP removal compared to relying solely on traditional anaerobic treatment. However, these options involve a significant degree of modification to WWTPs, and their implementation solely to reduce MP discharges may prove impractical in most cases. Adopting these systems is more likely to be successful when integrated with further polishing of other water quality parameters. This allows for more valued uses for the treated wastewater, such as water reuse and controlled groundwater recharge. It is important to keep in mind that any increase in effluent quality is accompanied by an increase in treatment cost and in the overall impacts associated with producing the necessary inputs for these post-treatments. In other words, the abatement of MP discharge might create environmental burdens beyond the walls of the treatment plant. Therefore, the endpoints for effluent discharge must be carefully assessed, aiming for an overall environmental optimum rather than the lowest achievable. A comprehensive benefit analysis is necessary.

## 9.3 CONCLUDING REMARKS

Sources of MPs in the environment are diverse. They can come from domestic wastewater, hospitals, commercial and industry effluents, farms, and several run-offs. This ample origin makes them diverse, complex, and difficult to monitor. Contamination of water and domestic sewage by MPs has adverse effects, such as resistance in pathogenic bacteria and reduced diversity of bacteria and algae in aquatic environments. These substances cause changes in the behavior and morphology of aquatic biota and interfere with the hormonal system of wildlife and humans. Even with the partial elucidation of the consequences and toxicity of exposure to MPs, most countries do not have legislation to impose a maximum limit for their concentration in water bodies. Currently, regulations have been adopted for a few micro-contaminants in some countries and regions only in the European Union.

Advanced physical–chemical treatment technologies, such as adsorption and membrane processes, have been demonstrated to be promising choices for MP removal; however, they imply high operation costs and formation of by-products. A biological anaerobic process of conversion of organic matter has, therefore, received attention from several research groups trying to assess if it is a viable option to convert MPs in domestic sewage. The development of anaerobic reactors for the removal of MPs from wastewater faces several challenges, including a selection of suitable microorganisms, organic load control, MP monitoring, reactor scaling-up, and cost–benefit. Studies on MP removal in anaerobic systems are emerging and a great deal of work should be carried out to evaluate whether conventional anaerobic reactors applied to domestic wastewater treatment under usual operating conditions are able to effectively remove contaminants of emerging concern. Moreover, in order to comply with the requirements of removing organic matter, nutrients, and MPs, beyond generating energy (biogas) in a robust, safe, and sustainable way, the study of new operating strategies and reactors configurations are required.

## REFERENCES


Alvarino T., Lema J., Omil F. and Suárez S. (2018). Trends in organic micropollutants removal in secondary treatment of sewage. *Reviews in Environmental Science and Bio/Technology*, **17**, 447–469, https://doi.org/10.1007/s11157-018-9472-3

Aquino S. F. de, Brandt E. M. F. and Chernicharo C. A. de L. (2013). Remoção de fármacos e desreguladores endócrinos em estações de tratamento de esgoto: revisão da literatura. *Engenharia Sanitaria e Ambiental*, **18**, 187–204, https://doi.org/10.1590/S1413-41522013000300002

Arcanjo G. S., dos Santos C. R., Cavalcante B. F., Moura G. de A., Ricci B. C., Mounteer A. H., Santos L. V. S., Queiroz L. M. and Amaral M. C. (2022). Improving biological removal of pharmaceutical active compounds and estrogenic activity in a mesophilic anaerobic osmotic membrane bioreactor treating municipal sewage. *Chemosphere*, **301**, 134716, https://doi.org/10.1016/j.chemosphere.2022.134716

Arrubla J. P., Cubillos J. A., Ramírez C. A., Arredondo J. A., Arias C. A. and Paredes D. (2016). Pharmaceutical and personal care products in domestic wastewater and their removal in anaerobic treatment systems: septic tank – up flow anaerobic filter. *Ingenieria e Investigación*, **36**, 70–78, https://doi.org/10.15446/ing.investig.v36n1.53076

Bailey J. E. and Ollis D. F. (1986). Biochemical Engineering Fundamentals. 2nd edn, McGraw Hill, New York.

Beltrán E. M., Pablos M. V., Fernández Torija C., Porcel M. Á. and González-Doncel M. (2020). Uptake of atenolol, carbamazepine and triclosan by crops irrigated with reclaimed water in a Mediterranean scenario. *Ecotoxicology and Environmental Safety*, **191**, 110171, https://doi.org/10.1016/j.ecoenv.2020.110171

Bernegossi A. C. (2019). Efeito Tóxico da Cafeína Sobre o Ciclo de Vida de Chironomus sancticaroli (Chironomidae, Diptera) e Daphnia magna (Daphniidae, Cladocera). Universidade de São Paulo, São Carlos.

Berthod L., Roberts G., Sharpe A., Whitley D. C., Greenwood R. and Mills G. A. (2016). Effect of sewage sludge type on the partitioning behaviour of pharmaceuticals: a meta-analysis. Environmental Science.

Brandt E. M. F., de Queiroz F. B., Afonso R. J. C. F., Aquino S. F. and Chernicharo C. A. L. (2013). Behaviour of pharmaceuticals and endocrine disrupting chemicals in simplified sewage treatment systems. *Journal of Environmental Management*, **128**, 718–726, https://doi.org/10.1016/j.jenvman.2013.06.003

Burke R. D., Todd S. W., Lumsden E., Mullins R. J., Mamczarz J., Fawcett W. P., Gullapalli R. P., Randall W. R., Pereira E. F. R. and Albuquerque E. X. (2017). Developmental neurotoxicity of the organophosphorus insecticide chlorpyrifos: from clinical findings to preclinical models and potential mechanisms. *Journal of Neurochemistry*, **142**, 162–177, https://doi.org/10.1111/jnc.14077

Butkovskyi A., Leal L. H., Rijnaarts H. H. M. and Zeeman G. (2015). Fate of pharmaceuticals in full-scale source separated sanitation system. *Water Research*, **85**, 384–392, https://doi.org/10.1016/j.watres.2015.08.045

Butkovskyi A., Sevenou L., Meulepas R. J. W., Leal L. H., Zeeman G. and Rijnaarts H. H. M. (2018). Micropollutant removal from black water and grey water sludge in a UASB–GAC reactor. *Water Science and Technology*, **77**, 1137–1148, https://doi.org/10.2166/wst.2017.640

Carey D. E., Zitomer D. H., Kappell A. D., Choi M. J., Hristova K. R. and McNamara P. J. (2016). Chronic exposure to triclosan sustains microbial community shifts and alters antibiotic resistance gene levels in anaerobic digesters. *Environmental Science: Processes & Impacts*, **18**, 1060–1067, https://doi.org/10.1039/C6EM00282J

Carneiro R. B., Sabatini C. A., Santos-Neto Á. J. and Zaiat M. (2019). Feasibility of anaerobic packed and structured-bed reactors for sulfamethoxazole and ciprofloxacin removal from domestic sewage. *Science of the Total Environment*, **678**, 419–429, https://doi.org/10.1016/j.scitotenv.2019.04.437

Carneiro R. B., Gonzalez-Gil L., Londoño Y. A., Zaiat M., Carballa M. and Lema J. M. (2020). Acidogenesis is a key step in the anaerobic biotransformation of organic micropollutants. *Journal of Hazardous Materials*, **389**, 121888, https://doi.org/10.1016/j.jhazmat.2019.121888

Carneiro R. B., Gomes G. M., Zaiat M. and Santos-Neto Á. J. (2022). Two-phase (acidogenic–methanogenic) anaerobic fixed bed biofilm reactor enhances the biological domestic sewage treatment: perspectives for recovering bioenergy and value-added by-products. *Journal of Environmental Management*, **317**, 115388, https://doi.org/10.1016/j.jenvman.2022.115388

Chen H., Zha J., Liang X., Li J. and Wang Z. (2014). Effects of the human antiepileptic drug carbamazepine on the behavior, biomarkers, and heat shock proteins in the Asian clam *Corbicula fluminea*. *Aquatic Toxicology*, **155**, 1–8, https://doi.org/10.1016/j.aquatox.2014.06.001

Chen W.-H., Wong Y., Huang T., Chen W.-H. and Lin J. (2019). Removals of pharmaceuticals in municipal wastewater using a staged anaerobic fluidized membrane bioreactor. *International Biodeterioration & Biodegradation*, **140**, 29–36, https://doi.org/10.1016/j.ibiod.2019.03.008

Corcoll N., Casellas M., Huerta B., Guasch H., Acuña V., Rodríguez-Mozaz S., Serra-Compte A., Barceló D. and Sabater S. (2015). Effects of flow intermittency and pharmaceutical exposure on the structure and metabolism of stream biofilms. *Science of the Total Environment*, **503–504**, 159–170, https://doi.org/10.1016/j.scitotenv.2014.06.093

de Graaff M. S., Vieno N. M., Kujawa-Roeleveld K., Zeeman G., Temmink H. and Buisman C. J. N. (2011). Fate of hormones and pharmaceuticals during combined anaerobic treatment and nitrogen removal by partial nitritation-anammox in vacuum collected black water. *Water Research*, **45**, 375–383, https://doi.org/10.1016/j.watres.2010.08.023

Dutta K., Lee M. Y., Lai W. W. P., Lee C. H., Lin A. Y. C., Lin C. F. and Lin J. G. (2014). Removal of pharmaceuticals and organic matter from municipal wastewater using two-stage anaerobic fluidized membrane bioreactor. *Bioresource Technology*, **165**, 42–49, https://doi.org/10.1016/j.biortech.2014.03.054

Escher B. I. and Fenner K. (2011). Recent advances in environmental risk assessment of transformation products. *Environmental Science & Technology*, **45**(9), 3835–3847, https://doi.org/10.1021/es1030799

European Commission. (2023). Water Framework Directive [WWW Document]. https://environment.ec.europa.eu/topics/water/water-framework-directive_en#implementation

Falås P., Wick A., Castronovo S., Habermacher J., Ternes T. A. and Joss A. (2016). Tracing the limits of organic micropollutant removal in biological wastewater treatment. *Water Research*, **95**, 240–249, https://doi.org/10.1016/j.watres.2016.03.009

Fent K., Weston A. and Caminada D. (2006). Ecotoxicology of human pharmaceuticals. *Aquatic Toxicology*, **76**, 122–159, https://doi.org/10.1016/j.aquatox.2005.09.009

Fernandes A. S., Mello F. V. C., Thode Filho S., Carpes R. M., Honório J. G., Marques M. R. C., Felzenszwalb I. and Ferraz E. R. A. (2017). Impacts of discarded coffee waste on human and environmental health. *Ecotoxicology and Environmental Safety*, **141**, 30–36, https://doi.org/10.1016/j.ecoenv.2017.03.011

Fischer K. and Majewsky M. (2014). Cometabolic degradation of organic wastewater micropollutants by activated sludge and sludge-inherent microorganisms. *Applied Microbiology and Biotechnology*, **98**, 6583–6597, https://doi.org/10.1007/s00253-014-5826-0

Ghattas A. K., Fischer F., Wick A. and Ternes T. A. (2017). Anaerobic biodegradation of (emerging) organic contaminants in the aquatic environment. *Water Research*, **116**, 268–295, https://doi.org/10.1016/j.watres.2017.02.001

Gibs J., Heckathorn H. A., Meyer M. T., Klapinski F. R., Alebus M. and Lippincott R. L. (2013). Occurrence and partitioning of antibiotic compounds found in the water column and bottom sediments from a stream receiving two wastewater treatment plant effluents in northern New Jersey, 2008. *Science of the Total Environment*, **458–460**, 107–116, https://doi.org/10.1016/j.scitotenv.2013.03.076

Golub M. S., Wu K. L., Kaufman F. L., Li L.-H., Moran-Messen F., Zeise L., Alexeeff G. V. and Donald J. M. (2010). Bisphenol A: developmental toxicity from early prenatal exposure. *Birth Defects Research, Part B: Developmental and Reproductive Toxicology*, **89**, 441–466, https://doi.org/10.1002/bdrb.20275

Gonzalez-Gil L., Mauricio-Iglesias M., Carballa M. and Lema J. M. (2018a). Why are organic micropollutants not fully biotransformed? A mechanistic modelling approach to anaerobic systems. *Water Research*, **142**, 115–128, https://doi.org/10.1016/j.watres.2018.05.032

Gonzalez-Gil L., Mauricio-Iglesias M., Serrano D., Lema J. M. and Carballa M. (2018b). Role of methanogenesis on the biotransformation of organic micropollutants during anaerobic digestion. *Science of the Total Environment*, **622–623**, 459–466, https://doi.org/10.1016/j.scitotenv.2017.12.004

Gonzalez-Gil L., Krah D., Ghattas A. K., Carballa M., Wick A., Helmholz L., Lema J. M. and Ternes T. A. (2019). Biotransformation of organic micropollutants by anaerobic sludge enzymes. *Water Research*, **152**, 202–214, https://doi.org/10.1016/j.watres.2018.12.064

Gonzalez-Rey M. and Bebianno M. J. (2014). Effects of non-steroidal anti-inflammatory drug (NSAID) diclofenac exposure in mussel *Mytilus galloprovincialis*. *Aquatic Toxicology*, **148**, 221–230, https://doi.org/10.1016/j.aquatox.2014.01.011

Government of Canada. (2023). Water Quality – Reports and Publications [WWW Document]. https://www.canada.ca/en/health-canada/services/environmental-workplace-health/reports-publications/water-quality.html

Granatto C. F., Grosseli G. M., Sakamoto I. K., Fadini P. S. and Varesche M. B. A. (2021). Influence of cosubstrate and hydraulic retention time on the removal of drugs and hygiene products in sanitary sewage in an anaerobic expanded granular sludge bed reactor. *Journal of Environmental Management*, **299**, 113532, https://doi.org/10.1016/j.jenvman.2021.113532

Gros M., Petrović M. and Barceló D. (2007). Wastewater treatment plants as a pathway for aquatic contamination by pharmaceuticals in the Ebro River Basin (northeast Spain). *Environmental Toxicology and Chemistry*, **26**, 1553, https://doi.org/10.1897/06-495R.1

Gutiérrez J. E. V., Camargo F. P., Sakamoto I. K. and Varesche M. B. A. (2022). Expanded granular sludge bed reactor technology feasibility for removal of nonylphenol ethoxylate in co-digestion of domestic sewage and commercial laundry wastewater: taxonomic characterization and biogas production. *Process Safety and Environmental Protection*, **161**, 556–570, https://doi.org/10.1016/j.psep.2022.03.055

Harb M., Lou E., Smith A. L. and Stadler L. B. (2019). Perspectives on the fate of micropollutants in mainstream anaerobic wastewater treatment. *Current Opinion in Biotechnology*, **57**, 94–100, https://doi.org/10.1016/j.copbio.2019.02.022

Hu X. (2018). The Most Widely Used Pesticide, One Year Later – Science in the News [WWW Document]. *Harvard University, Science Policy Blog*. https://sitn.hms.harvard.edu/flash/2018/widely-used-pesticide-one-year-later/

Jelic A., Gros M., Ginebreda A., Cespedes-Sánchez R., Ventura F., Petrovic M. and Barcelo D. (2011). Occurrence, partition and removal of pharmaceuticals in sewage water and sludge during wastewater treatment. *Water Research*, **45**, 1165–1176, https://doi.org/10.1016/j.watres.2010.11.010

Kasprzyk-Hordern B., Dinsdale R. M. and Guwy A. J. (2009). The removal of pharmaceuticals, personal care products, endocrine disruptors and illicit drugs during wastewater treatment and its impact on the quality of receiving waters. *Water Research*, **43**, 363–380, https://doi.org/10.1016/j.watres.2008.10.047

Keen O. S., Baik S., Linden K. G., Aga D. S. and Love N. G. (2012). Enhanced biodegradation of carbamazepine after UV/$H_2O_2$ advanced oxidation. *Environmental Science & Technology*, **46**, 6222–6227, https://doi.org/10.1021/es300897u

Kennes-Veiga D. M., Gónzalez-Gil L., Carballa M. and Lema J. M. (2022). Enzymatic cometabolic biotransformation of organic micropollutants in wastewater treatment plants: a review. *Bioresource Technology*, **344**, 126291, https://doi.org/10.1016/j.biortech.2021.126291

Kihlman B. A. (1974). Effects of caffeine on the genetic material. *Mutation Research/Fundamental and Molecular Mechanisms of Mutagenesis*, **26**, 53–71, https://doi.org/10.1016/S0027-5107(74)80036-9

Kümmerer K., Al-Ahmad A. and Mersch-Sundermann V. (2000). Biodegradability of some antibiotics, elimination of the genotoxicity and affection of wastewater bacteria in a simple test. *Chemosphere*, **40**, 701–710, https://doi.org/10.1016/S0045-6535(99)00439-7

Ku P., Wu X., Nie X., Ou R., Wang L., Su T. and Li Y. (2014). Effects of triclosan on the detoxification system in the yellow catfish (*Pelteobagrus fulvidraco*): expressions of CYP and GST genes and corresponding enzyme activity in phase I, II and antioxidant system. *Comparative Biochemistry and Physiology, Part C: Toxicology & Pharmacology*, **166**, 105–114, https://doi.org/10.1016/j.cbpc.2014.07.006

Lee T. H. Y., Chuah J. and Snyder S. A. (2022). Occurrence of emerging contaminants in southeast Asian environments: present status, challenges, and future prospects. *ACS ES&T Water*, **2**, 907–931, https://doi.org/10.1021/acsestwater.1c00453

Li S., Ren Y., Fu Y., Gao X., Jiang C., Wu G., Ren H. and Geng J. (2018). Fate of artificial sweeteners through wastewater treatment plants and water treatment processes. *PLoS ONE*, **13**, e0189867.

Luo Y., Guo W., Ngo H. H., Nghiem L. D., Hai F. I., Zhang J., Liang S. and Wang X. C. (2014). A review on the occurrence of micropollutants in the aquatic environment and their fate and removal during wastewater treatment. *Science of the Total Environment*, **473–474**, 619–641, https://doi.org/10.1016/j.scitotenv.2013.12.065

Macêdo W. V., Duarte Oliveira G. H. and Zaiat M. (2021). Tetrabromobisphenol A (TBBPA) anaerobic biodegradation occurs during acidogenesis. *Chemosphere*, **282**, 130995, https://doi.org/10.1016/j.chemosphere.2021.130995

McCurry D. L., Bear S. E., Bae J., Sedlak D. L., McCarty P. L. and Mitch W. A. (2014). Superior removal of disinfection byproduct precursors and pharmaceuticals from wastewater in a staged anaerobic fluidized membrane bioreactor compared to activated sludge. *Environmental Science & Technology Letters*, **1**, 459–464, https://doi.org/10.1021/ez500279a

Metcalfe C. D., Metcalfe T. L., Kiparissis Y., Koenig B. G., Khan C., Hughes R. J., Croley T. R., March R. E. and Potter T. (2001). Estrogenic potency of chemicals detected in sewage treatment plant effluents as determined by in vivo assays with Japanese medaka (*Oryzias latipes*). *Environmental Toxicology and Chemistry*, **20**, 297–308, https://doi.org/10.1002/etc.5620200210

Mladenov N., Dodder N. G., Steinberg L., Richardot W., Johnson J., Martincigh B. S., Buckley C., Lawrence T. and Hoh E. (2022). Persistence and removal of trace organic compounds in centralized and decentralized wastewater treatment systems. *Chemosphere*, **286**, 131621, https://doi.org/10.1016/j.chemosphere.2021.131621

Mora-Cabrera K., Peña-Guzmán C., Trapote A. and Prats D. (2021). Use of combined UASB+eMBR treatment for removal of emerging micropollutants and reduction of fouling. *Aqua Water Infrastructure, Ecosystems and Society*, **70**, 984–1001, https://doi.org/10.2166/aqua.2021.058

do Nascimento J. G. da S., Silva E. V. A., dos Santos A. B., da Silva M. E. R. and Firmino P. I. M. (2021). Microaeration improves the removal/biotransformation of organic micropollutants in anaerobic wastewater treatment systems. *Environmental Research*, **198**, 111313, https://doi.org/10.1016/j.envres.2021.111313

Oliveira G. H. D., Santos-Neto A. J. and Zaiat M. (2016). Evaluation of sulfamethazine sorption and biodegradation by anaerobic granular sludge using batch experiments. *Bioprocess and Biosystems Engineering*, **39**, 115–124, https://doi.org/10.1007/s00449-015-1495-3

Peña-Guzmán C., Ulloa-Sánchez S., Mora K., Helena-Bustos R., Lopez-Barrera E., Alvarez J. and Rodriguez-Pinzón M. (2019). Emerging pollutants in the urban water cycle in Latin America: a review of the current literature. *Journal of Environmental Management*, **237**, 408–423, https://doi.org/10.1016/j.jenvman.2019.02.100

Petersen K., Heiaas H. H. and Tollefsen K. E. (2014). Combined effects of pharmaceuticals, personal care products, biocides and organic contaminants on the growth of *Skeletonema pseudocostatum*. *Aquatic Toxicology*, **150**, 45–54, https://doi.org/10.1016/j.aquatox.2014.02.013

Pirete L. de M., Camargo F. P., Dornelles H. S., Granatto C. F., Sakamoto I. K., Grosseli G. M., Fadini P. S., Silva E. L. and Varesche M. B. A. (2022). Biodegradation of diclofenac and ibuprofen in fluidized bed reactor applied to sanitary sewage treatment in acidogenic and denitrifying conditions. *Journal of Water Process Engineering*, **49**, 102964, https://doi.org/10.1016/j.jwpe.2022.102964

Queiroz F. B., Brandt E. M. F., Aquino S. F., Chernicharo C. A. L. and Afonso R. J. C. F. (2012). Occurrence of pharmaceuticals and endocrine disruptors in raw sewage and their behavior in UASB reactors operated at different hydraulic retention times. *Water Science and Technology*, **66**, 2562–2569, https://doi.org/10.2166/wst.2012.482

Rajagopal R., Choudhury M., Anwar N., Goyette B. and Rahaman M. (2019). Influence of pre-hydrolysis on sewage treatment in an up-flow anaerobic sludge blanket (UASB) reactor: a review. *Water (Basel)*, **11**, 372.

Ren H., Tröger R., Ahrens L., Wiberg K. and Yin D. (2020). Screening of organic micropollutants in raw and drinking water in the Yangtze River Delta, China. *Environmental Sciences Europe*, **32**, 67, https://doi.org/10.1186/s12302-020-00342-5

Reyes-Contreras C., Matamoros V., Ruiz I., Soto M. and Bayona J. M. (2011). Evaluation of PPCPs removal in a combined anaerobic digester-constructed wetland pilot plant treating urban wastewater. *Chemosphere*, **84**, 1200–1207, https://doi.org/10.1016/j.chemosphere.2011.06.003

Ribeiro A. R., Nunes O. C., Pereira M. F. R. and Silva A. M. T. (2015). An overview on the advanced oxidation processes applied for the treatment of water pollutants defined in the recently launched Directive 2013/39/EU. *Environment International*, **75**, 33–51, https://doi.org/10.1016/j.envint.2014.10.027

Richard J., Boergers A., vom Eyser C., Bester K. and Tuerk J. (2014). Toxicity of the micropollutants bisphenol A, ciprofloxacin, metoprolol and sulfamethoxazole in water samples before and after the oxidative treatment. *International Journal of Hygiene and Environmental Health*, **217**, 506–514, https://doi.org/10.1016/j.ijheh.2013.09.007

Rizzo L., Malato S., Antakyali D., Beretsou V. G., Đolić M. B., Gernjak W., Heath E., Ivancev-Tumbas I., Karaolia P., Lado Ribeiro A. R., Mascolo G., McArdell C. S., Schaar H., Silva A. M. T. and Fatta-Kassinos D. (2019). Consolidated vs new advanced treatment methods for the removal of contaminants of emerging concern from urban wastewater. *Science of the Total Environment*, **655**, 986–1008, https://doi.org/10.1016/j.scitotenv.2018.11.265

Rogowska J., Cieszynska-Semenowicz M., Ratajczyk W. and Wolska L. (2020). Micropollutants in treated wastewater. *Ambio*, **49**, 487–503, https://doi.org/10.1007/s13280-019-01219-5

Sanganyado E. (2022). Policies and regulations for the emerging pollutants in freshwater ecosystems: challenges and opportunities. In: Emerging Freshwater Pollutants: Analysis, Fate and Regulations, T. Dalu and N. T. Tavengwa (eds.), Amsterdam: Elsevier, pp. 361–372, https://doi.org/10.1016/C2019-0-04698-5

Sawaya C., El Khoury C., Ramadan L., Deeb R. and Harb M. (2022). Effects of influent municipal wastewater microbial community and antibiotic resistance gene profiles on anaerobic membrane bioreactor effluent. *Water Reuse*, **12**, 304–318, https://doi.org/10.2166/wrd.2022.018

Smith A. L., Shimada T. and Raskin L. (2017). A comparative evaluation of community structure in full-scale digesters indicates that two-phase digesters exhibit greater microbial diversity than single-phase digesters. *Environmental Science*, **3**, 304–311.

Speece R. E. (1996). Anaerobic Biotechnology for Industrial Wastewaters. Archae Press, Nashville, Tennessee.

Stadler L. B., Ernstoff A. S., Aga D. S. and Love N. G. (2012). Micropollutant fate in wastewater treatment: redefining 'removal'. *Environmental Science & Technology*, **46**, 10485–10486, https://doi.org/10.1021/es303478w

The United States Environmental Protection Agency. (2023). National Primary Drinking Water Regulations [WWW Document]. https://www.epa.gov/ground-water-and-drinking-water/national-primary-drinking-water-regulations#Organic

Torresi E., Escolà Casas M., Polesel F., Plósz B. G., Christensson M. and Bester K. (2017). Impact of external carbon dose on the removal of micropollutants using methanol and ethanol in post-denitrifying moving bed biofilm reactors. *Water Research*, **108**, 95–105, https://doi.org/10.1016/j.watres.2016.10.068

Tran N. H., Reinhard M. and Gin K. Y. H. (2018). Occurrence and fate of emerging contaminants in municipal wastewater treatment plants from different geographical regions – a review. *Water Research*, **133**, 182–207, https://doi.org/10.1016/j.watres.2017.12.029

Vassalle L., García-Galán M. J., Aquino S. F., Afonso R. J. de C. F., Ferrer I., Passos F. and R Mota C. (2020). Can high rate algal ponds be used as post-treatment of UASB reactors to remove micropollutants? *Chemosphere*, **248**, 125969, https://doi.org/10.1016/j.chemosphere.2020.125969

Verlicchi P., Al Aukidy M. and Zambello E. (2012). Occurrence of pharmaceutical compounds in urban wastewater: removal, mass load and environmental risk after a secondary treatment – a review. *Science of the Total Environment*, **429**, 123–155, https://doi.org/10.1016/j.scitotenv.2012.04.028

Wang G., Shen J., Wei S., Cai D. and Liu J. (2022). Identification of heavy metals and organic micropollutants in drinking water sources in typical villages and towns in northeast China. *Molecules*, **27**, 8033, https://doi.org/10.3390/molecules27228033

Wentzel M. C. and Ekama G. A. (1997). Principles in the modelling of a biological wastewater treatment plants. In: Microbial Community Analysis: The Key to the Design of Biological Wastewater Treatment Systems, T. E. Cloete, and N. Y. O. Muyima (eds), Cambridge: IWA Publishing, pp. 73–82.

Wijekoon K. C., McDonald J. A., Khan S. J., Hai F. I., Price W. E. and Nghiem L. D. (2015). Development of a predictive framework to assess the removal of trace organic chemicals by anaerobic membrane bioreactor. *Bioresource Technology*, **189**, 391–398, http://dx.doi.org/10.1016/j.biortech.2015.04.034

Wolff D., Krah D., Dötsch A., Ghattas A. K., Wick A. and Ternes T. A. (2018). Insights into the variability of microbial community composition and micropollutant degradation in diverse biological wastewater treatment systems. *Water Research*, **143**, 313–324, https://doi.org/10.1016/j.watres.2018.06.033

Yang Y., Fu J., Peng H., Hou L., Liu M. and Zhou J. L. (2011). Occurrence and phase distribution of selected pharmaceuticals in the Yangtze Estuary and its coastal zone. *Journal of Hazardous Materials*, **190**, 588–596, https://doi.org/10.1016/j.jhazmat.2011.03.092

Zi L. (2022). China's plan to clean up 'new pollutants'. [WWW Document]. *China Dialogue*. https://chinadialogue.net/en/pollution/chinas-plan-to-clean-up-new-pollutants/

Zirlewagen J., Licha T., Schiperski F., Nödler K. and Scheytt T. (2016). Use of two artificial sweeteners, cyclamate and acesulfame, to identify and quantify wastewater contributions in a karst spring. *Science of the Total Environment*, **547**, 356–365, https://doi.org/10.1016/j.scitotenv.2015.12.112

doi: 10.2166/9781789063479_0275

# Chapter 10
# Process monitoring and control

*Jorge Rodríguez* and Wasim Ahmed*

*Department of Chemical Engineering, Khalifa University, PO Box 127788 Abu Dhabi, United Arab Emirates*
**Correspondence: jorge.rodriguez@ku.ac.ae*

**ABSTRACT**

Anaerobic treatment of municipal wastewater has its own peculiarities given the low organic loading, volatile suspended solid content, and other specifics; however, monitoring and control approaches for sewage anaerobic biological systems share most principles as for industrial wastewater anaerobic treatment. The main key point for stable nominal operation in anaerobic bioreactors is to maintain sufficient active methanogenic biomass in reactors. The loss of methanogenic biomass can occur due to hydraulic washout, insufficient supply of methanogenesis substrates, inhibitions or toxicity including temperature shock. In this chapter, general key aspects when thinking on anaerobic treatment control are presented and discussed, including the most important variables for effective monitoring. Both conventional and more advanced strategies for process control are discussed with an emphasis on the process dynamics and the utilization of the available information from instrumentation to the best possible way.

**Keywords:** advanced control, anaerobic reactor monitoring, anaerobic treatment control, instrumentation, model-based control.

## 10.1 MONITORING AND CONTROL: MOTIVATION AND GOALS

Similar to water treatment systems, the control of anaerobic processes is driven by a combination of forces and technology factors. These include the need to enhance operational efficiency to reduce both capital and operational expenses related to energy and chemicals (Olsson & Ingildsen, 2020). Good control involves maintaining consistent operation and quality of effluents as well as to proactively responding to unexpected events or disruptions. These include hydraulic and organic overloads, issues related to toxicity, sudden temperature or pH fluctuations, and more. Effective control measures not only serve to address these challenges but can also provide valuable insights into understanding the dynamics of the process for troubleshooting purposes.

Although anaerobic treatment is typically utilized to treat warm, high-strength industrial wastewater containing high concentrations of biodegradable organic matter, direct anaerobic treatment of domestic wastewater is becoming attractive as it can change a wastewater treatment plant from energy consuming to energy producing. However, anaerobic sewage treatment has its own limitations;

it is limited by hydrodynamic constraints including flow fluctuations, rather than organic conversion capacity. A high fraction of particulate chemical oxygen demand (COD) in influents favours flocculent rather than granular sludge in upflow anaerobic sludge blanket (UASB) reactors and the composition in general displays a larger variability when treating sewage wastewater. Although the monitoring and control principles applicable to anaerobic reactors treating domestic wastewater do not differ much from those treating industrial wastewater, these specific limitations should be considered when designing monitoring and control of anaerobic sewage treatment plants.

Good monitoring and control practices for anaerobic reactors in domestic wastewater treatment offer benefits such as enhanced process efficiency, increased biogas production, efficient waste treatment, odour control, prevention of process upsets, regulatory compliance and operational cost savings. Adopting robust monitoring and control systems will optimize the performance and sustainability of processes. It is also essential to prevent process upsets, to ensure regulatory compliance and to achieve operational cost savings. By adopting robust monitoring and control systems, process performance and sustainability can be maximized.

When approaching the problem of controlling anaerobic reactors it is important to be aware of their characteristic dynamic behaviour. An anaerobic digestion (AD) process is a multistep process in which normally only either the first or the last step can become limiting. These first and last steps are the hydrolysis of particulate matter and methanogenesis, respectively (Figure 10.1). This implies that the intermediate step of acidification, which produces components potentially inhibitory to methanogenesis is actually the fastest step. This has an immediate implication that if acidification proceeds too quickly the system can be flooded with acids crossing the pH inhibition threshold for methanogens. If methanogens cease to operate, the accumulation of acids is never eliminated and the system methanogenic capabilities can collapse permanently. This is an inherently unstable system once the pH inhibition threshold has been crossed and therefore will require an active control action to bring it back into the stable operation range.

### 10.1.1 Anaerobic reactor operation under nominal conditions

Adequate operational conditions in anaerobic reactors must be maintained to provide favourable reactor environmental conditions such that the ecological balance of key AD microorganisms, especially methanogens, is maintained (de Lemos Chernicharo, 2007). Under normal conditions and in the presence of sufficient active methanogenic biomass, intermediate volatile fatty acids (VFAs),

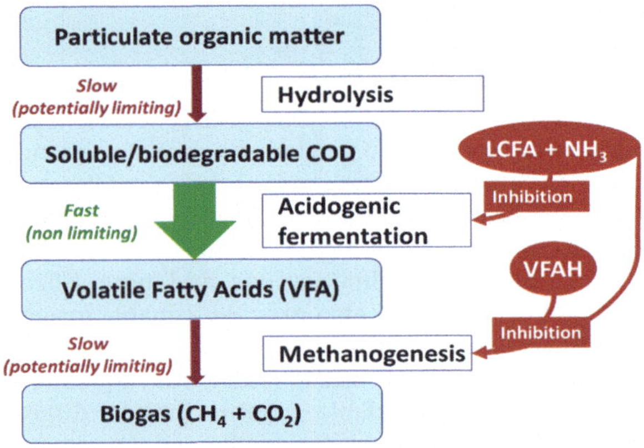


**Figure 10.1** AD is a multistep process. Methanogenesis can be inhibited by the accumulation of intermediate products of the acidogenic fermentation and other components. Effective control action must reject the disturbances leading to that accumulation over the threshold of instability.

generated during an intermediate fast acidification process, are consumed as quickly as they are generated. Under these conditions the limiting process step is methanogenesis. For the production of VFAs and their methanization to remain balanced it is necessary to ensure that sufficient active methanogenic biomass remains in the reactor. Adequate biomass retention is the key to ensure a feasible treatment capacity, especially for low-strength wastewater. As illustrated in the previous chapters, bioreactors such as UASBs and their later developments can deliver this capability. In particular, an expanded granular sludge bed reactor could offer better removal efficiency due to better contact between sludge and wastewater, making it suitable for dilute substrates such as domestic wastewater (Seghezzo *et al.*, 1998).

The design of monitoring and control systems should aim not only to ensure stable and robust operation under normal conditions but also to optimize the process during start-up or recovery from unexpected upsets as swiftly as possible. The normal operation of an anaerobic digester is characterized by a stable microbial methanogenic activity leading to high yield of methanization of the influent biological oxygen demand (BOD).

### 10.1.2 Anaerobic reactor operation under dynamic conditions

In response to process and operational changes that may occur at any stage, including sudden or unforeseen events, as well as during start-up and recovery phases, it is crucial to take timely and appropriate actions. These operational measures are aimed at preventing process upsets, such as the accumulation of acids or disruptions in the microbial community, and ensuring the continued smooth operation of the system.

A critical yet challenging operation phase that can determine the entire progression of anaerobic treatment systems is the start-up (Sbarciog *et al.*, 2012). In terms of operational strategy, the feeding regimen and inoculation procedures are pivotal factors during this phase. During start-up, a carefully planned feeding regimen becomes essential to support the growth and adaptation of the microbial consortium to changing or unfamiliar environments, all while avoiding destabilization. Consequently, across both experimental and full-scale studies, it is a common practice to gradually increase organic loading rate (OLR) of the feed, considering the current state of the digester and any required inoculation procedures.

## 10.2 MONITORING ANAEROBIC TREATMENT PROCESSES

Adequate operational conditions of an implemented anaerobic reactor are essential to create and maintain favourable reactor environmental conditions so that an ecological balance of key anaerobic microorganisms, especially methanogens, is maintained (de Lemos Chernicharo, 2007). The microbial consortia present in an anaerobic reactor can work effectively only under a specific range of conditions and they are sensitive to significant process changes that could eventually cause process failures if not adequately monitored.

### 10.2.1 Key variables and mechanisms affecting anaerobic treatment process operation and stability

Maintaining adequate balance of anaerobic microorganisms requires consideration of both physical and chemical characteristics of the reactor environment since these affect microbial growth (de Lemos Chernicharo, 2007). In this regard, important factors include:

- *Biomass retention*: a significant factor that increases treatment efficiency, shortens treatment period and/or reduces reactor size requirements is the density of microorganisms available for treatment (Taricska *et al.*, 2009). Also, the form of the microbial biomass, existing on their own or as aggregates, has a significant effect on microorganism survival, transfer of nutrients and consequently, on the overall efficiency of an anaerobic treatment process (de Lemos Chernicharo, 2007).

- *Nutrients:* a number of macro- and micro-nutrients are essential for maintenance and correct functioning of microorganisms. Hence, the reactor medium should have essential nutrients in sufficient quantities. For methanogens, these include N, P, S, K, Ca and Mg as macronutrients, and Fe, Ni, Co, Mo, Zn, Mn and Cu as micronutrients.
- *Temperature*: this is an important physical factor for microbial growth, favourable thermodynamics of several reactions and substrate diffusion rates. While an optimal temperature range for anaerobic treatment is between 30 and 40°C, anaerobic treatment rates at lower temperatures can decrease by ~11%/°C below the ideal temperature range (Lubbe & Haandel, 2019). This presents challenges for anaerobic processes operating at lower temperatures (e.g. domestic wastewater treatment in milder climate regions).
- *pH:* this affects enzymatic activity and toxicity of a number of compounds. In an anaerobic treatment process, VFAs and carbonic acid (from generated $CO_2$) are the main sources affecting pH. For methanogens in general, the optimal pH range is between 6.6 and 7.4 while inhibition occurs for pH below 6.0 or above 8.3. Acidogens are much less sensitive to pH and can survive even at very low pH (as low as 4.5).
- *Alkalinity*: this is the buffer capacity of an anaerobic treatment system especially in the form of bicarbonate ($HCO_3^-$) via ionic speciation reactions. The buffer capacity due to alkalinity helps in avoiding pH changes by regulating $H^+$ ion concentration in the reactor medium. Sufficient alkalinity will help in neutralization of acidic VFA intermediates and hence help in pH buffering in case of VFA accumulation. Overall, alkalinity is important for evaluation of pH and indication of acidification conditions inside a reactor.
- *Toxic compounds*: toxicity of anaerobic microorganisms towards a number of compounds is one of the main reasons for non-generalized use of an anaerobic treatment process. Controlling or eliminating such compounds is necessary to minimize or prevent inhibition to microorganisms, especially methanogens which can be easily inhibited by these compounds. The level of toxicity depends on the concentration of inhibitory compounds and their impact on microorganisms (reversible or irreversible). The chemical species that can be inhibitory to microorganisms include salts (primarily cations such as $Na^+$, $Ca^{2+}$, etc.), ammonia/ammonium ($NH_3$/ $NH_4^+$), hydrogen sulphide ($H_2S$) and heavy metals (chromium, nickel, zinc, copper, etc.). For sewage anaerobic treatment, possible toxic compounds in the feed include oxygen, $NO_x$, $NH_4^+$ /$NH_3$ and sulphide although they may be likely at non-inhibitory levels (Lubbe & Haandel, 2019).

Researchers have identified and suggested many key process parameters known to act as diagnosis indicators of process stability: pH, biogas/methane flow, dissolved $H_2$ concentration, alkalinity and VFA concentration (García-Gen *et al.*, 2015). For an anaerobic treatment process, VFA levels and alkalinity are important indicators to assess process stability while microbial community profiling can give a direct indication of the process status (performance and stability). Methane production is also one of the important indicators to indirectly access stability in AD systems.

### 10.2.2 Variable selection based on availability and practicality

For efficient anaerobic bioreactor operation, activities that evaluate process efficiency, operational stability and sludge characteristics (amount and quality) are important. During practical anaerobic reactor operation, some of the important process variables and parameters that need to be verified and monitored include (de Lemos Chernicharo, 2007):

- Influent flow rates
- Physicochemical and microbiological characteristics of influent wastewater
- Efficiency and operational problems of the reactor
- Biogas characteristics (amount and quality)
- Produced sludge characteristics (amount and quality)

**Table 10.1** List of monitoring variables suggested for assessing different factors during anaerobic reactor operation.

| | Process Assessment Factors | | |
|---|---|---|---|
| | **Treatment Efficiency** | **Reactor Stability** | **Sludge Quantity and Quality** |
| **Monitoring variables** | • Suspended solids<br>• COD<br>• BOD<br>• Biogas production<br>• Pathogenic organisms | • Temperature<br>• pH<br>• Alkalinity<br>• VFAs<br>• Variations in biogas composition | • Total solids<br>• Total volatile solids<br>• Sludge mass<br>• Specific methanogenic activity<br>• Fraction of biodegradable solids in sludge<br>• Sludge settleability |

COD, chemical oxygen demand; BOD, biological oxygen demand; VFAs, volatile fatty acids.

A list of process variables that should be generally monitored during steady-state and/or stable operating conditions is provided in Table 10.1 in order to assess process efficiency, stability and quality of resulting sludge. A number of these listed variables are also necessary for anaerobic treatment process monitoring in general irrespective of the operational phase. Assessing the levels of dissolved organic matter (DOM) using BOD measurements is important for domestic wastewater since there is heterogeneity in DOM quality and levels in such wastes (Goffin *et al.*, 2018). This affects the biological treatment efficiency which is highly sensitive to soluble organic matter fractions.

According to monitoring equipment availability and practical limitations, the measuring frequency of the variables given in Table 10.1 can range from daily to once in a month. The majority of these variables are limited to being measured and monitored offline only, which is problematic for instrumentation and control purposes.

### 10.2.3 Current AD process-monitoring practices

Owing to the long hydraulic retention times (HRTs) or solid retention times (SRTs) used in anaerobic reactor systems and the resource limitations in available monitoring equipment/techniques (particularly in case of microbial biomass and heterogeneous waste characterization), offline measurements are practically feasible and have long been used for monitoring the process. Offline measurements such as COD, total organic carbon, BOD, biomethane potential tests and organic matter biochemical analysis have been commonly utilized in anaerobic treatment systems (Jimenez *et al.*, 2015). Methane production (part of biogas) is one of the common variables measured and monitored for assessing anaerobic reactor output performance.

In case of anaerobic reactor-monitoring practices at a commercial scale, Table 10.2 provides a summary of variables monitored across five of the existing plants in Europe according to the report by The Wales Centre of Excellence for Anaerobic Digestion *et al.* (2012) as an example. Although a variety of process variables are being monitored in these full-scale systems, only a few of the variables are measured online (pH, temperature, biogas flow rate, biogas composition, biogas yield and partial pressures) while a majority are measured offline and less frequently.

### 10.2.4 Practical limitations in anaerobic reactor monitoring and control

A small number of variables typically monitored online, particularly in commercial-scale anaerobic treatment systems, presents a limitation for the development of effective operational control strategies, for which the variables monitored online to effectively assess the process state is a critical component. For an anaerobic treatment process, VFA levels and alkalinity are important indicators to assess process stability while microbial community profiling could in some cases provide some indication of the process status (performance and stability). On a commercial scale, there are still major hurdles to overcome when monitoring such variables. VFA and alkalinity are generally monitored offline and less frequently in case of full-scale AD plants while monitoring for microbial community diversity and/

**Table 10.2** Summary of variables monitored across some of the existing anaerobic treatment plants in Europe.

| Anaerobic Treatment Plant Component or Stream Analysed | Monitoring Variables | Measurement Frequency | Sampling |
|---|---|---|---|
| Feedstock | Volatile and total solids | On arrival, few times a week or monthly | Manual and on-site/in lab |
| | Temperature | Continuous | Online and on-site |
| | pH | Monthly or few times a year | Online/manual and on-site/in lab |
| | Potential biogas yield | Monthly | Manual and in lab |
| | Alkalinity | Monthly or few times a year | Manual and in lab |
| | C:N:P:S analysis | Monthly or few times a year | Manual and in lab |
| | Carbs, lipids, proteins, heavy metals, light metal ions | Less often (every year or two) | Manual and in lab |
| Digester | Total solids | Weekly or monthly | Manual and in lab |
| | Temperature | Continuous | Online and on-site |
| | pH | Continuous or weekly | Online/manual and on-site |
| | OLR | Weekly, monthly or few times a year | Manual and in lab or calculated |
| | VFAs | Weekly, monthly or few times a year | Manual and in lab |
| | C:N:P:S ratio | Every 4 months | Manual and in lab |
| | Hydraulic and solid retention times | Monthly | Manual and in lab |
| Biogas | Gas flow rate | Continuous or few times a week | Online and on-site/telemetry |
| | Biogas yield | Continuous, daily or few times a week | Online and on-site/telemetry |
| | Biogas composition ($CH_4$, $CO_2$, $NH_3$, $H_2S$, other trace gases, impurities) | Continuous or few times a week | Online and on-site/telemetry |
| | Partial pressures ($CH_4$, $CO_2$, $H_2$) | Continuous or few times a day | Online/manual and on-site |
| Digestate | Volatile and total solids | Few times a year (2–4) | Manual and in lab |
| | Temperature | Continuous | Online and on-site |
| | pH | Continuous or few times a year | Online/manual and on-site/in lab |
| | Nutrients and trace elements | Few times a year (2–4) | Manual and in lab |
| | Heavy metals | Yearly | Manual and in lab |
| | N total | Before spreading, weekly or few times a year | Manual and in lab |
| | Pathogens (e.g. *Salmonella*, *Enterobacteria*, helminth eggs) | Few times a year (1–4) | Manual and in lab |
| | VFAs | Biweekly or few times a year | Manual and in lab |

C:N:P:S, carbon:nitrogen:phosphorus:sulphur; VFAs, volatile fatty acids; N, nitrogen.
*Source:* Information adapted from The Wales Centre of Excellence for Anaerobic Digestion *et al.* (2012).

or abundance hardly exists or is not common. On the contrary, for anaerobic treatment of domestic wastewater/sewage, controlling some variables such as temperature is challenging although it is a readily available variable for online monitoring. For conventional suspended biomass treatment units, the low strength of such wastewater and the consequent low methane production from its anaerobic treatment could not provide sufficient heating to drive the process towards increased treatment rates in the optimal, higher temperature ranges (Lubbe & Haandel, 2019). A different behaviour is expected from high-rate systems, as extensively discussed in the previous chapters.

At the research level, novel techniques have been developed for online monitoring or rapid measurements. Existing studies have shown the potential and success of online-monitoring techniques for VFAs, alkalinity and other parameters (COD, total alkalinity, volatile suspended solids (VSS)) (Boe & Angelidaki, 2012; Falk *et al.*, 2015; Molina *et al.*, 2009; Morel *et al.*, 2005; Nielsen *et al.*, 2007; Steyer *et al.*, 2002). Techniques have been generally tested on lab- and pilot-scale systems, but tests with full-scale systems have been performed as well (as in the study by Nielsen *et al.*, 2007). With regards to practical considerations, the current limitations of such techniques include high cost (e.g. spectral techniques) and uncertainty in application to other case scenarios and process conditions since these techniques are generally calibration based. Another alternative strategy to deal with online-monitoring limitations is the development of model-based sensors (software sensors, observers, state estimators, etc.) that are able to predict the values of difficult to measure variables using variables that are easily measured or available during monitoring. Some of the popular approaches include Kalman filters and Luenberger observers (de Lemos Chernicharo, 2007). For AD processes in general, a number of studies on software sensors and state estimators/observers to predict unmeasured process variables have been published. For interested readers, de Lemos Chernicharo (2007) provide a review of such studies. Due to the predictive nature, such sensors have a major drawback of being limited to specific process conditions and operational ranges.

From a practical viewpoint, availability of skilled personnel, monitoring costs and opinions of technical consultants are the other major factors, besides equipment availability, affecting the extent of monitoring (e.g. number of variables measured and frequency) implemented in real-life anaerobic treatment plants (Drosg, 2013). However, when considering monitoring costs, one should also evaluate the economic losses resulting from insufficient monitoring (Drosg, 2013).

## 10.3 CONTROL OF ANAEROBIC TREATMENT PROCESSES

Despite several advantages of an anaerobic process as a treatment technology, a major concern in full-scale application is that it can become unstable under operational condition variations (Bernard *et al.*, 2001). The microbial consortia present in an anaerobic treatment system can work effectively under a certain range of conditions and are sensitive to significant process changes that could eventually cause process failures if not monitored. Hence, control of such conditions within safe and optimal ranges is essential for a stable and efficient performance.

Numerous studies have been conducted on proposal and implementation of control schemes for anaerobic reactors in general. In this regard, several control schemes have been proposed and implemented to monitor anaerobic reactors: on/off, standalone or combinations of proportional (P), integral (I) and derivative (D) controls, adaptive control, cascade control, neural network-based control, fuzzy logic, non-linear adaptive, non-linear model predictive control (NMPC) and many other expert control systems (García-Gen *et al.*, 2015; Jimenez *et al.*, 2015). A majority of these control strategies used HRT or OLR as a manipulated variable (MV) (García-Gen *et al.*, 2015).

### 10.3.1 Control strategies for nominal anaerobic reactor conditions

Operating under nominal conditions involves maintenance of stable and steady conditions with a given methane production rate. Under nominal steady-state conditions, anaerobic digester operation generally proceeds with minimal control action required (McCarty, 1964). During nominal operation the main task of the control system is the rejection/attenuation of disturbances. Some of the key anaerobic reactor disturbances that have been the basis of control strategies in existing studies involve acidification, alkalinity changes and organic overloads, together with the inflow of toxic components. As such, the variables targeted for control (control variables (CVs)) in existing studies include pH, alkalinity, VFA concentrations and methane production rate while manipulating dilution rate and/or added alkali/bicarbonate (MVs) (Jimenez *et al.*, 2015). The existence of several different validated control strategies for anaerobic reactors in the literature implies that there is diversity in control strategies for different objectives and applications, irrespective of the anaerobic reactor operation phase (steady state/nominal, start-up, recovery, re-start-up, etc.) and feed characteristics. Some researchers have considered targeting CVs for faster indication and management of process destabilization whereas others considered maximizing methane production while avoiding organic overloads.

In general, acidification has been the key phase to track and a number of studies have utilized different and/or early indicator variables for effective control. This could be more apparent for anaerobic reactors treating high-strength wastes than low-strength streams such as sewage or domestic wastewater. The study of McCarty (1964) represents one of the earliest guides on anaerobic treatment fundamentals, including heuristics and suggestions for corrective action against process imbalances. These included suggestions on using operational actions (e.g. adding lime or sodium bicarbonate) to maintain alkalinity and/or pH within a stable range (bicarbonate alkalinity in the range of 2500–5000 mg/L) (McCarty, 1964).

#### 10.3.1.1 Hydrogen-based control

An alternative and less conventional approach is based on the monitoring of another fast-response variable, which is hydrogen in the gas phase if available in terms of adequate instrumentation for it. Rodríguez *et al.* (2006) proposed an easy to use controller for stable maintenance of the concentration of $H_2$ in the gas phase. A controller operates by indirectly regulating the concentration of COD in the effluent, which reflects the methanization process. This is proposed by controlling the hydrogen concentration in the biogas. Theoretical analysis indicated that, under steady-state conditions, there is a functional connection between the concentration of COD in the effluent and the hydrogen in the biogas (Ruiz, 2005). This relationship can be directly determined through experiments for a specific substrate, reactor set-up and operational conditions when a real system operates in a stable state. This functional connection implies that setting a desired level for the hydrogen concentration in the biogas is essentially the same as establishing a target for the COD concentration in the effluent.

The controller is purposefully designed to maintain the hydrogen concentration in the biogas, and consequently, the COD concentration in the effluent, at a predefined set-point value. The control strategy is described in Equation (10.1), indicating how the dilution rate should change over time. Typically, a discrete implementation of the controller is the expected approach. The adjustment in the dilution rate is directly proportional to its current value and influenced by two factors: $f_{H_2}$ and $f_{CH_4}$. The factor $f_{H_2}$ depends on the current hydrogen concentration, whereas $f_{CH_4}$ is related to the current methane production rate:

$$\frac{dD}{dt} = KDf_{H_2}f_{CH_4}$$

The controller modifies the dilution rate ($D$) according to the values of hydrogen concentration in the biogas (ppm $H_2$) and methane production rate ($Q_{CH_4}$). Figure 10.2 shows the relationship between the controller factors and these monitored variables.

The quick reaction of hydrogen to disturbances, such as overloads, guarantees that the system is swiftly corrected before it reaches an unstable state. The controller is purposefully designed to enhance the process by maximizing methane production as long as the hydrogen concentration remains at a low level. This controller capitalizes on the fact that hydrogen, acting as an intermediary

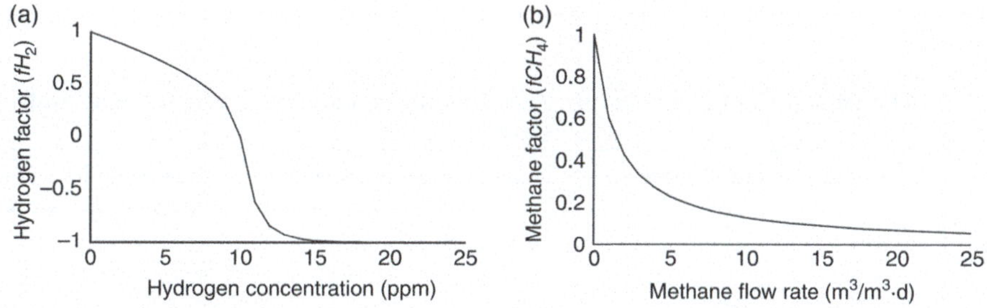


**Figure 10.2** Empirically derived functions for the hydrogen factor vs. hydrogen concentration in the gas (for an example desired ppm $H_{2sp}$ of 10 ppm) (a) and for the methane factor as a function of the methane production rate (b). It is worth nothing how the methane factor never attains zero meaning that the controller always pushes the system gently towards higher methane production rates if the capacity allows for it (*source*: Rodríguez *et al.*, 2006).

in AD, accumulates quickly in the biogas if almost any form of destabilization occurs. It serves as a well-established indicator of impending unstable conditions, particularly overloads. Consequently, it has the potential to be a highly effective early warning signal when monitored in real time.

### 10.3.2 Control strategies for dynamic AD conditions

For any given anaerobic treatment system, start-up can be a crucial phase as it can determine the entire progression of the system and this issue has not received much attention (Sbarciog & Vande Wouwer, 2014). Another significant issue is slow recovery of anaerobic reactors from process instability or other imbalances. The conditions required during these stages are usually different from those of an anaerobic reactor in nominal operation. Hence, in addition to the system monitoring during normal operation, efficient control and optimization during the anaerobic reactor start-up and recovery can be beneficial and economical in safely driving the system towards an optimal operation (Sbarciog & Vande Wouwer, 2014).

Some studies exist that focused on proposing and designing controllers for anaerobic reactors during the start-up phase. Table 10.3 provides a summary of such studies. Some of these studies also considered strategies for re-start-up, recovery and process upsets (e.g. overload). It should be noted that the substrates of interest in these studies have primarily been industrial or sludge wastes; high-strength wastes that make the start-up phase challenging due to organic overload risks and destabilization if not managed effectively. For anaerobic treatment of sewage, start-up could be less challenging owing to the inherent presence of microbial populations and buffering capacity of these liquid wastes (Lubbe & Haandel, 2019).

The existing control systems listed in Table 10.3 range from simpler configurations (e.g. proportional) to more advanced ones such as optimal controllers (e.g. NMPC implementation by Ahmed & Rodríguez, 2020) and an expert system (as in Barnett & Andrews, 1992). Overall, promising results have been observed from implementation of such model-based advanced control strategies for anaerobic reactor start-up and process upset recoveries in the existing studies. This indicates the potential for generally adopting such strategies for effective anaerobic treatment process control and optimization during start-up and/or process upsets.

### 10.3.3 Advanced control configurations

#### *10.3.3.1 Optimum control using model-predictive control in anaerobic treatment systems (model- based controllers)*

Model-predictive control (MPC) is an advanced control strategy that has proven to be particularly useful for difficult multivariable or multiple-input multiple-output control problems (Seborg *et al.*, 2011). The basic idea of MPC involves optimizing future behaviour of a system by using model predictions and,

**Table 10.3** Summary of existing studies on control strategies for anaerobic reactors during dynamic conditions (start up, recovery and upsets).

| | **Ahmed and Rodríguez (2020)** | **Austermann-Haun *et al.* (1994)** | **Barnett and Andrews (1992)** |
|---|---|---|---|
| **Aims** | Automatic, optimal start-up | Start-up enhancement | Diagnosing hydraulic, organic, toxic and $NH_3$ upsets |
| **Controller type** | NMPC | Automatic system | Expert system (fuzzy-logic-based) |
| **CVs** | (1) VFAs (effluent), (2) conc. of acetoclastic methanogens (reactor), and (3) $CH_4$ production rate | pH | pH, $CH_4$ flow rate and variables related to upsets considered |
| **MVs** | Inflow rates for (1) organic substrate, (2) dilution water and (3) added alkali | Feed flow rate | Influent flow rate, sludge recycle, dilution and acid/base addition |
| **AD system tested** | • Virtual UASB (9.75 L at 35°C)<br>• Substrate: synthetic (glucose) (set-up based on literature) | • Real-life AFFR (9.5 L at 35°C)<br>• Substrate: synthetic (VFAs) | • Virtual two-stage, continuous AD process<br>• Substrate: sludge (WW) |
| | **García-Diéguez *et al.* (2010)** | **Holubar *et al.* (2003)** | **Renard *et al.* (1991)** |
| **Aims** | Stable and efficient re-start-up and overload operation | Optimal start-up (non-adapted seed) and optimal recovery (shock loadings) | Stable start-up and steady-state performance |
| **Controller type** | Variable gain | Optimal control | Adaptive (model-based) |
| **CVs** | $H_2$ and $CH_4$ gas flow rate (effluent) | Biogas flow rate and composition | Propionate concentration |
| **MVs** | Feed flow rate | OLR | Feed dilution rate |
| **AD system tested** | • Virtual USBFR (1150 L at 37°C)<br>• Substrate: synthetic winery WW | • Real-life CSTR (20 L at 30°C)<br>• Substrate: primary and surplus sludge mix (MWWTP) | • Real-life CSTR (60 L at 35°C)<br>• Substrate: citrocol |
| | **Rozzi *et al.* (1994)** | **Sbarciog *et al.* (2012)** | **Puñal *et al.* (2001)** |
| **Aims** | Stability during start-up and overload | Optimal start-up | Diagnosing hydraulic and organic overload upsets |
| **Controller type** | Proportional | Optimal control | Expert system (fuzzy-logic-based) |
| **CVs** | Bicarbonate alkalinity (effluent) | $CH_4$ production rate | Acidification control (biogas flow rate, $CH_4$ and CO biogas composition) |
| **MVs** | Alkali concentration (influent) | Dilution rate | Feeding (and recycling) rate, and nutrients and alkali dosing |
| **AD system tested** | • Real-life hybrid AD (10 L at 37°C)<br>• Substrate: olive mill effluent | • Virtual CSTR<br>• Organic substrate (details not reported) | • Data from different AD operational situations (exact details not reported) |

AD, anaerobic digester/digestion; AFFR, anaerobic fixed film reactor; CSTR, continuous stirred tank reactor; MWWTP, municipal wastewater treatment plant; NMPC, non-linear model-predictive control; OLR, organic loading rate; USBFR, upflow sludge-bed filter reactor; VFAs, volatile fatty acids; WW, waste water.

together with system output measurements, determining the appropriate control input actions to be implemented (Grüne & Pannek, 2011; Seborg *et al.*, 2011). One of the characteristic features of MPC that contributes to its success is the consideration of constraints on inputs and/or outputs (Grüne & Pannek, 2011; Seborg *et al.*, 2011). Also, provided that an accurate model is available, predictions can help in early detection of potential problems (Seborg *et al.*, 2011). A critical factor, however, for MPC success lies in the availability of a reliable and accurate model.

MPC implementation in existing anaerobic treatment studies has shown successful performance for optimal control of anaerobic treatment systems, including robustness against disturbances (e.g. overload). Table 10.4 provides a summary of some of the existing studies on implementing model-predictive controllers in anaerobic reactors. Overall, a number of studies related to MPC implementation for anaerobic treatment systems exist in literature while few of these conducted real-life validation tests. Also, the substrates of interest in these MPC studies have been primarily high-strength wastewaters and solid wastes. Nevertheless, these studies have demonstrated the success and flexibility of MPC strategy for controlling anaerobic treatment processes.

Many of the studies listed in Table 10.4 have shown the success of MPC at optimizing the anaerobic co-digestion scheme; an anaerobic co-treatment strategy involving two or more substrates. Such a co-treatment scheme is beneficial to offset the limitations with anaerobic treatment of low-strength wastewaters such as sewage/domestic wastewater. Anaerobically co-treating sewage/domestic waste with substrates of desirable characteristics (e.g. higher COD, nutrient content, etc.) can mitigate low COD and suboptimal temperature operation issues with the domestic wastewater treatment alone. Implementing MPC for anaerobic co-treatment of sewage is an attractive approach to decide on effective operational strategy (e.g. ratio of co-substrate blending) for stable and optimal process performance, including optimal control during the start-up phase.

In a recent study by Ahmed and Rodríguez (2020), an NMPC system was constructed to optimize the start-up of anaerobic reactors treating soluble easily degradable organic materials (see Figure 10.3). Two different configurations were proposed for this system. For effective and cost-efficient control during the start-up process, a minimal set of practical CVs was chosen, which includes: (1) the quality of an effluent in terms of acetate equivalent COD, (2) the level of acetoclastic methanogenic biomass within the reactor and (3) the rate of methane production (only for one of the NMPC configurations). MVs for this process encompass the volumetric flow rates of organic substrate, dilution water and potentially concentrated alkali addition. In order to utilize the selected CVs, which are technically and economically feasible to measure or estimate, a custom-tailored AD model was specifically designed. This model plays a crucial role as a predictive component within an NMPC system.

A majority of the MPC studies for anaerobic treatment systems have implemented NMPC in particular which is justified and necessary due to the highly non-linear nature of models describing the complex anaerobic process. The existing studies on MPC have generally focused on ensuring stable, steady-state conditions for anaerobic reactors and as such, studies on MPC implementation specifically for the anaerobic reactor start-up phase are only a few. The operational conditions for anaerobic reactor start-up can be different that need to be optimized for faster, stable and efficient start-up to ensure successful onset of an anaerobic treatment process. In this regard, implementing an MPC strategy, inherently being an optimal control strategy, is effective for optimal anaerobic reactor control including start-up.

#### 10.3.3.2 Meta-heuristics-based control: fuzzy logic and hybrid controllers

Meta-heuristics algorithms have been utilized in control designs besides being primarily applied for modelling/optimization applications. For anaerobic treatment applications, a number of studies exist that successfully designed such controllers including fuzzy-logic controllers and hybrid controllers. Fuzzy-logic control is based on application of expert knowledge that can help in optimizing processes under dynamic conditions (Robles *et al.*, 2018). Moreover, fuzzy-logic-based control is advantageous for non-linear systems and does not require significant amount of data and/or an extensive/

**Table 10.4** Summary of existing studies on MPC strategies for anaerobic treatment systems.

| Authors | AD System | MPC Structure | Objective Function Terms | Constraints | Prediction Model |
|---|---|---|---|---|---|
| Ahmed and Rodríguez (2020) | • Substrates: synthetic (glucose) | • CV(s): VFAs (effluent), conc. of acetoclastic methanogens (reactor), and $CH_4$ production rate (as addition for one scenario)<br>• MV(s): inflow rates of organic substrates, dilution water and added alkali<br>• Sampling period = 12 h<br>• Control horizon = 2 days<br>• Prediction horizon = 4 days<br>• (virtual testing) | • CV set-point errors<br>• Penalty for alkali addition | • Maximum allowable OLR<br>• Maximum allowable IA/TA ratio<br>• Allowable range of feed flow, dilution water and added alkali<br>• Allowable range of total input flows | AM2 (simplified and modified) |
| Gaida *et al.* (2011) | • Co-digestion<br>• Substrates: maize silage, grass and manure | • MV(s): feed flows<br>• Sampling period = 2 days<br>• Control horizon = 2 days<br>• Prediction horizon = 100 days (virtual testing) | • Net income | • pH limitation<br>• Maximum allowable FOS/TAC<br>• Maximum allowable total solids<br>• Minimum $CH_4$ biogas composition | ADM1 |
| Gaida *et al.* (2012) | • Co-digestion<br>• Co-substrates: maize silage and manure (including solid fractions) | • MV(s): feed flows<br>• Control horizon = 7 days<br>• Prediction horizon = 250 days (virtual testing) | • Net income | • Allowable range of feed flows for each co-substrate<br>• pH limitation<br>• Maximum allowable VFA/TA ratio<br>• Maximum allowable total solids<br>• Minimum $CH_4$ biogas composition | ADM1 |
| Haugen *et al.* (2014) | • Substrate: dairy waste | • CV(s): $CH_4$ production rate, VFA concentration<br>• MV(s): feed flow rate<br>• Sampling period = 36 or 72 min[a]<br>• Control horizon = 1 or 2 days[a]<br>• Prediction horizon = 1 or 2 days[a] (virtual and real-life testing) | • CV set-point error<br>• Control input (MV) changes<br>• Terminal set-point error cost (2nd scenario) | • Maximum allowable VFA concentration<br>• Allowable range of feed flows | Modified Hill model |

**Table 10.4** Summary of existing studies on MPC strategies for anaerobic treatment systems. (*Continued*)

| Authors | AD System | MPC Structure | Objective Function Terms | Constraints | Prediction Model |
|---|---|---|---|---|---|
| Kil *et al.* (2017) | • Substrates: soluble and particulate substrates (2 test scenarios) | • CV(s): biogas flow<br>• MV(s): feed dilution rate<br>• Control horizon = 5 days<br>• Prediction horizon = 5 days (virtual testing) | • Biogas flow set-point error<br>• Control input (MV) changes | • Maximum allowable effluent COD<br>• Maximum allowable VFA/alkalinity ratio | AM2 (reduced and modified) |
| Mauky *et al.* (2016) | • Co-digestion<br>• Substrates: cattle manure, maize silage, grass silage, ground wheat grain | • CV(s): biogas storage fill level (related to biogas production)<br>• MV(s): co-substrates feed flows<br>• Sampling period = 30 min<br>• Control horizon = 1 day<br>• Prediction horizon = 7 days (real-life testing) | • Biogas storage fill level<br>• Control input (MV) changes | • Gas storage and utilization limitations<br>• Allowable range of feed flows<br>• Co-substrate type utilization | ADM1 (simplified and modified) |
| Ordace *et al.* (2012) | • Substrates: primary and thickened sludge | • EPSAC-MPC type controller<br>• CV(s): biogas production<br>• MV(s): feed flows<br>• Sampling period = 15 min<br>• Control horizon = 15 min<br>• Prediction horizon = 12.5 h (virtual testing) | $CH_4$ production set-point error | – | 2nd-order AD model[b] |
| Xue *et al.* (2015) | • Co-digestion<br>• Substrates: maize silage, swine manure | • CV(s): inhibitory compound levels (inhibition coefficients)<br>• MV(s): feed flows<br>• Sampling period = 20 days<br>• Control horizon = 20 days<br>• Prediction horizon = 60 days (virtual testing | • Inhibition coefficient set-point error<br>• Control input (MV) changes | • pH limitation<br>• Maximum allowable effluent COD<br>• Allowable range of feed flows | ADM1 (reduced) |

AD, anaerobic digester/digestion; ADM1, anaerobic model 1; CV, control variable; EPSAC-MPC, extended prediction self-adaptive control; FOS/TAC, volatile organic acids/total inorganic carbon; IA, intermediate alkalinity; MPC, model-predictive control/controller; MV, manipulated variable; OLR, organic loading rate; TA, total alkalinity; VFA, volatile fatty acids.

[a]Two testing scenarios.

[b]Identified using data from ADM1 predictions.

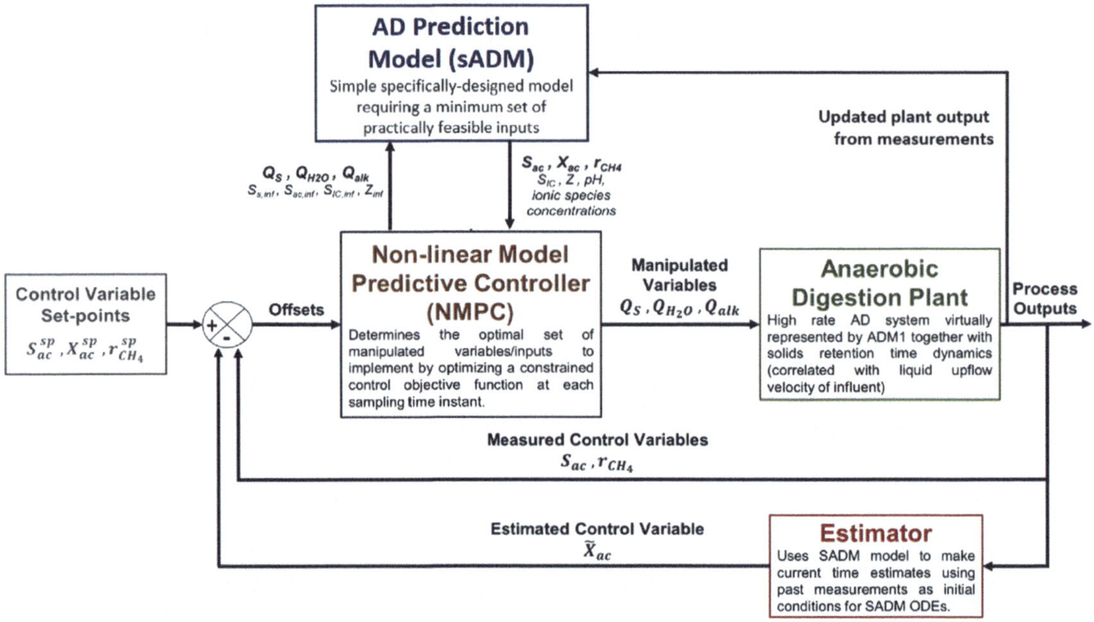


**Figure 10.3** Proposed NMPC system architecture for optimal start-up control of an anaerobic reactor treating soluble substrates (*source*: adapted from Ahmed & Rodríguez, 2020).

mechanistic model (Robles *et al.*, 2018). Some of the earlier studies, such as those by Barnett and Andrews (1992) and Puñal *et al.* (2001), showed the success of fuzzy-logic controllers in AD systems. These studies were particularly focused on controlling processes under disturbances such as upsets and feed overloads (summarized in Table 10.3).

Fuzzy-logic controllers have also been experimentally validated with anaerobic treatment systems. In a study by Robles *et al.* (2017), a fuzzy-logic controller was validated against a pilot-scale digester treating winery wastewater. The system was automated with a fuzzy-logic controller for regulating methane gas flow rate by manipulating feed flow rate while incorporating variable VFA levels as control input to avoid acidification. An on–off pH controller was utilized additionally that would regulate pH via alkali addition. In the presence of set-point changes and moderate level of disturbances (equipment-related), the controller successfully exhibited robust performance. In another similar study by Robles *et al.* (2018), a fuzzy-logic controller was experimentally validated against a fixed-bed anaerobic reactor treating winery wastewater. The control strategy aimed at controlling methane flow rate and effluent VFA concentration by manipulating feed flow but the focus was also on maximizing methane flow rate. Despite some process upsets before automating the experimental system, the controller was able to drive the process towards favourable and optimal conditions. A pilot-scale biogas reactor treating solid wastes was switched from manual mode operation to automated operation via fuzzy-logic controller by Scherer *et al.* (2022). The control scheme involved regulating pH, methane composition in biogas and specific gas production rate by manipulating OLR. Compared to manual operation strategy, the controller was able to operate the process at higher OLR without acidification issues.

The potential of hybrid controllers for anaerobic treatment systems has been realized in existing studies. A new class of controller was proposed and then virtually tested against an anaerobic treatment system as a case study by Chairez (2013). The proposed controller involved combining fuzzy-logic control with a differential neural network (DNN) to deal with limited process information and uncertainty in non-linear systems. For the anaerobic treatment process case scenario, the proposed controller successfully exhibited set-point tracking of biomass and fermentation product concentrations using predictions from the implemented DNN (state estimator) for the virtual testing of an anaerobic treatment system as a simple case scenario. A combined control design involving fuzzy logic and adaptive model-predictive control (AMPC) was implemented by Ghanavati *et al.* (2021) and virtually tested against an anaerobic treatment system. The controller aimed at regulating methane flow rate and effluent VFA levels as well as maximizing methane production by manipulating the feed flow. Auto-regressive moving average model (input–output model) was used for predictions in the AMPC while ADM1 was used as the plant model. Overall, the controller provided stable and near optimal process performance (methane production) under feed composition change and model mismatch scenarios. In a study by Anand *et al.* (2021), a combined fuzzy logic and proportional-integral-derivative (PID) controller was proposed to regulate an anaerobic reactor with a dual-heating system configuration (internal and external). The strategy aimed at regulating digester temperature via manipulation of the mass flows of the heating fluid. In comparison to a standalone PID controller, the proposed fuzzy–PID controller showed superior performance in terms of peak overshoots, rise time and settling time of the controlled variable (temperature) relative to set point. Moreover, the combined controller was able to deal with disturbances in feed temperatures.

Overall, implementing meta-heuristics-based approaches is a potential solution for tackling some of the challenges when modelling, optimizing and/or controlling anaerobic treatment systems (such as being highly non-linear systems in terms of modelling and having limited information due to the limitations in online monitoring for control purposes).

## 10.4 OUTLOOK

Effective and sufficient monitoring procedures are important for successful operation of an anaerobic treatment process. Anaerobic reactor failure resulting from poor or inadequate monitoring leads to significant financial losses since the reactor has to be started up again with new inoculum followed by significantly long waiting times (several months) to re-establish stable conditions and full loading capacity (Drosg, 2013). Moreover, monitoring efforts should be particularly highest for sensitive phases such as start-ups (Drosg, 2013).

A small number of variables monitored online in commercial-scale anaerobic treatment systems presents a limitation on the development of effective control strategies, where the number and type of variables monitored online to effectively assess the status of a process is a critical component. For challenging anaerobic reactor phases (start-up, recovery, etc.), VFA levels and alkalinity are important and ideal indicators to monitor. Also, microbial profiling can give a direct indication of the process status (performance and stability). Accounting for microbial diversity in management of anaerobic treatment processes is an emerging area of interest with high potential and can lead to a new standard in the field of anaerobic process control and optimization (Jimenez *et al.*, 2015). However, on a commercial scale, there are still major issues to overcome when monitoring such variables. VFA and alkalinity are generally monitored offline and less frequently in case of full-scale anaerobic treatment plants whereas monitoring for microbial community diversity and/or abundance hardly exists or is not common. With technological advancements over the years, it might be possible to overcome existing challenges in anaerobic treatment instrumentation and with a positive outlook, it is likely that more of the novel measurement techniques will be developed in the future that might be cost effective and applicable for commercial-scale applications.

For nominal conditions, effective operational control is less challenging than during the critical phases of start-up and recovery. Aiming for optimal operation during such critical phases can help increasing the practical feasibility of anaerobic treatment technology at large scale. In this regard, opting for optimal control strategies is an ideal strategy. More studies on optimal control strategies, such as the implementation of such strategies for low-strength wastewaters, for start-up and/or optimal performance in general are still needed for progress and developments in this field.

## REFERENCES


Ahmed W. and Rodríguez J. (2020). A model predictive optimal control system for the practical automatic start-up of anaerobic digesters. *Water Research*, **174**, 115599, https://doi.org/10.1016/j.watres.2020.115599

Anand K., Mittal A. P. and Kumar B. (2021). Modelling and simulation of dual heating of substrate with centralized temperature control for anaerobic digestion process. *Journal of Cleaner Production*, **325**, 129235, https://doi.org/10.1016/j.jclepro.2021.129235

Austermann-Haun U., Seyfried C. F., Zellner G. and Diekmann H. (1994). Start-up of anaerobic fixed film reactors: technical aspects. *Water Science and Technology*, **29**(10–11), 297–308, https://doi.org/10.2166/wst.1994.0773

Barnett M. W. and Andrews J. F. (1992). Expert system for anaerobic-digestion-process operation. *Journal of Environmental Engineering*, **118**(6), 94994, https://doi.org/10.1061/(ASCE)0733-9372(1992)118:6(949)

Bernard O., Hadj-Sadok Z., Dochain D., Genovesi A. and Steyer J.-P. (2001). Dynamical model development and parameter identification for an anaerobic wastewater treatment process. *Biotechnology and Bioengineering*, **75**(4), 424–438, https://doi.org/10.1002/bit.10036

Boe K. and Angelidaki I. (2012). Pilot-scale application of an online VFA sensor for monitoring and control of a manure digester. *Water Science and Technology*, **66**(11), 2496–2503, https://doi.org/10.2166/wst.2012.498

Chairez I. (2013). Differential neuro-fuzzy controller for uncertain nonlinear systems. *IEEE Transactions on Fuzzy Systems*, **21**(2), 369–384, https://doi.org/10.1109/TFUZZ.2012.2215875

de Lemos Chernicharo C. A. (2007). Anaerobic Reactors (Vol. 6). Water Intelligence Online. https://doi.org/10.2166/9781780402116

Drosg B. (2013). Process monitoring in biogas plants. IEA Bioenergy (ISBN: 978-1-910154-03-8). https://www.ieabioenergy.com/wp-content/uploads/2013/12/Technical-Brochure-process_montoring.pdf

Falk H. M., Reichling P., Andersen C. and Benz R. (2015). Online monitoring of concentration and dynamics of volatile fatty acids in anaerobic digestion processes with mid-infrared spectroscopy. *Bioprocess and Biosystems Engineering*, **38**(2), 237–249, https://doi.org/10.1007/s00449-014-1263-9

Gaida D., Luis A., Brito S., Wolf C., Back T., Bongards M. and Mcloone S. (2011). Optimal control of biogas plants using nonlinear model predictive control. 22nd IET Irish Signals and Systems Conference.

Gaida D., Wolf C., Back T. and Bongards M. (2012). Nonlinear model predictive substrate feed control of biogas plants. 20th Mediterranean Conference on Control & Automation (MED), pp. 652–657. https://doi.org/10.1109/MED.2012.6265712

García-Diéguez C., Molina F., Fernández E. and Roca E. (2010). Control of re-startup of anaerobic USBF reactors after short stops. *Industrial & Engineering Chemistry Research*, **49**(10), 4748–4755, https://doi.org/10.1021/ie901164t

García-Gen S., Rodríguez J. and Lema J. M. (2015). Control strategy for maximum anaerobic co-digestion performance. *Water Research*, **80**, 209–216, https://doi.org/10.1016/j.watres.2015.05.029

Ghanavati M. A., Vafa E. and Shahrokhi M. (2021). Control of an anaerobic bioreactor using a fuzzy supervisory controller. *Journal of Process Control*, **103**, 87–99, https://doi.org/10.1016/j.jprocont.2021.05.010

Goffin A., Guérin S., Rocher V. and Varrault G. (2018). Towards a better control of the wastewater treatment process: excitation–emission matrix fluorescence spectroscopy of dissolved organic matter as a predictive tool of soluble BOD5 in influents of six Parisian wastewater treatment plants. *Environmental Science and Pollution Research*, **25**(9), 8765–8776, https://doi.org/10.1007/s11356-018-1205-1

Grüne L. and Pannek J. (2011). Nonlinear Model Predictive Control, Springer, London, pp. 43–66, https://doi.org/10.1007/978-0-85729-501-9_3

Haugen F., Bakke R. and Lie B. (2014). State estimation and model-based control of a pilot anaerobic digestion reactor. *Journal of Control Science and Engineering*, **2014**, 1–19, https://doi.org/10.1155/2014/572621

Holubar P., Zani L., Hager M., Fröschl W., Radak Z. and Braun R. (2003). Start-up and recovery of a biogas-reactor using a hierarchical neural network-based control tool. *Journal of Chemical Technology & Biotechnology*, **78**(8), 847–854, https://doi.org/10.1002/jctb.854

Jimenez J., Latrille E., Harmand J., Robles A., Ferrer J., Gaida D., Wolf C., Mairet F., Bernard O., Alcaraz-Gonzalez V., Mendez-Acosta H., Zitomer D., Totzke D., Spanjers H., Jacobi F., Guwy A., Dinsdale R., Premier G., Mazhegrane S. and … Steyer J.-P. (2015). Instrumentation and control of anaerobic digestion processes: a

review and some research challenges. *Reviews in Environmental Science and Bio/Technology*, **14**(4), 615–648, https://doi.org/10.1007/s11157-015-9382-6
Kil H., Li D., Xi Y. and Li J. (2017). Model predictive control with on-line model identification for anaerobic digestion processes. *Biochemical Engineering Journal*, **128**, 63–75, https://doi.org/10.1016/j.bej.2017.08.004
Lubbe J. van der and Haandel A. van (2019). Anaerobic Sewage Treatment: Optimization of Process and Physical Design of Anaerobic and Complementary Processes. IWA Publishing, London.
Mauky E., Weinrich S., Nägele H.-J., Jacobi H. F., Liebetrau J. and Nelles M. (2016). Model predictive control for demand-driven biogas production in full scale. *Chemical Engineering & Technology*, **39**(4), 652–664, https://doi.org/10.1002/ceat.201500412
McCarty P. L. (1964). Anaerobic Waste Treatment Fundamentals – Part two – Environmental Requirements and Control. Public Works, pp. 123–126.
Molina F., Ruiz-Filippi G., Garcia C., Lema J. M. and Roca E. (2009). Pilot-scale validation of a new sensor for on-line analysis of volatile fatty acids and alkalinity in anaerobic wastewater treatment plants. *Environmental Engineering Science*, **26**(3), 641–649, https://doi.org/10.1089/ees.2007.0308
Morel E., Santamaria K., Perrier M., Guiot S. R. and Tartakovsky B. (2005). Multi-wavelength fluorometry for anaerobic digestion process monitoring. *Water Science and Technology*, **52**(1–2), 465–471, https://doi.org/10.2166/wst.2005.0554
Nielsen H., Uellendahl H. and Ahring B. (2007). Regulation and optimization of the biogas process: propionate as a key parameter. *Biomass and Bioenergy*, **31**(11–12), 820–830, https://doi.org/10.1016/j.biombioe.2007.04.004
Olsson G. and Ingildsen P. (2020). Process control. In: Biological Wastewater Treatment – Principles, Modelling and Design, 2nd edn, G. Chen, M. van Loosdrecht, G. Ekama, and D. Brdjanovic (eds), IWA Publishing, London, UK.
Ordace A., Ionescu C. M., Vannecke T. P. W., Nascu I. and De Keyser R. (2012). Predictive control of anaerobic digestion of wastewater sludge. A feasibility study. 16th International Conference on System Theory, Control and Computing (ICSTCC), pp. 1–7.
Puñal A., Rodríguez J., Franco A., Carrasco E. F., Roca E. and Lema J. M. (2001). Advanced monitoring and control of anaerobic wastewater treatment plants: diagnosis and supervision by a fuzzy-based expert system. *Water Science and Technology*, **43**(7), 191–198, https://doi.org/10.2166/wst.2001.0422
Renard P., Van Breusegem V., Nguyen M.-T., Naveau H. and Nyns E.-J. (1991). Implementation of an adaptive controller for the startup and steady-state running of a biomethanation process operated in the CSTR mode. *Biotechnology and Bioengineering*, **38**(8), 805–812, https://doi.org/10.1002/bit.260380802
Robles A., Latrille E., Ruano M. V. and Steyer J. P. (2017). A fuzzy-logic-based controller for methane production in anaerobic fixed-film reactors. *Environmental Technology*, **38**(1), 42–52, https://doi.org/10.1080/09593330.2016.1184321
Robles A., Capson-Tojo G., Ruano M. V., Latrille E. and Steyer J.-P. (2018). Development and pilot-scale validation of a fuzzy-logic control system for optimization of methane production in fixed-bed reactors. *Journal of Process Control*, **68**, 96–104, https://doi.org/10.1016/j.jprocont.2018.05.007
Rodríguez J., Ruiz G., Molina F., Roca E. and Lema J. M. (2006). A hydrogen-based variable-gain controller for anaerobic digestion processes. *Water Science and Technology*, **54**(2), 57–62, https://doi.org/10.2166/wst.2006.486
Rozzi A., Di Pinto A. C., Limoni N. and Tomei M. C. (1994). Start-up and operation of anaerobic digesters with automatic bicarbonate control. *Bioresource Technology*, **48**(3), 215–219, https://doi.org/10.1016/0960-8524(94)90149-X
Ruiz G. (2005). Monitoring and Advanced Control of Anaerobic Reactors. PhD thesis, Universidade de Santiago de Compostela, Spain.
Sbarciog M. and Vande Wouwer A. (2014). Start-up of multi-species anaerobic digestion systems: extrapolation of the single species approach. *Mathematical and Computer Modelling of Dynamical Systems*, **20**(1), 87–104, https://doi.org/10.1080/13873954.2013.817443
Sbarciog M., Loccufier M. and Vande Wouwer A. (2012). An optimizing start-up strategy for a bio-methanator. *Bioprocess and Biosystems Engineering*, **35**(4), 565–578, https://doi.org/10.1007/s00449-011-0629-5
Scherer P., Schreiber A., Arthur R., Antonczyk S. and Vollmer G.-R. (2022). Doubling the space–time yield of a pilot biogas reactor with swine manure and cereal residues by a closed loop feedback control based on an automated fuzzy logic control system. *Processes*, **10**(12), 2511, https://doi.org/10.3390/pr10122511
Seborg D., Edgar T., Mellicamp D. and Doyle III, F. (2011). Process Dynamics and Control. 3rd edn, John Wiley & Sons, New Jersey.
Seghezzo L., Zeeman G., van Lier J. B., Hamelers H. V. M. and Lettinga G. (1998). A review: The anaerobic treatment of sewage in UASB and EGSB reactors. *Bioresource Technology*, **65**(3), 175–190, https://doi.org/10.1016/S0960-8524(98)00046-7

Steyer J. P., Bouvier J. C., Conte T., Gras P., Harmand J. and Delgenes J. P. (2002). On-line measurements of COD, TOC, VFA, total and partial alkalinity in anaerobic digestion processes using infra-red spectrometry. *Water Science and Technology*, **45**(10), 133–138, https://doi.org/10.2166/wst.2002.0310

Taricska J. R., Long D. A., Chen J. P., Hung Y.-T. and Zou S.-W. (2009). Anaerobic digestion. In Biological Treatment Processes, Humana Press, New Jersey, pp. 589–634. https://doi.org/10.1007/978-1-60327-156-1_14

The Wales Centre of Excellence for Anaerobic Digestion, Landes Energie Verein Steiermark, & Vienna University of Technology (2012). European case studies of anaerobic digestion plants showcasing their monitoring practices. Bio-Methane Regions. https://www.severnwye.org.uk/fileadmin/Resources/SevernWye/Projects/Biomethane_Regions/Downloads/BMR_D_5_1_Best_Practice_Monitoring_FINAL_a_Resubmission_Final7.pdf

Xue L., Li D. and Xi Y. (2015). Nonlinear model predictive control of anaerobic digestion process based on reduced ADM1. 10th Asian Control Conference (ASCC), pp. 1–6. https://doi.org/10.1109/ASCC.2015.7244539

doi: 10.2166/9781789063479_00293

# Chapter 11

# Life-cycle assessment, carbon-footprint and techno-economic analysis

S. Estévez[1]*, V. Hernández[2], F. Hernández-Sancho[2], G. Feijoo[1] and M. T. Moreira[1]

[1]CRETUS, Department of Chemical Engineering, University of Santiago de Compostela, Santiago de Compostela, Spain
[2]Department of Economic Structure, University of Valencia, Valencia, Spain
*Correspondence: sofia.estevez.rivadulla@usc.es

**ABSTRACT**

Anaerobic treatment of domestic wastewater has been recognized as a feasible technology to reduce energy demand from the valorization of streams with high-organic load. However, there are many factors that influence the environmental and economic profiles of the process, such as the effluent pollutant load, operating temperature and methane emissions in the treated effluent and in the air. Therefore, all these parameters must be optimized taking into account not only the technical performance (i.e. biogas production), but also environmental and economic implications. This chapter presents a dual approach to highlight the characteristics of anaerobic treatment of domestic wastewater with life-cycle assessment and techno-economic analysis and how to approach these methodologies in diverse case studies.

**Keywords:** domestic anaerobic treatment, life-cycle assessment (LCA), techno-economic analysis (TEA), wastewater treatment (WWT).

## 11.1 SUSTAINABILITY AND CIRCULARITY IN WASTEWATER TREATMENT PLANTS

The 2030 Agenda for Sustainable Development sets out a series of ambitious objectives for the global community. These sustainable development goals include targets for access to safe water and sanitation and improved water management. To date, these are challenges that have proven difficult to meet, partly not only because they are complex, but also due to the global context, with political, social, economic uncertainty and environmental adversities. This calls for redoubled efforts and carefully selected approaches to achieve transformational change. Especially, drinking water scarcity, climate change and resource depletion are driving a paradigm shift in the wastewater treatment sector, reinforced by social awareness and new legislative changes. Because of this, there has been a more proactive attitude regarding the transformation of public perception of wastewater treatment plants from disposal to waste valorization facilities or 'biorefineries'. In Europe, the goal is not only compliance with the already outdated Directive 91/271/EEC, but also try to incorporate the precepts supported by the Circular Economy Action Plan published in 2020, the Regulation (EU) 2020/852 (also known as 'EU Taxonomy') in relation to sustainable development, the Regulation (EU) 2020/741 for minimum quality in water reuse and with the forthcoming wastewater treatment directive (revised proposal in October 2022). Quality thresholds for wastewater reuse have also been contemplated by non-European countries and worldwide organizations, such as the World Health Organization, Environmental Protection Agency (EPA), Australian Guideline for Water Recycling (AGWR) and International Organization for Standardization (ISO). In terms of the implementation of circular economy, other initiatives, platforms and organizations around the world can also be cited, such as African Circular economy Network, Economy and Social Commission for Asian and the Pacific, Asia Circular Economy Association and the Platform for Accelerating the Circular Economy.

In this context of sustainability and circularity, the strategic direction of the sector is moving towards the design of energy-neutral facilities, the application of more restrictive criteria for the removal of micropollutants, the improvement of reclaimed water quality and the improvement of sanitation in areas that had been left behind. Better access to water and sanitation, water management and governance and the multiple benefits they bring, can contribute significantly to positive transformation in these environments.

Because of these, strategies and technologies have been put forward to reduce, reclaim, reuse, recycle, recover and rethink and thus maintain material flows within the production–consumption chain. The most prominent resource of a wastewater treatment plant (WWTP) is wastewater, which could be transformed into a multitude of co-products such as liquid or solid fertilizers from digestate or as struvite, hydroxyapatite, k-struvite or ammonium sulphate, polymers (polyhydroxyalkanoates (PHAs), celluloses and polyhydroxybutyrates (PHBs)), biomass (microalgae and biochar) and energy (hydrogen, biomethane, biodiesel, bioethanol and biogas) and also reclaimed water to be reused (Singh *et al.*, 2022). Among them, one of the most recurrent products is biogas given the large technological maturity of anaerobic digestion (AD) processes and its well-known function as storable energy vector. Although AD has been typically supporting sewage sludge management in developed countries because the continuous challenges in the search for politically and economically acceptable treatment routes, its applicability goes beyond, as algal biomass, food waste, sewage, industrial feedstocks and other high-organic strength streams have been widely used in the co-digestion of organic waste (European Environment Agency, 2022). Anaerobic treatment offers an opportunity for the transformation of organic matter present in industrial and domestic wastewaters and reduced sludge production compared to a conventional activated sludge (CAS) system. Apart from this, anaerobic treatment improves the profile of facilities in relation to energy self-sufficiency by converting the produced biogas into heat and/or electricity (Stazi & Tomei, 2018). The system also responds more quickly to the addition of substrates after shutdown periods. In line with the above benefits, it entails a reduction in reactor size, a feature associated with its higher loading rate (Zieliński *et al.*, 2023). However, these are only some of the reported advantages, as there are also environmental (avoidance of carbon dioxide emissions) and economic (sale of energy) benefits to consider. To obtain a broader view of the performance of this process across these two pillars of sustainability, this chapter presents both a critical review of the literature to highlight the potential environmental and economic benefits of anaerobic domestic treatment and a practical case study. Given the limited publication rate of research articles in the fields of life-cycle assessment (LCA) and techno-economic analysis for anaerobic secondary treatment, prospective challenges for technologies were also analysed by addressing a broad systematic revision of the literature on the topic of anaerobic treatment of domestic wastewater.

## 11.2 ANAEROBIC TREATMENT FROM AN ENVIRONMENTAL PERSPECTIVE

Anaerobic treatment is a very versatile process whose application has been widely studied in the wastewater treatment sector, especially for the management of sludge. However, depending on the operating conditions the required energy can totally or partially offset the energy balance of the facility. Sustainable sewage treatment plants (STPs) operated with anaerobic processes must, then, rely on technologies capable of maintaining a positive balance between the desired quality of emissions and the recovered resources. Therefore, the design and operation of STPs must be framed within a context in which environmental impacts are comprehensively and continuously assessed. One of the most recognized methodologies for this purpose is LCA, which not only takes into account direct–indirect emissions, but it is also useful for the identification of internal process weaknesses and technological benchmarking through its life cycle.

It consists of four well-defined stages: definition of the scope and system boundaries, development of the life-cycle inventory (LCI), life-cycle impact assessment (LCIA) and interpretation of the results (ISO, 2006a, 2006b). The first stage comprises the formulation of the initial hypothesis, the level of detail of the system under study, the data search needs, the definition of system boundaries and functional unit (FU). The so-called 'system boundaries' include all elements of technology, process or product under assessment. On the contrary, the FU is a quantitative definition of the system, as it is used as a reference for inputs and outputs and allows comparison of processes (ISO, 2006a). The second stage or LCI involves the quantification of the quantities of materials and energy, emissions and waste calculated with reference to the previously selected FU. All the data collected is then used and translated into environmental burdens during the impact assessment (third stage), which classifies emissions into their respective impact categories. These categories are global indicators (e.g. climate change, eutrophication or resource depletion), which provide insight into the status of an environmental aspect. Finally, in the last stage, the results of the LCI and the LCIA are interpreted, and conclusions of the study are provided.

In this sense, the first question to be answered is whether the anaerobic treatment constitutes a good technological alternative compatible with a good quality state of ecosystems, wildlife and humans when installed in domestic wastewater treatment plants. As mentioned in Section 11.1, the technology can be implemented both in water and sludge lines. In the first case, the most powerful strategy is the direct reduction of energy demand by replacing typical aeration systems with mixing devices. It is worth noting that aeration electricity accounts for 45–75% of the overall cost of wastewater treatment plants (Kong *et al.*, 2021). In the second case, the use of anaerobic treatment in the sludge line allows higher methane yields as a source of bioenergy while allowing the concentration of phosphorus or nitrogen in a solid digestate that can be used in agriculture. In terms of environmental performance, anaerobic treatment, if all the produced methane is recovered, has the potential to reduce greenhouse gas (GHG) emissions. The best performance of an anaerobic process is directly related to the concentration of wastewater. Considering only direct emissions, a conventional aerated sludge process is only outperformed when the influent has a biochemical oxygen demand (BOD) concentration above 500 mg/L. This is related to emissions to the environment of dissolved methane contained in the anaerobically treated effluents, which can be as high as the methane recovered (Cakir & Stenstrom, 2005). Methane is a strong GHG with a global warming potential (GWP) of 28 (IPCC, 2021). Thus, GHG emissions associated with the anaerobic treatment of diluted sewage could be higher that the aerobic counterpart. Problems associated with the presence of dissolved methane in the anaerobically treated effluents were discussed in Chapter 7. For indirect emissions, the selection of post-treatment technologies, reactor configuration and wastewater salinity play key roles. The environmental profile is expected to be positive for anaerobic systems, at least for medium-strength influents (Smith *et al.*, 2014).

Concerning the environmental benefits of anaerobic treatment for sewage sludge three publications can be highlighted claiming its best environmental performance. Blanco *et al.* (2016) and Arias *et al.* (2020a) tried to quantify it with an LCA focused on the introduction of a digester within the sludge line of a domestic WWTP. The implementation of this technology led to a noteworthy impact reduction in the environmental profile of the facility of ~85 and 10% for each of the previously named studies, respectively. Awad *et al.* (2019) also supported the conclusions found in both publications, although the benefits achieved for the climate change category were much smaller (0.3% difference). It should be noted that, when an AD process is implemented, the dewatering characteristics of the sludge improve and, therefore, a lower amount of chemicals (polyelectrolytes) is required in the following dewatering stages. In addition, the energy received from the biogas transformation provides the plant with a certain degree of autonomy from the national grid. However, and as it was underlined for domestic wastewater, the results achieved would depend on the characteristics of influent and technologies.

### 11.2.1 Is technology selection affecting the environmental profile?

The environmental benchmarking between anaerobic treatment technologies and other technologies cannot yet be offered in its entirety. The reason is the difficulties encountered in the following two aspects: the implementation of the technology within wastewater treatment plants and the lack of comparability between LCA studies, as they have different objectives, scopes, FUs, system boundaries and so on. Due to different methodological decisions made in each publication related to the topic, a coherent comparison between technologies cannot be made among research studies published by different authors. Accordingly, the comparison is restricted to scenarios proposed within the same published manuscript. In this sense, currently the maximum number of scenarios evaluated in the same manuscript is around five–six, although it is usual to study three–four scenarios in a comparative way. In addition, the diversity of studies is greater when anaerobic treatment is applied for sludge valorization compared to when it is referring to the water line. Consequently, research for domestic treatment has been directed towards the comparison of biological technologies, whereas the ongoing study of the scientific community for sewage sludge has resulted in some interesting findings for incineration, PHA production, hydrothermal carbonization (HTC) systems, pyrolysis and AD with microalgae processes for side-streams and lagoon biodigesters.

Chronologically, the first research addressing the aerobic versus anaerobic debate as secondary treatment from an LCA perspective was published by Smith *et al.* (2014) trying to compare an anaerobic membrane bioreactor (AnMBR) with a -rate activated sludge (HRAS) system. Other technologies were evaluated, not only in terms of technological comparison, but also taking into account different domestic wastewater composition, temperature and sludge-handling practices. An AnMBR produced the highest environmental impact in all categories analysed for medium-strength wastewater (430 mg/L chemical oxygen demand (COD)) and was related to two aspects: high-energy requirements for membrane scouring by biogas sparging and dissolved methane emissions. At higher strength of wastewater (800 mg/L COD), an AnMBR was able to outperform a CAS system due to higher energy production. Two years later, Pretel *et al.* (2016) also share insights on the environmental performance of AnMBRs. Due to the problematic encountered with respect to diffuse methane emissions in the previous study, they proposed an improvement of the system with a degassing unit in order to capture the methane dissolved in effluents for additional energy production or to be used as organic matter for denitrification in the downstream post-treatment unit. As a result, they obtained the best profile for AnMBRs for moderate–high-strength domestic wastewater in all categories analysed except for eutrophication. However, many more parameters affect the environmental profile. Pretel *et al.* (2013) attempted to provide insight into energy demand, biogas recovery, nutrient recovery and sludge removal factors for an AnMBR. Although it is true that higher operating temperatures resulted in higher methane production, this energy was not sufficient to tackle the heating requirements to operate. To address the environmentally needed improvements for AnMBR technologies in domestic wastewater treatment, Harclerode *et al.* (2020) evaluated eight new scenarios. They considered primary treatment, membrane fouling, sulphide removal, phosphorus removal, dissolved methane management and waste management. Oxidative biological sulphide removal was shown to be more sustainable than coagulation, vacuum degassing tanks were shown to be better than hollow fibre contactors for methane removal and primary sedimentation was shown to be better than simple fine screening. Among these, sulphide and phosphorus removal offered the greatest improvements, with reductions of up to 70% in all impact categories.

The study of Sills *et al.* (2016) was no longer related to AnMBRs. Their environmental assessment was rather carried out for an anaerobic baffled reactor (ABR), for a combination of a tricking filter with ABR, tricking filter and ABR with constructed wetland (CW) and for an ABR with CW. Of the four scenarios, the ABR followed by tricking filter resulted in the best profile. However, the benefits of bioelectricity production in the climate change category for ABRs were again hampered by dissolved methane in anaerobic effluents. This is because ~95% of GWP was affected by this specific direct emission. It appears that, regardless of the technology used, the environmental competitiveness of anaerobically operated technologies as secondary treatment is related to the degassing of methane in effluents and the concentration of nitrogenous compounds. Upflow anaerobic sludge blanket (UASB) with wetland post-treatment and CAS were also compared in a study by Laitinen *et al.* (2017), who found a strong relationship between energy demand and climate change. The clear winner of their analysis was the CW-based treatment. UASB + CW was also the core topic of the study by de Sampaio Lopes *et al.* (2014) but the interpretation of the results did not include a comparative assessment of technologies. However, the contribution analysis provided pointed out the importance of sodium hypochlorite solution control during the disinfection stage. There are no longer restrictions for the joint application of UASB with CWs as a post-treatment step, Patel and Singh (2022) carried out a comparative LCA of the stand-alone performance of a UASB reactor with activated sludge processes, sequential batch reactors and CWs separately. Overall, UASBs show a better profile in the categories analysed compared to CAS and SBR (sequencing batch reactor) processes, except in eutrophication. This is due to the negligible nutrient removal from the anaerobic treatment, whereas SBR, despite being the highest contributor in the other categories, shows the lowest eutrophication potential (Table 11.1).

Considering the AD of domestic sewage sludge, two groups of technologies were studied: non-biological and biological-based. Within the first family, Chen *et al.* (2022) attempted to provide a comparative view for sewage sludge management of direct incineration and a combination of AD with incineration, whereas Medina-Martos *et al.* (2020) focused on the integration of the above-mentioned technology with a newly developed HTC system. The combination of HTC and AD resulted in a reduction of global warming potential by 75%. The improved performance compared to stand-alone AD was due to the recovery of hydrochar for heat production, which has a renewable biogenic origin. Studies by Li and Feng (2018) and Li *et al.* (2017a) are other examples on pyrolysis and co-incineration (Table 11.2).

For biological systems, AD was combined with a microalgae treatment. Benefits can be obtained from exhaust gas treatment with microalgae and dewatering of digestate from primary and secondary sludge. Because co-digestion would result in increased biogas and reduced energy for aeration in a secondary treatment reactor (the algae pond acts as a treatment process for ammonia emissions and prevents its return to the water line), implementation of the technology improves the environmental profile in relation to climate change, ozone layer depletion, freshwater eutrophication and water consumption (Tua *et al.*, 2021). However, microalgae production does not necessarily need to occur in the sludge line and advantages can also be taken from the substitution of the CAS process.

It is noteworthy that the energy sector seems to benefit from technological advances in AD. A future optimized perspective of a two-stage system including stages of a dark fermentation process and PHA accumulation can reduce the environmental impact associated with climate change by up to 41%. However, AD as a single solution remains as of today the most advantageous alternative for wastewater valorization, as three gaps remain to be filled in two-stage systems: the microbial growth demands an increase, the need of higher accumulation yields for polyhydroxybutyrate production and larger organic loading rates or amount of feedstock processed inside the reactor (Asunis *et al.*, 2021).

### 11.2.2 Prospective environmental–technological challenges

One of the recurrent applications of LCA is the identification of critical points and a comparison among products/scenarios performing the same function, especially in large-scale operation, and also in the early-design phases when it is possible to identify those stages with the greatest potential impact and on time implementation of changes that can result in a significant environmental performance improvement. Therefore, this section presents a critical review of the environmental consequences of different variables and technologies related to anaerobic treatment. So far, published studies have discussed the concentration of volatile solids, the impact of upstream and downstream processes, the temperature and co-digestion with other organic waste streams.

**Table 11.1** Methods and FUs of anaerobic domestic wastewater treatment processes for domestic wastewater and related.

| Type of Resource | LCA Method | FU | References |
|---|---|---|---|
| Domestic wastewater, blackwater, urine and greywater | ReCiPe | 26,000 inhabitants and 6510 jobs/year | Lehtoranta *et al.* (2022) |
| Domestic wastewater | IMPACT 2002+ | 1 $m^3$ of treated wastewater and 1 kg of sludge | Patel and Singh (2022) |
| Domestic wastewater | ReCiPe | 1 $m^3$ of water available for consumption or 1 $m^3$ of wastewater | Boldrin *et al.* (2022) |
| Domestic wastewater | TRACI | 5 million gallons of wastewater | Harclerode *et al.* (2020) |
| Blackwater | ReCiPe | 1 $m^3$/day of wastewater treated | Estévez *et al.* (2022a) |
| Blackwater | ReCiPe | 1 $m^2$ of urban garden | Estévez *et al.* (2022b) |
| Blackwater and kitchen waste | ReCiPe | Blackwater (in $m^3$) and kitchen waste (in kg) generated in a four-person household/year | Prado *et al.* (2020) |
| Domestic wastewater and sewage sludge | CML | 1 t of TS | Cañote *et al.* (2021) |
| Domestic wastewater and food waste | TRACI | 5 million gallons/day | Becker *et al.* (2017) |
| Domestic wastewater and food waste | ReCiPe | 2000 inhabitants/day | Lijó *et al.* (2017) |
| Domestic wastewater and sewage sludge | IPCC | 1000 $m^3$ of influent wastewater | Laitinen *et al.* (2017) |
| Domestic wastewater | TRACI | 1 $m^3$ wastewater | Shoener *et al.* (2016) |
| Domestic wastewater and sewage sludge | IMPACT 2002+ | 2 MGD of domestic wastewater, assuming a plant lifetime of 30 years | Sills *et al.* (2016) |
| Domestic wastewater | CML | 1 $m^3$ of treated wastewater | Pretel *et al.* (2016) |
| Domestic wastewater | TRACI | 5 million gallons/day of wastewater | Smith *et al.* (2014) |
| Domestic wastewater | CML | 1 $m^3$ of treated effluent | de Sampaio Lopes *et al.* (2014) |
| Domestic wastewater | CML | 1 $m^3$ of treated wastewater | Pretel *et al.* (2013) |

#### 11.2.2.1 Solid concentration

There is direct relationship among feedstock concentration, the conversion efficiency of the organic matter and environmental impact. However, the conclusions drawn by several authors regarding to the best techno-environmental performance do not seem to go in the same direction. The reason may be related to LCA methodological approaches adopted by each author, but also to the use of feedstocks with different solid concentrations (despite considering similar VS/TS (volatile solids/total solids) ratios in the studies) and technological configurations. Research conducted in the scientific community on this topic has been applied mainly to the most concentrated streams, sludge. This is because domestic wastewater usually contains a solid concentration of <0.1% and, therefore, the benefits of stream concentration are not sufficiently relevant unless the wastewater is mixed with food waste (Becker *et al.*, 2017; Lijó *et al.*, 2017). Recent LCA studies have demonstrated that the environmental feasibility of anaerobic treatment depends on both the biodegradability of organic matter and the concentration of solids (Chen *et al.*, 2022; Li & Feng, 2018; Li *et al.*, 2017a). Although all of them mainly focused on sludge treatment, lessons can be learnt for domestic wastewater treatment. For example, the environmental profile would improve with increasing volatile solid concentration

**Table 11.2** Methods and FUs of anaerobic domestic wastewater treatment processes for sewage sludge and related.

| Type of Resource | LCA Method | FU | References |
|---|---|---|---|
| Sewage sludge | ReCiPe | 1 $m^3$ of biogas | Singh *et al.* (2022) |
| Sewage sludge | IPCC | 500 t of raw sewage sludge (5% TS) | Chen *et al.* (2022) |
| Sewage sludge and food waste | CML-IA baseline | 1 kg of digested sludge | Satayavibul and Ratanatamskul (2021) |
| Sewage sludge with algae biomass | ILCD, CED, WC and LO | 1000 $m^3$ of influent wastewater | Tua *et al.* (2021) |
| Sewage sludge | ReCiPe and CML | 1 t of mixed sludge | Arias *et al.* (2021) |
| Sewage sludge | Environmental footprint | 1 $m^3$ of urban wastewater | Brockmann *et al.* (2021) |
| Sewage sludge | ReCiPe | 1 $m^3$ of treated wastewater and 1 $m^3$ of produced methane | Lanko *et al.* (2020) |
| Sewage sludge | Usetox | 1000 kg wet mixed sludge | Medina-Martos *et al.* (2020) |
| Sewage sludge and food waste | Not specified | 1 $m^3$ wastewater | Morelli *et al.* (2020) |
| Sewage sludge, cow manure, forage waste and returned dairy products | CML | 1 t of wet manure | Adghim *et al.* (2020) |
| Sewage sludge digestate | ILCD/PEF | 1 kg of biopolymer | Vogli *et al.* (2020) |
| Sewage sludge | eFootprint | 1000 kg of dry biomass | Wang *et al.* (2020) |
| Sewage sludge | CML and ReCiPe | 1 $m^3$ of treated wastewater | Arias *et al.* (2020b) |
| Sewage sludge | CML and ReCiPe | 1 t of mixed sludge | Arias *et al.* (2020a) |
| Sewage sludge | CML | 1 $m^3$ of treated wastewater | Awad *et al.* (2019) |
| Sewage sludge | Not specified | 1 t of mixed sludge | Cartes *et al.* (2018) |
| Sewage sludge | CML | 1 t TS of thickened sludge | Li and Feng (2018) |
| Sewage sludge | CML | 1000 kg of mixed sludge | Yoshida *et al.* (2018) |
| Sewage sludge | CML | 1 $m^3$ of raw wastewater | Colzi Lopes *et al.* (2018) |
| Sewage sludge, septage and high-strength organic waste | ReCiPe and TRACI | 1 MGD | Morelli *et al.* (2018) |
| Sewage sludge | ReCiPe | 1 t of TS of sludge entering the sewage sludge treatment line | Gourdet *et al.* (2017) |
| Sewage sludge | Not specified | 1 dry tonne of sludge | Heimersson *et al.* (2017) |

(*Continued*)

**Table 11.2** Methods and FUs of anaerobic domestic wastewater treatment processes for sewage sludge and related (*Continued*).

| Type of Resource | LCA Method | FU | References |
|---|---|---|---|
| Sewage sludge and food waste | CML | Annual management of municipal waste and sewage | Edwards *et al.* (2017) |
| Sewage sludge | CML | 1 t of TS | Li *et al.* (2017b) |
| Sewage sludge | CML | 1 t of TS | Li *et al.* (2017a) |
| Sewage sludge | Not specified | 1 dry tonne of biosolids | Alvarez-Gaitan *et al.* (2016) |
| Sewage sludge | ReCiPe, EDIP and CML | 1 $m^3$ of influent wastewater | Piao *et al.* (2016) |
| Sewage sludge | Not specified | 1 day for a plant with capacity of 70,000 inhabitants | Tomei *et al.* (2016a) |
| Sewage sludge | ReCiPe | 1 person equivalent as 60 g of $BOD_5$/day | Blanco *et al.* (2016) |
| Sewage sludge | Not specified | Daily inflow to the WWTP | Gianico *et al.* (2015) |
| Sewage sludge | CM | 1 t of TS | Mills *et al.* (2014) |
| Sewage sludge | ReCiPe and Usetox | 10,000 $m^3$ wastewater per day | Heimersson *et al.* (2014) |
| Sewage sludge supernatant | CML | 1 kg $PO_4^{3-}$ eq. removed | Rodriguez-Garcia *et al.* (2014) |
| Sewage sludge | IPCC | 500 $m^3$ liquid raw sewage sludge | Cao and Pawłowski (2013) |
| Sewage sludge | Not specified | Amount of sludge generated/year | Remy *et al.* (2013) |
| Sewage sludge and food waste | CML | 1000 t of OFMSW (Organic Fraction of Municipal Solid Waste) and 2000 t of sewage sludge (dry matter (DM) 17%) | Righi *et al.* (2013) |

regardless of reactor configuration. However, the results (methane production) might be negligible and not be able to offset the environmental impacts of other inputs that cross system boundaries.

This is the case of Li *et al.* (2017a), which has obtained a very small (1.6–7.1% among scenarios) modification in the environmental profile when the organic content of the sludge varied from 70 to 40%. Other publications can be highlighted: Chen *et al.* (2022) demonstrated for sludge that incineration is a good option instead of AD for recoverable biodegradable waste with a low-organic content (VS/TS <55%), and Li and Feng (2018) showed that pyrolysis improve as much as its organic matter content does (the results were 4.6 times higher when achieving 70% compared to the baseline value of 50%). Finally, Li *et al.* (2017b) analysed processes with operation at low (3–6%) and high (10–15%) total solid concentrations with the same organic matter fraction (i.e. VS/TS of 70%). In the latter case, environmental benefits are obtained for high-solid concentration technologies at thermophilic temperature and for low-solid concentration, at mesophilic temperature.

#### 11.2.2.2 Upstream and downstream processes

The system boundaries of an LCA should be defined in accordance with the goals of the study. Therefore, the analysis of the environmental impacts of anaerobic treatment may be complemented by other processes or technologies used to facilitate organic matter solubilization, enhanced nutrient recovery or removal or compliance with legislation. The analysis of wider system boundaries has been mainly applied for the sewage line, but two studies have been reported for anaerobic secondary treatment. The first one is the research of Harclerode *et al.* (2020) who studied the differences between the use of primary sedimentation and screening only, included subsequent treatment with dissolved methane removal and nutrient removal and sludge management with AD and lime stabilization. The other one is the study of Laitinen *et al.* (2017) that incorporated a CW as post-treatment for a UASB reactor. Indeed, the substitution of secondary treatment technologies from aerobic to anaerobic can lead to the creation of new technological challenges within the sludge line, which indirectly contribute to the overall environmental sustainability of the system. This technological replacement has been analysed through several studies. For example, Brockmann *et al.* (2021) compared the performance of a CAS system with oxygen photogranulation and Arias *et al.* (2020a) provided information on the benefits of HRASs followed by an integrated film-activated sludge. In anaerobic treatment technologies, the change in environmental impact could be due to both a decrease in excess sludge produced and a change in its quality. This would influence not only the size of the equipment, but also the selection of the most appropriate technologies for its management. Within this topic, the study by Arias *et al.* (2021) is noteworthy as they have proposed a comparison between a chemical and thermal pre-treatment for sewage sludge management after the subsequent AD. Although the absence of these technologies implies a lower energy demand, a worse environmental profile in terms of climate change was found. The best performance in this category corresponded to the chemical approach due to higher biogas production and lower energy demand. However, a higher impact was recorded in other categories such as terrestrial acidification, particulate formation and terrestrial eco-toxicity due to indirect emissions generated for the production of these chemicals. Wang *et al.* (2020) also compared stand-alone AD with the AD process with heat treatment and a new alkali/acid pre-treatment. The latter treatment shows less impact on climate change, although other categories such as acidification and eutrophication potentials favoured the implementation of the thermal technology. Despite this, the authors claim that alkaline treatment is comparable to thermal treatment and that the small differences are due to sodium chloride consumption and higher energy demand.

#### 11.2.2.3 Temperature

Temperature is another parameter that must be considered within the process because there is a strong relationship with the reaction yield and with the energy requirements. When it comes to the valorization of wastewater for energy purposes, the energy consumed must be compensated by that resulting from the manufacture of the co-products. In domestic anaerobic treatment, the only product analysed so far with LCAs is the biogas and, thus, only an energy perspective has been considered in the literature. The temperature was addressed for membrane bioreactors (MBRs) at 15, 20 and 33°C in Smith *et al.* (2014) and Pretel *et al.* (2013). However, for temperatures up to 33°C, energy production from urban wastewater is not able to compensate for the heating needs of the system. For this reason, it is necessary to optimize the biogas yield, but only if the balance between production and consumption is positive. Studies on anaerobic sludge digestion processes (more data available) have shown that higher temperatures do not always lead to better environmental performance, and, in many situations, such improvement is only possible for some impact categories and FUs.

In this regard, Lanko *et al.* (2020) demonstrated the effects of temperature for sewage management through the analysis of three options: mesophilic, thermophilic and temperature-phased AD. The latter alternative showed the best environmental performance of the three in all the categories analysed (i.e. toxicity, land occupation, eutrophication and fossil resource depletion, among others) but climate change on which mesophilic digestion slightly outperforms. However, the authors found inconsistencies among different FUs. By changing the FU from 1 $m^3$ of the treated wastewater to 1 $m^3$ of the produced methane, the previously stated environmental results were rearranged and led to a better profile of the thermophilic reactor for all categories except for climate change where mesophilic operation is still more favourable. With a similar approach, Li *et al.* (2017b) presented the environmental differences among five processes (conventional AD, high-solid AD, AD with thermal pre-treatment hydrolysis, thermophilic AD and thermophilic high-solid AD). Regardless of the solid concentration (suitable for both high- and low-solid contents), the best environmental results were obtained with high-solid thermophilic AD (44% improvement for a VSS (volatile suspended solids) concentration of 70%). It is true that this technology requires more energy for heating the feed sludge, agitation and transport, but its efficiency is also higher and therefore more biogas is recovered. It seems that, for the same solid concentration, thermophilic technologies provide the best profile.

On the contrary, domestic anaerobic wastewater treatment still lacks LCA studies referring to the production of non-energy bioproducts. In contrast, the sewage sludge digestion may produce VFAs (volatile fatty acids) as a result. However, the production of non-energy products reduces the energy self-sustainability of facilities because they cannot be converted in electricity or heat to be recycled back to systems. Therefore, the efficiency of heating devices, energy loss prevention (adiabatic systems are preferable) and energy-environmental optimization become essential. For example, Elginoz *et al.* (2020) reported a 10% decrease in environmental load by improving the efficiency of the heating device by 10%. Furthermore, in many impact categories (i.e. terrestrial acidification, eutrophication, eco-toxicity and global warming, among others) temperature reduction showed a better profile (between 13 and 36%) despite the lower performance associated with a temperature decrease.

#### 11.2.2.4 Co-digestion of streams

Temperature and organic matter content are commonly parameters in the evaluation of anaerobic technologies, as indicated in sections 11.2.2.1 and 11.2.2.3. In this sense, co-digestion of other biodegradable waste streams with domestic wastewater would allow process optimization as higher volatile solid concentrations, C/N ratios and removal efficiencies can be achieved. Within this context, Becker *et al.* (2017) have demonstrated decreased impact on GWP with increased treatment of food waste with domestic wastewater. They have evaluated three technologies: a conventional (CAS), an anaerobic (AnMBR) and a newly developed aerobic treatment (HRAS). Their study could not draw conclusive results, as the environmental profiles of each were very similar and the uncertainty analysis showed small confidence intervals (probability within which the results of the environmental profile can be found). However, they highlight the potential of AnMBRs due to the recent development of the technology and the comparability of the net energy balance with HRAS-AD (technology that Arias *et al.* (2020a) stated as environmentally friendly compared to CAS for municipal wastewater).

Although not for domestic wastewater, other publications can be cited for sewage sludge. Morelli *et al.* (2020) have reported that there was a clear reduction (between 46 and 108% depending on the impact category) of the environmental impact with the implementation of co-digestion in wastewater treatment plants. Edwards *et al.* (2017) compared the same co-digestion strategy with a segregated treatment where food waste ended up in landfill. As a result, the co-digestion scenario outperformed (given the additional bioenergy generated) the segregated treatment in climate change potential (represented as 53–71% of the impact of the business-as-usual or landfill scenario for food waste).

## 11.3 TECHNO-ECONOMIC VIABILITY OF THE AD

Techno-economic assessment (TEA), cost–benefit analysis (CBA), life-cycle costing (LCC), cost–effectiveness analysis, cost–utility analysis and cost method of accounting are some examples of methodologies for the analysis of economic aspects of products, technologies and/or processes. Among them, TEA, CBA and LCC are some of the most frequently mentioned methodologies in the scientific community. Although LCC is the oldest methodology, they have been recently developed in parallel leaving to a vague distinction among them. CBA mainly focuses on projects and policies with the monetization of costs and benefits and aims at financial profitability. Although it is advantageously an autonomous tool unveiling whether a selected system under study has attributable welfare, CBA is considered a 'black-box' methodology. It simply ignores what happens inside the system in order to confirm its economic viability. Because of this problematic, TEA emerged as a tool simultaneously integrating the implications of technological aspects in the economy of a process or project as it systematically examines the interrelationships among them. There is, thus, a combination of engineering decision taking aiming at process optimization with the economic changes associated with them. In this regard, TEA separates from the vision provided for CBA as is not seem only a tool for investment. Therefore, TEA analysis should always go accompanied by a technological description and definition of technology-readiness levels. Despite implying and improvement for industrial processes of its use compared to CBA, TEA still lacks an appropriate regulatory foundation. In this regard, a new ISO standard is being developed (ISO/WD TS 14076) covering not only techno-economic analyses at any size or scale but also incorporating the environmental impacts. However, the inclusion of TEA in the legislation is far behind LCC which has already being recognized in directives (2014/24/EU and 2014/25/EU) and in standards (ISO 15663:2000 and 15686:2017).

Then, considering Tables 11.3 and 11.4, why scientists have decided to use TEA instead of LCC in their studies in the topic of anaerobic treatment? Generally, LCC has been associated with a product approach and a cost analysis though all life-cycle stages whereas TEA has been typically limited to an inherent investor perspective with gate-to-gate system boundaries (around the factory). However, the truth is that both TEA and LCC have their origin altogether in LCA methodology and both of them could be suitably used interchangeably in some contexts. It is then that LCC has a wider applicability whereas the scope of TEA is narrowing more and more to the stand-alone analysis of technologies and facilities. It is also very associated with the use of LCC the monetization of externalities (noise, environmental pollution, social effects and vibration, among others) which could partially be internalized (such as GHG emission taxes or waste-disposal costs) in the analysis. This could be appreciated for the studies dealing with anaerobic wastewater treatment. Out of the 33 studies shown in Tables 11.3 and 11.4 for TEA, only one has included environmental externalities whereas two of nine of the LCC studies were considering them. And yet, the sector seems to prefer the use of TEA instead of LCC (~72% of the studies were carried out with TEA). On the contrary, only one publication classified its study as CBA. Regarding the type of wastewater, the percentage of economic analyses focused on stand-alone analysis of sewage sludge was higher (57.6%) than for co-digestion and for other flows. The use of anaerobic treatment for domestic wastewater has only been addressed by five publications.

### 11.3.1 Techno-economic emerging challenges

Although the environmental analysis for anaerobic treatment of the state-of-the art focused primarily on parameter optimization (i.e. temperature and solid concentration) and upstream and downstream

**Table 11.3** Methodology and tools used for the economic analysis of anaerobic technologies in domestic treatment for sewage sludge.

| Type of Resource | Methodology | Accounting Tools | Reference |
|---|---|---|---|
| Sewage sludge | TEA | OPEX, CAPEX, PBP and TAEC | Gholamian *et al.* (2023) |
| Sewage sludge mixed with lignocellulosic biomass | TEA | OPEX, CAPEX, NPV and IRR | Ebrahimi *et al.* (2023) |
| Sewage sludge | TEA | PBP, IRR, NPV and SIR | El-Qanni *et al.* (2022) |
| Sewage sludge | TEA | OPEX | He *et al.* (2022) |
| Sewage sludge | TEA | OPEX, CAPEX and NPV | Vinardell *et al.* (2022) |
| Sewage sludge | TEA | OPEX and PBP | Díaz *et al.* (2021) |
| Sewage sludge, food waste and fish sludge | TEA | OPEX, CAPEX, ROI and NPV | Fernando-Foncillas and Varrone (2021) |
| Food waste and sewage sludge | TEA | OPEX, CAPEX and NPV | Vinardell *et al.* (2021) |
| Sewage sludge and organic municipal solid waste | TEA | OPEX, CAPEX, NPV, IRR and PBP | El Ibrahimi *et al.* (2021) |
| Sewage sludge and food waste | LCC | OPEX and CAPEX | Andreasi Bassi *et al.* (2021) |
| Sewage sludge | TEA | OPEX, CAPEX and PBP | Bahreini *et al.* (2020) |
| Sewage sludge | TEA | OPEX and CAPEX | Medina-Martos *et al.* (2020) |
| Sewage sludge | LCC | OPEX, CAPEX, TAEC and ROI | Roldán *et al.* (2020) |
| Sewage sludge | LCC | OPEX, CAPEX, PBP and MARR | Cuéllar-Franca *et al.* (2019) |
| Sewage sludge | TEA | OPEX and CAPEX | Ranganathan and Savithri (2019) |
| Sewage sludge and organic municipal solid waste | LCC | NPV and IRR | Francini *et al.* (2019) |
| Sewage sludge and organic waste | CBA | OPEX, CAPEX and ROI | Thomsen *et al.* (2018) |
| Sewage sludge | TEA | OPEX and CAPEX | Dussan and Monaghan (2018) |
| Sewage sludge | LCC | OPEX and CAPEX | Tarpani and Azapagic (2018) |
| Sewage sludge | TEA | OPEX, CAPEX, NPV, IRR and PBP | Rus *et al.* (2017) |
| Sewage sludge | TEA | OPEX and CAPEX | García-Gutiérrez *et al.* (2016) |
| Sewage sludge | TEA | OPEX | Tomei *et al.* (2016a) |
| Sewage sludge | TEA | OPEX | Tomei *et al.* (2016b) |
| Sewage sludge | TEA | OPEX and CAPEX | Gianico *et al.* (2015) |
| Sewage sludge | TEA | OPEX, CAPEX and TAEC | Garrido-Baserba *et al.* (2015) |
| Sewage sludge | LCC | OPEX | Xu *et al.* (2014) |

*Note:* CBA, cost–benefit analysis; CAPEX, capital expenditure; IRR, internal rate of return; LCC, life-cycle costing; MARR, minimum acceptable rate of return; NPV, net present value; OPEX, operational expenditure; PBP, payback period; ROI, return of investment; SIR, saving to investment ratio; TAEC, total annual equivalent cost; TEA, techno-economic assessment.

**Table 11.4** Methodology and tools used for the economic analysis of anaerobic technologies in domestic treatment for wastewater resources.

| Type of Resource | Methodology | Accounting Tools | Reference |
|---|---|---|---|
| Domestic wastewater | TEA | OPEX, CAPEX and NPV | Sanchez *et al.* (2023) |
| Domestic wastewater, blackwater, greywater and urine | LCC | OPEX and CAPEX | Lehtoranta *et al.* (2022) |
| Domestic wastewater | TEA | OPEX and CAPEX | Vinardell *et al.* (2020) |
| Blackwater | LCC | OPEX, CAPEX and NPV | Estévez *et al.* (2022b) |
| Blackwater and urine | LCC | OPEX and CAPEX | Postacchini *et al.* (2022) |
| Domestic wastewater | TEA | OPEX and CAPEX | Shoener *et al.* (2016) |
| Domestic wastewater | TEA | OPEX, CAPEX and NPV | Sills *et al.* (2016) |

*Note:* CBA, cost–benefit analysis; CAPEX, capital expenditure; IRR, internal rate of return; LCC, life-cycle costing; MARR, minimum acceptable rate of return; NPV, net present value; OPEX, operational expenditure; PBP, payback period; ROI, return of investment; SIR, saving to investment ratio; TAEC, total annual equivalent cost; TEA, techno-economic assessment.

technologies, the economic studies were mainly aimed at creating a roadmap around anaerobic processes and how the process could overcome cost-effectiveness constraints. Therefore, the study of complementary technologies for anaerobic systems plays an important role in the quality (and thus the price) and marketability improvement of co-products.

#### 11.3.1.1 Upstream and downstream

Although the economic studies centred on anaerobic processes implemented within the sector of wastewater treatment have thermal pre-treatment for sewage (i.e. hydrolysis and carbonization), domestic anaerobic treatment as a secondary process has been investigated by the implementation of forward osmosis and reverse osmosis as pre-treatments. Within this context, Vinardell *et al.* (2020) have analysed the feasibility of forward and reverse osmosis to concentrate diluted solutions. Given that the wastewater pre-concentration accounts for more than 74% of the total budget, pre-treatment worsens the costs associated with the treatment. Therefore, the stand-alone AnMBR operation is cheaper than the scenario with such pre-treatment, those being half of the scenario with pre-concentration. Because of the better economic performance of forward osmosis–AnMBR, Vinardell *et al.* (2022) studied the influence of the draw solution used during the operation of the system because it affects the membrane fluxes and the salinity (inhibiting anaerobic bacteria in the subsequent reactor). The most economically favourable draw solute was dependent on the membrane type but in any of the case studies proposed by the authors the economic impact from these substances was moderate. The hotspot still was the capital investment for pre-concentration. Although the forward and reverse osmosis for water reclamation is still not a viable configuration, pre-concentration of sewage sludge with membrane remains a promising approach.

The techno-economic analysis for domestic anaerobic wastewater treatment has been poorly addressed and the literature has been focused on sewage sludge. Two approaches can be found in the literature: AD is one of the multiple options for sludge treatment and AD is the main process discussed, but it is followed by other treatment. Garrido-Baserba *et al.* (2015), Tarpani and Azapagic (2018) and Xu *et al.* (2014) are examples of economic analysis for sewage treatment.

#### 11.3.1.2 Feasibility of reactor configuration

Despite the large variability of reactors capable of operating at total solid concentration lower than 3% (CSTRs or continuous stirred tank reactors, UASBs, AFBs or anaerobic fluidized bed reactors, ECGS or expanded granular sludge bed reactors, MBRs, among others), the techno-economic analysis of domestic anaerobic treatment has only paid attention to the operation of MBRs and ABRs.

For example, Sills *et al.* (2016) compared the performance of ABRs with other systems such as the combination of an aerobic tricking filter with AD. This reactor provided a lower cycle cost (~40% in agreement with the results of its net current value) because of both better capital and operating costs. It should be highlighted also the lower solid production of this system, which reduced the costs associated with disposal from 25 to 7%. Regarding the implementation of MBRs, membrane purchase seems to be the largest hotspot (over 49%) of the analysis followed by chemical addition for cleaning (between 0 and 26% depending on the flowrate) (Sanchez *et al.*, 2023). Vinardell *et al.* (2020) also agreed on a higher contribution of capital expenditures (CAPEX) in membrane reactors (between 63 and 77%).

#### 11.3.1.3 Influence of co-digestion on economic profile

The organic matter concentration of domestic wastewater is not expected to be as high as, for example, that of molasses, cheese whey or sewage, among others. What can be expected from co-digestion of organic waste and municipal wastewater? To clarify this issue, Vinardell *et al.* (2021) proposed a comparison of the following scenarios: (1) anaerobic secondary treatment with an AnMBR and with another AD reactors as side-stream for sewage sludge and food co-digestion, (2) the CAS system, (3) anaerobic secondary treatment but without co-digestion in the side-stream with AD and (4) another including nutrient recovery from the centrate of the side-stream AD. Although the analysis centred in the sludge line, the highest costs were also achieved by domestic WWTPs with an aerated secondary treatment because a larger amount of sludge with poor biodegradability is produced. The implementation of an AD AnMBR and co-digestion significantly increased the revenues (triple) despite the increase of the costs up to a 44%. Among the case studies with anaerobic co-digestion, the greatest benefits were obtained with the configuration in which the nutrients are treated in the mainstream. Co-digestion of food waste in a conventional domestic WWTP is not only preferable in terms of the facility itself, but also in comparison to the separate treatment of both types of waste.

#### 11.3.1.4 Nutrient recovery efficiencies

The nutrient recovery can be performed both in the water and sludge lines and brings multiple benefits beyond the transformation of pollutants into marketable products. This strategy also supports the increase of quality of effluents and prevents the uncontrolled precipitation, which increases the performance efficiencies of unitary operations. However, nutrient recovery poses a challenge because it may compromise the economic feasibility of the processes. When implemented in the sludge line, the question to be solved is whether the precipitation of phosphorus should be performed before or after AD. Considering the results achieved by Roldán *et al.* (2020), there is 1% difference between both options, which ensures the viability of the processes when the LCC analysis is expressed by an amount of phosphorus recovered. However, a change in the FU per $m^3$ unit led to a drastic change in the direction of the results because only one of the newly proposed alternatives was viable. This is related, thus, to the lower maintenance costs associated with the uncontrolled phosphorus precipitation and higher biogas production of the alternative considering the phosphorus precipitation before AD.

When anaerobic treatment is applied in the water line, costs are saved compared to aerobic systems, as aeration requirements are avoided. However, anaerobic processes are not capable of removing nutrients such as phosphorus and nitrogen by themselves and thus the effluent do not comply with European standards. Therefore, resource recovery technologies should be installed in the facility to maintain the emission thresholds. Because of this, the economic feasibility of the entire system is compromised because in many cases the methane production of the water line does not offset the higher CAPEX and operational expenditure (OPEX) associated with it (even for concentrations higher than 1100 mg/L). It should be considered that the biogas from the side-stream AD treating the sewage sludge from the main anaerobic treatment located in the water line may not also be sufficient to provide a profitable system (Vinardell *et al.*, 2021).

#### 11.3.1.5 Valorization of the raw biogas

The valorization pathway selected for the raw biogas produced during anaerobic treatment has a strong influence on the economic viability of the facilities, as each of the co-products is co-related to a market price and to a specific post-treatment cost. The selection of the best technology has not been an easy task within this topic as each of the researchers compared different technologies with each other, and all drawn different conclusions. The knowledge in this section comes specifically from publications on AD of domestic sewage sludge treatment. However, and as the product obtained from the anaerobic secondary treatment of the domestic wastewater is also biogas, the recommendations and guidelines from the environmental studies of sewage sludge anaerobic treatment can also be applied. Volpi *et al.* (2023) recommended the use of biogas for the production of PHA, Purwanta *et al.* (2022) for co-firing in boilers, Alfonso-Cardero *et al.* (2021) for electricity production and Fuess and Zaiat (2018) for biomethane production.

Besides upgrading and transformation of the raw biogas, the $CO_2$ in the biogas can also be transformed into a vast number of co-products. In this regard, Cuéllar-Franca *et al.* (2019) compared four technological configurations for the production of fuels: the capture of the $CO_2$ by adsorption with mono-ethanolamine (MEA) from the raw biogas, the capture of $CO_2$ with MEA from the flue gases of combusted raw biogas, combustion of raw biogas and direct use of the $CO_2$ in the flue gases and capture of $CO_2$ from raw biogas, combustion and final use of flue gases. The highest capital and operating costs arise from the $CO_2$ capture by absorption from the flue gas, with a difference of ~13% compared to the best scenario ($CO_2$ capture from unsweetened biogas). Despite being the most expensive process, it has the most cost-effective design. This is because the liquid fuel has a higher price.

## 11.4 ANALYSIS OF WASTEWATER TREATMENT TECHNOLOGIES AND PROCESSES WITH LCA AND LCC

### 11.4.1 LCA approach for wastewater scenarios: a case study

Despite the growing interest in the application of LCA over the last decade, the complexity involved in collecting data throughout the entire life cycle, especially for novel practitioners, remains a threshold to overcome. This becomes more evident across sectors, as the focus (i.e. process or product) and logic (existence of cross-functionality) of the analysis vary depending on the objective of the study. To resolve some issues related to the applicability of LCA in the field of wastewater, Corominas *et al.* (2020) proposed a practical guide with a hypothetical approach for the sector. In this sense, this chapter aims to perform an LCA for domestic anaerobic treatment with experimental data as a starting point. Therefore, the main focus of the example provided is more on the early stages of research and development than on decision making (product purchasing or development of environmental legislation) or marketing (eco-design).

In this sense, the environmental profile of a sequential granular UASB bioreactor has been analysed taking into account the effects of changing temperature and hydraulic retention time. For this purpose, laboratory experimental data were obtained from the study of Stazi *et al.* (2022). Four scenarios were constructed based on two aspects: data availability and comparability between scenarios (Table 11.5). Effluent composition, operating temperature, product characterization (purity and amount of biogas), hydraulic retention times and pump run times are some of the raw data collected. Because two variables were analysed, the scenarios were named accordingly. 'T' refers to temperature whereas 'H' refers to hydraulic retention time. The numbers just after each letter correspond to the temperature in degrees Celsius and the hydraulic retention time in hours. Thus, the stage operating at 35°C and 22 h will be referred to as T35H22.

**Table 11.5** Data collection from the experimental study of Stazi *et al.* (2022).

| Parameters | Units | Scenario T35H22 | Scenario T25H22 | Scenario T35H14 | Scenario T35H09 |
|---|---|---|---|---|---|
| Working volume of the reactor | L | 0.90 | 0.90 | 0.90 | 0.90 |
| Number of pumps | – | 3.00 | 3.00 | 3.00 | 3.00 |
| Feeding volume/cycle | L | 0.50 | 0.50 | 0.50 | 0.50 |
| Feeding time | min | 12.00 | 12.00 | 12.00 | 12.00 |
| Sedimentation time | min | 30.00 | 30.00 | 30.00 | 30.00 |
| Effluent discharge time | min | 8.00 | 8.00 | 8.00 | 8.00 |
| COD in the influent | mg/L | 500.00 | 500.00 | 500.00 | 500.00 |
| Total nitrogen (TN) in the influent | mg/L | 53.00 | 53.00 | 53.00 | 53.00 |
| Total phosphorus (TP) in the influent | mg/L | 5.00 | 5.00 | 5.00 | 5.00 |
| Temperature of operation | °C | 35.00 | 25.00 | 35.00 | 35.00 |
| Reaction time | min | 670.00 | 670.00 | 420.00 | 230.00 |
| HRT | h | 22.00 | 22.00 | 14.00 | 9.00 |
| COD in the effluent | mg/L | 18.50 | 39.00 | 27.00 | 34.00 |
| Nitrogen in the effluent as $NH_3$ | mg/L | 60.50 | 52.50 | 59.00 | 59.50 |
| Phosphorus in the effluent as $PO_4^-$ | mg/L | 7.75 | 8.10 | 7.25 | 7.10 |
| Total dissolved solids (TSS) in the effluent | mg/L | 15.00 | 13.50 | 15.00 | 15.00 |
| Produced biogas | $m^3$/kg COD removed | 0.24 | 0.21 | 0.22 | 0.18 |
| Methane in biogas | % | 75.50 | 53.00 | 75.50 | 75.50 |
| Concentration of methane in the effluent | mg/L | 7.09 | 7.34 | 7.09 | 7.09 |

T35H09, 35°C of temperature and 9 h of hydraulic residence time; T35H14, 35°C of temperature and 14 h of hydraulic residence time; T25H22, 25°C of temperature and 22 h of hydraulic residence time; T35H22, 35°C of temperature and 22 h of residence.

Regardless of the sector or technology to be analysed, LCA should always proceeds according to ISO 14040 and 14044 standards, so the analysis of environmental aspects should be carried out following their four stages.

*Goal and scope*: One of the main assumptions when approaching anaerobic wastewater treatment is whether to consider the digestate/effluent as a discharge, a stream to be further treated, or a co-product. Generally, effluents from AD processes do not meet the discharge thresholds set by European legislation and, therefore, the removal of some compounds such as nitrogen or phosphorus should be considered. This feature could be managed by an extension of system boundaries to incorporate the respective treatment technologies. However, many other questions must be answered before decisions can be made: Is the analysis focused on reactor optimization? Is the process decentralized and space limited? Does the country where the technology is implemented have restrictive emission limits? On the contrary, the circular economy concept is driving the recovery of nutrients from wastewater streams. However, direct reuse of liquid and solid digestate is sometimes limited by crop type, proper fertigation practices, nutrient concentration in effluent, presence of metals and pathogens (which are only removed at higher temperatures), distance from facility to point of application, community acceptance and economic feasibility (Helmecke *et al.*, 2020). The complexity of the treatment paradigm detailed above is something that must be reflected by LCA practitioners through the definition of system boundaries. As one of the key aspects, system boundaries should predefine the set of unit processes, inputs and outputs for total emissions accounting and is therefore fundamental in the creation of the LCI.

The system boundaries should consider three dimensions: geographical, temporal and technical (Li *et al.*, 2014). Among them, the last one has a particular importance among the wastewater scientific community, corresponding to data collection for new technologies. However, the other two are rarely defined due to the limitations of the environmental characterization factor in the databases. For example, the Eco-invent database only covers 20,000 datasets, far fewer than the 204 million organic substances, alloys, mixtures, polymers and salts accounted for in the Chemical Abstract Service (2023) record. Moreover, many of them are not country or region specific and the inventory refers to a global or European approach. When studying a sequential granular UASB reactor, the system boundaries can therefore only be threefold: cradle-to-gate, gate-to-gate and cradle-to-gate with expansion/substitution/boundary partitioning allocation. In the first approach, emissions are accounted for from raw material extraction to UASB effluent.

In the second case, according to International Reference Life Cycle Data (ILCD) guidelines, direct foreground process emissions, such as methane or carbon dioxide, would be included. The latter perspective is the most difficult to define and a full consensus among LCA practitioners has not yet been achieved. Depending on how emissions are accounted for, three main schools of thoughts can be distinguished as: attributional, consequential and socioeconomic (Moretti *et al.*, 2020). The pure attributional approach advocates that environmental impact should only be described by physical causalities entering and leaving through system boundaries. In contrast, the pure consequential approach aims at the description of the change of the environmental profile in response to a given modification in process technologies or the assessment of the future marginal energy supply. However, not many of the studies referring to anaerobic processes in wastewater treatment (11% according to the results in Tables 11.1 and 11.2) present how these emissions were reported. So, how to approach the limits of the cradle-to-gate system with expansion/substitution for sequential UASB reactors? One option could be to compare the scenarios with an attributional perspective considering that all viable alternatives provide the same number of products. This implies 'expansion' of the system boundaries to introduce other productive processes such as fertilizer or energy industries. 'Substitution' is, however, more typical of the consequential approach and is implemented through the avoidance of emissions from such production processes. Table 11.6 summarizes the decisions made at the goal and scope stage to carry out the case study described in this chapter.

Secondary treatment of domestic wastewater becomes, with the implementation of an anaerobic UASB, a multifunctional system capable of reducing direct pollution to water resources and simultaneously providing marketable products. Therefore, the complete definition of the system boundaries also depends on the function selected and the strategy followed to achieve the objective of the study. The functions of anaerobic wastewater treatment can be oriented at three different levels: micro-level, macro-level and accounting decision support (European Commission *et al.*, 2011). The first is closely related to specific products, the second to the comparison of technological scenarios, material strategies and policy options and the third is interested in the documentation of what has happened or will happen based on decisions already taken. An example of a micro-level approach in the case of AD for the example provided in this chapter could be the comparison of the electricity and liquid bio-fertilizer produced versus products already on the world market.

As far as the process or macro-level analysis is concerned, multiple strategies can be cited: technological optimization by analysing the weak points and identifying the flows with the highest environmental impact, forecasting and comparing technologies, affecting the remaining unit operations of a facility and the overall impact through technology implementation, geographical coverage and system response to specific environmental conditions (especially in the case of open-air reactors, as the profile may change between regions for the same technological approach), as well as the use of specific characterization factors. All these strategies belong to one of the following groups: planning, design and operation/optimization. Within the first group one could include decision-making studies regarding the selection of treatment alternatives, the second group refers mainly to the elucidation of environmental hotspots and the anticipation of impacts and the third to the creation of roadmaps to ameliorate environmental impacts. Therefore, it should be considered whether the objective of the LCA is to provide guidance on planning, design or operation. Because the research objective of Stazi

**Table 11.6** Summary of the attributes of the LCA for the case study.

| Attribute | Description |
|---|---|
| Initial hypothesis | (1) Two analyses were used for comparison: (a) the effluent of the UASB is directly discharged and (b) the effluent is directly reused as fertilizer source.<br>(2) The biogas has been valorized into heat and electricity in a heat and power unit with an efficiency of 45% for heat and 35% for electricity.<br>(3) Because of the scale of the facility, a conservative value of 0.4 m distance between equipment has been assumed to estimate the electricity consumption in the pumps. Readers should be aware that the distance may change in accordance with the location/ geographical aspects of the area of implementation and with the scale-up.<br>(4) Except for the energy demand of the pumps (whose efficiency is expected to be higher at higher design scales), it is assumed that the results of the remaining parameters do not differ from other scales.<br>(5) The reactor was sized considering the up-flow velocity indicated by Stazi *et al.* (2022) and a height/diameter ratio of 3.<br>(6) The pH control could not be included because in laboratory experiments the amount of chemicals (sodium bicarbonate) consumed is not usually recorded and the initial and final pH remain unchanged.<br>(7) Average data estimated from the upper and lower limits of the operating conditions for each of the scenarios analysed. |
| Technical system boundaries | Cradle-to-gate for the hypothesis (a) and cradle-to-gate with system expansion for hypothesis (b) |
| Geographical system boundaries | Continental/global level |
| Temporary system boundaries | Not applicable |
| Emission accounting approach | Attributional |
| Strategy group | Operational, optimization |
| Level/scale of implementation | Macro-level or process-oriented |
| FU | 1 $m^3$ of wastewater treated |

*et al.* (2022) for the technology was the optimization of temperature and hydraulic retention time in a domestic wastewater treatment reactor, the case study provided can be framed within operation/ optimization.

Finally and as mentioned earlier in Section 11.2, the first stage (called objective and scope) of the LCA should also include the identification of the FU. As can be seen in Tables 11.1 and 11.2 in Section 11.1, the most recurrent FU in AD for wastewater treatment is the volume of treated wastewater or '1 $m^3$ of treated wastewater'. This FU has also been selected by many other authors focused on the analysis and comparison of technologies to be implemented as anaerobic secondary treatment. Examples are the studies of: Boldrin *et al.* (2022), de Sampaio Lopes *et al.* (2014), Harclerode *et al.* (2020), Laitinen *et al.* (2017), Patel and Singh (2022), Pretel *et al.* (2013), Pretel *et al.* (2016), Sills *et al.* (2016) and Smith *et al.* (2014). However, this FU is only representative of one of the system functions: wastewater treatment. The product approach should be considered with other FUs such as: 1 $m^3$ of biogas, 1 kWh or 1 kg of methane produced, 1 kg of nutrients recovered and 1 $m^2$ of soil fertilized.

*Life-cycle inventory*: In LCA, two LCI perspectives can be differentiated depending on the sources of data collection. In process-based inventories, data have been collected with a 'bottom-up' approach and use primary input–output data associated with the foreground system process. On the contrary, input–output inventories follow a 'top-down' approach and have been collected from statistical data.

In the latter case, emissions are estimated from the selling prices of consumables/processes used and are therefore more suitable for regional/national scales than for industrial/facility scales. Although the data from Stazi *et al.* (2022) cannot be considered primary due to its bibliographic origin, it is certain that similar LCAs can be performed from laboratory data that have not already been published. Consequently, the nature of the study forces the foreground LCI to be a process-based inventory compiled from a 'bottom-up' approach. In addition to the application of mass and energy balances, software databases (for this case SimaPro®) were used to fill in gaps related to background processes, such as electricity and heat production. The LCI should be completed throughout the operation, maintenance, construction and demolition phases of a process/technology lifetime. However, the case study was only exemplified for the operational phase and the others were outside the system boundaries. For this reason, the LCI is constituted by direct and indirect emissions but only related to the treatment process.

Therefore, the emissions for anaerobic treatment in a UASB can be divided into direct and indirect. Direct emissions are considered within the boundaries of a gate-to-gate or inside-the-fence system, as they can be measured or modelled within a foreground system (Li *et al.*, 2022). As a secondary biological treatment, emissions come from the microbial metabolic activities of the biomass involved in the unit operations or from the limited removal of the target pollutant. Estimation of such emissions can be performed with direct laboratory measurements, with IPCC (Intergovernmental Panel on Climate Change) guidelines, with characterization factors exposed in the scientific literature with similar objectives and from process modelling. However, the latter options may resort to overestimation or underestimation of such emissions, as inhibitory effects and other operational particularities of a system are not included. Therefore, only methane and carbon dioxide emissions were considered for the LCA example provided in this chapter. A comparison of scenarios has been shown according to commonly measured parameters at a laboratory scale. In addition, many of the remaining emissions were not considered relevant in other publications and were therefore assumed to have an impact of <5% (cut-off criterion) (Laurent *et al.*, 2020). On the contrary, and depending on the composition achieved by the technologies, some of these components, such as methane, may rather be considered as valuable output. This is because biogas first undergoes a combustion process to produce electricity and/or heat.

Indirect emissions are usually defined for broader system boundaries, such as cradle-to-gate and cradle-to-grave, as emissions upstream (inputs from the Technosphere) and downstream (outputs to the Technosphere) of the target process or product may be considered background (system but not under the direct control or decisive influence of the producer of the good) processes and are often predefined in regulatory databases. Indirect and direct emissions were included in the case study. It should be noted that depending on the hypothesis and initial boundaries, the data may be reorganized differently. Table 11.7 specifically defines effluent nitrogen and phosphorus as emissions and not as feasible outputs.

*Life-cycle impact assessment*: This third stage of LCA aims to transform with impact assessment modelling factors the inventory in Table 11.7 into results that can be understood in terms of environmental impact or damage. It is further subdivided into the following stages: selection and definition of impact categories, classification, characterization, normalization and weighting. Unlike other LCA methodologies, such as carbon or water footprinting, LCA is a multidimensional methodology. Therefore, a multi-criteria analysis must be provided through the investigation of the results of indicators or categories. Many methods are currently available from which categories can be selected: USEtox, ReCiPe, IMPACT 2002+, TRACI, EDIP, CML, MEEUP, EPS, IPCC, Eco-indicator and LIME (European Commission *et al.*, 2011). The first pre-selection of the method should be performed on the level at which the impact categories should act, namely midpoint or endpoint. This implies that the results will be shown for different parts of the environmental value chain. Furthermore, the selection of the categories does not follow a strict rule; however, the usual practice is to select the method based on the categories of interest. The other option could be to choose the categories independently of the method. However, attributes such as completeness of

**Table 11.7** LCI per FU (1 $m^3$ of wastewater treated) for the case study defined from the data of Stazi *et al.* (2022).

| LCI Materials | Units | Scenario T35H22 | Scenario T25H22 | Scenario T35H14 | Scenario T35H09 |
|---|---|---|---|---|---|
| Inputs from Technosphere | | | | | |
| Feeding pump | $kWh/m^3$ | 0.41 | 0.41 | 0.41 | 0.41 |
| Recirculation pump | $kWh/m^3$ | 22.99 | 22.99 | 14.41 | 7.89 |
| Effluent pump | $kWh/m^3$ | 0.27 | 0.27 | 0.27 | 0.27 |
| Heat | $kWh/m^3$ | 18.98 | 5.92 | 19.01 | 19.08 |
| Biogas blower | $kWh/m^3$ | $2.51 \times 10^{-3}$ | $2.67 \times 10^{-3}$ | $2.21 \times 10^{-3}$ | $1.60 \times 10^{-3}$ |
| Outputs to Technosphere | | | | | |
| Electricity | $kWh/m^3$ | $2.93 \times 10^{-1}$ | $1.77 \times 10^{-1}$ | $2.69 \times 10^{-1}$ | $2.11 \times 10^{-1}$ |
| Heat | $kWh/m^3$ | 0.00 | 0.00 | 0.00 | 0.00 |
| Outputs to the Nature | | | | | |
| Methane ($CH_4$) | $kg/m^3$ | $7.09 \times 10^{-3}$ | $7.34 \times 10^{-3}$ | $7.09 \times 10^{-3}$ | $7.09 \times 10^{-3}$ |
| Carbon dioxide ($CO_2$) | $kg/m^3$ | $1.97 \times 10^{-1}$ | $1.70 \times 10^{-1}$ | $1.81 \times 10^{-1}$ | $1.42 \times 10^{-1}$ |
| COD | $kg/m^3$ | $1.85 \times 10^{-2}$ | $3.90 \times 10^{-2}$ | $2.70 \times 10^{-2}$ | $3.40 \times 10^{-2}$ |
| Nitrogen as ammonia ($NH_3$) | $kg/m^3$ | $6.05 \times 10^{-2}$ | $5.25 \times 10^{-2}$ | $5.90 \times 10^{-2}$ | $5.95 \times 10^{-2}$ |
| Phosphorus as phosphate ($PO_4^-$) | $kg/m^3$ | $7.75 \times 10^{-3}$ | $8.10 \times 10^{-3}$ | $7.25 \times 10^{-3}$ | $7.10 \times 10^{-3}$ |
| Dissolved solids (TSS) | $kg/m^3$ | $1.50 \times 10^{-2}$ | $1.35 \times 10^{-2}$ | $1.50 \times 10^{-2}$ | $1.50 \times 10^{-2}$ |

scope, robustness and uncertainty, reproducibility, transparency or stakeholder acceptance would vary among the selected categories. For example, the ILCD manual 'Recommendations for Life Cycle Impact Assessment in the European context' provides guidelines for pre-selection of methods based on the most common impact categories (climate change, ozone depletion, human toxicity, respiratory particulates/inorganics, photochemical ozone formation, ionizing radiation impacts, acidification, eutrophication, eco-toxicity, land use and resource depletion).

However, the guidelines were published in 2010 and many methods have since been updated as ReCiPe 2016. The selection of impact categories for the case study aimed at representativeness of the following aspects: background energy consumed, implications of biogas production in reducing resource depletion, relevance of direct emissions control and water resource pollution. For these reasons, the categories analysed were climate change, ozone depletion, eutrophication and resource depletion. During the classification stage, the LCI data are assigned to each of the previously selected impact categories. All of these categories can be found in CML and ReCiPe, two of the most widely used methods in the field of AD (see Tables 11.1 and 11.2). After selection of the LCIA method, the connection between the LCI and the environmental impact/harm is performed during the classification and characterization stages. Environmental loadings are assigned to each LCIA data element and category using factors. Both stages can be supported by the use of software such as SimaPro, EASETECH and OpenLCA, among many others. In particular, the first three LCIA stages for the environmental comparison of the UASB granular sequential reactor scenarios have been performed with SimaPro version 9.3 for the ReCiPe 2016 Midpoint (H) V1.07/World (2010) H method. Normalization and weighting, the optional ICLC stages were not included within the study.

*Interpretation of the results*: All of the above phases of LCA are interrelated. The objective of this analysis is not only to provide a clear message to the readers about the environmental expectations of the analysed process, product or technology, but it is also relevant for the identification of missing data and errors. Therefore, the LCA is iterative, and the results obtained at this stage lead to the modification or redesign of the assumptions adopted in the previous stages. When interpreting the results, the first step should be a data and sensitivity analysis, because the results depend on the decisions made throughout the design process. This should be especially important when using data from unreliable sources, numerical data that falls within a range, assumptions, and when alternative production routes can be modelled. Referring to the case study proposed for Stazi *et al.* (2022), ranges of data can be found regarding pollutant removal efficiency, assumptions were used for reactor sizing and distances between pumps and other equipment and the European profile was used to power electrical devices (the more renewable the profile, the lower the environmental impact). To check whether some of these parameters represent a significant change in the results, the sensitivity analysis should be linked to a contribution statement. The hotspots or elements of the process system that have the greatest impact on the overall profile should be identified. Depending on the details of the contribution analysis, the results could highlight the species of substances impacting the environment, the processes involved (i.e. energy or chemicals) or sections of a facility (i.e. primary treatment, secondary treatment). Finally, for comparative analysis, the results of different scenarios can be contrasted with each other. Figure 11.1 shows the visualization of the contribution of one of the four scenarios analysed, as well as the comparison among all scenarios considering hypothesis (a) of the analysis (Table 11.6).

Table 11.6 highlights the two initial hypotheses of the LCA study conducted: discharge of the effluent directly into the environment and valorization of the effluent compounds as liquid fertilizer. Under the first hypothesis, five background processes and six polluting substances emitted to nature are responsible for the current distribution of the environmental profile in Figure 11.1a. Among them, the use of electricity in the pump needed for agitation and recirculation of the stream inside the reactor is the main contributor in three (between 54.8 and 78.6% for FRS or fossil resource scarcity and SOD or stratospheric ozone depletion, respectively) of the four impact categories under study. The ME (marine eutrophication), on the

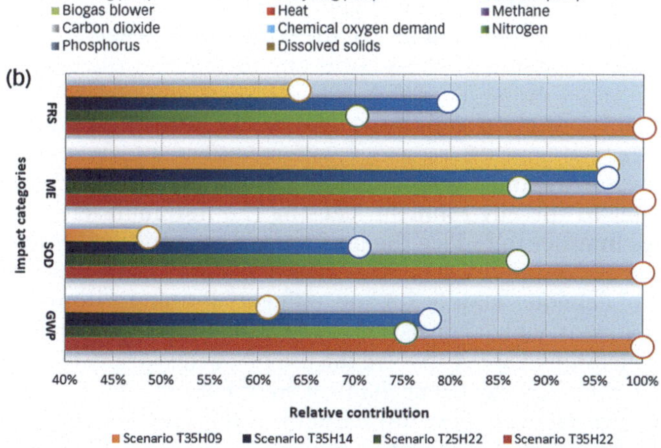


**Figure 11.1** Environmental relative contribution profile for the four scenarios of the case study analysed: (a) Process contribution for the scenario T35H22 and (b) scenario comparison. T35H09, 35°C and HRT of 9 h; T35H14, 35°C and HRT of 14 h; T25H22, 25°C and HRT of 22 h; T35H22, 35°C and HRT of 22 h.

contrary, is affected 96.2% by nitrogen emissions to water resources. With the exception of the marine eutrophication category, heat consumption is the second most important process in the profile with a minimum and maximum representativeness of 19.1 and 43.6% for SOD and FRS. This is due to two reasons: the five listed background processes are energy related, with an LCI underlining a much higher influence of electricity demand on the recirculation pump, and the unit process (expressed in impacts per kWh) of electricity is more polluting than the selected heating process. In this context, effluent recirculation is the main hotspot of the process when the distance between units is 0.4 m. A reduction of the length of the pump suction and discharge piping to 0.1 m would lead to completely different results, as the electricity demand would no longer be the main hotspot. Unlike temperature, hydraulic retention time and influent concentrations, the distance between equipment is not an intensive variable. Thus, full-scale anaerobic treatment systems may have pump arrangements and distances differing from those indicated within this chapter. For such reason, the results of the herein described LCA have been proposed as an example for the identification of environmental weak points from an early stage of design. The procedure could be repeated for larger scale technologies or facilities but considering that not all the parameters from the inventory are going to relate proportionally to the scaling-up of the facility. Another option to decrease the environmental impact of electricity use would be a modification of the hydraulic retention time. This is due to the relationship between the running time of the pump used in effluent recycling and the energy demand. To this effect, a comparison of three scenarios is shown in Figure 11.1b. Scenarios T35H09, T35H14 and T35H22 demonstrate how a shorter hydraulic residence time is accompanied by a decrease in environmental impact by an order of magnitude of up to 51.5%, from 22 to 9 h. Lower operating temperature also translates into better results (up to 30%). The answer to which of the two parameters (temperature or HRT (hydraulic retention time)) is more relevant for the profile with the same degree of modification cannot be obtained with this case study, as the hypotheses were built on experimental data. However, it appears that for modifications ~30% the results depend on the distance from the recycle pump to the reactor. For example, three of the categories analysed are favourable to T35H14 for 0.4 m, whereas the same hypothesis is worse than T25H22 for 0.1 m.

When comparing initial hypotheses, A with B, the qualitative LCA results are similar. In both cases, T35H22 is the worst scenario in all impact categories and T35H09 is the best. Accordingly, the results are congruent regardless of the adopted scenario. The decrease in the environmental profile between the most and least polluting scenarios is, however, 11.8% higher for ME under scenario B. This means that T35H09 is still a better alternative in terms of lower effluent nutrient composition and thus under-recovery. The reason for this is the higher affect of a direct emission compared to fertilizer production from other sources (a comparability of the scenarios was performed considering that all of them could provide the same amount of phosphorus and nitrogen fertilizer).

### 11.4.2 How to approach the techno-economic analysis of anaerobic domestic wastewater treatment

The objective of wastewater treatment plants is to eliminate the pollutants present in the water in order to return the water to the environment, causing the minimum environmental impact, or to use it for other purposes in order to reduce the pressure on conventional water sources. To carry out wastewater treatment, a set of physicochemical treatments are necessary; the degree of treatment required for wastewater depends mainly on the effluent discharge limits. Most facilities have a conventional type of treatment, which consists of pre-treatment, primary and secondary treatment. In turn, various processes mentioned above involve the use of a wide variety of assets and can be grouped into civil construction, piping and electromechanical equipment.

As in any other industrial process, these infrastructures require a series of costs: energy, personnel, reagents, maintenance and so on, for the operation of facilities. In this sense, as with investment costs, these can vary depending on multiple variables that must be considered. Precisely, one of the first phases in any investment project is the feasibility analysis. The study makes possible to assess the profitability and financial sustainability of a project in the long term. This is particularly important when, as is the case in the wastewater treatment sector, there are different treatment technologies available. Another important aspect of feasibility analyses is that they enable decision makers to assess the environmental, social and legal implications of a project. These assessments can help prevent risks and ensure that the project complies with applicable legal and environmental regulations.

A wastewater treatment project is viable when, in addition to technically complying with the legal criteria regarding the quality of the influent, it is sustainable in the long term, which implies identifying a priori all the aspects associated with the process that may jeopardize the operation itself in short- and medium-term future. From the techno-economic point of view, some treatment technologies may require higher operating costs but with a lower investment or, on the contrary, technologies requiring higher economic investment that minimize operating costs. However, any economic analysis must take into account the unique characteristics associated with the site of the facility, local energy costs, specific quality and regulatory requirements, for example. In this sense, the economic analysis of the treatment process can include, in a second stage, the quantification of the benefits generated, or, in other words, the resources produced in the process. In the case of anaerobic treatment, in addition to generating water resources for irrigation, it also generates an alternative source of energy in the form of biogas that is converted into heat or electricity to save energy resources, thereby reducing GHG emissions. With respect to sludge generation, compared to other treatments, it is lower, which reduces the costs associated with sludge disposal. This stabilized sludge can be used as a fertilizer to enrich and improve soil characteristics. In summary, understanding the wastewater treatment infrastructure as a source of resources, which can be quantified, makes it possible to integrate the benefits generated into an overall decision balance (Figure 11.2).

In scenarios where there are different technical alternatives, cycle cost analysis can help to identify which option is economically more viable taking into account the entire investment period. Life-cycle cost, also known by its acronym LCC, is used to evaluate project costs. LCC is a method of economic analysis used to evaluate the total cost of ownership of a product or service over its entire life cycle. This method is based on life-cycle thinking and takes into account all costs associated with a product or service, from acquisition to disposal. The main objective of LCC is to provide a complete and accurate picture of the total costs associated with a product or service, to help companies and individuals make more informed procurement decisions. LCC calculations can be used on any piece of equipment or treatment system to determine the cost of procurement, operation, maintenance or disposal over its lifetime. There are currently a wide range of models used to calculate the total cost over the lifetime of a product. Regardless of the model used, they all share a common objective: to provide an accurate estimate of total pump system costs over time, expressed in today's currency value. In this sense, the result of an LCC can be used to compare different products or services and choose those with a lower total cost.

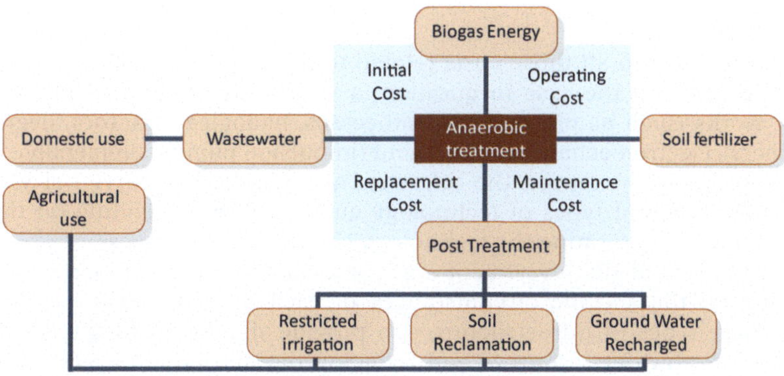


**Figure 11.2** Anaerobic wastewater treatment costs and resource generation.

Naturally, the weighting of certain factors in an LCC analysis will depend on local circumstances. For example, in countries with low-energy prices or for stations that run infrequently, energy costs may not be a major factor. Similarly, maintenance costs will not be a major factor in locations where labour is inexpensive. The advantage of an LCC analysis is that it lets the user focus on factors that matter most for a specific treatment system and situation. LCC can help in making informed decisions in the design and planning of the project. By evaluating the total costs associated with different design and planning options, the most cost-effective and sustainable option in the long term can be chosen. In addition to the great usefulness of LCC in the design phase of the project, it is also particularly useful throughout the operation period of the project. By evaluating the total costs associated with the project, areas where resource consumption such as water and energy can be reduced, and waste-disposal costs can be minimized, can be identified. In this way, improvements can be made that increase the sustainability of the project.

LCC is divided into several steps, including identifying all costs associated with a product or service, estimating the future cost and projecting the cost over the life cycle. In general, we can simplify the following four costs:

(1) Acquisition costs (CAPEX). These are all costs associated with the initial investment plus research and development costs, corresponding to the engineering and construction, testing, transfer and integrated logistical support activities incurred to incorporate an asset into an organization.

(2) Operating costs (OPEX). They correspond to the variable costs derived from the operation of the system in accordance with the planned degree of activity (which may be hours/year).

  (2.1) Maintenance costs. These reflect the consumption of resources derived from preventive and corrective maintenance, basic infrastructure, spare parts and associated consumption (not considered in the acquisition), repairs, modifications and/or modernizations, to ensure the availability of the system to fulfil its mission.

  (2.2) Repair costs. These costs relate to technical failures or equipment breakdowns. They may require the intervention of a technician or the replacement of damaged components.

To summarize, the investment and/or acquisition costs are called CAPEX (capital expenditures) and the operating and maintenance costs are called OPEX (operational expenditures). The costs associated with fixed assets (CAPEX) take into account the useful life for which the infrastructures have been designed. Thus, they represent the capacity of the facilities to generate a profit over time. The costs associated with maintenance and operation tasks (OPEX) ensure the correct functioning of the facilities, optimizing their use and avoiding wear and tear and deterioration.

#### *11.4.2.1 Capital expenditures*

Investment costs correspond to all those costs related to the acquisition of the necessary assets and the start-up of the project. In the case in question, a wastewater treatment plant mainly involves land, piping, civil works (such as physical unit processes, biological reactors, degritting units and settlers, among others), electromechanical equipment (impulsion pumps, submersible pumps, blowers, centrifuges, etc.) and piping. Moreover, the infrastructure required to carry out water reclamation processes may include different types of technology and can vary depending on the quality of the reclaimed water, which will determine the technology chosen for the process.

It is important to note that the investment costs associated with reclaimed water infrastructure can vary depending on the specific circumstances of each project. However, despite the initial investment costs, these types of infrastructure can provide significant long-term benefits, such as reducing demand for freshwater resources and reducing the discharge of wastewater into surface waters, leading to improved environmental and public health outcomes. The percentage distribution of the investment cost for wastewater treatment and reclaimed water process infrastructure can vary depending on a number of factors, such as the type and size of the facility, the technology used, the geographic location and local regulatory conditions.

In the case of an STP, construction costs can represent ~70–80% of the total investment cost. Within construction costs, the cost of treatment equipment (e.g. biological and chemical treatment technology) and construction costs (e.g. installation of piping, construction of settling tanks and digesters) are the two major components (EPA, 2016). The remaining costs include the cost of land acquisition, engineering, design, licensing and permitting costs and administration and supervision costs. In the case of reclaimed water processing infrastructure, the cost of treatment equipment can represent the majority of the total investment cost. Advanced water treatment technology, such as reverse osmosis and membrane filtration, can be expensive to install and operate. Other costs include the cost of constructing and maintaining pipelines, storage tanks, pumping stations and other equipment associated with the distribution of reclaimed water. There are also costs associated with the management and oversight of the reclaimed water programme, including quality monitoring and regulatory compliance.

Another aspect to be taken into account with regards to investment costs is the useful life of the assets that make up these infrastructures in order to establish their depreciation. It should be borne in mind that the expected life of the different assets (infrastructures, electromechanical equipment and piping) may be defined by operational variables such as operating hours, or fixed variables such as the age of the element. However, there are references that allow these maximum useful life parameters to be established approximately.

A period of 30 years is considered appropriate for civil works and the rest of the first establishment costs, as although it may be necessary to remodel some equipment beforehand, the civil works are perfectly usable for a new installation. Shorter periods are established for the rest of the equipment, and in many cases not because they will continue to operate with good performance at the end of this period, but because of technical obsolescence, given that technical progress may make it advisable to replace them with others that are more efficient or better adapted to the real needs of the installation. With these criteria, the most frequently adopted periods are:

- Mechanical equipment: 12 years
- Membranes: 8 years
- Electrical equipment: 15 years
- Instrumentation and control: 12 years
- Piping: 15 years

The investment required to build any STP depends very much on its size. The specific investment (investment required for each $m^3$/day of production) decreases as the size of the plant increases, which means that the scale factor plays an important role in the investment. In addition to the scale factor, other aspects such as the quality of the influent and effluent plays an important role, which will determine the type of treatment required. The investment costs of a wastewater treatment system have been studied by numerous authors, many of whom, through the development of cost functions, provide a model to estimate the investment costs for different types of treatment. These authors include the work of Friedler and Pisanty (2006) who establishes a cost function for secondary treatment and advanced secondary treatment with nitrification or Singhirunnusorn and Stenstrom (2010) who develops cost functions for four types of technologies: activated sludge, oxidation tank, aerated lagoons and oxidation ponds. On the contrary, Tsagarakis *et al.* (2003) focus on the analysis of the investment, operation and maintenance costs of different secondary treatments whose main difference is aeration. They conclude that activated sludge treatments can be the most economical above a certain size. On the contrary, Rodríguez-Miranda *et al.* (2015) estimates the investment costs of wastewater treatment by differentiating between investment costs associated with primary treatment and secondary treatment.

With respect to anaerobic wastewater treatment, one of the most used types of treatment is anaerobic sludge blanket reactors. Sludge blanket reactors are a type of anaerobic treatment in which wastewater is passed through a floating 'blanket' of suspended sludge particles. As the anaerobes in the sludge digest the organic components of the wastewater, they multiply and accumulate into larger granules that settle at the bottom of the reactor tank and can be recycled for future cycles. The treated effluent flows up and out of the unit. Biogas resulting from the degradation process is collected through collection hoods throughout the treatment cycle. With respect to investment costs, Tchobanoglus *et al.* (2003), based on a sample of infrastructures with similar characteristics, suggest the following cost function:

$$IC = Q_d \times 35{,}877 \times Z^{-0.43} \tag{11.1}$$

where $IC$ is the investment cost (€), $Q_d$ is the design flow rate and $Z$ is the equivalent inhabitants (design).

In this sense, it is important to point out the influence of economies of scale on investment costs, so that a larger infrastructure would imply a lower unit cost according to the design flow rate (Figure 11.3). These results coincide with studies by other authors relating investment and operating costs to economies of scale in the wastewater treatment sector (Hernández-Chover *et al.*, 2018).

#### 11.4.2.2 Operational expenditures

In the urban water cycle sector, a large number of equipment and infrastructures are necessary to carry out the process, in this sense, operation and maintenance (O&M) costs may exceed the initial investment costs. The initial investment in a project can be significant, but it is often only a fraction of the total costs that must be considered over the entire life cycle of the project. O&M costs can include expenses such as energy required to operate the project, equipment replacement and repair, labour costs and material costs. It is important to take these costs into account when assessing the feasibility and profitability of a project. An effective approach to minimize O&M costs is to carefully plan and design the project from the outset, with the goal of minimizing maintenance and operational requirements and maximizing energy efficiency. In addition, implementing preventive and predictive maintenance programmes can help reduce long-term maintenance costs by detecting and addressing issues before they become serious problems.

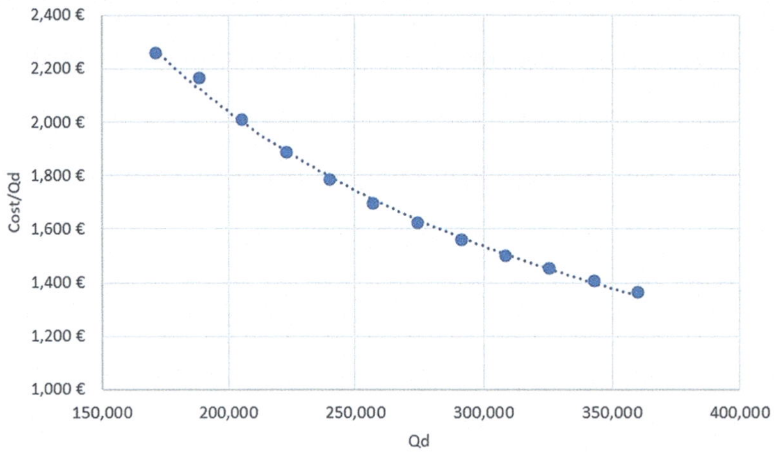


**Figure 11.3** Investment costs and design flow rate (€/m³) for an UASB reactor.

Operating and maintenance costs can be divided into two main groups:

(a) Fixed costs, independent of the treatment flow (€/day), such as:
- Operating and maintenance personnel costs
- Electromechanical maintenance costs
- Monitoring and control, health and safety and administrative costs

(b) Variable costs, depending on the flow treated (€/m$^3$). These are
- Electricity costs
- Costs of chemical reagents
- Replacement costs of membranes or other treatment elements
- Waste management

Any analysis that aims to evaluate the operating costs associated with the wastewater treatment process will include technical and economic aspects that may affect the process. With respect to the technical aspects, the technology used can influence in a higher or lower pollutant collection and consequently in the formation of costs. At the same time, there are other variables that must be evaluated because of the influence they can have on the variation of the economic costs of the process. In the previous section, we have observed the influence of economies of scale on the investment costs of anaerobic treatment, so that the larger the size of the infrastructure, according to the design capacity, the lower the costs. With respect to operating costs, Hernández-Chover *et al.* (2018) confirmed that they have a similar behaviour, that is, larger infrastructure dimensions will generate a lower unit cost in terms of treatment (also shown in Figure 11.4).

There are other aspects that can influence the operational costs of these infrastructures: Sala-Garrido *et al.* (2012) showed that WWTPs located in tourist areas are affected by seasonality and variability of pollutant loads, which implies that these infrastructures operate at full capacity during part of the year, generating higher operational costs and possible problems with effluent quality.

As in the previous section, an economic function is proposed to project the costs associated with the O&M of these infrastructures:

$$OC = 64{,}286 \times Z^{-0.445} \tag{11.2}$$

where $OC$ is the operational cost (€) and $Z$ is the equivalent inhabitants treated.

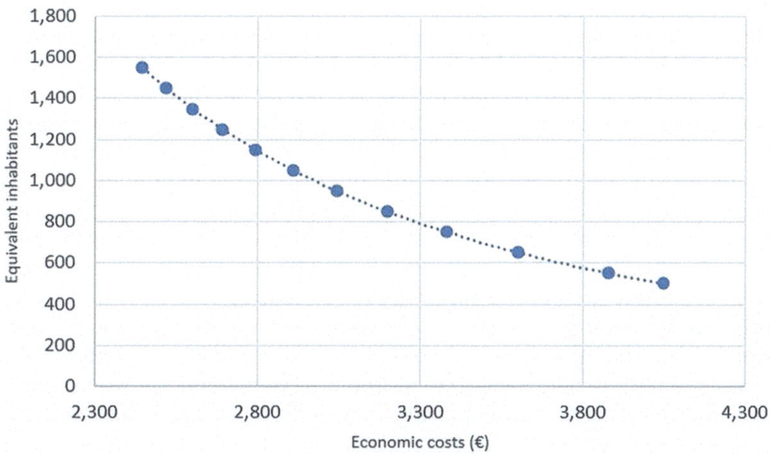


**Figure 11.4** Operational costs per equivalent inhabitant treated for an UASB reactor.

The results indicate that the operational costs of this type of technology are influenced by the presence of economies of scale, so that a greater number of treated inhabitants would reduce the relative costs of the treatment process. In addition, it should be taken into account that higher organic loads would generate greater amounts of resources, such as energy in the form of biogas, sludge that can be valorized and water.

In this section, we have synthesized the economic costs associated with this type of facility, suggesting economic functions that model the investment and operating costs of an anaerobic treatment plant. The LCC methodology divides the costs according to their nature, so that the proposed function would be as follows:

$$\mathrm{LCC} = C_{\mathrm{ic}} + C_{\mathrm{in}} + C_{\mathrm{e}} + C_{\mathrm{o}} + C_{\mathrm{m}} + C_{\mathrm{s}} + C_{\mathrm{env}} + C_{\mathrm{d}}$$

where $C_{\mathrm{ic}}$ = initial costs. Initial costs relate to the cost of purchasing systems, piping and all mechanical and electrical equipment as well the cost of engineering, testing and inspection, including any spare parts and training; $C_{\mathrm{in}}$ = installation and commissioning costs. These costs can include civil work, foundations, connection of piping, electrical wiring and instrumentation. They also cover the cost of setting and grouting of equipment on the foundations, provisions for flushing as well as performance evaluations at start-up. The installation and commissioning of monitoring and control equipment is also included in this item. Installation time can be minimized or eliminated by selecting a pre-programmed variable speed drive that requires a minimum of configuration settings; $C_{\mathrm{e}}$ = energy costs. These costs can include civil work, foundations, connection of piping, electrical wiring and instrumentation. They also cover the cost of setting and grouting of equipment on the foundations, provisions for flushing as well as performance evaluations at start-up. The installation and commissioning of monitoring and control equipment is also included in this item. Installation time can be minimized or eliminated by selecting a pre-programmed variable speed drive that requires a minimum of configuration settings; $C_{\mathrm{o}}$ = operational costs. Operational costs cover the labour costs for normal operation of the pumping system. This includes, for example, normal wear and tear, system supervision and keeping the station clean. Operational costs do not include costs attributable to energy or maintenance of the treatment water system. An LCC analysis can be a good tool to see how fast the investment in a new supervision system will pay back;

$C_{\mathrm{m}}$ = maintenance and repair costs. Such costs relate directly to the total number of hours spent on maintenance and the cost of spare parts, including planned and unplanned maintenance; $C_{\mathrm{s}}$ = downtime costs. This category relates mainly to unexpected downtime but may also be due to a loss of production or even loss of trust from a customer. Downtime costs can be minimized by using maintenance contracts that ensure regular service to maximize uptime and shorten response time in the event of emergencies. Monitoring and control solutions can also create early warnings that help to prevent downtime; $C_{\mathrm{env}}$ = environmental costs. These include costs for dealing with spills, environmental inspections and contaminant disposal during the lifetime of the water treatment system. Such costs are often set by local regulatory authorities and vary from country to country. The disposal of used parts and materials is also included. For information about what materials are used and their effects on the environment, see the Environmental Product Declaration; $C_{\mathrm{d}}$ = decommissioning costs. Decommissioning costs usually include the disposal of the water treatment system and auxiliary services as well as restoration of the local environment. The decommissioning costs seldom vary for similar solutions and are often excluded from an LCC calculation.

An LCC analysis can be used to determine the total cost for the system over its lifetime. When conducting a complete analysis, it is necessary to gather and enter data for all eight categories in the formula. An LCC analysis can also be used to examine how beneficial an investment can be, meaning that only factors that are of relevance for the analysis need to be included. Making two analyses – one with the investment and one without – and comparing the results will show the payback time for the investment. When comparing different systems, the relevant data should be entered for the same categories. The comparison of technological alternatives should include, in addition to the technical aspects, the resources that each technology is capable of generating. The valorization of the resources generated can influence the reduction of the economic costs associated with both investment and operation. Thus, technologies with higher investment or operating requirements may be more viable options due to the resources they can generate over other alternatives.

## ACKNOWLEDGEMENTS

The authors express their gratitude to the HP-NANOBIO project (PID2019-111163RB-I00) granted by the Spanish Ministry of Science and Innovation, to the SPOTLIGHT project (PDC2021-121240-I00) funded by MCIN/AEI/10.13039/501100011033 and European Union NextGenerationEU/PRTR and the BIORECER project (No. 101060684) funded by HORIZON-CL6-2021-ZEROPOLLUTION-01. S.E. also acknowledges the Spanish Ministry of Science, Innovation and Universities for financial support (Grant reference PRE2020-092074). S.E., M.T.M. and G.F. belong to Cross-disciplinary Research in Environmental Technologies (CRETUS Research Center, Spain, ED431E 2018/01) and the Galician Competitive Research Group (GRC ED431C 2021/27).

## REFERENCES

Adghim M., Abdallah M., Saad S., Shanableh A., Sartaj M. and El Mansouri A. E. (2020). Comparative life cycle assessment of anaerobic co-digestion for dairy waste management in large-scale farms. *Journal of Cleaner Production*, **256**, 120320, https://doi.org/10.1016/J.JCLEPRO.2020.120320

Alfonso-Cardero A., Pagés-Díaz J., Contino F., Rajendran K. and Lorenzo-LLanes J. (2021). Process simulation and techno-economic assessment of vinasse-to-biogas in Cuba: deterministic and uncertainty analysis. *Chemical Engineering Research and Design*, **169**, 33–45, https://doi.org/10.1016/J.CHERD.2021.02.031

Alvarez-Gaitan J. P., Short M. D., Lundie S. and Stuetz R. (2016). Towards a comprehensive greenhouse gas emissions inventory for biosolids. *Water Research*, **96**, 299–307, https://doi.org/10.1016/J.WATRES.2016.03.059

Andreasi Bassi S., Boldrin A., Frenna G. and Astrup T. F. (2021). An environmental and economic assessment of bioplastic from urban biowaste. The example of polyhydroxyalkanoate. *Bioresource Technology*, **327**, 124813, https://doi.org/10.1016/J.BIORTECH.2021.124813

Arias A., Behera C. R., Feijoo G., Sin G. and Moreira M. T. (2020a). Unravelling the environmental and economic impacts of innovative technologies for the enhancement of biogas production and sludge management in wastewater systems. *Journal of Environmental Management*, **270**, 110965, https://doi.org/10.1016/J.JENVMAN.2020.110965

Arias A., Feijoo G. and Moreira M. T. (2020b). What is the best scale for implementing anaerobic digestion according to environmental and economic indicators? *Journal of Water Process Engineering*, **35**, 101235, https://doi.org/10.1016/J.JWPE.2020.101235

Arias A., Feijoo G. and Moreira M. T. (2021). Benchmarking environmental and economic indicators of sludge management alternatives aimed at enhanced energy efficiency and nutrient recovery. *Journal of Environmental Management*, **279**, 111594, https://doi.org/10.1016/J.JENVMAN.2020.111594

Asunis F., De Gioannis G., Francini G., Lombardi L., Muntoni A., Polettini A., Pomi R., Rossi A. and Spiga D. (2021). Environmental life cycle assessment of polyhydroxyalkanoates production from cheese whey. *Waste Management*, **132**, 31–43, https://doi.org/10.1016/J.WASMAN.2021.07.010

Awad H., Gar Alalm M. and El-Etriby H. K. (2019). Environmental and cost life cycle assessment of different alternatives for improvement of wastewater treatment plants in developing countries. *Science of the Total Environment*, **660**, 57–68, https://doi.org/10.1016/J.SCITOTENV.2018.12.386

Bahreini G., Elbeshbishy E., Jimenez J., Santoro D. and Nakhla G. (2020). Integrated fermentation and anaerobic digestion of primary sludges for simultaneous resource and energy recovery: impact of volatile fatty acids recovery. *Waste Management*, **118**, 341–349, https://doi.org/10.1016/J.WASMAN.2020.08.051

Becker A. M., Yu K., Stadler L. B. and Smith A. L. (2017). Co-management of domestic wastewater and food waste: a life cycle comparison of alternative food waste diversion strategies. *Bioresource Technology*, **223**, 131–140, https://doi.org/10.1016/j.biortech.2016.10.031

Blanco D., Collado S., Laca A. and Díaz M. (2016). Life cycle assessment of introducing an anaerobic digester in a municipal wastewater treatment plant in Spain. *Water Science and Technology*, **73**(4), 835–842, https://doi.org/10.2166/wst.2015.545

Boldrin M. T. N., Formiga K. T. M. and Pacca S. A. (2022). Environmental performance of an integrated water supply and wastewater system through life cycle assessment – a Brazilian case study. *Science of the Total Environment*, **835**, 155213, https://doi.org/10.1016/J.SCITOTENV.2022.155213

Brockmann D., Gérand Y., Park C., Milferstedt K., Hélias A. and Hamelin J. (2021). Wastewater treatment using oxygenic photogranule-based process has lower environmental impact than conventional activated sludge process. *Bioresource Technology*, **319**, 124204, https://doi.org/10.1016/J.BIORTECH.2020.124204

Cakir F. Y. and Stenstrom M. K. (2005). Greenhouse gas production: a comparison between aerobic and anaerobic wastewater treatment technology. *Water Research*, **39**(17), 4197–4203, https://doi.org/10.1016/J.WATRES.2005.07.042

Cañote S. J. B., Barros R. M., Lora E. E. S., dos Santos I. F. S., Silva A. P. M., Piñas J. A. V., Cañote A. L. B., Silva d. C. e. and L H. (2021). Life cycle assessment of upflow anaerobic sludge blanket sludge management and activated sludge systems aiming energy use in the municipality of Itajubá, Minas Gerais, Brazil. *Journal of Material Cycles and Waste Management*, **23**(5), 1810–1830, https://doi.org/10.1007/s10163-021-01253-0

Cao Y. and Pawłowski A. (2013). Life cycle assessment of two emerging sewage sludge-to-energy systems: evaluating energy and greenhouse gas emissions implications. *Bioresource Technology*, **127**, 81–91, https://doi.org/10.1016/J.BIORTECH.2012.09.135

Cartes J., Neumann P., Hospido A. and Vidal G. (2018). Life cycle assessment of management alternatives for sludge from sewage treatment plants in Chile: does advanced anaerobic digestion improve environmental performance compared to current practices? *Journal of Material Cycles and Waste Management*, **20**(3), 1530–1540, https://doi.org/10.1007/s10163-018-0714-9

Chemical Abstract Service (2023). CAS Registry. https://www.cas.org/

Chen R., Yuan S., Chen S., Ci H., Dai X., Wang X., Li C., Wang D. and Dong B. (2022). Life-cycle assessment of two sewage sludge-to-energy systems based on different sewage sludge characteristics: energy balance and greenhouse gas-emission footprint analysis. *Journal of Environmental Sciences*, **111**, 380–391, https://doi.org/10.1016/J.JES.2021.04.012

Colzi Lopes A., Valente A., Iribarren D. and González-Fernández C. (2018). Energy balance and life cycle assessment of a microalgae-based wastewater treatment plant: a focus on alternative biogas uses. *Bioresource Technology*, **270**, 138–146, https://doi.org/10.1016/J.BIORTECH.2018.09.005

Corominas L., Byrne D. M., Guest J. S., Hospido A., Roux P., Shaw A. and Short M. D. (2020). The application of life cycle assessment (LCA) to wastewater treatment: a best practice guide and critical review. *Water Research*, **184**, 116058, https://doi.org/10.1016/J.WATRES.2020.116058

Cuéllar-Franca R., García-Gutiérrez P., Dimitriou I., Elder R. H., Allen R. W. K. and Azapagic A. (2019). Utilising carbon dioxide for transport fuels: the economic and environmental sustainability of different Fischer–Tropsch process designs. *Applied Energy*, **253**, 113560, https://doi.org/10.1016/J.APENERGY.2019.113560

de Sampaio Lopes T. A., Matos Queiroz L. and Kiperstok A. (2014). Environmental performance of full-scale a wastewater treatment plant applying life cycle assessment. *Revista Ambiente e Agua*, **9**(3), 445–, https://doi.org/10.4136/1980-993X

Díaz I., Díaz-Curbelo A., Ignacio Matute K., Fdz-Polanco M. and Pérez-Elvira S. I. (2021). Influence of the operating conditions of the intermediate thermal hydrolysis on the energetic efficiency of the sludge treatment process. *Bioresource Technology*, **333**, 125114, https://doi.org/10.1016/J.BIORTECH.2021.125114

Dussan K. and Monaghan R. F. D. (2018). Integrated thermal conversion and anaerobic digestion for sludge management in wastewater treatment plants. *Waste and Biomass Valorization*, **9**(1), 65–85, https://doi.org/10.1007/s12649-016-9812-x

Ebrahimi M., Ramirez J. A., Outram J. G., Dunn K., Jensen P. D., O'Hara I. M. and Zhang Z. (2023). Effects of lignocellulosic biomass type on the economics of hydrothermal treatment of digested sludge for solid fuel and soil amendment applications. *Waste Management*, **156**, 55–65, https://doi.org/10.1016/J.WASMAN.2022.11.020

Edwards J., Othman M., Crossin E. and Burn S. (2017). Anaerobic co-digestion of municipal food waste and sewage sludge: a comparative life cycle assessment in the context of a waste service provision. *Bioresource Technology*, **223**, 237–249, https://doi.org/10.1016/J.BIORTECH.2016.10.044

Elginoz N., Atasoy M., Finnveden G. and Cetecioglu Z. (2020). Ex-ante life cycle assessment of volatile fatty acid production from dairy wastewater. *Journal of Cleaner Production*, **269**, 122267, https://doi.org/10.1016/J.JCLEPRO.2020.122267

El Ibrahimi M., Khay I., El Maakoul A. and Bakhouya M. (2021). Techno-economic and environmental assessment of anaerobic co-digestion plants under different energy scenarios: a case study in Morocco. *Energy Conversion and Management*, **245**, 114553, https://doi.org/10.1016/J.ENCONMAN.2021.114553

El-Qanni A., Alsayed M., Alsurakji I. H., Najjar M., Odeh D., Najjar S., Hmoudah M., Zubair M., Russo V. and Di Serio M. (2022). A technoeconomic assessment of biological sludge dewatering using a thermal rotary dryer: a case study of design applicability, economics, and managerial feasibility. Biomass Conversion and Biorefinery, https://doi.org/10.1007/s13399-022-03480-3

EPA (2016). Clean Watersheds Needs Survey 2012. Report to Congress. https://www.epa.gov/sites/default/files/2015-12/documents/cwns_2012_report_to_congress-508-opt.pdf

Estévez S., Feijoo G. and Moreira M. T. (2022a). Environmental synergies in decentralized wastewater treatment at a hotel resort. *Journal of Environmental Management*, **317**, 115392, https://doi.org/10.1016/J.JENVMAN.2022.115392

Estévez S., González-García S., Feijoo G. and Moreira M. T. (2022b). How decentralized treatment can contribute to the symbiosis between environmental protection and resource recovery. *Science of the Total Environment*, **812**, 151485, https://doi.org/10.1016/J.SCITOTENV.2021.151485

European Commission, Joint Research Centre and Institute for Environment and Sustainability (2011). International Reference Life Cycle Data System (ILCD) Handbook – Recommendations for Life Cycle Impact Assessment in the European context. First edition. EUR 24571 EN. Luxemburg: Publication Office of the European Union. ISBN: 978-92-79-17451-3.

European Environment Agency (2022). Beyond water quality sewage treatment in a circular economy.

Fernando-Foncillas C. and Varrone C. (2021). Potential of the sewage sludge valorization in Scandinavia by co-digestion with other organic wastes: a techno-economic assessment. *Journal of Cleaner Production*, **324**, 129239, https://doi.org/10.1016/J.JCLEPRO.2021.129239

Francini G., Lombardi L., Freire F., Pecorini I. and Marques P. (2019). Environmental and cost life cycle analysis of different recovery processes of organic fraction of municipal solid waste and sewage sludge. *Waste and Biomass Valorization*, **10**(12), 3613–3634, https://doi.org/10.1007/s12649-019-00687-w

Friedler E. and Pisanty E. (2006). Effects of design flow and treatment level on construction and operation costs of municipal wastewater treatment plants and their implications on policy making. *Water Research*, **40**(20), 3751–3758, https://doi.org/10.1016/j.watres.2006.08.015

Fuess L. T. and Zaiat M. (2018). Economics of anaerobic digestion for processing sugarcane vinasse: applying sensitivity analysis to increase process profitability in diversified biogas applications. *Process Safety and Environmental Protection*, **115**, 27–37, https://doi.org/10.1016/J.PSEP.2017.08.007

García-Gutiérrez P., Jacquemin J., McCrellis C., Dimitriou I., Taylor S. F. R., Hardacre C. and Allen R. W. K. (2016). Techno-economic feasibility of selective $CO_2$ capture processes from biogas streams using ionic liquids as physical absorbents. *Energy and Fuels*, **30**(6), 5052–5064, https://doi.org/10.1021/acs.energyfuels.6b00364

Garrido-Baserba M., Molinos-Senante M., Abelleira-Pereira J. M., Fdez-Güelfo L. A., Poch M. and Hernández-Sancho F. (2015). Selecting sewage sludge treatment alternatives in modern wastewater treatment plants using environmental decision support systems. *Journal of Cleaner Production*, **107**, 410–419, https://doi.org/10.1016/J.JCLEPRO.2014.11.021

Gholamian E., Mehr A. S., Yari M. and Carton J. G. (2023). Dynamic simulation and techno-economic assessment of hydrogen utilization in dual fuel (hydrogen/biogas) micro gas turbine systems for a wastewater treatment plant. *Process Safety and Environmental Protection*, **169**, 220–237, https://doi.org/10.1016/J.PSEP.2022.10.045

Gianico A., Bertanza G., Braguglia C. M., Canato M., Laera G., Heimersson S., Svanström M. and Mininni G. (2015). Upgrading a wastewater treatment plant with thermophilic digestion of thermally pre-treated secondary sludge: techno-economic and environmental assessment. *Journal of Cleaner Production*, **102**, 353–361, https://doi.org/10.1016/J.JCLEPRO.2015.04.051

Gourdet C., Girault R., Berthault S., Richard M., Tosoni J. and Pradel M. (2017). In quest of environmental hotspots of sewage sludge treatment combining anaerobic digestion and mechanical dewatering: a life cycle assessment approach. *Journal of Cleaner Production*, **143**, 1123–1136, https://doi.org/10.1016/J.JCLEPRO.2016.12.007

Harclerode M., Doody A., Brower A., Vila P., Ho J. and Evans P. J. (2020). Life cycle assessment and economic analysis of anaerobic membrane bioreactor whole-plant configurations for resource recovery from domestic wastewater. *Journal of Environmental Management*, **269**, 110720, https://doi.org/10.1016/J.JENVMAN.2020.110720

He C., Fang K., Wang W., Wang Q., Luo J., Ma J., Xue X., Gao F., Sun K., Liu M. and Wang K. (2022). Techno-economic feasibility of 'membrane-based pre-concentration + post-treatment' systems for municipal wastewater treatment and resource recovery. *Journal of Cleaner Production*, **375**, 134113, https://doi.org/10.1016/J.JCLEPRO.2022.134113

Heimersson S., Harder R., Peters G. M. and Svanström M. (2014). Including pathogen risk in life cycle assessment of wastewater management. 2. Quantitative comparison of pathogen risk to other impacts on human health. *Environmental Science and Technology*, **48**(16), 9446–9453, https://doi.org/10.1021/es501481m

Heimersson S., Svanström M., Cederberg C. and Peters G. (2017). Improved life cycle modelling of benefits from sewage sludge anaerobic digestion and land application. *Resources, Conservation and Recycling*, **122**, 126–134, https://doi.org/10.1016/J.RESCONREC.2017.01.016

Helmecke M., Fries E. and Schulte C. (2020). Regulating water reuse for agricultural irrigation: risks related to organic micro-contaminants. *Environmental Sciences Europe*, **32**(1), 4, https://doi.org/10.1186/s12302-019-0283-0

Hernández-Chover V., Bellver-Domingo Á and Hernández-Sancho F. (2018). Efficiency of wastewater treatment facilities: the influence of scale economies. *Journal of Environmental Management*, **228**, 77–84, https://doi.org/10.1016/J.JENVMAN.2018.09.014

IPCC (2021). Climate Change 2021: The Physical Science Basis. Contribution of Working Group I to the Sixth Assessment Report of the Intergovernmental Panel on Climate Change.

ISO (2006a). EN ISO 14040:2006. Environmental management – Life cycle assessment – Principles and framework. http://www.cscses.com/uploads/2016328/20160328110518251825.pdf

ISO (2006b). EN ISO 14044:2006. Environmental management – Life cycle assessment – Requirements and guidelines. https://www.iso.org/standard/38498.html

Kong Z., Li L., Wu J., Wang T., Rong C., Luo Z., Pan Y., Li D., Li Y., Huang Y. and Li Y. Y. (2021). Evaluation of bio-energy recovery from the anaerobic treatment of municipal wastewater by a pilot-scale submerged anaerobic membrane bioreactor (AnMBR) at ambient temperature. *Bioresource Technology*, **339**, 125551, https://doi.org/10.1016/J.BIORTECH.2021.125551

Laitinen J., Moliis K. and Surakka M. (2017). Resource efficient wastewater treatment in a developing area – climate change impacts and economic feasibility. *Ecological Engineering*, **103**, 217–225, https://doi.org/10.1016/J.ECOLENG.2017.04.017

Lanko I., Flores L., Garfí M., Todt V., Posada J. A., Jenicek P. and Ferrer I. (2020). Life cycle assessment of the mesophilic, thermophilic, and temperature-phased anaerobic digestion of sewage sludge. *Water (Switzerland)*, **12**(11), 1–20, https://doi.org/10.3390/w12113140

Laurent A., Weidema B. P., Bare J., Liao X., Maia de Souza D., Pizzol M., Sala S., Schreiber H., Thonemann N. and Verones F. (2020). Methodological review and detailed guidance for the life cycle interpretation phase. *Journal of Industrial Ecology*, **24**(5), 986–1003, https://doi.org/10.1111/jiec.13012

Lehtoranta S., Malila R., Särkilahti M. and Viskari E. L. (2022). To separate or not? A comparison of wastewater management systems for the new city district of Hiedanranta, Finland. *Environmental Research*, **208**, 112764, https://doi.org/10.1016/J.ENVRES.2022.112764

Li H. and Feng K. (2018). Life cycle assessment of the environmental impacts and energy efficiency of an integration of sludge anaerobic digestion and pyrolysis. *Journal of Cleaner Production*, **195**, 476–485, https://doi.org/10.1016/J.JCLEPRO.2018.05.259

Li T., Zhang H., Liu Z., Ke Q. and Alting L. (2014). A system boundary identification method for life cycle assessment. *International Journal of Life Cycle Assessment*, **19**(3), 646–660, https://doi.org/10.1007/s11367-013-0654-5

Li H., Jin C. and Mundree S. (2017a). Hybrid environmental and economic assessment of four approaches recovering energy from sludge with variant organic contents. *Journal of Cleaner Production*, **153**, 131–138, https://doi.org/10.1016/J.JCLEPRO.2017.03.167

Li H., Jin C., Zhang Z., O'Hara I. and Mundree S. (2017b). Environmental and economic life cycle assessment of energy recovery from sewage sludge through different anaerobic digestion pathways. *Energy*, **126**, 649–657, https://doi.org/10.1016/J.ENERGY.2017.03.068

Li L., Wang X., Miao J., Abulimiti A., Jing X. and Ren N. (2022). Carbon neutrality of wastewater treatment – a systematic concept beyond the plant boundary. *Environmental Science and Ecotechnology*, **11**, 100180, https://doi.org/10.1016/J.ESE.2022.100180

Lijó L., Malamis S., González-García S., Moreira M. T., Fatone F. and Katsou E. (2017). Decentralised schemes for integrated management of wastewater and domestic organic waste: the case of a small community. *Journal of Environmental Management*, **203**, 732–740, https://doi.org/10.1016/J.JENVMAN.2016.11.053

Medina-Martos E., Istrate I. R., Villamil J. A., Gálvez-Martos J. L., Dufour J. and Mohedano Á.F. (2020). Techno-economic and life cycle assessment of an integrated hydrothermal carbonization system for sewage sludge. *Journal of Cleaner Production*, **277**, 122930, https://doi.org/10.1016/J.JCLEPRO.2020.122930

Mills N., Pearce P., Farrow J., Thorpe R. B. and Kirkby N. F. (2014). Environmental & economic life cycle assessment of current & future sewage sludge to energy technologies. *Waste Management*, **34**(1), 185–195, https://doi.org/10.1016/J.WASMAN.2013.08.024

Morelli B., Cashman S., Cissy Ma X., Garland J., Turgeon J., Fillmore L., Bless D. and Nye M. (2018). Effect of nutrient removal and resource recovery on life cycle cost and environmental impacts of a small-scale water resource recovery facility. *Sustainability (Switzerland)*, **10**(10), 3546, https://doi.org/10.3390/su10103546

Morelli B., Cashman S., Ma X., Turgeon J., Arden S. and Garland J. (2020). Environmental and cost benefits of co-digesting food waste at wastewater treatment facilities. *Water Science and Technology*, **82**(2), 227–241, https://doi.org/10.2166/wst.2020.104

Moretti C., Corona B., Edwards R., Junginger M., Moro A., Rocco M. and Shen L. (2020). Reviewing ISO compliant multifunctionality practices in environmental life cycle modeling. *Energies*, **13**(14), 3579. https://doi.org/10.3390/en13143579

Patel K. and Singh S. K. (2022). A life cycle approach to environmental assessment of wastewater and sludge treatment processes. *Water and Environmental Journal*, **36**(3), 412–424, https://doi.org/10.1111/wej.12774

Piao W., Kim Y., Kim H., Kim M. and Kim C. (2016). Life cycle assessment and economic efficiency analysis of integrated management of wastewater treatment plants. *Journal of Cleaner Production*, **113**, 325–337, https://doi.org/10.1016/J.JCLEPRO.2015.11.012

Postacchini M., Di Giuseppe E., Eusebi A. L., Pelagalli L., Darvini G., Cipolletta G. and Fatone F. (2022). Energy saving from small-sized urban contexts: integrated application into the domestic water cycle. *Renewable Energy*, **199**, 1300–1317, https://doi.org/10.1016/J.RENENE.2022.09.063

Prado L. O., Souza H. H. S., Chiquito G. M., Paulo P. L. and Boncz M. A. (2020). A comparison of different scenarios for on-site reuse of blackwater and kitchen waste using the life cycle assessment methodology. *Environmental Impact Assessment Review*, **82**, 106362, https://doi.org/10.1016/J.EIAR.2019.106362

Pretel R., Robles A., Ruano M. V., Seco A. and Ferrer J. (2013). Environmental impact of submerged anaerobic MBR (SAnMBR) technology used to treat urban wastewater at different temperatures. *Bioresource Technology*, **149**, 532–540, https://doi.org/10.1016/J.BIORTECH.2013.09.060

Pretel R., Robles A., Ruano M. V., Seco A. and Ferrer J. (2016). Economic and environmental sustainability of submerged anaerobic MBR-based (AnMBR-based) technology as compared to aerobic-based technologies for moderate-/high-loaded urban wastewater treatment. *Journal of Environmental Management*, **166**, 45–54, https://doi.org/10.1016/J.JENVMAN.2015.10.004

Purwanta Bayu A. I., Mellyanawaty M., Budiman A. and Budhijanto W. (2022). Techno-economic analysis of reactor types and biogas utilization schemes in thermophilic anaerobic digestion of sugarcane vinasse. *Renewable Energy*, **201**, 864–875, https://doi.org/10.1016/J.RENENE.2022.10.087

Ranganathan P. and Savithri S. (2019). Techno-economic analysis of microalgae-based liquid fuels production from wastewater via hydrothermal liquefaction and hydroprocessing. *Bioresource Technology*, **284**, 256–265, https://doi.org/10.1016/J.BIORTECH.2019.03.087

Remy C., Lesjean B. and Waschnewski J. (2013). Identifying energy and carbon footprint optimization potentials of a sludge treatment line with life cycle assessment. *Water Science and Technology*, **67**(1), 63–73, https://doi.org/10.2166/wst.2012.529

Righi S., Oliviero L., Pedrini M., Buscaroli A. and Della Casa C. (2013). Life cycle assessment of management systems for sewage sludge and food waste: centralized and decentralized approaches. *Journal of Cleaner Production*, **44**, 8–17, https://doi.org/10.1016/J.JCLEPRO.2012.12.004

Rodriguez-Garcia G., Frison N., Vázquez-Padín J. R., Hospido A., Garrido J. M., Fatone F., Bolzonella D., Moreira M. T. and Feijoo G. (2014). Life cycle assessment of nutrient removal technologies for the treatment of anaerobic digestion supernatant and its integration in a wastewater treatment plant. *Science of the Total Environment*, **490**, 871–879, https://doi.org/10.1016/J.SCITOTENV.2014.05.077

Rodríguez-Miranda J. P., García-Ubaque C. A. and Penagos-Londoño J. C. (2015). Analysis of the investment costs in municipal wastewater treatment plants in Cundinamarca. *Dyna*, **82**(192), 230–238, http://dx.doi.org/10.15446/dyna.v82n192.44699

Roldán M., Bouzas A., Seco A., Mena E., Mayor and Barat R. (2020). An integral approach to sludge handling in a WWTP operated for EBPR aiming phosphorus recovery: simulation of alternatives, LCA and LCC analyses. *Water Research*, **175**, 115647, https://doi.org/10.1016/J.WATRES.2020.115647

Rus E., Mills N., Shana A., Perrault A., Fountain P., Thorpe R. B., Ouki S. and Nilsen P. J. (2017). The intermediate thermal hydrolysis process: results from pilot testing and techno-economic assessment. *Water Practice and Technology*, **12**(2), 406–422, https://doi.org/10.2166/wpt.2017.031

Sala-Garrido R., Molinos-Senante M. and Hernández-Sancho F. (2012). How does seasonality affect water reuse possibilities? An efficiency and cost analysis. *Resources, Conservation and Recycling*, **58**, 125–131, https://doi.org/10.1016/J.RESCONREC.2011.11.002

Sanchez L., Vinardell S., Charreton J., Heran M. and Lesage G. (2023). Assessing the impact of granular anaerobic membrane bioreactor intensification on treatment performance, membrane fouling and economic balance. *Journal of Environmental Chemical Engineering*, **11**(2), 109369, https://doi.org/10.1016/J.JECE.2023.109369

Satayavibul A. and Ratanatamskul C. (2021). Life cycle assessment of a novel zero organic-waste model using the integrated anaerobic digester and oxidation-ditch membrane bioreactor for high-rise building application. *Waste and Biomass Valorization*, **12**(10), 5425–5436, https://doi.org/10.1007/s12649-021-01418-w

Shoener B. D., Zhong C., Greiner A. D., Khunjar W. O., Hong P. Y. and Guest J. S. (2016). Design of anaerobic membrane bioreactors for the valorization of dilute organic carbon waste streams. *Energy and Environmental Science*, **9**(3), 1102–1112, https://doi.org/10.1039/c5ee03715h

Sills D. L., Wade V. L. and DiStefano T. D. (2016). Comparative life cycle and technoeconomic assessment for energy recovery from dilute wastewater. *Environmental Engineering Science*, **33**(11), 861–872, https://doi.org/10.1089/ees.2016.0153

Singhirunnusorn W. and Stenstrom M. K. (2010). A critical analysis of economic factors for diverse wastewater treatment processes. *Journal of Environmental Engineering and Management*, **20**(4), 263–268, https://www.researchgate.net/publication/267382358

Singh E., Mishra R., Kumar A., Shukla S. K., Lo S. L. and Kumar S. (2022). Circular economy-based environmental management using biochar: driving towards sustainability. *Process Safety and Environmental Protection*, **163**, 585–600, https://doi.org/10.1016/J.PSEP.2022.05.056

Smith A. L., Stadler L. B., Cao L., Love N. G., Raskin L. and Skerlos S. J. (2014). Navigating wastewater energy recovery strategies: a life cycle comparison of anaerobic membrane bioreactor and conventional treatment systems with anaerobic digestion. *Environmental Science and Technology*, **48**(10), 5972–5981, https://doi.org/10.1021/es5006169

Stazi V. and Tomei M. C. (2018). Enhancing anaerobic treatment of domestic wastewater: state of the art, innovative technologies and future perspectives. *Science of the Total Environment*, **635**, 78–91. https://doi.org/10.1016/j.scitotenv.2018.04.071

Stazi V., Annesini M. C. and Tomei M. C. (2022). Anaerobic domestic wastewater treatment in a sequencing granular UASB bioreactor: feasibility study of the temperature effect on the process performance. *Journal of Environmental Chemical Engineering*, **10**(5), 108512, https://doi.org/10.1016/J.JECE.2022.108512

Tarpani R. R. Z. and Azapagic A. (2018). Life cycle costs of advanced treatment techniques for wastewater reuse and resource recovery from sewage sludge. *Journal of Cleaner Production*, **204**, 832–847, https://doi.org/10.1016/J.JCLEPRO.2018.08.300

Tchobanoglus G., Burton F. and Stensel H. D. (2003). Wastewater engineering: treatment and reuse. *Water Works Association Journal*, **95**(5), 201.

Thomsen M., Romeo D., Caro D., Seghetta M. and Cong R. G. (2018). Environmental-economic analysis of integrated organic waste and wastewater management systems: a case study from Aarhus City (Denmark). *Sustainability (Switzerland)*, **10**(10), 3742, https://doi.org/10.3390/su10103742

Tomei M. C., Bertanza G., Canato M., Heimersson S., Laera G. and Svanström M. (2016a). Techno-economic and environmental assessment of upgrading alternatives for sludge stabilization in municipal wastewater treatment plants. *Journal of Cleaner Production*, **112**, 3106–3115, https://doi.org/10.1016/J.JCLEPRO.2015.10.017

Tomei M. C., Mosca Angelucci D. and Levantesi C. (2016b). Two-stage anaerobic and post-aerobic mesophilic digestion of sewage sludge: analysis of process performance and hygienization potential. *Science of the Total Environment*, **545–546**, 453–464, https://doi.org/10.1016/J.SCITOTENV.2015.12.053

Tsagarakis K. P., Mara D. D. and Angelakis A. N. (2003). Application of cost criteria for selection of municipal wastewater treatment systems.

Tua C., Ficara E., Mezzanotte V. and Rigamonti L. (2021). Integration of a side-stream microalgae process into a municipal wastewater treatment plant: a life cycle analysis. *Journal of Environmental Management*, **279**, 111605, https://doi.org/10.1016/J.JENVMAN.2020.111605

Vinardell S., Astals S., Mata-Alvarez J. and Dosta J. (2020). Techno-economic analysis of combining forward osmosis-reverse osmosis and anaerobic membrane bioreactor technologies for municipal wastewater treatment and water production. *Bioresource Technology*, **297**, 122395, https://doi.org/10.1016/J.BIORTECH.2019.122395

Vinardell S., Dosta J., Mata-Alvarez J. and Astals S. (2021). Unravelling the economics behind mainstream anaerobic membrane bioreactor application under different plant layouts. *Bioresource Technology*, **319**, 124170, https://doi.org/10.1016/J.BIORTECH.2020.124170

Vinardell S., Blandin G., Ferrari F., Lesage G., Mata-Alvarez J., Dosta J. and Astals S. (2022). Techno-economic analysis of forward osmosis pre-concentration before an anaerobic membrane bioreactor: impact of draw solute and membrane material. *Journal of Cleaner Production*, **356**, 131776, https://doi.org/10.1016/J.JCLEPRO.2022.131776

Vogli L., Macrelli S., Marazza D., Galletti P., Torri C., Samorì C. and Righi S. (2020). Life cycle assessment and energy balance of a novel polyhydroxyalkanoates production process with mixed microbial cultures fed on pyrolytic products of wastewater treatment sludge. *Energies*, **13**(11), 2706, https://doi.org/10.3390/en13112706

Volpi M. P. C., Fuess L. T. and Moraes B. S. (2023). Economic performance of biogas production and use from residues co-digestion in integrated 1G2G sugarcane biorefineries: better electricity or biomethane? *Energy Conversion and Management*, **277**, 116673, https://doi.org/10.1016/J.ENCONMAN.2023.116673

Wang S., Yu S., Lu Q., Liao Y., Li H., Sun L., Wang H. and Zhang Y. (2020). Development of an alkaline/acid pre-treatment and anaerobic digestion (APAD) process for methane generation from waste activated sludge. *Science of the Total Environment*, **708**, 134564, https://doi.org/10.1016/J.SCITOTENV.2019.134564

Xu C., Chen W. and Hong J. (2014). Life-cycle environmental and economic assessment of sewage sludge treatment in China. *Journal of Cleaner Production*, **67**, 79–87, https://doi.org/10.1016/J.JCLEPRO.2013.12.002

Yoshida H., ten Hoeve M., Christensen T. H., Bruun S., Jensen L. S. and Scheutz C. (2018). Life cycle assessment of sewage sludge management options including long-term impacts after land application. *Journal of Cleaner Production*, **174**, 538–547, https://doi.org/10.1016/J.JCLEPRO.2017.10.175

Zieliński M., Kazimierowicz J. and Dębowski M. (2023). Advantages and limitations of anaerobic wastewater treatment – technological basics, development directions, and technological innovations. *Energies*, **16**(1), 83. https://doi.org/10.3390/en16010083

doi: 10.2166/9781789063479_0329

# Chapter 12

# Still open research questions on technologies, microplastics, and antibiotic resistance

J. C. de Araújo[1], H. A. Santos[1], C. Vijande[2], J. M. Garrido[2] and M. C. Tomei[3]*

[1]Department of Sanitary and Environmental Engineering, Federal University of Minas Gerais, Belo Horizonte 31270-010, Brazil
[2]CRETUS, Department of Chemical Engineering, Universidade de Santiago de Compostela, Santiago de Compostela, Spain
[3]Water Research Institute (IRSA), National Research Council (CNR), Roma, Italy
*Correspondence: concetta.tomei@irsa.cnr.it

**ABSTRACT**

In this final chapter, a summary of the still open questions to be solved for the scale-up of the technologies described in the previous chapters (including post-treatment options), for application in cold and temperate climate areas is provided. Special attention is dedicated to upflow anaerobic sludge blanket and anaerobic membrane bioreactors (AnMBRs), which are, even if with different peculiarities, the most promising solutions. In addition to anaerobic mainstream treatment technologies, recent research demand focuses also on two important emerging topics, which are of interest not only for anaerobic wastewater treatment, but also for conventional activated sludge plants, that is, the fate of microplastics (MPs) in plants and in receiving water bodies and the antibiotic resistance spreading. For these two topics, which were not analyzed in detail in the previous chapters, we propose in this chapter a general presentation of the state of knowledge in relation to wastewater treatment plants (WWTPs) with particular focus on anaerobic systems and related research needs. Concerning MPs, in spite they are recognized as ubiquitous pollutants, there is a strong research demand on the standardization of sampling and analytical protocols and on powerful technologies able to improve their removal. Even though information about the fate and effect of MPs in WWTPs with anaerobic treatments is scarce, AnMBRs have demonstrated high removal rates, which suggest them as a promising technology. As regard to antibiotic resistance, WWTPs are one of the main sources of dissemination of antibiotic-resistant bacteria (ARBs), antibiotic-resistant genes (ARGs), and antibiotic residues into the environment. Even if the available data are referring mainly to warm climate regions, it is worth noting that anaerobic reactors, alone or in combination with aerobic post-treatment, can remove from 0.5 to 3.0 log units of ARBs and ARGs depending on the resistance bacteria or gene. Also in this case AnMBRs achieved the best performance while good results have been also obtained with combined treatment options (i.e., anaerobic–aerobic and anaerobic post-treatment). Research demand on antibiotic resistance is driven by the need of clarifying the fate of ARBs, ARGs, antibiotics, pathogens in the treatment line, and the effects of wastewater characteristics on the plant performance. Important challenge is the development and evaluation of more effective disinfection and treatment methods (such as ultrafiltration and advanced oxidative processes) able to reduce the antibiotic resistance spreading from WWTPs.

**Keywords:** anaerobic domestic wastewater treatment, antibiotic residues, antibiotic-resistant bacteria, antibiotic-resistant genes, high-rate bioreactors, microplastics.

## 12.1 ANAEROBIC DOMESTIC WASTEWATER TREATMENT IN A CIRCULAR ECONOMY

The concept of circular economy has been defined in many ways as well highlighted by Kirchherr *et al.* (2017), who, after a systematic analysis of the 114 circular economy definitions found in literature, proposed the following reference definition: 'circular economy is an economic system that replaces the "end-of-life" concept with reducing, alternatively reusing, recycling, and recovering materials in production/distribution and consumption processes. It operates at the micro level (products, companies, consumers), meso-level (eco-industrial parks) and macro level (city, region, nation and beyond), with the aim to accomplish sustainable development, thus simultaneously creating environmental quality, economic prosperity and social equity, to the benefit of current and future generations.'

An anaerobic process fits perfectly this complete and exhaustive definition, because it allows resource recovery, including energy, safe water, and nutrients, from valuable by-products of industries, but, and even more important, from wastes and wastewater. In the case of domestic wastewater (DWW) treatment, the recovery of energy and nutrients is accomplished from a huge 'dirty' matrix whose conventional treatment generally implies energy consumption and loss of the nutrients contained in the influent. In this context, the anaerobic process potentially represents the optimal mainstream treatment option for DWW. Unfortunately, in cold and moderate climate regions, this option is still challenging due to the low process efficiency at these temperatures for dilute streams such as DWW. In the previous chapters, several key aspects for the application of anaerobic processes as mainstream treatment of DWW have been highlighted as well as the related research needs. In this final chapter, a summary of the still open questions to be solved to promote the scale-up of the proposed technologies, including post-treatment options, for application in cold and temperate climate areas is provided. Special attention is dedicated to upflow anaerobic sludge blanket (UASB) and anaerobic membrane bioreactors (AnMBRs), which are, even if with different peculiarities, the most promising solutions.

In addition to technologies, recent research demand focuses also on two important emerging topics, which are of interest not only for anaerobic DWW treatment, but also for conventional wastewater treatment plants (WWTPs), that is, the fate of microplastics (MPs) and antibiotic resistance spreading, both of them not extensively investigated for anaerobic processes. For these two topics, which were not analyzed in detail in the previous chapters, we propose in this chapter a general presentation of the state of knowledge in relation to WWTPs with particular focus on anaerobic systems and related research needs.

## 12.2 MAINSTREAM TECHNOLOGIES FOR THE ANAEROBIC TREATMENT OF DWW

### 12.2.1 General aspects to be investigated for high-rate systems

The first key element for the feasibility of anaerobic processes is the technology: it has been demonstrated that conventional suspended biomass reactors cannot achieve the required performance in terms of effluent quality with reactor volumes and footprint suitable for practical application, thus high-rate bioreactors are required.

Types and related characteristics of high-rate systems have been presented in this book; they are generally at medium/high technology readiness level but there are common research issues to be investigated. The first aspect is the performance optimization to define the best set of operating parameters depending on the final destination of the produced effluent: many studies have been conducted with this objective and they provided consistent results demonstrating the feasibility of high-rate bioreactors for chemical oxygen demand (COD) removal in DWW. Concerning the nutrients, anaerobic effluents are characterized by high N and P contents, which, as optimal solution congruent with sustainability goals, should be recovered as much as possible. Their recovery is not always easy and feasible due to geographical, logistic, and legislation issues. As already mentioned, research efforts and management strategies should be devoted to modify the approach in dealing with urban wastewater to be considered more as a source of energy and valuable compounds than a liquid waste. When the nutrient recovery is not feasible, appropriate post-treatment is necessary, also the challenge in this case is the achievement of the good effluent quality with a minimum expense of energy in order to not having a massive impact on the recovered energy in the anaerobic step.

Another general critical aspect in the performance of anaerobic systems is the presence of dissolved methane ($dCH_4$) in the treated effluent whose recovery has to be maximized to avoid energy loss and reduce dangerous greenhouse gas emissions. Research efforts on strategies for $dCH_4$ recovery or reuse within the same treatment process are mandatory to approach energy-neutral anaerobic treatment, and to exploit the intrinsic merits of the process to be economically feasible and environmentally friendly. Several technologies are available for the dissolved $CH_4$ recovery, both through physical methods, such as aeration, gas stripping, and degassing membranes, and for its biological removal through down-flow hanging sponge reactors and the more recent proposed process based on denitrification and anaerobic $CH_4$ oxidation (N-Damo) (Stazi & Tomei, 2021). All the proposed technologies need further investigations to optimize the process performance at pilot and demonstration scales. In fact, their applicability has not yet been fully evaluated in terms of economic feasibility and process safety (Liu *et al.*, 2014). Other issues, which deserve greater care, would be the regular and accurate measurement of the concentration of $dCH_4$ in the effluent, instead of the generally assumed theoretical data referring to the thermodynamic equilibrium conditions, and the influence of some technical aspects, such as the technology itself, the operation mode, and the type of biomass (Cookney *et al.*, 2016). The most promising solution for $CH_4$ recovery, that is, membrane separation, requires additional research on the evaluation and related impact of wetting, fouling, and clogging phenomena. There are available control strategies for these phenomena but in most cases are energy-intensive and not economically feasible and research efforts should be devoted to these critical aspects.

### 12.2.2 UASB bioreactors

UASB reactors, described in detail in Chapter 2, represent an effective technology relatively easy to apply whose advantages for the anaerobic treatment of DWW in warm climate areas are well recognized. The extension of UASB application even in moderate/cold climate regions requires technological improvements as the maximization of the biomass activity with configurations able to concentrate the biomass inside the bioreactor. As reported in Chapter 4, promising results have been achieved with attached and granular biomass systems. These studies, as the most part of literature in this field, are conducted at lab scale and consistent work should be dedicated to their scale-up. Another interesting technical solution investigated in the two above mentioned studies is the operation in sequential mode, which is not common for an anaerobic process, but, given its flexibility, can be an effective solution for small wastewater treatment installations and to face the high seasonal load variations, which can occur in touristic zones.

### 12.2.3 Anaerobic membrane bioreactors

Membrane bioreactors represent the most powerful technology developed so far for enhancing the performance of biological processes and it is of particular relevance for the anaerobic process application even for treatment of low-strength wastewaters as the domestic ones. The use of filtration membranes would revolutionize treatment systems, eliminating practically all the colloidal compounds, germs, and suspended solids usually present in anaerobic effluents. Nevertheless, the anaerobic treatment of DWW, using submerged membrane technology, is today still a promise and not a reality. This is due, in practice, to the lower flux obtained in both laboratory and pilot scale AnMBR units, usually between 5 and 10 $L/m^2/h$, far from the values of 20 $L/m^2/h$ or higher, observed in full-scale aerobic MBRs. Chapter 3 is dedicated to this technology and related research needs are well highlighted. Besides the need of performance optimization and facing the well-known fouling phenomenon, the most important aspect for the anaerobic treatment of DWW is the 'the minimization of energy use' associated in practice with the stable operation at higher flux, which is mandatory to achieve a positive energy balance making an anaerobic process competitive with an aerobic one. Energy consumption and fouling phenomena are strictly related and it is challenging to find the optimal equilibrium solution allowing acceptable fouling levels and, at the same time, acceptable energy consumptions. Strategies and technological alternatives to work in this direction are described in detail in Chapter 3.

### 12.2.4 Post-treatments

As long as there is no breakthrough on AnMBR technology, the anaerobic treatment of urban wastewater will in practice relies on the use of UASB systems and their modifications as is presented in Chapter 2. Currently, there are multiple technologies available for the effective removal of biodegradable organic matter, Total Suspended Solids (TSS), and partial Total Nitrogen (TN) removal through aerobic post-treatment systems, including suspended biomass (activated sludge processes); biofilm (trickling filter and down-flow hanging sponges, moving bed biofilm reactors, etc.), and integrated fixed-film and activated sludge processes. Nowadays, the increase in the elimination of TN means, in practice, bypassing part of the raw sewage directly to the post-treatment system, with the consequent loss of biogas production. Thus, the removal of dissolved methane and nitrogen compounds as was mentioned in Chapter 5 should be improved, with more effective strategies potentially involving advanced microbial communities or engineering systems, using among others anammox or N-damo processes at ambient temperature.

The use of post-treatment systems is also required if the anaerobically treated wastewater has to be reused, as has been mentioned in Chapter 8. Water reuse offers a good opportunity for recovering not only water, but also the contained nutrients for agriculture or landscape irrigation, diminishing the chemical fertilizer requirements of crops. Water reuse could play a crucial role in both mitigating and adapting to the impacts of climate change. Although traditionally the sources of drinking water have come from aquifers and surface waters, it is more and more common, due to water stress, to use energy-intensive processes, such as desalination of seawater through reverse osmosis to produce drinking water. In Spain this represented the 5.9% of the total drinking water consumed in the country in 2020 (INE, 2022). Replacing the use of drinking water with reclaimed water, especially for landscape irrigation and other urban uses, could be therefore a way to reduce energy consumption and greenhouse gas emissions, associated with seawater desalination, which would be added to the potential savings obtained from the use of combined anaerobic–aerobic systems to produce such reclaimed water.

## 12.3 MICROPLASTICS IN CONVENTIONAL WWTPS AND ANAEROBIC SYSTEMS

MP pollution has become a highly relevant issue in recent years. MPs can be defined as plastic fragments smaller than 5 mm. Depending on their origin, they can be classified as primary or secondary. Primary MPs are the ones that can be found in cosmetic and medical products. In addition, MPs originated from abrasion of tires and fibers that are released during laundry are also commonly classified as primary MPs (Rossatto *et al.*, 2023). Secondary MPs are the ones that originate from physical, chemical, or biological fragmentation of larger plastic debris, such as plastics bags or bottles (Akdogan & Guven, 2019).

Besides their origin, MPs can also be classified based on their physical and chemical characteristics. Despite the lack of and standardized protocol for their identification leads to a great variability in the criteria used for their description, there is a general agreement in referring at least to the type of polymer, size, color, and shape. Thus, *microbeads/spheres*, *fibers*, *films*, *fragments/irregular particles*, or *foams* are commonly used terms (Hartmann *et al.*, 2019).

Due to this diversity of sources, MPs can enter natural systems through different pathways. On the one hand, primary MPs are prone to be released through household sewage discharge or application of sewage sludge containing synthetic fibers or sedimented MPs from personal care or household products. On the other, anthropogenic activities such as littering or municipal solid waste collection and disposal contribute to plastic pollution. These larger plastic wastes and the MPs formed from them may be introduced into aquatic environments by wind dispersal, surface runoff, or soil erosion (Akdogan & Guven, 2019).

Moreover, their characteristics make them a potential risk for ecosystems. Due to their small size, MPs can interfere with food chains, because they can be ingested by organisms in lower trophic levels such as zooplankton. Furthermore, their size in combination with their great specific surface area and hydrophobic surface enables them to act as carriers of other pollutants, such as persistent organic pollutants, heavy metals, polycyclic aromatic hydrocarbons, polychlorinated biphenyls, and even pathogens (Akdogan & Guven, 2019).

MPs are nowadays considered ubiquitous pollutants, as they have been found in every known environment, that is, seas, oceans, lakes, rivers, beaches, and marine soils, including the sediments of Arctic fresh water (Rossatto *et al.*, 2023). In this context, domestic WWTPs are now considered important for the release of MPs due to the diverse origins of influents, all of which could contain a significant amount of MPs. The removal rates of MPs in domestic WWTPs are above 90%, but considerable quantities of MPs are released to the surrounding receiving waters, ending in ocean systems, due to the daily discharge of large volumes of effluents. The discharges of MPs in effluent and sludge are considered foreign matter and constitute a contaminant and source of pollution to the receiving waters and soil systems (Liu *et al.*, 2022).

Numerous types of MPs have been detected in domestic WWTPs. The most common ones would be polyethylene (PE), polystyrene (PS), polyethylene terephthalate (PET), polypropylene (PP), polyamide fibers, and polyester fibers (Liu *et al.*, 2021). Other polymers such as acrylates, alkyds, polyurethane, polyvinyl alcohol, and polyvinyl chloride (PVC) have been reported too (Sun *et al.*, 2019).

### 12.3.1 Characterization techniques

Nowadays, one of the main challenges in the study of MPs is the lack of standardization in sampling, identification, and quantification methods, which explain the wide range of MP concentrations that can be found in literature among different studies. Although many researchers have reviewed the identification and characterization of MPs in marine environment and freshwater systems, less review papers have focused on the techniques used for the analysis of MPs in samples taken from domestic WWTPs. This can be explained by the complexity of wastewater and sludge, whit high contents of organic matter, suspended solids, and other contaminants (Liu *et al.*, 2022; Tirkey & Upadhyay, 2021).

In general, many protocols have been developed, with a significant number of steps and long times of processing. This diversity of techniques results in a great extent of non-comparable results, contributing to a lack of consistency in them. Most relevant steps that are always mentioned in literature are shown in Figure 12.1.

In addition, when working with MPs, some precautions must be taken. During the sample collection and pre-treatment, contaminations of samples might be induced from atmospheric fallout, the equipment and devices, used and even clothing of workers. Some recommendations would be using glass materials, cover the samples with aluminum foil, or wearing laboratory coats made of natural fabric. It has also been suggested to setup a blank control sample. Finally, the recovery of MPs after the extraction processes is recommended to be tested (Sun *et al.*, 2019).

As can be seen in literature, for each step of the sampling and processing, different options can be chosen:

*Sampling*: the first stage for the characterization and quantification of MPs is to collect wastewater and sewage sludge samples at different points within domestic WWTPs. For wastewater samples, the most common methods used include grab samples at different sampling points or using auto-samplers to collect samples at intervals over periods from a few hours to 24 h. Sludge samples can be directly collected from the primary clarifiers, the activated sludge recirculation lines, and the anaerobic digesters. Sample volumes or sludge quantities collected are a critical factor, and the first source of variability among authors. Sample volumes used for the MP quantification in the water line could be in the ranges of 10–100, 100–200, and 50–1000 L, respectively. The quantity of sludge samples used for MP analysis can vary from 5 to 100 g or 150 to 200 mL.

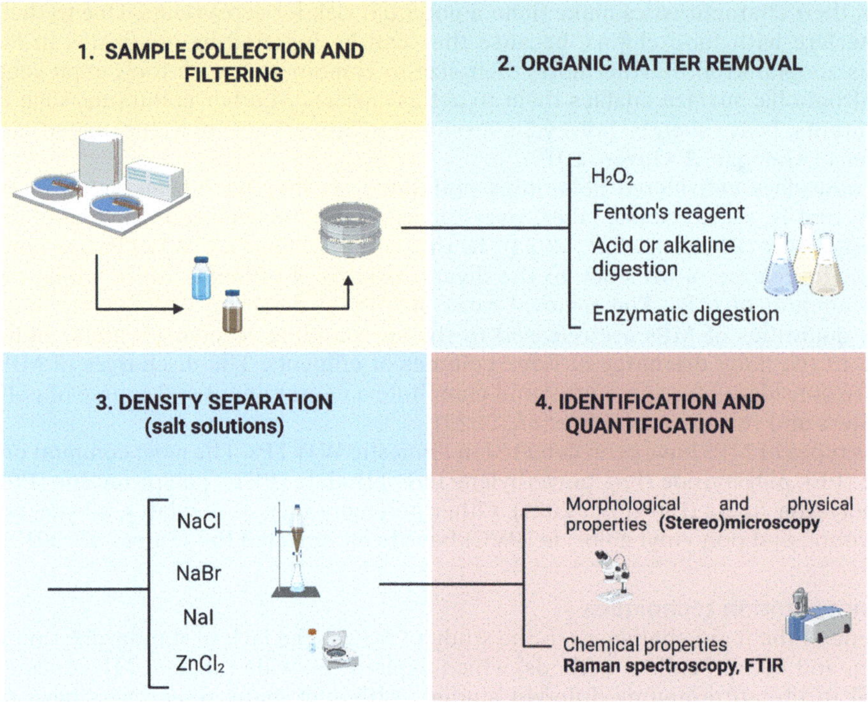


**Figure 12.1** Flowchart summarizing steps and techniques used for MP detection in domestic WWTPs. *Source*: adapted from Liu *et al.* (2022) and Sun *et al.* (2019). Figure created with Biorender.com.

The proper sample volumes or sludge quantities for MP characterization will depend on their concentration in the samples (Liu *et al.*, 2022).

*Filtering*: Collected samples are usually filtered using filters or sieves. So far, the mesh/pore sizes applied in the related studies have not been standardized and vary from ~1 to 500 μm, considerably affecting final results. This mesh-based size categorization may not be very accurate. Some studies have observed that some particles would not pass through the sieves even if sufficiently small due to their irregular shapes, and morphology of fibers enable them to pass longitudinally through smaller filters. In addition, the materials present in wastewater and sludge can cause blockages. For this reason, a common practice for sludge samples is to collect and transport them in a container to a laboratory for further sample processing (Sun *et al.*, 2019). Once reached the laboratory, sludge samples are commonly subjected to some form of pre-treatment, such as sonication (Liu *et al.*, 2022).

*Organic material removal*: it is necessary to remove all non-plastic organic substances in the samples, which comprises the major contents of solids recovered from domestic WWTPs. The clean-up of these materials may include various treatments and separation steps. Once again, the methodology applied varies widely among the different studies. The most cited methodologies usually include the use of acid or alkaline solutions, hydrogen peroxide, Fenton's reagent, or even enzymes such as cellulase, chitinase, or protease. However, NaOH and KOH have been reported to lead to insufficient reduction of organics in sludge samples and cause destruction to PE, PS, polycarbonate, PET, and PVC polymers, so they are not recommended for organic removal in WWTP sludge (Liu *et al.*, 2022; Tirkey & Upadhyay, 2021).

*Density separation*: the aim of this step is to segregate MPs from inorganic substances or sediment that are not eliminated with the digestion. The most used salt solution would be NaCl. However, polymers such as PET or PVC cannot be separated with this methodology. For this reason, other salt solutions have been suggested in several studies: NaI, NaBr, and $ZnCl_2$ (Tirkey & Upadhyay, 2021).

*Identification and quantification techniques*: Liu *et al.* (2022) classified the most used techniques for MP characterization into two main groups. On the one hand, morphological and physical characterization using microscopy or stereomicroscopy is widely extended as a methodology to pre-sort and even count MPs. However, relying solely on the visual identification could result in overestimation of the actual number of MPs, especially fibers. On the other, techniques focused on chemical properties of MPs are now widely used, commonly in combination with the visual identification. These techniques range from relatively fast spectroscopy techniques (Fourier transform infrared or Raman spectroscopy) to pyrolysis-gas chromatography/mass spectrometry or thermogravimetric analysis-differential scanning calorimetry.

### 12.3.2 Occurrence and removal rates

As previously mentioned, there is not a standardized protocol for sampling, identifying, and quantifying MPs in domestic WWTPs. Because of this, the data reported by different studies vary significantly. There is discrepancy even in the units used, with authors referring to the mass or the number of MPs, particles, or items, taking into account in particles that cannot be confirmed as either plastics or non-plastics (Michielssen *et al.*, 2016; Sheriff *et al.*, 2023).

Several review articles summarize the quantity and typology of MPs found. According to Sun *et al.* (2019) the amount of particles reported vary from 1 to 10,044 particles/L in influents and from 0 to 447 particles/L in effluents. Sheriff *et al.* (2023) mentioned quantities in influents from 0.28 to 8400 MPs/L and from 0.05 to 297 MPs/L in effluents. Other studies reported similarly mixed data (Liu *et al.*, 2021). The wide variations in MP concentration or quantity in the studies could be partially related to the aforementioned differences on sampling methodology, pre-treatment of the samples, and the analysis methods applied, as well as the characteristics of the facilities (Sun *et al.*, 2019). Other factors that must be considered are the size of the facility, the population served, and the source of the raw water (Liu *et al.*, 2021; Sheriff *et al.*, 2023).

Regarding sludge samples, studies focusing on MPs in the sewage sludge including their occurrence, transformation, and further mobilization are quite limited. According to Sun *et al.* (2019), the number of MPs in dry weight (DW) sludge could reach from around 1500 to 170,000 particles/kg DW. Nguyen *et al.* (2022) reported the number of MPs in sludge from domestic WWTPs worldwide ranges from >1000 to 301,400 particles/kg. These quantities are substantially higher than that in the liquid phase. The great variation in the number of MPs in sludge also emphasizes the importance of representative sampling and harmonized detection methods. On the contrary, size of MPs in the sewage sludge usually is significantly different from that in the raw water, with the average size being relatively large. In terms of the shape, synthetic fibers are the most common MPs found in sludge. Therefore, the continuous addition of MPs to soils due to the use of sewage sludge as fertilizer should not be ignored (Sun *et al.*, 2019).

Although no treatment technology has been specifically designed for the removal of MPs in wastewater, most of domestic WWTPs exhibit high removal rates. Above 90% of MPs that enter WWTPs are removed from water lines, being retained in the sludge via adsorption mechanisms. Hence, the type of treatment plays a key role in this removal rate (Sheriff *et al.*, 2023). The pre-treatment and primary treatment steps can effectively remove most MPs in wastewater. Grease and grit pre-treatment effectively removes MPs of larger size, whereas primary settling treatment achieve high efficiency in removing MPs. Due to this reason, the use of primary treatment systems would guarantee a high removal of MPs in domestic WWTPs, regardless of whether either an aerobic or anaerobic biological treatment can be used later. The secondary aerobic treatment, which usually comprises biological treatment and clarification, manage to further decrease the number of MPs to 0.2–14%. Finally, in general, WWTPs with tertiary treatment processes had lower MP quantities (0–51 particles/L) in effluents than those with primary or secondary treatment processes only (Sun *et al.*, 2019).

### 12.3.3 MPs in anaerobic systems

Regarding occurrence and fate of MPs in domestic WWTPs with anaerobic treatments, available information is scarce. Currently, there are no studies on the presence of MPs in full-scale facilities and on their removal rates. Only few studies have focused on anaerobic technologies in pilot-scale facilities. Moreover, different studies have tried to assess the influence of MPs on anaerobic granular sludge (AGS), which will be discussed in the next section.

Michielssen *et al.* (2016) assessed the efficiency of different unit processes at three domestic WWTPs in removing small anthropogenic litter (SAL), which includes MPs. Two of them were full-scale conventional facilities with different secondary and tertiary treatments, and the third one was a pilot-scale AnMBR with a side-stream configuration, placed after preliminary treatment of one of the other conventional facilities. The number of SALs in raw water and its removal rates for the three technologies were determined. The overall removal rates achieved ranged from 93.8 to 99.4%. Even though an important part of this removal took place in the preliminary treatment, AnMBR pilot process removed 99.1% of the remaining SAL, outperforming removal rates of final treatment in both plants with conventional configurations. This is due to the small pore size of the micro- or ultrafiltration membranes employed, usually between 0.4 and 0.05 μm. In addition, the unit processes employed affected also shape classes of SAL detected on final effluents. Some particles such as paint chips were not detected in the AnMBR effluent, unlike the other domestic WWTPs. Moreover, as compared to conventional systems, there was a 10-fold reduction in the number of fibers released by the AnMBR treatment. However, fibers still represented a greater percentage of SAL in effluents. It must be noted that in this study they did not focus on MPs exclusively, and part of these fibers could be natural fibers such as cotton. These results suggest that an AnMBR is an efficient technology for removing SAL and therefore, MPs.

In another study, Pittura *et al.* (2021) compared a full-scale WWTP with conventional activated sludge (CAS) treatment and a pilot-scale UASB coupled with an AnMBR. This study focused on determining MP occurrence and their removal rates. For an anaerobic configuration, they determined the presence of MPs in the pretreated influent, the UASB effluent, and the permeate, as well in the granular sludge. In this case, authors focused only on particles identified as MPs, classifying them in terms of their shape as fiber-shaped (MPFs) and particle-shaped (MPPs). The innovative configuration removed 94% of influent MPs compared with the conventional treatment scheme. The overall removal of 52.6% was achieved by the UASB and further 44% by the AnMBR. In terms of shape, the removal rate for MPPs achieved 100%, whereas only 87% of MPFs were removed. On the contrary, MPPs (particularly fragments and films) were identified in conventional WWTP final effluent. Hence, the main conclusion of this study was that the overall MP removal in a pilot-scale configuration was greater than that in a full-scale CAS scheme. The authors indicate that the high removal efficiency is due to the ultrafiltration unit, that was mostly responsible for the total abatement of MPPs. Regarding MP accumulation in sludge, the anaerobic sludge from UASB reactors presented a higher contribution of MPPs of lower size class (0.03–0.1 mm) than the sludge from primary in the conventional configuration.

In conclusion, anaerobic treatments emerge as a promising technology for removal of MPs from wastewater, especially membrane-based systems. This better performance of membrane bioreactor (MBR) technology to remove MPs compared to conventional systems has been highlighted by other authors, but in aerobic systems (Sun *et al.*, 2019). However, it should be noted that AnMBRs have different fouling and permeation characteristics compared to aerobic MBRs (Pittura *et al.*, 2021). Finally, particular attention should be paid to fibers because both studies mentioned emphasize that AnMBRs are not able to remove them from influents.

### 12.3.3.1 Effects of MPs on AGS systems

Several authors have studied the response of AGS to the exposure of different MPs. However, most of the studies carried out were laboratory-scale experiments. Thus, these laboratory-scale experiments were in part focused on determining the mechanisms of toxicity, so they applied conditions that could differ from real systems. A few authors have studied the fate and influence of MPs on pilot plants (Pittura *et al.*, 2021) or the effect of more than a type of MP at a time (Wei *et al.*, 2022). To date, there is no information about MP effect on full-scale anaerobic systems for municipal wastewater treatment. Most relevant studies in this field and their main results are summarized in Table 12.1.

In general, all authors conclude that MP addition has a negative effect on the performance of AGS systems, affecting extracellular polymeric substance (EPS) production, granule size and integrity, and methane production. It is important to note that the size of MPs and their concentration are relevant factors, and they must be considered when drawing conclusions. In addition, exposure time might play an important role in the effects observed, as well as the type of studied MPs.

These negative effects can be explained by two main mechanisms. On the one hand, some studies have reported that the leachates from different types of MPs inhibit microbial growth and affect

**Table 12.1** Principal effects of MP addition on AGS in different experiments.

| Experimental Setup | Type of MP | MP Concentrations/ Quantities | Main Results | References |
|---|---|---|---|---|
| Laboratory-scale batch experiments | PS (particle size from 0.5 to 150 μm) | 75 mg/L | EPS secretion inhibition. Damage in AGS integrity and cell viability and methane production inhibited with increasing particle size | Wang *et al.* (2022a) |
| Laboratory-scale BEAD reactors | PE (average particle size: 40–48 μm) | 0.5–10 mg/L | Higher concentrations (10 mg/L) inhibited methane production. Granule breakage and decrease in cell viability. Changes in microbial populations | Wang *et al.* (2022b) |
| Laboratory-scale UASB reactors operated continuously for 120 days | PET (average size: 1.27 mm) | 15–300 MPs/L | Low concentration (15 MPs/L) did not affect AGS whereas high concentrations (75–300 MPs/L) significantly inhibited AGS activity, with granule breakage and EPS production inhibition. Changes in microbial populations | Zhang *et al.* (2020a) |
| Pilot-scale UASB reactors treating municipal wastewater | PP | 5–50 PP/g TS | Methanogenic activity tolerated up to a concentration of 18 MPs/g TS<br>50 PP/g TS caused a remarkable inhibition on methane production | Pittura *et al.* (2021) |
| Laboratory-scale batch experiments<br>Remediation strategy with hydrochar tested | PET, PS, PE, and PP (size: 150 μm) | 75 mg MP/L (36% of PET, 15% of PS, 42% of PE, and 8% of PP) | Inhibition of cumulative methane production. Granule size decreased. Changes in microbial populations | Wei *et al.* (2022) |
| Laboratory-scale UASB reactor operated for over 300 days with synthetic wastewater | PVC (size: 0.1 mm) | 0–150 MPs/L | Decrease in methane production and granule size. Cell viability was also affected. Changes in microbial population | Zhang *et al.* (2020b) |

PS: polystyrene; PE: polyethylene; PET: polyethylene terephthalate; PP: polypropylene; PVC: polyvinyl chloride; UASB: upflow anaerobic sludge blanket; EPS: extracellular polymeric substance.

microbial activities. Experimental results of the studies carried out with PVC, PET, and PS exposure demonstrated the negative influence of these leachates on methane production. PET can cause this negative effect through leaching of di-*n*-butyl-phthalate (Wei *et al.*, 2019; Zhang *et al.*, 2020a). PS leachates (Zhang *et al.*, 2020a) were strongly associated with dropping methane production on AGS, especially the leachates from larger particles. On the other, results from diverse studies suggest that MP addition could lead to the production of reactive oxygen species (ROS). Even in the anaerobic environment, sub-micromolar oxygen still remains in the medium, which can interact with abundant active sites on MP surface and generate ROS. Compared with the control, the exposure of PET-MPs at 15 MPs/L had no impact on the ROS production, but the presence of 75, 150, and 300 MPs/L enhanced ROS production (Wang *et al.*, 2022b; Zhang *et al.*, 2020a).

Going into further detail, MP addition seems to contribute to the alteration of granule size and integrity. Studies with PS particles and biomass from an UASB reactor (Wang *et al.*, 2022a) reported a decrease in granule size and breakage of them due to the MP addition. Moreover, the authors found a positive relationship between particle size and the damage caused. However, it is noteworthy that the PS concentrations might be higher than the ones found in a WWTP. In addition, another study with PE in a bio-electrochemical anaerobic (BEAD) reactor demonstrated alterations in AGS surface and a diminishment in particle size with MP concentration of 10 mg/L (Wang *et al.*, 2022a). Other studies with PET (Zhang *et al.*, 2020a) indicated the addition of 15 MPs/L had no effect in comparison with a control reactor, but higher quantities (from 75 to 300 MPs/L) resulted in a decrease of AGS particle size. Interestingly, in this study a lower number of MPs added (15 MPs/L) seemed to be related to an increase in EPS production, which is not observed with the higher ones. These results suggest that a low concentration of MPs can enhance EPS productions with a protective effect. In contrast, results from Zhang *et al.* (2020b) with PVC with 15 MPs/L did show negative effects.

These results are in consonance with the ones observed for AGS performance. Both methane production and COD removal show a descending trend when concentrations and size of MPs increase, regardless of the type of MP studied (Wang *et al.*, 2022a; Zhang *et al.*, 2020a). It is worth to mention that the lower quantities studied for PET had no remarkable effect on biomass performance in opposition to the results with PVC, as in the case of granule integrity. Experiments with PP in a pilot-scale UASB reactor only reported negative effects with MPs above 18 PP-MPs/g TS, but not with 0.5 and 5 PP-MPs/g TS (Pittura *et al.*, 2021).

In conclusion, the effect of MP addition on AGS reactor performance has been demonstrated by various studies. However, more research is needed, particularly in full-scale reactors. Regarding possible solutions, Wei *et al.* (2022) proposed the addition of hydrochar as an effective remediation strategy. Authors indicate that hydrochar amendment effectively mitigated the reductions in methane production, granule size, and cell viability and reduced the toxicity of MPs to microbial community.

### 12.3.4 Research needs on MPs

The first highlighted important aspect to be investigated is the definition of standardized protocols for sampling, identification, and analysis of MPs, which currently are not available thus making data analysis procedures extremely difficult. This is a mandatory start point to realize systematic studies on the fate of MPs in different water matrices and in WWTPs. Standardized methodologies can eliminate/reduce the high variability observed in MP concentration and removal data reported in the scientific literature (even for very similar WWTPs) opening the way for reliable comparative studies.

Concerning the anaerobic treatment, promising results have been obtained for MP removal, but these are based on few studies mainly focused on AnMBRs. As mentioned in Chapter 3, the most critical bottlenecks for MBRs requiring further investigations are fouling and clogging phenomena. In addition, in the case of MPs another important issue, which can cause clogging is constituted by fibers, so appropriate pre-treatment should be investigated in order to minimize the impact of fibers on the anaerobic treatment performance in AnMBRs. For the specific case of high-rate anaerobic bioreactors operated with granular biomass, which also demonstrated good performance in MP removal, additional research is required on strategies for performance enhancement, that is, by addition of sorbent media (i.e., biochar, hydrochar, etc.), which showed positive effects in preliminary studies.

Finally, as observed in general for the studies on anaerobic processes to be applied for DWW treatment in moderate climate regions, also for the fate of MPs there is a strong research demand on experimental validation at demonstration and full-scale level, which is necessary for providing reliable feasibility evaluation.

## 12.4 ANTIBIOTIC-RESISTANT BACTERIA AND ANTIBIOTIC-RESISTANT GENES: WHY INVESTIGATE ANTIMICROBIAL-RESISTANT ELEMENTS IN WASTEWATER?

Antibiotic resistance is a global threat to public health leading to 700,000 deaths worldwide annually, according to the World Health Organization (WHO). Based on a report (O'Neill, 2016) by an independent committee, 10 million deaths annually are expected by 2050, meaning that more people will die from this than from cancer.

Antimicrobial resistance (AMR) can be defined as the ability of microorganisms, bacteria, viruses, fungi, and parasites to withstand the effects of drugs used to contain infections, that is, antibiotics, antivirals, antifungal, and antiparasitic. In the case of bacteria, antibiotic resistance is a process in which said microorganisms can achieve a resistance mechanism, due to the acquisition of resistant genes from other bacteria, or due to the mutation of key genes during cell replication. From that moment on, antibiotics will only act on those susceptible microorganisms, not affecting those microorganisms that have acquired AMR.

Although the development of antibiotic resistance is a natural evolutionary process mediated by microorganisms, it has been accelerated by selective pressures due to anthropogenic activities (Bengtsson-Palme *et al.*, 2017). In view of the various factors that encourage the emergence and spread of this resistance, the excessive use and misuse of antibiotics by the world population and the subsequent spread of antibiotic-resistant genes (ARGs) among microorganisms can be highlighted (WHO, 2021). Antibiotic drugs enter the wastewater stream via human, animal, medical, and industrial waste, along with heavy metals at different concentrations according to their sources. These wastewater streams also contain enteric pathogens, coliforms, phages, antibiotic-resistant bacteria (ARBs), and ARGs, which then are combined during the treatment in domestic WWTPs (Hazra & Durso, 2022).

Domestic WWTPs aim to remove and control pollutants and conditions harmful to health, arising from chemical, physical, and biological nature. Although they present processes mainly directed to the removal of solids and organic matter, some removal of micropollutants, ARBs, and ARGs can occur (Machado *et al.*, 2023; Uluseker *et al.*, 2021), but still these micropollutants remain in effluents (Leroy-Freitas *et al.*, 2022; Machado *et al.*, 2023). In Brazil, for instance, most of the municipal WWTPs operate with secondary treatment and ~22% of the domestic WWTPs include some kind of effluent post-treatment, such as ultraviolet (UV) disinfection, maturation ponds, and chlorination.

Simultaneously to the indiscriminate use of antibiotics, WWTPs are considered hotspots for ARB proliferation and horizontal gene transfer of ARGs, being a major source of enriching and disseminating ARGs and ARBs to the environment (Bouki *et al.*, 2013; Rizzo *et al.*, 2013). This occurs because, in these environments, there is a combination of different ARBs and ARGs, close contact of cells, in addition to the great availability of nutrients, the presence of antibiotic residues, heavy metals, and other chemical compounds at subinhibitory concentration, creating favorable conditions for the spread of AMR through horizontal gene transfer. Therefore, it is possible to state that access to adequate and efficient DWW treatment systems constitutes one of the interfaces between basic sanitation and human health.

In line with the potential for spreading and monitoring the spread of AMR attributed to domestic WWTPs, there is a proposal from the European Union for member states to monitor AMR in domestic WWTPs receiving effluents generated by a population of more than 100,000 inhabitants (European Commission, 2022; Larsson *et al.*, 2022). This is a unique case because there are no effluent discharge limits (or standards) for the maximum allowed concentration of resistant genes or bacteria, as well as antibiotic residues in other countries. It is worth noting that, due to increasing population, urbanization, water stress, resource consumption, and water reuse plants, it is becoming unsustainable to treat wastewaters to only meet discharge limits. Therefore, it is important to produce valuable products (such as biosolids), recover nutrients, energy, and produce water having appropriate quality for reuse, and to protect the environment and public health. Thus, it is necessary to monitor chemical and biological micropollutants in wastewaters and their removal by different treatment options including anaerobic systems. It is also important to analyze the operational parameters and environmental factors that affect the occurrence, abundance, and removal efficiency of ARBs and ARGs by each wastewater treatment system (Baranchesme & Munir, 2018).

There are four types of mechanisms that bacteria have developed against antibiotics: (1) efflux pumps, which are proteins that excrete antibiotics from the cells (some examples of multidrug efflux genes are: mdtH, mdtN, mexB, mexD, mexF, Yang *et al.*, 2014); (2) inactivation of antibiotics by hydrolysis or by conversion of functional groups (examples of $\beta$-lactamase genes are blaTEM, blaCTX-M32, blaCTX-M15, among others); (3) target by pass (such as overproduction of the target compound/enzyme); and (4) target modification (modification of the antibiotic targets themselves). The most common antibiotic-resistant elements investigated in wastewaters are listed in Table 12.2. It can be noticed that the majority of the studies have detected and quantified class 1 integron integrase gene (intl). This gene has been monitored as an indicator of putative multiple antibiotic resistance (Zhang *et al.*, 2018). ARGs encoding resistance to sulfonamides (sul1, sul2, and sul3), macrolide (ermB), quinolone (qnrB), tetracycline (tetM), and $\beta$-lactams (blaTEM) are commonly monitored in wastewaters (Table 12.2), among others. Alexander *et al.* (2020), in a study investigating 23 WWTP effluents in Germany, reported that the most frequently detected ARGs in wastewater effluents, among 12 clinically relevant ARGs studied, were those that confer resistance to sulfonamides (sul1), macrolides (ermB), $\beta$-lactams (blaTEM), and tetracycline (tetM); the intermediates were blaCTX-M32, blaOXA48, blaCTX-M15, blaCMY-2 (conferring resistance to $\beta$-lactam antibiotics), and rare ARGs were mecA (responsible for methicillin resistance), blaNDM-1 (gene that produces carbapenemase $\beta$-lactamase, conferring resistance to carbapenem antibiotics), mcrl (mobilized colistin-resistant gene), and vanA (conferring resistance to vancomycin). They also reported that, based on the numbers of total ARG cell equivalents discharge per day, ARG dissemination from WWTP effluents, independently on the catchment area, was not related to the WWTP size. In fact, smaller WWTPs (for instance with a 26,000 population equivalent) release as many ARG cell equivalents per day (e.g., $2.97 \times 10^{14}$) as larger WWTPs ($4.76 \times 10^{13}$ with a 45,000 population equivalent). According to the authors, these results lead to the question as to what kind of wastewater treatment or treatment efficiency smaller WWTPs apply.

Depending on the goals of the study, resources, and analytical capabilities, 5–30 ARGs can be investigated by real-time quantitative polymerase chain reaction (qPCR), more than 170 ARGs can be investigated through high-capacity qPCR, and even more ARGs can be assessed via metagenomics.

Wang *et al.* (2020) in a review on the occurrence and fate of antibiotics, ARBs and ARGs in WWTPs (mainly activated sludge systems) in different geographical areas (Europe, America, Asia, and Africa), reported that the ARGs commonly observed in WWTPs were bla (blaCTXM, blaTEM), sul (sul1, sul2), tet (tetO, tetQ, tetW), and ermB genes, whereas the most frequently detected antibiotics were macrolides (clarithromycin, erythromycin/erythromycin-$H_2O$, azithromycin, roxithromycin), sulfonamides (sulfamethoxazole), trimethoprim, quinolones (ofloxacin, ciprofloxacin, norfloxacin), and tetracyclines (tetracycline). They observed that there was a positive correlation between antibiotics and ARGs commonly detected in domestic WWTPs, except for $\beta$-lactam antibiotics and bla genes. The bla genes were found frequently, despite $\beta$-lactam antibiotics being seldom detected owing to hydrolysis. In secondary treatment effluents, the concentration of trimethoprim was the highest (138 ng/L in median) and the concentration of other antibiotics remained lower than 80 ng/L, whereas the relative abundance of ARGs ranged from 2.9 to 4.6 logs (copies/mL, in median).

**Table 12.2** Types of antibiotics, ARBs, and ARGs commonly investigated in domestic and urban wastewater worldwide.

| Antibiotics | Bacteria Resistant to the Following Antibiotics | ARGs Encoding Resistant to | References |
|---|---|---|---|
| Ceftazidime, meropenem, ciprofloxacin, lincomycin, clindamycin, azithromycin, clarithromycin, erythromycin, sulfamethazine, sulfamethoxazole, trimethoprim, tetracycline, and chloramphenicol. | Amikacin, meropenem, ceftazidime, clindamycin, erythromycin, ciprofloxacin, co-trimoxazole (trimethoprim:sulfamethoxazole, tetracycline, vancomycin, and chloramphenicol) | $\beta$-Lactam antibiotics (blaNDM, blaKPC, blaSHV, blaCTX-M), amikacin [aac(6′)-Ib], sulfonamides and co-trimoxazole (sul1, sul2, dfrA), fluoroquinolone (qnrA, qnrB), macrolides (ermB), tetracycline (tetM, tetO), quinolones (qnrB, qnrA), integron (int1 gene) | Le *et al.* (2018) |
| Not investigated | Not investigated | Sulfonamide (sul2), macrolide (ermB), chloramphenicol resistance (cmlA and floR) | Chen *et al.* (2019) |
| 8 Antibiotics: erythromycin-$H_2O$ (ETM-$H_2O$), monensin (MON), clarithromycin (CTM), leucomycin (LCM), sulfamethoxazole (SMX), trimethoprim (TMP), sulfamethazine (SMZ), and sulfapyridine (SPD) | Not investigated | Sulfonamide (sul1, sul2, and sul3), tetracycline (tetG, tetM, tetO, and tetX), macrolide (ermB and ermC), chloramphenicol (cmlA and floR), and 16S rRNA (bacteria) | Chen *et al.* (2016) |
| Not investigated | Vancomycin, cephalexin, sulfadiazine, and erythromycin | vanA, ampC, sulI, and ereA | Yuan *et al.* (2014) |
| Not mentioned | Not investigated | 30 ARGs were investigated (tetA, tetB, tetE, tetG, tetH, tetS, tetT, tetX, sul1, sul2, qnrB, ermC) | Mao *et al.* (2015) |
| Not investigated | Not investigated | 9 ARGs (blaTEM, blaOXA-48, blaOXA-58, blaCTX-M-15, blaCTX-M-32, blaKPC-3, sul1, tetM, mcr-1), 16S rRNA gene and intI1 gene | Cacace *et al.* (2019) |
| 49 antibiotics (such as ciprofloxacin, ofloxacin, pipemidic acid, azithromycin, etc.) | Not investigated | 11 ARGs (such as sul1, sul2, qnrS, blaTEM, ermB, etc.) | Ávila *et al.* (2021) |
| Not investigated | ARBs not investigated. Taxonomic detection via qPCR of *E. coli*, *P. aeruginosa*, *K. pneumoniae*, *A. baumannii*, and enterococci. | 12 ARGs: sul1, ermB, blaTEM, tetM, mcr-1 (colistin resistance), mecA (methicillin resistance in staphylococci), blandm-1 (new deli $\beta$-lactamase), vanA (vancomycin resistance), blaCTX-M-15, blaCTX-M-32, blaOXA-48, blaCMY2 | Alexander *et al.* (2020) |

*(Continued)*

**Table 12.2** Types of antibiotics, ARBs, and ARGs commonly investigated in domestic and urban wastewater worldwide. (*Continued*)

| Antibiotics | Bacteria Resistant to the Following Antibiotics | ARGs Encoding Resistant to | References |
|---|---|---|---|
| Not investigated | Amoxicillin, sulfamethoxazole, cephalexin, ciprofloxacin, sulfadiazine, tetracycline | Many ARGs investigated via metagenomics and IntI1, 16S rRNA | Dias *et al.* (2022) |
| 14 antibiotics | Cultivable multiple-antibiotic-resistant bacteria | 178 genes were detected in the hospital wastewaters, among them intI1 and qnrD, intI2 and sul3, intI3 and tetX, Tn916/Tn1545 and sul2, and ISCR1 and sul3 | Wang *et al.* (2018) |
| Amoxicillin, cefaclor, cefprozil, cefdinir, levofloxacin, ciprofloxacin, azithromycin, clindamycin, clarithromycin, and triclocarban | ARBs were not investigated. 16S rRNA gene (total bacteria) and the human specific fecal HF183 bacteroides 16S rRNA genetic marker were determined | Class 1 integrase (intl1), class $\beta$-lactamase (blaCTX-M and blaTEM), erythromycin (ermB), fluoroquinolone (qnrS), sulfonamide (sul1 and sul2), tetracycline (tet(O)), methicillin (mecA), and vancomycin (vanA) | McConnell *et al.* (2018) |
| Many classes of antibiotics: macrolides (clarithromycin, erythromycin/erythromycin-$H_2O$, azithromycin, roxithromycin), sulfonamides (sulfamethoxazole), trimethoprim, quinolones (ofloxacin, ciprofloxacin, norfloxacin), tetracyclines (tetracycline) | Bacteria resistant to sulfonamide and tetracycline; *E. coli* and enterococci resistant to ampicillin, tetracycline, trimethoprim/sulfamethoxazole, cefotaxime, nitrofurantoin, erythromycin, and vancomycin, among others | Sul1, sul2, tetM, tetO, tetQ, tetW, tetH, tetZ, tetX, tetA, blaTEM, blaSHV, blaCTX-M, qnrB, qnrS, er, mC, ermB, ermF, mecA, int1 | Wang *et al.* (2020) (a review) |
| Many classes of $\beta$-lactam; sulfonamide, fluoroquinolones, tetracyclines, macrolides | Not mentioned | $\beta$-Lactams (ampR, blaTEM, blaampC, blaCMY-13, blaCTX-m, blaCTX-m1, blaCTX-m9, blaCTX-m12, blaCTXm32, blaFOX, blaOXA-10, blaOXA, blaOXA-46, blaOXA-58, blaSHV-5, blaVIM1, blaVIM11), macrolides (ermB, ermF, ereA, ereB, macB, mef, mphA), quinolones (gyrA, parC, gnrC, gnrD, gnrS), tetracyclines (tetA, tetB, tetB(P), tetC, tetE, tetG, tetM, tetO, tetQ, tetV, tetW, tetX, tetZ), sulfonamide sul1, sul2, sul3), multidrug efflux pump genes (mdtF, mdtG, mdtH, mdtN, mexB, mexD, mexF) | Uluseker *et al.* (2021) (a review) |
| Not investigated | Not investigated. 16S rRNA gene (for total bacteria) | Class 1 integrase (intl1), class $\beta$-lactamase (blaTEM), erythromycin (ermB), fluoroquinolone (qnrB), sulfonamide (sul1), tetracycline (tetA) | Leroy-Freitas *et al.* (2022) |

### 12.4.1 ARB and ARG reduction in WWTPs

Many studies have been conducted in different countries and WWTPs operated with CAS processes (as reported by Le *et al.*, 2018; Pazda *et al.*, 2019; Rafraf *et al.*, 2016; Uluseker *et al.*, 2021; Wang *et al.*, 2020). Nevertheless, very few studies have investigated ARB and ARG removal in WWTPs operated with anaerobic systems. In general, these studies reported that biological treatment systems (mainly conventional (CAS) or modified activated sludge (MAS)) reduce ARB and ARG abundance by 2–4 log units depending on operational conditions applied. Antibiotics removal varied and could be up to 70% depending on the class of antibiotics (Le *et al.*, 2018; McConnell *et al.*, 2018; Wang *et al.*, 2020; Wen *et al.*, 2016; Yang *et al.*, 2014).

Le *et al.* (2018) investigated the occurrence of 19 antibiotics, bacterial resistance to 10 antibiotics and 15 ARGs in a municipal WWTP comprised of CAS and MBR systems (including primary clarifier + aerobic/anoxic tanks + microfiltration membrane unit). They reported that physical (primary clarifier) and biological treatments (anoxic/aerobic tanks) played an important role in removing the majority of antibiotics (median removal efficiency for amoxicillin, azithromycin, ciprofloxacin, chloramphenicol, meropenem, minocycline, oxytetracycline, sulfamethazine, and vancomycin was >70% in both CAS or MBR, whereas trimethoprim and lincomycin persisted in CAS (<50% removal), ARGs (up to 4.2 log removal), and ARBs (5.0 log removal). On the contrary, the microfiltration membrane treatment completely removed ARBs, reduced ARG concentration (up to 4.8 log removal, and up to 7.1 log removal for the whole MBR train). The microfiltration membrane unit alone insignificantly reduced concentrations of antibiotic residues in comparison to the treatment in the secondary clarifier. They concluded that MBR system (comprised of primary clarifier + aerobic/anoxic tanks + microfiltration membrane unit) outperformed CAS in the elimination of ARBs, ARGs, and most detected antibiotics.

Some studies (Du *et al.*, 2015) have reported that aerobic membrane bioreactors (AeMBRs) can substantially reduce the concentration of ARGs (more than 5 orders of magnitude removal of tetG, tetX, tetW, and sul1 resistant genes) and complete removal of ARBs (Le *et al.*, 2018) from DWWs, mostly due to pore-size exclusion mechanism (pore size: 0.1–0.4 mm). Wang *et al.* (2020), in a review on the occurrence and fate of antibiotics, ARBs, and ARGs in municipal WWTPs (mainly CAS) worldwide, reported that ARG and ARB abundance was efficiently reduced in WWTPs (2 log reduction), and no obvious proliferation of ARGs and increase in *Escherichia coli* resistance rate were observed. However, they mentioned that ARGs (2.9–4.6 logs), ARBs (2.3–4.5 logs of *E. coli*/enterococci), and antibiotics (macrolides, quinolones, sulfonamides, trimethoprim, and tetracycline) were still present in the treated effluent. In this review, the authors concluded that because ARGs, ARBs, and antibiotics are not entirely removed from WWTPs (even after disinfection treatments, such as ozonation, chlorination, and UV) the key point to control the occurrence of AR from WWTPs is to reduce antibiotics consumption by both human medicine and animal breeding.

Uluseker *et al.* (2021) reviewed and summarized the current knowledge about AR removal efficiencies of different WWTP methods (studies mainly conducted in Europe, China, and Canada). They showed that CAS treatment, with aerobic and/or anaerobic reactors alone or in series, followed by post-treatment methods (such as UV, ozonation, and oxidation) removes considerably more ARGs and ARBs than activated sludge alone. They also examined AR in biosolids and discussed removal efficiency of different sludge treatment procedures. They concluded that advanced post-treatment methods such as UV, ozonation, and oxidation of effluents, and heat drying, lime stabilization, and pyrolysis of biosolids, remove considerably more ARGs and ARBs than activated sludge treatment alone, but there are disadvantages such as more complex operation and higher cost.

Studies investigating fate of ARGs in anaerobic digested sludge are more common (Tong & Wang, 2014; Wu *et al.*, 2016; Xu *et al.*, 2018; Yang *et al.*, 2014), but few studies have been found for anaerobic domestic WWTPs (as can be seen in Table 12.3).

As mentioned earlier, domestic WWTPs have been widely monitored in Europe, Asia, and North America regarding the occurrence and fate of antibiotic-resistant elements, such as ARGs, integrons,

**Table 12.3** Studies that investigated ARBs and ARGs in domestic WWTPs with an anaerobic process and/or one anaerobic system in the WWTP treatment train.

| Type of Sample | ARB/ARGs/Antibiotics | Type of WWTP/ Treatment Process | ARB and/or ARG Removal Efficiency | References |
|---|---|---|---|---|
| Sewage sludge | ARGs (tetA, tetB, tetC, tetW, tetM, tetQ, tetX, sul1, sul2, ermB, ermC, blaAOX, blaTEM), integrons (intl1, intl2) and 16S rRNA | 1 WWTP/anaerobic digestion | Highest ARG removal in AD was achieved with thermal hydrolysis at 140°C | Haffiez *et al.* (2022) |
| Sewage sludge | ARBs (tetracycline and $\beta$-lactam antibiotics-resistant bacteria) | 1 WWTP | ARB removal of 1.5–1.6 log units after AD | Tong and Wang (2014) |
| Sewage sludge | ARGs by metagenomic approach. 271 ARGs subtypes belonging to 18 ARG types were identified | 1 WWTP (activated sludge (AS) and AD) | AD removed 20.7% of ARGs from sludge. AS removed 99.8% of ARGs | Yang *et al.* (2014) |
| Sewage sludge | ARGs (tetA, tetC, tetM, tetO, tetX) and ARBs (tetracycline and $\beta$-lactam antibiotics-resistant bacteria) | 1 WWTP, Beijing, China | AD of sludge with microwave (MW-H), microwave (MW), and microwave-$H_2O_2$-alkaline (MW-$H_2O_2$). Removal varied from 0.55 to 5.04 logs, according to the conditions evaluated | Tong *et al.* (2016) |
| Sewage sludge | *tetA, tetG, tetX, tetO, tetW, sul1, sul2, sul3, ermB, ermF, blaTEM dfrA1, dfrA2*, and *intI1* | Thermophilic and mesophilic AD | Thermophilic digestion removed *tetA, tetG, tetX, sul1, sul2, ermB, dfrA1, dfrA2*, and *intI1* by 0.1–0.72 log units, whereas it increased *tetO, tetW, sul3, ermF*, and *blaTEM*, and showed no decrease in sul2. Mesophilic digestion was ineffective in removing ARGs, except for *ermB* and *blaTEM* | Wu *et al.* (2016) |
| Sewage sludge | ARGs (sul1, tetA, tetO, tetX) | Thermophilic AD | ARG removal >80% | Xu *et al.* (2018) |
| Hospital wastewater | Real wastewater + ciprofloxacin | UASB reactor | 41% ciprofloxacin reduction and 68% COD removal | Guney and Sponza (2016) |
| DWW | Sulfonamide, chloramphenicol, aminoglycoside, tetracycline, $\beta$-lactam-resistant genes | Anaerobic bioreactors (lab scale) | ARG removal of 62% | Christgen *et al.* (2015) |
| DWW | Many ARGs via metagenomics | AAS bioreactors, lab scale | ARG removal >85% | Christgen *et al.* (2015) |
| Municipal wastewater | ARBs (resistant to tetracycline and sulfadiazine) | UASB reactor | ARB reduction of 0.95–1.16 log units. | Yuan *et al.* (2016) |

*(Continued)*

**Table 12.3** Studies that investigated ARBs and ARGs in domestic WWTPs with an anaerobic process and/or one anaerobic system in the WWTP treatment train. (*Continued*)

| Type of Sample | ARB/ARGs/Antibiotics | Type of WWTP/ Treatment Process | ARB and/or ARG Removal Efficiency | References |
|---|---|---|---|---|
| DWW | ARGs (sul1, sul2, and sul3, tetG, tetM, tetO, and tetX, (ermB and ermC), (cmlA and floR), and 16S rRNA gene | Vertical flow constructed wetland | ARG removal of 63.9 and 84.0% | Chen *et al.* (2016) |
| Municipal wastewater | ARGs (tetA, tetO, tetW, sul1, sul2, blaCTX-M, blaTEM, blaSHV), int1, and 16S rRNA gene | 4 WWTPs, Harbin, China; 2 had an anaerobic step in its flowchart | ARG removal of 0.3–2.7 log units in the four WWTPs.<br>CAS removed 70% of antibiotic residues, 5.0 log units for ARBs, and 4.2 log units for ARGs | Wen *et al.* (2016) |
| Municipal wastewater | 19 antibiotics (amoxicillin, azithromycin, ciprofloxacin, chloramphenicol, meropenem, minocycline, oxytetracycline, sulfamethazine, vancomycin, etc.); 10 ARBs and 15 ARGs (blaKPC, blaNDM, blaSHV, ermB, intI1, sul1, tetO, among others) | 1 WWTP comprised of CAS (anoxic/ aerobic tank) and MBR | MBR system completely removed ARBs, ARGs (up to 4.8 log rate removal (LRV) for ARGs and up to 7.1 LRV for the whole MBR train), but microfiltration membrane alone did not reduce antibiotics in comparison to the treatment of the secondary clarifier | Le *et al.* (2018) |
| Municipal wastewater | ARGs (intl1, blaCTX-M, blaTEM, ermB, qnrS, sul1 and sul2, tetO, mecA, and vanA), 16S rRNA (total bacteria) and HF183 (human bacteroides fecal indicator) | Full-scale WWTP with biological nitrogen removal (BNR) reactors (anoxic, anaerobic, and aerobic reactors) and, UV disinfection step | ARG removal of 2 log copy/mL. WWTP was still releasing ARGs in the order of $2.8 \times 10^{14}$ copies/day | McConnell *et al.* (2018) |
| Municipal primary clarifier effluent | ARGs (ermB, tetO, sul1, int1) | 1 bench-scale AnMBR | ARG reduction of 3–4 log units | Kappell *et al.* (2018) |
| DWW and manure | ARGs (int1, sul1, sul2, ampC, blaOXA1, ermB, ermF, tetO, tetW, tp614, blaNDM-1) | Bench-scale AnMBR (4.5 L) | ARG removal of 99.95% during the stage with the greatest addition of manure was observed | Lou *et al.* (2020) |
| Domestic and industrial wastewater from WWTPs with an anaerobic treatment stage | ARGs (tet(A), tet(B), tet(C), tet(G), tet(L), tet(M), tet(O), tet(Q), tet(W), tet(X), sul1, sul2, and int1); antibiotic residues and ARBs (8 antibiotics: 3 tetracyclines (TCs), 4 sulfonamides, and 1 trimethoprim) | 2 WWTPs | ARG removal of 1 log, according to the different conditions evaluated | Li *et al.* (2016) |

(*Continued*)

**Table 12.3** Studies that investigated ARBs and ARGs in domestic WWTPs with an anaerobic process and/or one anaerobic system in the WWTP treatment train. (*Continued*)

| Type of Sample | ARB/ARGs/Antibiotics | Type of WWTP/ Treatment Process | ARB and/or ARG Removal Efficiency | References |
|---|---|---|---|---|
| DWW (raw and treated) | ARGs (blaCTX, blaOXA, blaTEM, and blaNDM-1), mobile genes (int1, int2, and int3) and 16S rRNA gene) and ARBs (ESBL-resistant bacteria; carbapenem-resistant bacteria) | 3 WWTPs. One has an anaerobic flow-through reactor in its flowchart | Removals of 1.5–3.0 log units were observed, according to the different conditions and microorganisms evaluated | Lamba and Ahammad (2017) |
| Municipal wastewater | 12 ARG types: aminoglycoside, $\beta$-lactams, chloramphenicol, fosmidomycin, macrolide–lincosamide–streptogramin, polymyxin, quinolone, rifamycin, sulfonamide, multidrug-resistant and tetracycline-resistant genes | 1 WWTP with conventional treatment (with anoxic–oxic–anoxic tanks) and MBR | MBR was effective at reducing ARG abundance. 5.3–7.4 log removal values were observed for *E. coli*, enterococci, and *P. aeruginosa* | Ng *et al.* (2019) |
| Municipal wastewater | ARGs (blaTEM, ermB, tetW, tetO, sul1, sul2, addD, and qnrS) and int1 | 4 full-scale MBRs (coupled with anaerobic–anoxic–oxic process) | ARG removal of 1.1–7.3 logs; ermB, sul1, and int1 were reduced by MBRs (1.5–7.3 log removal) | Li *et al.* (2019) |
| DWW | ARGs (sul2, ermB, cmlA, and floR) | Hybrid/integrated flow constructed wetland | ARG removal of 87.8–99.1% | Chen *et al.* (2019) |
| DWW | Addition of 250 μg/L of sulfamethoxazole, ampicillin, and erythromycin to the wastewater ARGs (sul1, tetO, ermF, int1) | AnMBR (lab-scale) | Antibiotic removal of 50–98% | Zarei-Baygi *et al.* (2020) |
| DWW | ARGs (aadA, blaOXA1, ermB, mexF, sul1, sul2, tetG, tetM, tetW, vanC03, vanXD), 16S rRNA, and intI1 | 2 domestic WWTPs. One has an anaerobic step in its flowchart | ARG removal of 74–100% | Sun *et al.* (2022) |

(*Continued*)

**Table 12.3** Studies that investigated ARBs and ARGs in domestic WWTPs with an anaerobic process and/or one anaerobic system in the WWTP treatment train. (*Continued*)

| Type of Sample | ARB/ARGs/Antibiotics | Type of WWTP/ Treatment Process | ARB and/or ARG Removal Efficiency | References |
|---|---|---|---|---|
| Municipal wastewater | ARGs (blaTEM, sul1, tetA), intI1, and 16S rRNA | 3 domestic WWTPs (pilot scale), two with UASB reactor followed by BTF, UASB followed by High rate algal pond (HRAP) | ARG removal of 1.0–3.5 logs depending on the system combination. In UASB/HRAP a reduction of 3.5 log units was observed for blaTEM. ARG removal of 0.3–0.9 logs was observed for the UASB | Santos (2021) |
| Municipal wastewater | ARBs (resistant to ampicillin, chloramphenicol, tetracycline, streptomycin, amoxicillin, sulfadiazine, sulfamethoxazole, trimethoprim, ciprofloxacin, erythromycin, sulfamethoxazole + trimethoprim combination) | 3 full-scale WWTPs, one with UASB reactor followed by BTF | ARB removal of 0.5, 1, and 2 order of magnitude in UASB/BTF, CAS, and MAS/UV | Machado *et al.* (2023) |
| Municipal wastewater | ARGs (sul1, tetA, blaTEM, ermB, qnrB), int1, and 16S rRNA | 3 full-scale WWTPs, one with UASB reactor followed by BTF | ARG removal of 0.2–2.0 log units. blaTEM, ermB, and qnrB removals were 1–2 log units in CAS, UASB/ BTF, and MAS/UV; intI1, sul1, and tetA, removals were 0.2–0.5 logs in the three WWTPs. UASB/BTF showed a similar performance to that of CAS and MAS/UV | Leroy-Freitas *et al.* (2022) |
| Decentralized wastewater treatment facility | ARGs (sul1, sul2), tetracycline-resistant gene (tetO), macrolide-resistant gene (ermB), one integron (int1), and 16S rRNA gene | 5 domestic WWTPs investigated in rural areas of China having anaerobic–oxic (AO) and AO–MBR processes | ARG removal varied from 0.4 to 2 log units, according to the conditions evaluated. AO–MBR achieved one-fold or higher ARG removal | Huang *et al.* (2023) |

Only studies treating domestic and/or urban wastewaters are included.

ARBs, and pathogens (Uluseker *et al.*, 2021). However, monitoring data about AR elements in municipal WWTPs applying anaerobic reactors such as UASB followed by biological trickling filters (UASB/BTFs), one of the main technologies used in warm climate regions (such as in Brazil), are scarce.

Christgen *et al.* (2015) investigated the fate of ARGs in anaerobic, aerobic, and anaerobic–aerobic sequential (AAS) bioreactors (lab-scale) treating DWW through metagenomic approaches. They reported that AAS and aerobic reactors were superior to anaerobic units in reducing ARG-like sequence abundances, especially aminoglycoside, tetracycline, and $\beta$-lactam-resistant genes, whereas sulfonamide and chloramphenicol ARG levels were unaffected by treatment. They concluded that AAS reactors are more promising for future applications because they can reduce more ARGs with lower-energy consumption (32% less energy), but all three treatment methods have limitations and need further studies.

Leroy-Freitas et al. (2022) investigated the abundance of integron (int1), ARGs (sul1, tetA, blaTEM, ermB, qnrB), and 16S rRNA in raw and treated wastewater of three full-scale WWTPs, using different treatment systems: CAS, UASB/BTF, and MAS/UV (modified activated sludge with UV disinfection stage). They observed that all WWTPs decreased the loads of genetic markers finally discharged to receiving water bodies and showed no evidence of being hotspots for AMR amplification in wastewater because the abundances of intI1 and ARGs within the bacterial population did not increase in the treated effluents. UASB/BTF showed a similar performance to that of the CAS and MAS/UV (ARGs removal of 1–2 log units), reinforcing the sanitary and environmental advantages of this biological treatment. Potential pathogenic population underwent a considerable decrease after the treatments; however, strong significant correlations with intI1 and ARGs revealed potential multidrug-resistant pathogenic bacteria (*Aeromonas*, *Arcobacter*, *Enterobacter*, *Escherichia–Shigella*, *Stenotrophomonas*, and *Streptococcus*) in the treated effluents, although in relatively reduced abundances.

Santos (2021) reported ARG reduction of 0.3–0.9 log units after treatment in a UASB reactor. When UASB reactor was combined with other post-treatment systems such as high-rate algal ponds (HRAPs) or BTFs, ARG removal of 1.0–3.5 orders of magnitude were observed depending on the system combination. In UASB/HRAPs, a reduction of 3.5 log units was observed for $\beta$-lactam-resistant gene (blaTEM).

Machado *et al.* (2023) investigating the same three WWTPs in the southeast part of Brazil as Leroy-Freitas *et al.* (2022) reported that MAS was effective in reducing ARB counts (by 2–3 log units), compared to CAS (1 log unit) and UASBs/BTFs (0.5 log unit). However, multidrug-resistant bacteria were still present in treated effluents despite the technology treatment applied. Yuan *et al.* (2016) reported ARB reductions of 0.95–1.16 log units in a UASB reactor.

ARGs and ARBs that ended up being accumulated in the sewage sludge can be removed to some extent by the anaerobic digestion of the sludge. Thermophilic anaerobic digestion has been shown to be more effective in the reduction of ARGs than mesophilic digestion (Wu *et al.*, 2016; Xu *et al.*, 2018) (Table 12.3).

Anaerobic digestion of the sewage sludge can remove ~20.7–80% of ARGs (Wu *et al.*, 2016; Xu *et al.*, 2018). In a recent review about the feasibility of anaerobic treatment including AD (of sludge and manure) in eliminating antibiotics and ARGs, Aziz *et al.* (2022) reported that AD at mesophilic and thermophilic temperatures were effective in eliminating ARGs including tetracyclines (tetA, tetO, tetW, tetX, tetC, tetG, tetL, tetM, tetQ), sulfonamides (sul1, sul2), macrolides (ermF, ermB, ermQ, mefA, mphB, ereA), fluoroquinolones (qnrS, aac(6′)-ib-cr, qnrA), trimethoprim (dfrA1, dfrA2), $\beta$-lactamase (blaTEM), aminoglycosides (aphA1, aphA2, aac(3)-II, aacA4, aadA, aadB, aadE, strA, strB), and mobile genetic elements (intI1, intI2, ISCR1, Tn916/1545). Nevertheless, contradictory results have been obtained for tetC, tetG, tetX, tetA, tetO, tetW, sul1, sul2, blaTEM, ermF, dfrA1, dfrA2, fluoroquinolones (qnrA), and intI1 genes, which are found to be resistant during the digestion period, and sometimes elevated concentrations have been observed after the anaerobic treatment.

They also reported that some resistant genes (sulfonamides – sul3, macrolides – ermX and mefA, and trimethoprim – dfrA5) are still not removed after AD.

In the study by Zarei-Baygi *et al.* (2020), operating an AnMBR for the treatment of DWW containing antibiotics (250 μg/L each of sulfamethoxazole, ampicillin, and erythromycin), they reported that ARG abundances (sul1, ermF, tetO) in the effluent increased upon initial antibiotic exposure to the system and then dropped immediately thereafter. The reactor removed 69–78% of sulfamethoxazole, 89–98% of ampicillin, and 40–58% of erythromycin. Some effluent ARGs (tetO and ermF) were minimal, whereas other genes such as sul1 and int1 were still present in the effluent and strongly correlated with several potentially pathogenic genera.

Aziz *et al.* (2022) concluded that AnMBRs are potentially the most efficient technology for removal of antibiotics (sulfamethoxazole, sulfadiazine, trimethoprim, clarithromycin, erythromycin, ciprofloxacin, ofloxacin, cefalexin, cephradine) and ARGs (sul1, sul2, tetO, tetW, ermF, ermB, blaNDM-1, blaCTX-M-15, blaoxa-48, blaoxa-1).

More details on the removal of antibiotics and ARGs are provided in Chapter 3 dedicated to AnMBRs.

In summary, anaerobic treatment technologies can reduce the burden of antibiotic resistance from the wastewaters and sludge (by reducing ARBs, ARGs, and antibiotic residues) and the removal efficiency will depend on reactor configuration, combination with other treatments, class of the antibiotic, and resistant genes.

### 12.4.2 Research needs on antibiotic resistance

Antibiotic resistance is a major threat to global public health. WWTPs applying biological treatment (involving aerobic and anaerobic processes) can reduce antibiotic residues, ARBs, and ARGs, but these resistant elements are not fully eliminated even after the disinfection treatment. In general, WWTPs operated with CAS processes can remove ~2–4 logs of ARGs and ARBs and up to 70% of antibiotics. On the contrary, anaerobic reactors can reduce from 1.0 to 3.0 log units of ARB and ARG abundance depending on the bacteria or resistant genes. AeMBRs and AnMBRs can outcompete other treatments by removing 3–5 log units of ARGs and up to 98% of antibiotics.

Future studies should focus on innovative/powerful technologies and operational strategies able to increase the removal of antibiotics, ARBs, and ARGs from municipal wastewater. To achieve this and facilitate the comparison of removal efficiencies and resistant levels between different WWTPs and countries, research is needed to select the most appropriate indicators of resistant bacteria and/or pathogens and ARGs, and standard methods for assessing them (Uluseker *et al.*, 2021). In the case of pathogenic bacteria, it could be a good choice to select, at least, microorganisms of the group ESKAPE, a group comprising six pathogens that show resistance to multiple classes of antibiotics: *Enterococcus faecium*, *Staphylococcus aureus*, *Klebsiella pneumoniae*, *Acinetobacter baumannii*, *Pseudomonas aeruginosa*, and *Enterobacter* spp.

Promising results obtained with combined treatments, that is, AAS processes or anaerobics in UASB bioreactors followed by post-treatment in HRAPs or BTFs should be consolidated with experiments at pilot/demonstration scale to achieve a full confirmation of their feasibility. The removal of ARBs, ARGs, antibiotics, and pathogens, as well as the influence of water quality parameters on the plant performance should be monitored and investigated. In addition, the development and evaluation of more effective disinfection and treatment methods (such as ultrafiltration and advanced oxidative processes) could contribute to reducing and/or eliminating the dissemination of AMR from WWTPs to the environment. The advantages and limitations of each treatment method and their combinations should also be investigated.

Finally, it is worth noting that to reduce antibiotic resistance spreading it is necessary to modify the current practice by reducing antibiotics consumption in medicine and animal breeding. This is because even the best technology cannot achieve the true 'zero discharge' and even minimal discharged amounts can contribute to the spreading.

## REFERENCES

Akdogan Z. and Guven B. (2019). Microplastics in the environment: a critical review of current understanding and identification of future research needs. *Environmental Pollution*, **254**, 113011, https://doi.org/10.1016/j.envpol.2019.113011

Alexander J., Hembach N. and Schwartz T. (2020). Evaluation of antibiotic resistance dissemination by wastewater treatment plant effluents with different catchment areas in Germany. *Scientific Reports*, **10**, 8952, https://doi.org/10.1038/s41598-020-65635-4

Ávila C., García-Galán M. J., Borrego C. M., Rodríguez-Mozaz S., García J. and Barceló D. (2021). New insights on the combined removal of antibiotics and ARGs in urban wastewater through the use of two configurations of vertical subsurface flow constructed wetlands. *Science of the Total Environment*, **755**, 142554, https://doi.org/10.1016/j.scitotenv.2020.142554

Aziz A., Sengar A., Basheer F., Farooqi I. H. and Isa M. H. (2022). Anaerobic digestion in the elimination of antibiotics and antibiotic-resistant genes from the environment – a comprehensive review. *Journal of Environmental Chemical Engineering*, **10**, 106423, https://doi.org/10.1016/j.jece.2021.106423

Baranchesme F. and Munir M. (2018). Strategies to combat antibiotic resistance in the wastewater treatment plants. *Frontiers in Microbiology*, **8**, 2603, https://doi.org/10.3389/fmicb.2017.02603

Bengtsson-Palme J., Kristiansson E. and Larsson D. J. (2017). Environmental factors influencing the development and spread of antibiotic resistance. *Federation of European Microbiological Societies*, **42**, 68–89.

Bouki C., Venieri D. and Diamadopoulos E. (2013). Detection and fate of antibiotic resistant bacteria in wastewater treatment plants: a review. *Ecotoxicology and Environmental Safety*, **91**, 1–9, https://doi.org/10.1016/j.ecoenv.2013.01.016

Cacace D., Fatta-Kassinos D., Manaia C. M., Cytryn E., Kreuzinger N., Rizzo L., Karaolia P., Schwartz T., Alexander J., Merlin C., Garelick H., Schmitt H., de Vries D., Schwermer C. U., Meric S., Ozkal C. B., Pons M.-N., Kneis D. and Berendonk T. U. (2019). Antibiotic resistance genes in treated wastewater and in the receiving water bodies: a pan-European survey of urban settings. *Water Research*, **162**, 320–330, https://doi.org/10.1016/j.watres.2019.06.039

Chen J., Ying G.-G., Wei X.-D., Liu Y.-S., Liu S.-S., Hu L.-X., He L.-Y., Chen Z.-F., Chen F.-R. and Yang Y.-Q. (2016). Removal of antibiotics and antibiotic resistance genes from domestic sewage by constructed wetlands: effect of flow configuration and plant species. *Science of the Total Environment*, **571**, 974–982, https://doi.org/10.1016/j.scitotenv.2016.07.085

Chen J., Deng W.-J., Liu Y.-S., Hu L.-X., He L.-Y., Zhao J.-L., Wang T.-T. and Ying G.-G. (2019). Fate and removal of antibiotics and antibiotic resistance genes in hybrid constructed wetlands. *Environmental Pollution*, **249**, 894–903, https://doi.org/10.1016/j.envpol.2019.03.111

Christgen B., Yang Y., Ahammad S. Z., Li B., Rodriguez D. C., Zhang T. and Graham D. (2015). Metagenomics shows that low-energy anaerobic–aerobic treatment reactors reduce antibiotic resistance gene levels from domestic wastewater. *Environmental Science & Technology*, **49**, 2577–2584, https://doi.org/10.1021/es505521w

Cookney J., Mcleod A., Mathioudakis V., Ncube P., Soares A., Jefferson B. and McAdam E. J. (2016). Dissolved methane recovery from anaerobic effluents using hollow fibre membrane contactors. *Journal of Membrane Science*, **502**, 141–150, https://doi.org/10.1016/j.memsci.2015.12.037

Dias M. F., Leroy-Freitas D., Machado E. C., da Silva Santos L., Leal C. D., da Rocha Fernandes G. and Araújo J. C. (2022). Effects of activated sludge and UV disinfection processes on the bacterial community and antibiotic resistance profile in a municipal wastewater treatment plant. *Environmental Science and Pollution Research*, **29**(24), 36088–36099, https://doi.org/10.1007/s11356-022-18749-3

Du J., Geng J., Ren H., Ding L., Xu K. and Zhang Y. (2015). Variation of antibiotic resistance genes in municipal wastewater treatment plant with a 2O-MBR system. *Environmental Science and Pollution Research*, **22**, 3715–3726, https://doi.org/10.1007/s11356-014-3552-x

European Commission. (2022). Proposal for a revised Urban Wastewater Treatment Directive. https://environment.ec.europa.eu/publications/proposal-revised-urban-wastewater-treatment-directive_en (accessed 24 October 2023)

Guney G. and Sponza D. T. (2016). Comparison of biological and advanced treatment processes for ciprofloxacin removal in a raw hospital wastewater. *Environmental Science and Pollution Research*, **37**, 3151–3167.

Haffiez N., Azizi S. M. M., Zakaria B. S. and Dhar B. R. (2022). Propagation of antibiotic resistance genes during anaerobic digestion of thermally hydrolyzed sludge and their correlation with extracellular polymeric substances. *Scientific Reports*, **12**(1), 6749, https://doi.org/10.1038/s41598-022-10764-1

Hartmann N. B., Hüffer T., Thompson R. C., Hassellöv M., Verschoor A., Daugaard A. E., Rist S., Karlsson T., Brennholt N., Cole M., Herrling M. P., Hess M. C., Ivleva N. P., Lusher A. L. and Wagner M. (2019). Are we speaking the same language? Recommendations for a definition and categorization framework for plastic debris. *Environmental Science & Technology*, **53**(3), 1039–1047, https://doi.org/10.1021/acs.est.8b05297

Hazra M. and Durso L. M. (2022). Performance efficiency of conventional treatment plants and constructed wetlands towards reduction of antibiotic resistance. *Antibiotics*, **11**(1), 114, https://doi.org/10.3390/antibiotics11010114

Huang D., Li B., Qiu Y. and Li J. (2023). Distribution and co-occurrence of antibiotic resistance genes and bacterial pathogens in the effluent of decentralized sewage treatment systems in China. *International Biodeterioration and Biodegradation*, **180**, 1–9, https://doi.org/10.1016/j.ibiod.2023.105596

INE, Spanish Statistical National Institute. (2022). Statistics on water supply and sanitation 2020 (in Spanish). https://www.ine.es/prensa/essa_2020.pdf (accessed 28 November 2023)

Kappell A., Kimbell L. K., Seib M. D., Carey D. E., Choi M. J., Kalayil T., Fujimoto M., Zitomer D. H. and McNamara P. J. (2018). Removal of antibiotic resistance genes in an anaerobic membrane bioreactor treating primary clarifier effluent at 20°C. *Environmental Science: Water Research & Technology*, **4**, 1783, https://doi.org/10.1039/C8EW00270C

Kirchherr J., Reike D. and Hekkert M. (2017). Conceptualizing the circular economy: an analysis of 114 definitions. *Resources, Conservation and Recycling*, **127**, 221–232, https://doi.org/10.1016/j.resconrec.2017.09.005

Lamba M. and Ahammad S. Z. (2017). Performance comparison of secondary and tertiary treatment systems for treating antibiotic resistance. *Water Research*, **127**, 172–182, https://doi.org/10.1016/j.watres.2017.10.025

Larsson D. G. J., Flach C. F. and Laxminarayan R. (2022). Sewage surveillance of antibiotic resistance holds both opportunities and challenges. *Nature Reviews Microbiology*, **21**, 213–214, https://doi.org/10.1038/s41579-022-00835-5

Le T., Ng C., Tran N., Chen H. and Gin H. (2018). Removal of antibiotic residues, antibiotic resistant bacteria and antibiotic resistance genes in municipal wastewater by membrane bioreactor systems. *Water Research*, **145**, 498–508, https://doi.org/10.1016/j.watres.2018.08.060

Leroy-Freitas D., Machado E. C., Torres-Franco A. F., Dias M. F., Leal C. D. and Araújo J. C. (2022). Exploring the microbiome, antibiotic resistance genes, mobile genetic element, and potential resistant pathogens in municipal wastewater treatment plants in Brazil. *Science of the Total Environment*, **842**, 156773, https://doi.org/10.1016/j.scitotenv.2022.156773

Li J., Cheng W., Xu L., Jiao Y., Baig S. A. and Chen H. (2016). Occurrence and removal of antibiotics and the corresponding resistance genes in wastewater treatment plants: effluents' influence to downstream water environment. *Environmental Science and Pollution Research*, **23**, 6826–6835, https://doi.org/10.1007/s11356-015-5916-2

Li B., Qiu Y., Li J., Liang P. and Huang X. (2019). Removal of antibiotic resistance genes in four full-scale membrane bioreactors. *Science of the Total Environment*, **653**, 112–119, https://doi.org/10.1016/j.scitotenv.2018.10.305

Liu Z. H., Yin H., Dang Z. and Liu Y. (2014). Dissolved methane: a hurdle for anaerobic treatment of municipal wastewater. *Environmental Science & Technology*, **48**(2), 889–890, https://doi.org/10.1021/es405553j

Liu W., Zhang J., Liu H., Guo X., Zhang X., Yao X., Cao Z. and Zhang T. (2021). A review of the removal of microplastics in global wastewater treatment plants: characteristics and mechanisms. *Environment International*, **146**, 106277, https://doi.org/10.1016/j.envint.2020.106277

Liu Y., Wang B., Pileggi V. and Chang S. (2022). Methods to recover and characterize microplastics in wastewater treatment plants. *Case Studies in Chemical and Environmental Engineering*, **5**, 100183, https://doi.org/10.1016/j.cscee.2022.100183

Lou E. G., Harb M., Smith A. L. and Stadler L. B. (2020). Livestock manure improved antibiotic resistance gene removal during co-treatment of domestic wastewater in an anaerobic membrane bioreactor. *Environmental Science: Water Research & Technology*, **6**(10), 2832–2842, https://doi.org/10.1039/D0EW00387E

Machado E. C., Freitas D. L., Leal C. D., De Oliveira A. T., Zerbini A., Chernicharo C. A. L. and Araújo J. C. (2023). Antibiotic resistance profile of wastewater treatment plants in Brazil reveals different patterns of resistance and multi resistant bacteria in final effluents. *Science of the Total Environment*, **857**, 159376, https://doi.org/10.1016/j.scitotenv.2022.159376

Mao D., Yu S., Rysz M., Luo Y., Yang F., Li F., Hou J., Mu Q. and Alvarez P. J. J. (2015). Prevalence and proliferation of antibiotic resistance genes in two municipal wastewater treatment plants. *Water Research*, **85**, 458–466, https://doi.org/10.1016/j.watres.2015.09.010

McConnell M. M., Hansen L. T., Jamieson R. C., Neudorf K. D., Yost C. K. and Tong A. (2018) Removal of antibiotic resistance genes in two tertiary level municipal wastewater treatment plants. *Science of the Total Environment*, **643**, 292–300, https://doi.org/10.1016/j.scitotenv.2018.06.212

Michielssen M. R., Michielssen E. R., Ni J. and Duhaime M. B. (2016). Fate of microplastics and other small anthropogenic litter (SAL) in wastewater treatment plants depends on unit processes employed. *Environmental Science: Water Research & Technology*, **2**(6), 1064–1073, https://doi.org/10.1039/C6EW00207B

Ng C., Tan B., Jiang X. T., Gu X., Chen H., Schmitz B. W., Haller L., Rathinam F., Charles Zhang T. and Gin K. (2019). Metagenomic and resistome analysis of a full-scale municipal wastewater treatment plant in Singapore containing membrane bioreactors. *Frontiers in Microbiology*, **10**, 1–13, https://doi.org/10.3389/fmicb.2019.00172

Nguyen M. K., Hadi M., Lin C., Nguyen H.-L., Thai V.-B., Hoang H.-G., Vo D.-V. N. and Tran H.-T. (2022). Microplastics in sewage sludge: distribution, toxicity, identification methods, and engineered technologies. *Chemosphere*, **308**, 136455, https://doi.org/10.1016/j.chemosphere.2022.136455

O'Neill J. (2016). Review on antimicrobial resistance: tackling drug-resistant infections globally: final report and recommendations. https://amr-review.org/sites/default/files/160525_Final%20paper_with%20cover.pdf (accessed 24 October 2023)

Pazda M., Kumirska J., Stepnowski P. and Mulkiewicz E. (2019). Antibiotic resistance genes identified in wastewater treatment plant systems – a review. *Science of the Total Environment*, **697**, 607, 134023, https://doi.org/10.1016/j.scitotenv.2019.134023

Pittura L., Foglia A., Akyol Ç., Cipolletta G., Benedetti M., Regoli F., Eusebi A. L., Sabbatini S., Tseng L. Y., Katsou E., Gorbi S. and Fatone F. (2021). Microplastics in real wastewater treatment schemes: comparative assessment and relevant inhibition effects on anaerobic processes. *Chemosphere*, **262**, 128415, https://doi.org/10.1016/j.chemosphere.2020.128415

Rafraf I. D., Lekunberri I., Sànchez-Melsió A., Aouni M., Borrego C. M. and Balcázar J. L. (2016). Abundance of antibiotic resistance genes in five municipal wastewater treatment plants in the Monastir Governorate, Tunisia. *Environmental Pollution*, **219**, 353–358, https://doi.org/10.1016/j.envpol.2016.10.062

Rizzo L., Manaia C., Merlin C., Schwartz T., Dagot C., Ploy M. C. and Fatta-Kassinos D. (2013). Urban wastewater treatment plants as hotspots for antibiotic resistant bacteria and genes spread into the environment: a review. *Science of the Total Environment*, **447**, 345–360, https://doi.org/10.1016/j.scitotenv.2013.01.032

Rossatto A., Arlindo M. Z. F., de Morais M. S., de Souza T. D. and Ogrodowski C. S. (2023). Microplastics in aquatic systems: a review of occurrence, monitoring and potential environmental risks. *Environmental Advances*, **13**, 100396, https://doi.org/10.1016/j.envadv.2023.100396

Santos L. S. (2021). Occurrence and removal of antibiotic resistance genes in simplified sewage treatment systems, in Belo Horizonte-Minas Gerais-Brazil. Master dissertation, Sanitary and Environmental Engineering Department, Federal University of Minas Gerais, Belo Horizonte, Brazil.

Sheriff I., Yusoff M. S. and Halim H. B. (2023). Microplastics in wastewater treatment plants: a review of the occurrence, removal, impact on ecosystem, and abatement measures. *Journal of Water Process Engineering*, **54**, 104039, https://doi.org/10.1016/j.jwpe.2023.104039

Stazi V. and Tomei M. C. (2021). Dissolved methane in anaerobic effluents: a review on sustainable strategies for optimization of energy recovery or internal process reuse. *Journal of Cleaner Production*, **317**, 128359, https://doi.org/10.1016/j.jclepro.2021.128359

Sun J., Dai X., van Loosdrecht M. C. and Ni B.-J. (2019). Microplastics in wastewater treatment plants: detection, occurrence and removal. *Water Research*, **152**, 21–37, https://doi.org/10.1016/j.watres.2018.12.050

Sun S., Geng J., Ma L., Sun X., Qi H., Wu Y. and Zhang R. (2022). Changes in antibiotic resistance genotypes and phenotypes after two typical sewage disposal processes. *Chemosphere*, **291**, 132833, https://doi.org/10.1016/j.chemosphere.2021.132833

Tirkey A. and Upadhyay L. S. B. (2021). Microplastics: an overview on separation, identification and characterization of microplastics. *Marine Pollution Bulletin*, **170**, 112604, https://doi.org/10.1016/j.marpolbul.2021.112604

Tong J. and Wang Y. Y. (2014). Distribution and removal of anaerobic antibiotic resistant bacteria during mesophilic anaerobic digestion of sewage sludge. *Huan Jing ke Xue = Huanjing Kexue*, **35**(10), 3814–3820.

Tong J., Liu J., Zheng X., Zhang J., Ni X., Chen M. and Wei Y. (2016). Fate of antibiotic resistance bacteria and genes during enhanced anaerobic digestion of sewage sludge by microwave pretreatment. *Bioresource Technology*, **217**, 37–43, https://doi.org/10.1016/j.biortech.2016.02.130

Uluseker C., Kaster K. M., Thorsen K., Basiry D., Shobana S., Jain M., Kumar G., Kommedal R. and Pala-Ozkok I. (2021). A review on occurrence and spread of antibiotic resistance in wastewaters and in wastewater treatment plants: mechanisms and perspectives. *Frontiers in Microbiology*, **12**, 717809, https://doi.org/10.3389/fmicb.2021.717809

Wang Q., Wang P. and Yang Q. (2018). Occurrence and diversity of antibiotic resistance in untreated hospital wastewater. *Science of the Total Environment*, **621**, 990–999, https://doi.org/10.1016/j.scitotenv.2017.10.128

Wang J., Chu L., Wojnárovits L. and Takács E. (2020). Occurrence and fate of antibiotics, antibiotic resistant genes (ARGs) and antibiotic resistant bacteria (ARB) in municipal wastewater treatment plant: an overview. *Science of the Total Environment*, **744**, 140997, https://doi.org/10.1016/j.scitotenv.2020.140997

Wang C., Wei W., Zhang Y.-T., Dai X. and Ni B.-J. (2022a). Different sizes of polystyrene microplastics induced distinct microbial responses of anaerobic granular sludge. *Water Research*, **220**, 118607, https://doi.org/10.1016/j.watres.2022.118607

Wang S., Wang X., Fessler M., Jin B., Su Y. and Zhang Y. (2022b). Insights into the impact of polyethylene microplastics on methane recovery from wastewater via bioelectrochemical anaerobic digestion. *Water Research*, **221**, 118844, https://doi.org/10.1016/j.watres.2022.118844

Wei W., Huang Q-S., Sun J., Dai X. and Ni B.-J (2019). Revealing the mechanisms of polyethylene microplastics affecting anaerobic digestion of waste activated sludge. *Environmental Science & Technology*, **53**, 16, 9604–9613, https://doi.org/10.1021/acs.est.9b0297

Wei W., Wang C., Shi X., Zhang Y.-T., Chen Z., Wu L. and Ni B.-J. (2022). Multiple microplastics induced stress on anaerobic granular sludge and an effectively overcoming strategy using hydrochar. *Water Research*, **222**, 118895, https://doi.org/10.1016/j.watres.2022.118895

Wen Q., Yang L., Duan R. and Chen Z. (2016) Monitoring and evaluation of antibiotic resistance genes in four municipal wastewater treatment plants in Harbin, Northeast China. *Environmental Pollution*, **212**, 34–40, https://doi.org/10.1016/j.envpol.2016.01.043

World Health Organization (WHO). (2021). Antimicrobial Resistance. WHO.int Geneva. Available at: https://www.who.int/news-room/fact-sheets/detail/antimicrobial-resistance (accessed 24 October 2023)

Wu Y., Cui E., Zuo Y., Cheng W., Rensing C. and Chen H. (2016). Influence of two-phase anaerobic digestion on fate of selected antibiotic resistance genes and class I integrons in municipal wastewater sludge. *Bioresource Technology*, **211**, 414–421, https://doi.org/10.1016/j.biortech.2016.03.086

Xu R., Yang Z. H., Wang Q. P., Bai Y., Liu J. B., Zheng Y. and Fan C. Z. (2018). Rapid startup of thermophilic anaerobic digester to remove tetracycline and sulfonamides resistance genes from sewage sludge. *Science of the Total Environment*, **612**, 788–798, https://doi.org/10.1016/j.scitotenv.2017.08.295

Yang Y., Li B., Zou S., Fang H. H. and Zhang T. (2014). Fate of antibiotic resistance genes in sewage treatment plant revealed by metagenomic approach. *Water Research*, **62**, 97–106, https://doi.org/10.1016/j.watres.2014.05.019

Yuan Q. B., Guo M. T. and Yang J. (2014). Monitoring and assessing the impact of wastewater treatment on release of both antibiotic-resistant bacteria and their typical genes in a Chinese municipal wastewater treatment plant. *Environmental Science: Processes & Impacts*, **16**(8), 1930–1937, https://doi.org/10.1039/C4EM00208C

Yuan Q. B., Guo M. T., Wei W. J. and Yang J. (2016). Reductions of bacterial antibiotic resistance through five biological treatment processes treated municipal wastewater. *Environmental Science and Pollution Research*, **23**, 19495–19503, https://doi.org/10.1007/s11356-016-7048-8

Zarei-Baygi A., Harb M., Wang P., Stadler L. B. and Smith A. L. (2020). Microbial community and antibiotic resistance profiles of biomass and effluent are distinctly affected by antibiotic addition to an anaerobic membrane bioreactor. *Environmental Science: Water Research & Technology*, **6**, 724–736, https://doi.org/10.1039/C9EW00913B

Zhang A. N., Li L.-G., Ma L., Gillings M. R., Tiedje J. M. and Zhang T. (2018). Conserved phylogenetic distribution and limited antibiotic resistance of class 1 integrons revealed by assessing the bacterial genome and plasmid collection. *Microbiome*, **6**: 130, 724–736, https://doi.org/10.1186/s40168-018-0516-2

Zhang Y.-T., Wei W., Huang Q.-S., Wang C., Wang Y. and Ni B.-J. (2020a). Insights into the microbial response of anaerobic granular sludge during long-term exposure to polyethylene terephthalate microplastics. *Water Research*, **179**, 115898, https://doi.org/10.1016/j.watres.2020.115898

Zhang Y.-T., Wei W., Sun J., Xu Q. and Ni B.-J. (2020b). Long-term effects of polyvinyl chloride microplastics on anaerobic granular sludge for recovering methane from wastewater. *Environmental Science & Technology*, **54**(15), 9662–9671, https://doi.org/10.1021/acs.est.0c02433

IWA Publishing's authorised EU representative for General Product Safety Regulations is Diane D'Arras, 15 rue Duret, 75116 Paris, France, e-mail: safety@iwap.co.uk.

Printed and bound by CPI Group (UK) Ltd, Croydon, CR0 4YY
12/05/2026
02108869-0013